COCAINE

Pharmacology, Physiology, and Clinical Strategies

Edited by

Joan M. Lakoski
University of Texas Medical Branch
Galveston, Texas

Matthew P. Galloway
Lafayette Clinic
Detroit, Michigan

Francis J. White
Lafayette Clinic
Detroit, Michigan

CRC Press
Boca Raton Ann Arbor London

Library of Congress Cataloging-in-Publication Data

Cocaine: pharmacology, physiology, and clinical strategies/edited
by Joan M. Lakoski, Matthew P. Galloway, Francis J. White.
p. cm.
Includes bibliographical references and index.
ISBN 0-8493-8813-9
1. Cocaine—Physiological effect. 2. Cocaine—Toxicology.
3. Cocaine habit. I. Lakoski, Joan M. II. Galloway, Matthew P.
III. White, Francis J.
[DNLM: 1. Cocaine—metabolism. 2. Cocaine—pharmacology.
3. Nervous System—drug effects. 4. Substance Dependence—therapy.
QV 113 C65928]
QP801.C68C64 1991
616.86′47—dc20
DNLM/DLC 91-15743
for Library of Congress CIP

Developed by: Telford Press

Direct all inquiries to CRC Press, Inc., 2000 Corporate Blvd., N.W., Boca Raton, Florida 33431.

International Standard Book Number 0-8493-8813-9

Library of Congress Number 91-15743
Printed in the United States 1 2 3 4 5 6 7 8 9 0

Contents

to

PAUL MAYERCIK, PEGGY GALLOWAY, MARINA E. WOLF

and

in memory of

E. EILEEN SHRAGER, PH.D.

PREFACE

It is both a privilege and a pleasure to have the opportunity to coalesce the research on cocaine that has been brought together in this volume. Basic and clinical research on this widely abused substance is a privilege as these efforts have been provided from the public's support of research funded via the federal government; this privilege confers considerable responsibility on its recipients to utilize both their knowledge and resources wisely. The pleasurable aspects of such a venture include the daily joys and frustrations of scientific pursuit carried out with the firm conviction that this new knowledge may serve the societal need to address the issues of drug abuse and dependence. We hope that this volume not only illustrates the current state of knowledge regarding the numerous effects of cocaine but will stimulate further research in the field of drug abuse.

Acknowledgment of the individuals who have given freely of their expertise and time in reviewing portions of this material must include Drs. Eric W. Black, Mark Kneupfer, and Oomen P. Mathew. We also wish to acknowledge the steady professionalism of our editor with The Telford Press, Pat Kelly. Finally, we offer our continued thanks to Dr. Charles Schuster, Dr. Roger M. Brown, and the dedicated staff at the National Institute on Drug Abuse for their unswerving commitment to facilitating the exciting science contained herein.

Joan M. Lakoski, Ph.D.
Matthew P. Galloway, Ph.D.
Francis J. White, Ph.D.

INTRODUCTION
THE NEED FOR RESEARCH WITH COCAINE AND OTHER ABUSED DRUGS

Societies have abused drugs throughout recorded history. The popularity of particular substances has changed from time to time, yet the phenomenon of abuse of chemicals by society has been continuous. In the 1960s marijuana and LSD were popular substances for abuse. Phencyclidine and related-type compounds were abused widely during the 1970s. In the 1980s we saw an increase of the abuse of cocaine. Nicotine, alcohol, opiates, and caffeine have a longer history of abuse. It is important for us to learn more about the general phenomenon of drug abuse so that society in the future will be able to more adequately respond to the consequences of the abuse of both old and new chemicals. Increased knowledge is needed in a number of different areas; including better methods for determining the incidence of abuse, education or prevention, basic and clinical biomedical research, and treatment. It is essential that we learn how to get the appropriate information to susceptible individuals at the best time to prevent the prevalence of drug abuse.

If we are to solve the problem of drug abuse we must learn how these chemicals interact with cellular components within the brain and understand the complex behavioral and sociological patterns that lead to drug abuse. We need better methodologies to study the interaction of abused substances with biological systems and on the opposite end of the spectrum better methodologies are needed to understand the needs of people so that they will not have to rely on the reward mechanisms stimulated by a chemical.

Knowledge of the molecular effects of cocaine and other drugs of abuse on biological systems is necessary if we are to understand how these chemicals alter neuronal function. However, these molecular events do not occur in isolation; therefore, it is important to know how these chemicals alter brain function at the molecular subcellular and neuronal level as well as how these alterations effect the rest of the brain. A considerable amount of research is needed to determine the best way to medically treat people who abuse drugs. These are complex issues which require understanding at multiple levels of investigation. Drug abuse usually is characterized by abuse of multiple substances; therefore, research encompasses the effects of the drug itself and the appropriate treatment modality for treating people who abuse each as well as multiple chemicals. Due to illegality of the act and for many other reasons, the prevalence of abuse of each substance is often very difficult to determine. Few problems pose as many and as complicated and inter-related components of research as drug abuse.

Even though cocaine has been abused for a very long time, the marked increase and prevalence of its abuse during the 1980s increased the awareness of our lack of knowledge about the effect of this substance on biological systems. It is clear that many lives have seriously been altered by the abuse of what chemically is a relatively simple molecule. The increased abuse of methamphetamine ''ICE'' is another indication of the continuous abuse of new chemicals.

It is intriguing and interesting to note that the widespread abuse of cocaine was predictable based on knowledge generated from basic biomedical and behavioral experiments; that is, cocaine was identified some years ago as one of the most rewarding substances tested in self-injection studies in subhuman primates. It was predicated that, if cocaine were available at a cost which allowed it to be abused by many people and if a route of administration were found which allowed cocaine to produce more rapid effects, the incidence of abuse would increase dramatically. Unfortunately, this is exactly what has happened in the past decade. Considerable research as described above is needed before we will be able to respond quickly and accurately to increased abuse of other substances.

The value of drug abuse research is best appreciated when one remembers that information generated from drug abuse research contributes much to our understanding of normal function and diseases of the brain. Clearly the identification and characterization of the endorphins and other endogenous opioid peptides was the result of a long effort by researchers to understand the addiction potential of heroin and its analogues. It is now becoming increasingly obvious that these endogenous chemicals play a role in normal function of the brain and that alterations in their activity and abnormalities in their function can be implicated in various disease states.

William Dewey, Ph.D.
Virginia Commonwealth University
Richmond, Virginia

1

Pharmacokinetics, Metabolism, and Disposition of Cocaine

LOUIS SHUSTER
Department of Pharmacology
Tufts University School of Medicine
Boston, Massachusetts

INTRODUCTION

Two driving forces are focusing renewed attention on the pharmacokinetics and metabolism of cocaine:

1. The widespread abuse of cocaine by addicts, involving self-administration by different routes with varying efficiency and side effects.
2. The realization that the oxidative pathway of cocaine metabolism has the potential for causing tissue damage by formation of reactive end products.

There is great variability in the responses of different individuals to cocaine. At least part of this variability can be attributed to differences in pharmacokinetics and metabolism. The administration of cocaine by inhalation of the free base can produce high blood levels in a very short time and lead to new pathological lesions such as "crack lung" and cardiac infarction. This chapter summarizes the fragmentary state of our present knowledge of cocaine disposition and points out additional information that is sorely needed in order to understand what happens when cocaine is introduced into the body.

ABSORPTION AND DISTRIBUTION

Cocaine is a lipid-soluble tertiary amine with a pKa of 8.6. Its absorption through cell membranes should, to a large extent, be determined by ionization. At pH 7.4, according to the Henderson-Hasselbalch equation, only about 6% of cocaine hydrochloride would be present as the uncharged base which should readily diffuse through cell membranes. The inhalation of a finely dispersed aerosol of cocaine base, such as is produced in ''free-basing'' or the smoking of crack (Snyder *et al.*, 1988), should yield higher blood levels than the injection or sniffing of cocaine hydrochloride. The absorption of cocaine is slowed by the intense vasoconstriction produced when cocaine blocks re-uptake of norepinephrine by adrenergic nerve endings in the walls of blood vessels in skin and mucous membranes.

When various doses of cocaine hydrochloride in aqueous solution were administered intranasally to human volunteers, peak blood levels were observed after 40 to 90 min. For crystalline cocaine hydrochloride, this time was 35 min. The variability may in part reflect different degrees of vasoconstriction (Javaid *et al.* 1978; Wilkinson *et al.*, 1980; Bromley and Hayward, 1988).

Van Dyke *et al.* (1978) gave cocaine hydrochloride by mouth in gelatin capsules in order to avoid absorption from oropharyngeal membranes. No cocaine could be detected in plasma until after 30 min. This lag period is believed to indicate poor absorption of cocaine which is almost completely ionized at the acidic pH of gastric juice. Peak plasma levels were observed between 50 and 90 min, and were comparable to those achieved after the sniffing of cocaine hydrochloride.

The oral absorption of cocaine from chewed coca leaves is facilitated by mixing the leaves with alkaline wood ashes. Plasma levels of several hundred nanograms per milliliter have been measured after 3 h of leaf chewing by Peruvian Indians (Paly *et al.*, 1979).

Cocaine is very rapidly absorbed when cocaine paste (''free base'') is smoked in cigarettes. Paly *et al.* (1982) measured plasma concentrations of up to 462 ng/ml after 3 min of smoking. Similar rapid rates of absorption have been described when cocaine hydrochloride was smoked in a glass pipe heated at 260°F (Cook *et al.* 1985). The half-time for absorption was 1.1 min.

The i.v. injection of cocaine hydrochloride produces peak plasma levels within 15 min in rats (Nayak *et al.*, 1976) and within 1 to 5 min in humans (Javaid *et al.*, 1978; Cook *et al.*, 1985) and sheep (Khan *et al.*, 1987).

As might be expected, absorption from subcutaneous sites is much slower, peak plasma levels being achieved only after 4 h in rats (Nayak *et al.*, 1976) and after 45 min in dogs (Woods *et al.*, 1951).

Due to its high lipid solubility, cocaine reaches concentrations in the tissues that may be as much as 10 to 20 times higher than plasma levels (Lemberger and Rubin, 1986). Thus, 30 min after a single i.v. injection of cocaine into rats, the highest concentrations were found in the spleen, kidney, and brain. The concentrations in brain were about nine times that in plasma. With chronic administration (twice daily for 3 weeks), the highest concentrations were found in fat (Nayak *et al.*, 1976).

PHARMACOKINETICS OF ELIMINATION

Table 1 presents some pharmacokinetic parameters that have been determined in several species. The volume of distribution, in the range of 2–4 l/kg, indicates that

Table 1
Pharmacokinetic Parameters for Cocaine

Species	Route	Dose (mg/kg)	$t_{1/2}$ (min)	Vd (1/kg)	Clearance	Bioavailability (%)	Ref.
Man	Intranasal	0.38	75				Wilkinson *et al.* (1980)
	Oral	2.0 and 3.0[a]	47				
	Smoking paste		38			70% of oral and nasal	Paly *et al.* (1982)
	i.v.	0.23–0.29	87	2.7	1.83 1/min total	100	Cook *et al.* (1985)
	Intranasal	1.2–1.5	85			80	Cook *et al.* (1985)
	Smoking	0.4–0.6	88			57	Cook *et al.* (1985)
	i.v.	0.4	90				Jacob *et al.* (1984)
	Intranasal	2.0	90				
	i.v.	[b]	48	1.97	2.034 1/min		Ambre *et al.* (1988)
	Inhaled			1.6	1.2 1/h/kg	60	Barnett *et al.* (1981)
	i.v.	1.4–2.9	52, 82	1.2–1.7	0.6–1.4 1/h/kg		Barnett *et al.* (1981)
	i.v.	0.42–0.62	48	1.96	2.06 1/min		Chow *et al.* (1985)
	i.v.	0.23–0.46	16–87		2.45 1/min		Javaid *et al.* (1978)
	Intranasal	0.23–1.4	34–54				Javaid *et al.* (1978)
	Intranasal	0.2–1.5	72–91				Van Dyke *et al.* (1982)
	Oral	2.0[a]	54				Van Dyke et al. (1982)
	Intranasal	2.0	78				Bromley and Hayward (1977)
	Intranasal	2.9[c]	98				Dvorchick *et al.* (1977)
Sheep	i.v.	1–4	8.6–10.3	3.5–4.0	0.29 1/min/kg		Kahn *et al.* (1987)
Monkey			72–78				Misra *et al.* (1976)
Dog			72				Misra *et al.* (1976)
Rat	I.P.	10	68	2.93	3.3 1/h (calc)		Stewart *et al.* (1978)
	S.C.	8	48–60				Nayak *et al.* (1976)
	i.v.	8	15–20				Nayak *et al.* (1976)
Mouse	I.P.		15–20				Benuck *et al.* (1987)

[a] Administered in capsules.
[b] Continuous infusion at varying rate.
[c] Applied by a cocaine-soaked cotton strip for 10 min.

there is appreciable sequestration in tissues. Clearance is almost entirely due to metabolic breakdown. Renal clearance amounts to, at most, a few percent of the total clearance.

The clearance rate of cocaine in sheep exceeds total cardiac output. This unusual finding is not due to a rapid rate of hydrolysis in the blood. Kahn *et al.* (1987) multiplied the rate of hydrolysis in sheep blood *in vitro* by the blood volume. The resulting clearance value of 0.16 ml/min/kg is a negligible fraction of total systemic clearance of 290 ml/min/kg. The explanation proposed by Kahn *et al.* (1987) is that there is rapid first pass clearance of cocaine in the lungs of sheep. This phenomenon is not seen in humans.

The rate of elimination of cocaine should not vary with the route of administration once the elimination phase has been attained. When the drug is administered subcutaneously or intranasally, absorption may continue for an appreciable period after elimination has begun. For this reason, the apparent elimination half-time of intranasal cocaine in humans is about twice that seen after i.v. administration.

The bioavailability of cocaine by the oral route is as great as after intranasal administration. Peak plasma levels are just as high after equal doses, and the subjective "high" appears to be even greater (Van Dyke *et al.*, 1978). These findings suggest that some of the activity of intranasal cocaine may be due to material that is swallowed. Compared to the i.v. route, the bioavailability of inhaled or smoked cocaine is about 60 to 80%.

METABOLISM

There are two main metabolic pathways for the disposition of cocaine: (1) hydrolysis by esterases in blood and tissues, accounting for 80 to 90% of total elimination and (2) oxidation by mixed function oxidases, mainly in the liver.

Deesterification

There are two ester groups in the cocaine molecule (Figure 1) and both are theoretically susceptible to hydrolysis catalyzed by esterases in the body. However, the main product of esterase action appears to be ecgonine methyl ester, produced by cleavage of the benzoyl residue. Both liver and serum esterases hydrolyse cocaine to ecgonine methyl ester (Stewart, 1979). The formation of benzoylecgonine when cocaine is incubated with serum (Taylor *et al.*, 1976) is probably the result of nonenzymatic hydrolysis (Stewart *et al.*, 1979; Garrett and Seyda, 1983). Blood and serum samples generally become alkaline on standing, with a pH of 8–9, because of the loss of CO_2 (Fletcher and Hancock, 1981).

The rate of nonenzymatic hydrolysis of cocaine has been examined in several laboratories (Table 2). Additional estimates of nonenzymatic rates of hydrolysis can be inferred from experiments in which the esterase activity of serum was either absent or was abolished by inhibitors. For example, Kahn *et al.* (1987) found that the half-life ($t^{1/2}$) of cocaine (120 μg/ml) in heparinized sheep blood at 37°C was 3.1 h. The addition of sodium fluoride to inhibit esterase activity increased this

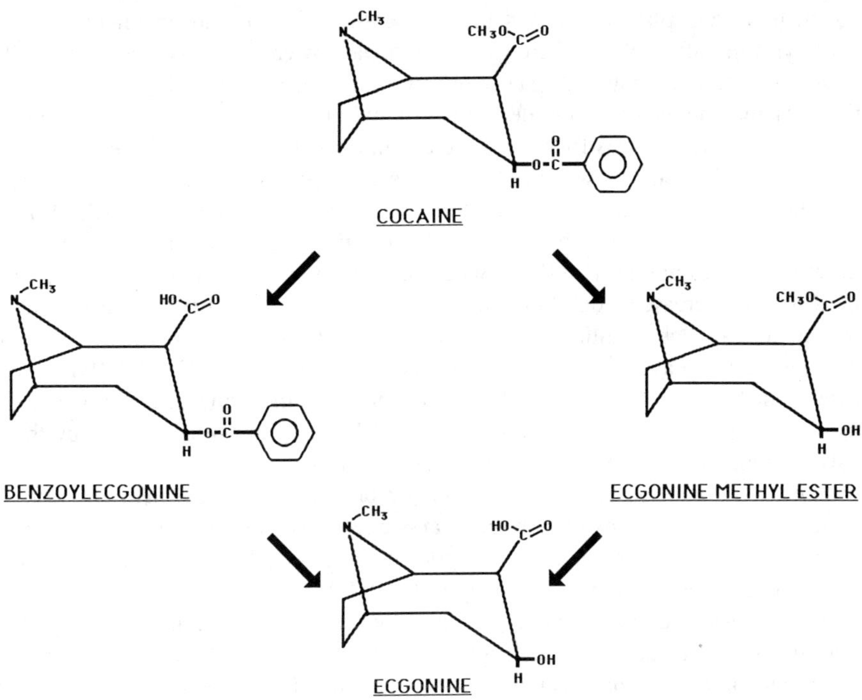

Figure 1. Hydrolysis of cocaine to ecgonine methyl ester, benzoylecgonine, and ecgonine.

Table 2
Rates of Nonenzymatic Hydrolysis of Cocaine to Benzoylecgonine

Concentration	pH	Temperature (°C)	Incubation time (h)	Percent Hydrolyzed	$t_{1/2}$ (h)	Ref.
	7.25	40			5	Garret and Seyda (1983)
	8.0	25	1	17	4.1	Fletcher and Hancock (1981)
	9.0	25	1	28	2.5	Fletcher and Hancock (1981)
	9.4	25	1	41	1.7	Fletcher and Hancock (1981)
	7.4	37	4	8		Stewart *et al.* (1978)

value to 5 to 8 h. The $t^{1/2}$ of 100 ng/ml cocaine in normal human plasma was reported to be 30 min. With plasma from patients who were unusually sensitive to succinylcholine because of a deficiency of plasma esterase, there was no loss of cocaine in 2 h under these conditions (Jatlow *et al.*, 1976).

The rate of hydrolysis of cocaine by serum esterases is strongly influenced by the concentration of the substrate. The apparent half-life of (299 μg/ml) cocaine at

40° in fresh dog plasma was 5.4 h, a value similar to the nonenzymatic rate of hydrolysis in buffer. When the concentration was lowered to 1 μg/ml, the half-life in dog plasma was about 2 h. These observations suggest that at higher concentrations plasma esterases are saturated and most hydrolysis is nonenzymatic. In support of this suggestion, it was found that the extent of hydrolysis of cocaine to benzoyl ecgonine in dog plasma was 90% when the cocaine concentrations was 299 μg/ml, but only 50% when the concentration was 30 μg/ml (Garrett and Seyda, 1983). These observations may explain the apparent 19% decrease in clearance that was observed by Kumor *et al.* (1988) when the rate of i.v. infusion of cocaine into humans was increased from 25 mg/h to 65 mg/h. In general, there is still very little information available on the pharmacokinetics of cocaine when very high doses are administered repeatedly over long periods of time. Weiss and Gawin (1988) have measured detectable amounts of benzoylecgonine in the urine of patients as long as 22 d after the cessation of cocaine intake. The daily intake of cocaine by these patients ranged from 8 to 30 g/d for several months.

Benzoylecgonine is metabolically much more stable than cocaine. Benzoylecgonine was hydrolyzed by human serum at one twentieth the rate of cocaine (Stewart *et al.*, 1979). When labeled benzoylecgonine was injected into rats, approximately 90% was excreted unchanged (Misra *et al.*, 1976).

In summary, the above findings suggest that there is appreciable nonenzymatic hydrolysis of cocaine to benzoylecgonine under *in vivo* conditions when the concentration of cocaine is greater than 1000 ng/ml. The enzymatic hydrolysis of cocaine to ecgonine methyl ester is most prominent when serum concentrations are a few hundred nanograms per milliliter. At higher concentrations there is apparently competitive inhibition by the products of hydrolysis. The action of esterases on benzoylecgonine and/or the nonenzymatic hydrolysis of ecgonine methyl ester would produce ecgonine, which together with benzoylecgonine is excreted in the urine.

In man, urinary excretion accounts for 85 to 90% of an administered dose of cocaine. Of this amount, only between 1 and 5% is unchanged cocaine. The remainder consists mostly of benzoylecgonine and ecgonine methyl ester (Jatlow, 1988; Fish and Wilson, 1969; Inaba *et al.*, 1978).

In rats, the urinary excretion amounts to approximately 50% of the administered dose of cocaine. Only 1% is free cocaine. About 22% could be accounted for as metabolites in the feces, a value that reflects the extent of biliary excretion in animals (Nayak *et al.*, 1976).

Oxidation

The oxidative pathway of cocaine metabolism illustrated in Figure 2 is relatively minor compared to the extent of enzymatic and nonenzymatic hydrolysis. However, it has important toxicological implications.

N-Demethylation

The initial step in the oxidation of cocaine by the mixed function oxidase system is N-demethylation to yield norcocaine and formaldehyde. This reaction was first demonstrated *in vitro* with a preparation of rat liver microsomes by Axelrod (1956).

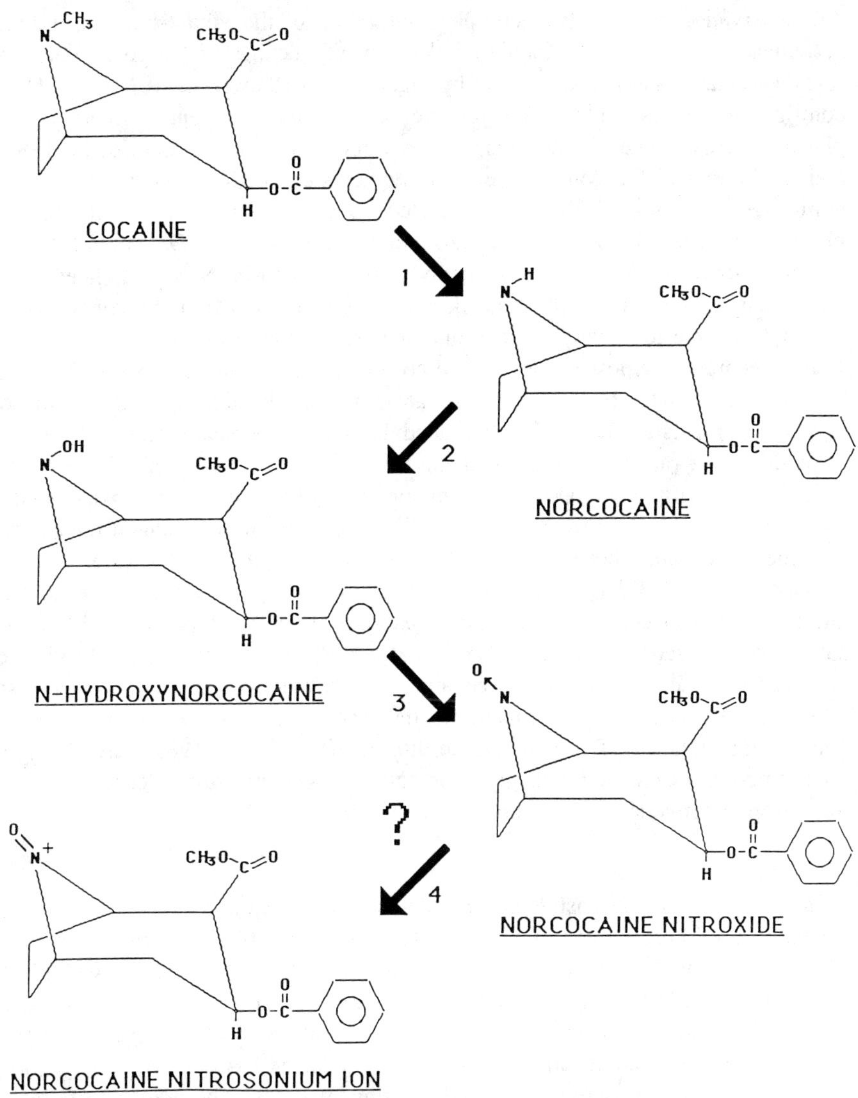

Figure 2. Oxidative pathway of cocaine metabolism. (Reproduced from Shuster (1988) by permission of the National Institute on Drug Abuse.)

The extent of N-demethylation *in vivo* can be established by measuring the exhalation of $^{14}CO_2$ by an animal that has been injected with N-methyl-labeled cocaine. In the rat, Stewart *et al.* (1978) observed that after a dose of 30 μmol (approximately 10 mg) per kg, 14% of the label was exhaled as $^{14}CO_2$ in 2 h. These results duplicated those reported earlier by Englert *et al.* (1976). A smaller proportion, down to 6%, was obtained from lower doses. Possible explanations for this observation include the incomplete saturation of the mixed-function oxidase system at low doses, and increased delivery of cocaine to the liver when the cardiovascular system is stimulated by higher doses (Stewart *et al.*, 1978). Other contributing factors could be the relative decrease in the extent of hydrolysis by plasma esterases when higher doses are administered (Garrett and Seyda, 1983), and the biological fixation of labeled formaldehyde and CO_2. Inaba *et al.* (1978) administered 10 mg of ^{14}C-methyl-labeled cocaine to two human subjects and observed a cumulative excretion of 2.4 and 6.2% as labeled CO_2 within 5 h.

Benzoylecgonine and ecgonine methyl ester seem to be N-demethylated much less readily than cocaine itself. Nayak *et al.* (1976) did not find any conversion of benzoylecgonine to benzoylnorecgonine or norecgonine in the rat.

It is theoretically possible to oxidize cocaine metabolically to cocaine N-oxide. A similar reaction has been described for atropine and related compounds (Phillipson *et al.*, 1976). Misra *et al.* (1979) examined the fate of chemically prepared cocaine-*N*-oxide in the rat. They found that the injection of 50 mg/kg of this derivative labeled with tritium yielded, as the major metabolite, ecgonine-*N*-oxide methyl ester. There were also small amounts (1% or less) of the unchanged compound, cocaine, norcocaine, benzoylecgonine, ecgonine, and ecgonine-*N*-oxide.

Kloss *et al.* (1983) have shown that the FAD monooxygenase of liver can catalyze the conversion of cocaine to cocaine-*N*-oxide. The latter compound could be metabolized to norcocaine by the microsomal cytochrome P450 system. In view of the very low *in vivo* conversion of cocaine-*N*-oxide to norcocaine reported by Misra *et al.* (1979), it appears that this sequential mechanism is not the major route for the N-demethylation of cocaine. According to Kloss *et al.* (1983), the direct cytochrome-P450-catalyzed demethylation pathway acccounts for at least 75% of the microsomal conversion of cocaine to norcocaine.

N-Hydroxylation

Misra *et al.* (1979) postulated that *N*-hydroxynorcocaine might be an oxidized metabolite of norcocaine, in view of other instances of the formation of toxic hydroxylamines from secondary alkyl amines and *N*-alkyl-aryl amines. With the development of HPLC assays for *N*-hydroxynorcocaine, evidence has been obtained for the formation of this metabolite both *in vitro* and *in vivo* (Shuster *et al.*, 1983). The conversion of norcocaine to N-hydroxynorcocaine has the usual requirements that are normally attributed to mixed-function oxidases. The addition of sodium fluoride *in vitro* or the injection of diazinon *in vivo* increases the yield of *N*-hydroxynorcocaine, presumably by inhibiting esterase activity.

One unusual feature of *N*-hydroxynorcocaine is its ability to cyclize spontaneously under physiological conditions to a dihydroisoxazole (Figure 3). The formation of this product probably results from a base-catalyzed reaction between the hydroxylamine residue of *N*-hydroxynorcocaine and the methyl ester at position 2. The half-time for this reaction at pH 7.4 and 37°C is 17 min (Shuster *et al.*, 1983).

Kloss *et al.* (1983) found that flavin monooxygenase could catalyze the conversion of norcocaine to *N*-hydroxynorcocaine at pH 8.3, as determined by qualitative thin layer chromatography. Shuster *et al.* (1983) concluded from experiments with the

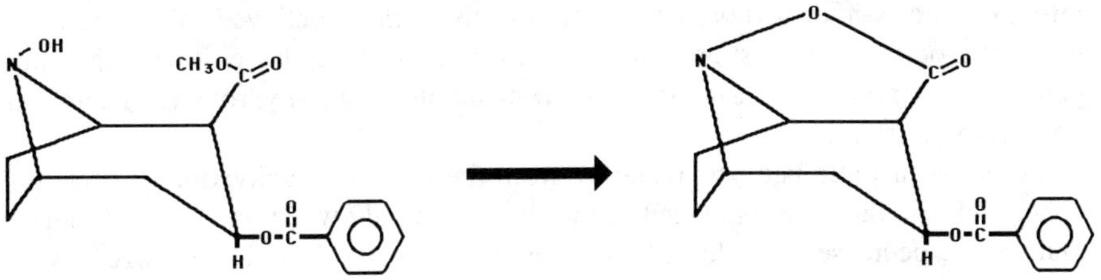

Figure 3. Spontaneous conversion of N-hydroxynorcocaine to a dihydro-isoxazole.

cytochrome P450 inhibitors carbon monoxide, SKF-525 A, and *n*-octylamine that the maximal contribution of the flavin monooxygenase system to the N-hydroxylation of norcocaine is no more than 25%. Pretreatment of mice with phenobarbital decreased the amount of *N*-hydroxynorcocaine recovered from the liver after the injection of norcocaine, presumably because such treatment increased the activity of plasma and microsomal esterases (Shuster *et al.*, 1983).

After the injection of mice with cocaine, appreciable levels of *N*-hydroxynorcocaine were detected in the brain by Benuck *et al.* (1988). The source of this metabolite appears to be the liver because intracisternal injection of cocaine did not lead to the formation of *N*-hydroxynorcocaine in the brain (Benuck *et al.*, 1989).

Most of the tritiated *N*-hydroxynorcocaine that was injected into rats was excreted in the urine as *N*-hydroxybenzylecgonine and *N*-hydroxyecgonine methyl ester. The same products were also obtained when *N*-hydroxynorcocaine was incubated with rat liver microsomes (Misra *et al.*, 1979).

Formation of Norcocaine Nitroxide

In 1976, Nayak *et al.* postulated that norcocaine might be oxidized *in vivo* to the nitroxide free radical which they described as "highly reactive". This free radical can be synthesized chemically by oxidizing norcocaine with m-chloroperbenzoic acid (Nayak *et al.*, 1976). Evidence for enzymatic formation of the nitroxide by liver microsomes *in vitro* has been obtained by Evans (1981) and Rauckman *et al.* (1982). The nitroxide is also formed by brain microsomes (Kloss *et al.*, 1984). These workers used electron spin resonance spectroscopy (ESR) to detect radical formation.

It was originally suggested by several workers that norcocaine nitroxide was responsible for the hepatic necrosis observed when mice are injected with cocaine. Hepatotoxicity requires metabolic activation and is prevented by inhibitors of the mixed function oxidase system, as well as by sulfhydryl compounds such as cysteine and cysteamine (Thompson *et al.*, 1979). The active metabolite also appears to bind covalently to liver proteins (Evans, 1983). However, norcocaine nitroxide turned out to be a stable free radical, similar to spin labels such as 2,2,6,6-tetramethylpiperidine-1-oxyl (TEMPO). It does not react with glutathione or liver proteins (Rauckman *et al.*, 1982). For this reason, Kloss *et al.*, (1984) have suggested that cocaine-induced hepatic necrosis may result from depletion of glutathione secondary to the consumption of NADPH in a futile redox cycle between norcocaine nitroxide and N-hydroxynorcocaine. This futile cycle is believed to generate both hydrogen peroxide and superoxide. Rosen *et al.* (1982) have shown that lipid peroxides are produced during the *in vitro* metabolism of *N*-hydroxynorcocaine and norcocaine nitroxide.

Studies of hepatic necrosis resulting from the metabolic activation of acetaminophen indicate that liver damage becomes manifest only when the levels of reduced glutathione decrease by at least 80% (Mitchell *et al.*, 1973). The change actually observed in mice with cocaine-induced liver damage may be as little as 20% (Thompson *et al.*, 1979). Overnight starvation by itself can lower reduced glutathione levels in the liver by 50% or more without causing any hepatotoxicity (Strubelt *et al.*, 1981). Furthermore, any theory based on glutathione depletion does not account for the covalent binding of activated cocaine to liver proteins (Evans, 1983).

Possible Formation of a Nitrosonium Ion

For these reasons, Charkoudian and Shuster (1985) have suggested that the ultimate hepatotoxin may be a metabolite of norcocaine nitroxide. They postulated that the nitrosonium ion of norcocaine might be formed by a one-electron oxidation of the nitroxide (Figure 2). The nitrosonium ion was produced by chemical oxidation as well as by electrochemical oxidation. It was obtained in the form of red crystals that were stable in air but decomposed very rapidly when dissolved in water or in solutions of cysteine or glutathione. The incubation of the nitrosonium ion of the model nitroxide TEMPO with erythrocytes produced severe membrane damage, as indicated by hemolysis (Charkoudian and Shuster, 1985).

The shuttling of model compounds such as TEMPO between the nitrosonium form and the nitroxide allows nitrosonium ions to act as potent catalysts for the oxidation of alcohols, aldehydes, and amines (Semmelhack *et al.*, 1983; Semmelhack and Schmidt, 1983). Nitrosonium ions react with hydroxide ions to produce hydrogen peroxide (Endo *et al.*, 1984) and with cell membrane lipids to yield malondialdehyde (Charkoudian and Shuster, 1985). One way in which nitrosonium ions might be generated metabolically could be by cytochrome P450 acting as a peroxidase together with lipid hydroperoxides.

The involvement of lipid hydroperoxides could explain some puzzling aspects of cocaine-induced hepatotoxicity. Liver damage from cocaine is prevented by adrenalectomy or by the injection of adrenergic α blockers (Shuster *et al.*, 1988; James *et al.*, 1987). The injection of catecholamines restores the capacity for liver damage in adrenalectomized mice and potentiates it in intact animals (Shuster *et al.*, 1988). Furthermore, only those cocaine analogs that can block the re-uptake of catecholamines appear to be hepatotoxic (Thompson *et al.*, 1979). For example, pseudococaine, which is metabolized like cocaine (Misra and Pontani, 1977) and can give rise to a nitroxide when incubated with microsomes (Charkoudian and Shuster, unpublished), is not hepatotoxic. These observations have led to the hypothesis that the adrenergic effects of cocaine are required for liver damage (Shuster *et al.*, 1988). The role of catecholamines may be to stimulate hormone-sensitive lipases to release free fatty acids which give rise to fatty acid hydroperoxides.

MINOR OXIDIZED METABOLITES

Nayak *et al.* (1976) mentioned suggestive evidence for the formation in the rat of small amounts of cocaine metabolites that were hydroxylated in the *para* position of the aromatic ring and at positions 6 and 7 of the pyrolidine ring. Hydroxylated metabolites have been identified in rat urine by Jindal and Lutz (1986) using gas chromatography-mass spectrometry. There was also tentative identification of metabolites with methylated hydroxyl groups, such as hydroxymethoxycocaine and dimethoxyhydroxy cocaine. Presumably these hydroxylated metabolites are produced by the action of mixed-function oxidases. While hydroxylated metabolites are often conjugated to glucuronic acid, there appears to be little or no evidence for the excretion of such glucuronides in urine (Nayak *et al.*, 1976).

ISSUES DESERVING FURTHER INVESTIGATION

There have not yet been systematic studies of cocaine metabolism in organs other than the liver, such as the kidney and lung. There is an active mixed-function oxidase system in nasal mucosa that is capable of metabolizing cocaine to norcocaine (Dahl and Hadley, 1983). When labeled cocaine is injected intravenously into mice, it is selectively concentrated and bound in the olfactory mucosa as well as the liver (Brittelo, 1988). Rapid first pass metabolism of cocaine in sheep lungs was detected by Kahn *et al.* (1987), but the nature of the metabolic steps involved is unknown.

There is a need for a clearer definition of which esterases and which isozymes of cytochrome P450 are involved in the metabolism of cocaine. These enzymes show considerable diversity between sexes and strains of animals, as does cocaine metabolism (Thompson *et al.*, 1984).

Another area that merits more attention is the pharmacokinetics and metabolism of cocaine analogs such as the phenyltropane derivatives (Clark *et al.*, 1973) and tropacocaine, as well as enantioners and isomers of cocaine itself.

Considerable useful information could be obtained by examining the metabolism of cocaine in perfused livers and by the use of suspensions and cultures of hepatocytes.

The preparation of cocaine labeled with ^{15}N or ^{13}C would greatly facilitate the application of NMR spectroscopy to studies of cocaine metabolism. This technique should allow one to follow changes in a whole range of cocaine metabolites without isolation from an incubation mixture. With the proper coil detector, it could also be used for *in vivo* studies.

A great deal still remains to be learned about the disposition of a drug that has been used by millions of people and for hundreds of years.

REFERENCES

J. Ambre, G. Smith, and K. Smith. Methylecgonine, an important metabolite of cocaine. *Clin. Pharm. Ther.* **29**: 231 (1981).

J. Ambre, S. Belnap, J. Nelson, T. Ruo, S.G. Shin, and A.J. Atkinson. Acute tolerance to cocaine in humans. *Clin. Pharm. Ther.* **44**: 1 (1988).

J. Axelrod, Possible mechanism of tolerance to narcotic drugs. *Science* **124**: 263, (1956).

E.B. Brittebo. Binding of cocaine in the liver, olfactory mucosa, eye and fur of pigmented mice. *Toxicol. Appl. Pharmacol.* **96**: 315 (1988).

G. Barnett, R. Hawks, and R. Resnick. Cocaine pharmacokinetics in humans. *J. Ethnopharmacol.* **3:** 353 (1981).

M. Benuck, M.E.A. Reith, and A. Lajtha. Presence of the toxic metabolite *N*-hydroxy-norcocaine in brain and liver of the mouse. *Biochem. Pharmacol.* **36**: 1169 (1988).

M. Benuck, A. Lajtha, and M.E.A. Reith. Pharmacokinetics of systemically administered cocaine and locomotor stimulation in mice. *J. Pharmacol. Exp. Ther.* **243**: 114 (1987).

M. Benuck, M.E.A. Reith, H. Sershen, H.L. Wiener, and A. Lajtha. Oxidative metabolism of cocaine: comparison of brain and liver. *Proc. Soc. Exp. Biol. Med.* **190**: 7 (1989).

L. Bromley, and A. Hayward. Cocaine absorption from the nasal mucosa. *Anaesthesia* **43**: 356 (1988).

M.J. Chow J.J. Ambre, T.I. Rho, A.J. Atkinson, D.J. Bowsher, and M.W. Fischman. Kinetics of cocaine distribution, elimination, and chronotropic effects. *Clin. Pharmacol. Ther.* **38**: 318 (1985).

J.C. Charkoudian, and L. Shuster. Electrochemistry of norcocaine nitroxide and related compounds: implications for cocaine hepatoxicity. *Biochem. Biophys. Res. Commun.* **130**: 1044 (1985).

C.E. Cook, A.R. Jeffcoat, and M. Perez-Reyes. Pharmacokinetic studies of cocaine and phencyclidine in man in: *Pharmacokinetics and Pharmacodynamics of Psychoactive Drugs*. G. Barnett and C.N. Chang, eds., Biomedical Publications, Foster City, CA, p. 48 (1985).

A.R. Dahl, and W.M. Hadley. Formaldehyde production promoted by rat nasal cytochrome P-450-dependent monooxygenases with nasal decongestants, essences, solvents, air pollutants, nicotine and cocaine as substrates. *Toxicol. Appl. Pharmacol.* **67**: 200 (1983).

B.H. Dvorchik, S.H. Miller, and W.P. Graham. Gas chromatographic determination of cocaine in whole blood and plasma using a nitrogen-sensitive flame ionization detector. *J. Chromatography* **135**: 141 (1977).

T. Endo, T. Miyazawa, S. Shiihashi, and M. Okawara. Oxidation of hydroxide ion by immonium oxide. *J. Am. Chem. Soc.* **106**: 3877 (1984).

L.F. Englert, V.S. Estevez, and B.T. Ho. Demethylation in rats chronically treated with cocaine. *Res. Commun. Chem. Pathol. Pharmacol.* **13**: 555 (1976).

M. A. Evans. Role of protein binding in cocaine-induced hepatic necrosis. *J. Pharmacol. Exp. Ther.* **224**: 73 (1983).

M. A. Evans. Microsomal activation of N-hydroxycocaine to a reactive nitroxide. *The Toxicologist* **1**: 1 (1981).

F. Fish, and W.D.C. Wilson. Excretion of cocaine and its metabolites in man. *Pharmacol* **21**: 135 (1969).

S.M. Fletcher, and V.S. Hancock. Potential errors in benzoylecgonine and cocaine analysis. *J. Chromatog.* **206**: 193 (1981).

E.R. Garrett, and K. Seyda. Prediction of stability in pharmaceutical preparations. XX. Stability evaluation and bioanalysis of cocaine and benzoylecgonine by high-performance liquid chromatography. *J. Pharmaceutical Sci.* **72**: 258 (1983).

T. Inaba, D.J. Stewart, and W. Kalow. Metabolism of cocaine in man. *Clin. Pharmacol. Ther.* **78**: 547 (1978).

P. Jacob, B.A. Elias-Baker, R.T. Jones, and N.L. Benowitz. Determination of cocaine in plasma by automated gas chromatography. *J. Chromatogr.* **306**: 173 (1984).

R.C. James, M.A. Schiefer, S.M. Roberts, and R.D. Harbison. Antagonism of cocaine-induced hepatotoxicity by the alpha adrenergic antagonists phentolamine and yohimbine. *J. Pharm. Exp. Ther.* **242**: 726 (1987).

P. Jatlow, P.L. Barash, C. Van Dyke, and R. Byck. Impaired hydrolysis of cocaine in plasma from succinylcholine sensitive individuals. *Clin. Res.* **24**: 255A (1976).

P.I. Jatlow, C. Van Dyke, P. Barash, and R. Byck. Measurement of benzoylecgonine and cocaine in urine, separation of various cocaine metabolites using reversed-phase high-performance liquid chromatography. *J. Chromatogr.* **152**: 115 (1978).

P. Jatlow. Cocaine: analysis, pharmacokinetics, and metabolic disposition. *Yale J. Biol. Med.* **61**: 105 (1988).

J.I. Javaid, M.W. Fischman, C.R. Schuster, H. Dekirmenjian, and J.M. Davis. Cocaine plasma concentration: relation to physiological and subjective effects in humans. *Science* **202**: 227 (1978).

S.P. Jindal and T. Lutz. Ion cluster techniques in drug metabolism: use of a mixture of labeled and unlabeled cocaine to facilitate metabolite identification. *J. Anal. Toxicol.* **10**: 150 (1986).

M. Kahn, P.K. Gupta, R. Cristie, A. Nangia, H. Winter, F.C. Lam, and D.G. Perrier. Determination of pharmacokinetics of cocaine in sheep by liquid chromatography. *J. Pharmaceutical Sci.* **76**: 39 (1987).

M.W. Kloss, G.M. Rosen, and E.J. Rauckman. N-Demethylation of cocaine to norcocaine: evidence for participation by cytochrome P-450 and FAD-containing monooxygenase. *Mol. Pharmacol.* **23**: 482 (1983).

M.W. Kloss, G.M. Rosen, and E.J. Rauckman. Cocaine-mediated hepatoxicity: a critical review. *Biochem. Pharmacol.* **33**: 169 (1984).

K. Kumor, M. Sherer, L. Thompson, E. Cone, J. Mahaffey, and J.H. Jaffe. Lack of cardiovascular tolerance during intravenous cocaine infusions in human volunteers. *Life Sci.* **42**: 2063 (1988).

L. Lemberger and A. Rubin. *Physiologic Disposition of Drugs of Abuse.* Spectrum Publications, New York, p. 311 (1986).

A.L. Misra. Disposition and transformation of cocaine. *Cocaine: Chemical Biological Clinical, Social and Treatment Aspects.* S.J. Mule, ed., CRC Press, Boca Raton, FL, p. 71 (1976).

A.L. Misra, P.K. Nayak, R. Bloch, and S.J. Mule. Estimation and disposition of [^{3}H] benzoylecgonine and pharmacological activity of some cocaine metabolites. *J. Pharm. Pharmacol.* **27**: 784 (1975).

A.L. Misra, R.B. Pontani, and N.L. Vadlamani. Metabolism of norcocaine, *N*-hydroxy norcocaine and cocaine-*N*-oxide in the rat. *Xenobiotica* **9**: 189 (1979).

A.L. Misra and R.B. Pontani. Disposition and metabolism of Ψ-cocaine (dextrococaine) in the rat. *Drug Metabolism and Disposition* **5**: 556 (1977).

A.L. Misra, R.B. Pontani, and S.J. Mule. [^{3}H]-Norcocaine and [^{3}H]-pseudococaine: effect of N-demethylation and C_2-epimerization of cocaine on its pharmacokinetics in the rat. *Experientia* **32**: 895 (1976).

A.L. Misra, P.K. Nayak, R. Bloch, and S.J. Mule. Estimation and disposition of [^{3}H]-benzoylecgonine and pharmacological activity of some cocaine metabolites. *J. Pharm. Pharmacol.* **27**: 784 (1975).

A.L. Misra, V.V. Giri, M.N. Patel, V.R. Alluri, R.B. Pontani, and S.J. Mule. Disposition of [^{3}H]-benzoylnorecgonine (cocaine metabolite) in the rat. *Res. Commun. Chem. Path. Pharmacol.* **13**: 579 (1976).

J.R. Mitchell, J.D. Jollow, W.Z. Potter, J.R. Gillette, and B.B. Brodie. Acetaminophen-induced hepatic necrosis. IV. Protective role of glutathione. *J. Pharmacol. Exp. Ther.* **187**: 211 (1973).

S.J. Mule, G.A. Casella, and A.L. Misra. Intracellular disposition of [^{3}H]-norcocaine, [^{3}H]-cocaine, ^{3}H-norcocaine, [^{3}H]-benzoylecgonine and [^{3}H]-benzoylnorecgonine in the brain of rats. *Life Sci.* **19**: 1585 (1976).

P.K. Nayak, A.L. Misra, and S.J. Mule. Physiological disposition and biotransformation of [^{3}H]cocaine in acutely and chronically treated rats. *J. Pharmacol. Exp. Ther.* **196**: 566 (1976).

D. Paly, C. Van Dyle, P. Jatlow, F. Cabieses, and R. Byck. Cocaine plasma concentrations in coca chewers. *Clin. Pharm. Ther.* **25**: 240 (1979).

D. Paly, P. Jatlow, C. Van Dyke, F. Raul Jeri, and R. Byck. Plasma cocaine concentrations during cocaine paste smoking. *Life Sci.* **30**: 30 (1982).

M. Perez-Reyes, S. Di Guiseppi, G. Ondrusek, A.R. Jeffcoat, and C.E. Cook. Free-base cocaine smoking. *Clin. Pharm.* **32**: 459 (1982).

J.D. Phillipson, S.S. Handa, and J.W. Gorrod. Metabolic N-oxidation of atropine, hyoscine and the corresponding nor-alkaloids by guinea pig liver microsomal preparations. *J. Pharm. Pharmacol.* **20**: 687 (1976).

E. J. Rauckman, G. M. Rosen, and J. Cavagnaro. Norcocaine nitroxide: A potential hepatotoxic metabolite of cocaine. *Mol. Pharmacol.* **21**: 458 (1982).

E.A. Reith, M. Benuck, and A. Lajtha. Cocaine disposition in the brain after continuous or intermittent treatment and locomotor stimulation in mice. *J. Pharmacol. Exp. Ther.* **243**: 28 (1987).

G. M. Rosen, M. W. Kloss, and E. J. Rauckman. Initiation of *in vitro* lipid peroxidation by N-hydroxynorcocaine and norcocaine nitroxide. *Mol. Pharmacol.* **22**: 529 (1982)

M.F. Semmelhack and C.R. Schmid. Nitroxyl-mediated electrooxidation of amines to nitriles and carbonyl compounds. *J. Am. Chem. Soc.* **105**: 6732 (1983).

M.F. Semmelhack, C.S. Chou, and D.A. Cortes. Nitroxy-mediated electrooxidation of alcohols to aldehydes and ketones. *J. Am. Chem. Soc.* **105**: 4492 (1983).

L. Shuster, E. Casey, and S.S. Welankiwar. Metabolism of cocaine and norcocaine to N-hydroxynorcocaine. *Biochem. Pharmacol.* **32**: 3045 (1983).

L. Shuster, C.A. Garhart, J. Powers, Y. Grunfeld, and G. Kanel. Hepatotoxicity of cocaine, in: *Mechanisms of Cocaine Abuse and Toxicity. NIDA Research Monograph* 88, D. Clouet, K. Asghar, and R. Brown, eds., National Institute on Drug Abuse, Rockville, MD p. 250 (1988).

D.J. Stewart, T. Inaba, M. Lucassen, and W. Kalow. Cocaine Metabolism: cocaine and norcocaine hydrolysis by liver and serum esterases. *Clin. Pharmacol. Ther.* **25**: 464 (1979).

D.J. Stewart, T. Inaba, and W. Kalow. N-Demethylation of cocaine in the rat and in isolated rat hepatocytes: comparison with aminopyrine demethylation. *J. Pharmacol. Exp. Ther.* **207**: 171 (1978).

C.A. Snyder, R.W. Wood, J.F. Graefe, A. Bowers, and K. Margar. ''Crack smoke'' is a respirable aerosol of cocaine base. *Pharmacol. Biochem. Behavior* **29**: 93 (1988).

O. Strubelt, E. Dost-Kempf, C.P. Siegers, M. Younes, M. Volpel, V. Preuss, and J.G. Dreckmann. The influence of fasting on the susceptibility of mice to hepatotoxic injury. *Toxicol. Appl. Pharmacol.* **60**: 66 (1981).

D. Taylor, V. S. Estevez, L. F. Englert, and B. T. Ho. Hydrolysis of carbon-labeled cocaine in human serum. *Res. Commun. Chem. Pathol. Pharmacol.* **14**: 249 (1976).

M.L. Thompson, L. Shuster, and K. Shaw. Cocaine-induced hepatic necrosis in mice. The role of cocaine metabolism. *Biochem. Pharmacol.* **28**: 2389 (1979).

M.L. Thompson, L. Shuster, and E. Casey. Sex and strain differences in response to cocaine. *Biochem. Pharmacol.* **33**: 1229 (1984).

C. Van Dyke, P. Jatlow, J. Ungerer, P.G. Barash, and R. Byck. Oral cocaine: plasma concentrations and central effects. *Science* **200**: 211 (1978).

C. Van Dyke, J. Ungerer, P. Jatlow, P. Barash, and R. Byck. Intranasal cocaine: dose relationships of psychological effects and plasma levels. *Int. J. Psychiatry Med.* **12**: 2127 (1982).

T.G. Vitti and R.L. Boni. Metabolism of cocaine, in: *Pharmacokinetics and Pharmacodynamics of Psychoactive Drugs.* G. Barnett and C.N. Chang. eds., Biomedical Publications, Foster City, CA, p. 427 (1985).

R.D. Weiss and F.H. Gawin. Protracted elimination of cocaine metabolites in long-term, high-dose cocaine abusers. *Am. J. Med.* **85**: 879 (1988).

L. A. Woods, F. G. McMahon, and M. H. Seevers. Distribution and metabolism of cocaine in the dog and rabbit. *J. Pharmacol. Exp. Ther.* **101**: 200 (1951).

2

Cocaine-Induced Hepatotoxicity

STEPHEN M. ROBERTS, ROBERT C. JAMES, AND RAYMOND D. HARBISON
Center for Environmental & Human Toxicology
University of Florida
Gainesville, Florida

INTRODUCTION

Perhaps one of the most intriguing puzzles in the study of the pharmacology and toxicology of cocaine lies in its effects on the liver. Studies in laboratory animals reveal that a dose of cocaine, depending upon the circumstances, may be innocuous or result in devastating liver damage. Research in a number of laboratories over the last 10–12 years has attempted to understand the basis for the hepatotoxcity of cocaine, the doses and conditions under which it occurs, and its meaning in the context of the human use and abuse of cocaine.

There are reports in the older literature of chronic liver dysfunction among coca leaf chewers and some case reports in the newer literature of liver injury from cocaine in drug abusers. However, most of the knowledge concerning hepatotoxic effects of cocaine has come from careful investigations in animals. The first such report of cocaine hepatotoxicity in mice appeared 100 years ago (Ehrlich, 1890). A later report, in 1939, also described abnormal histopathology in mice treated with repeated doses of cocaine (MacLachlan and Hodge, 1939). Despite these published observations, there was apparently little interest in the hepatic effects of cocaine until the late 1970s, when two independent laboratories presented new studies of cocaine hepatotoxicity and results suggesting that its toxicity is intimately tied to its hepatic metabolism.

It is interesting to note that the hepatotoxic potential of cocaine is somewhat unique; that is, it is not shared by other local anesthetic drugs nor other substances of abuse. Studies of possible liver effects of compounds similar to cocaine indicate that the structure of the cocaine molecule is extremely important in its hepatotoxic action (Thompson *et al.*, 1979; Freeman and Harbison, 1981). The structural requirements for hepatotoxicity from cocaine are so rigid that even subtle alterations

Table 1
Hepatotoxicity of Cocaine and Structurally Related Compounds

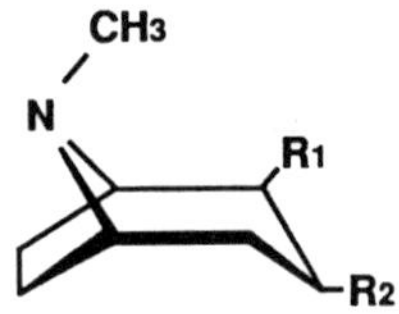

Common name	R_1	R_2	Hepatotoxic?
Cocaine	$-COOCH_3$	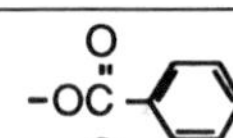	Yes
Pseudococaine[a]	$-COOCH_3$	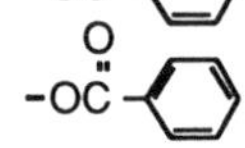	No
Tropacocaine	–H	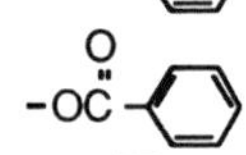	No
Tropine[b]	–H	–OH	No
—	$-COOCH_3$	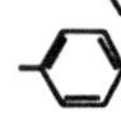	No
—	$-COOCH_3$	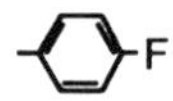	No
—	$-COOCH(CH_3)_2$	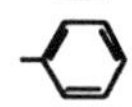	No
—	$-COOCH(CH_3)_2$	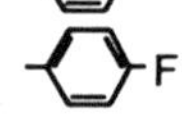	Yes
Atropine	–H	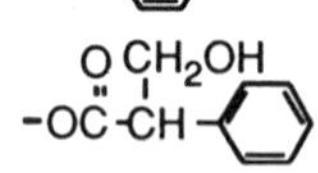	No

Note: Based on data in Thompson *et al.*, 1979 and Freeman and Harbison, 1981.

[a] Carboxymethyl group in equitorial position, not axial as in cocaine molecule.
[b] Tropine N-oxide is also not hepatotoxic.

in its structure or conformation usually result in a nonhepatotoxic compound (Table 1).

MANIFESTATIONS OF COCAINE HEPATOTOXICITY IN LABORATORY ANIMALS

Histopathology

The most detailed histopathologic studies of cocaine-induced liver injury have been conducted in the DBA/2Ha mouse, a strain in which extensive liver injury can be produced with a single dose of cocaine (Kloss *et al.*, 1982; Gottfried *et al.*, 1986). Dilation of rough endoplasmic reticulum, glycogen depletion, and evidence of peroxidation of microsomal lipids have been observed in centrilobular hepatocytes as early as 1 h following cocaine administration in this strain. Mitochondrial changes, progressing from focal membrane disruption to swelling and disruption of cisternae,

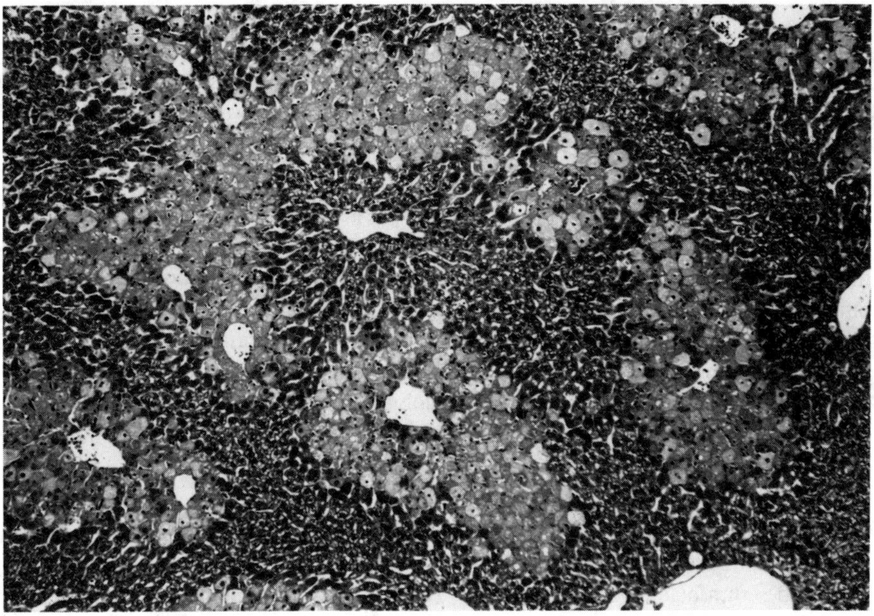

Figure 1. Centrilobular necrosis in a naive DBA/2Ha male mouse 24h after treatment with a single dose of 50 mg/kg cocaine, i.p. Hematoxylin-eosin stain.

occurred in the centrilobular region beginning approximately 2 h after cocaine administration. Progressive changes were also noted in centrilobular peroxisomes, including changes in morphology and density. Irreparable injury and necrosis were visible in the centrilobular region within 6–8 h after cocaine administration, with lesser injury extending into the midzonal region. The periportal hepatocytes, in contrast, showed no necrosis and only mild fatty metamorphosis. An example of the hepatic lesion produced by cocaine is shown in Figure 1.

A number of studies have provided histopathologic evidence suggesting that the intralobular site of the hepatic necrosis may be altered by pretreatment with hepatic enzyme-inducing drugs such as phenobarbital, and perhaps by the duration or extent of cocaine treatment. For example, the administration of four daily doses of cocaine to B6AF1/J mice results in necrosis which appears first in the midzonal region and later extends into the centrilobular region (Shuster *et al.*, 1977; Thompson *et al.*, 1979). When mice of the same strain are administered a single hepatotoxic dose of cocaine following phenobarbital pretreatment, the lesion appears in the periportal rather than midzonal or centrilobular regions. Hepatic necrosis from a single dose of cocaine in phenobarbital-induced ICR mice also appears in the periportal region, and a similar pattern of necrosis can be produced through extended dosing of cocaine in naive ICR mice (Evans and Harbison, 1978; Freeman and Harbison, 1981; Evans, 1983). The cocaine metabolite norcocaine also produces zonal hepatocellular necrosis in this mouse strain — centrilobular in naive mice and periportal in phenobarbital-induced mice (Freeman and Harbison, 1981a). Not all hepatic enzyme-inducing agents are associated with the appearance of the cocaine lesion in the periportal region. Centrilobular necrosis is seen in B6AF1/J mice following hepatic enzyme induction through housing on pine bedding (Thompson *et al.*, 1979) and in ICR mice induced through pretreatment with ethanol (Smith *et al.*, 1981).

The variability in hepatic lesion produced by cocaine under different conditions is unusual. A clear description of this apparent shifting lesion is hampered by the absence of a systematic examination of the phenomenon. Simultaneous variables of hepatic enzyme inducer pretreatment and cocaine dosing regimen (single vs. multiple doses) in existing studies often confound the determination of the contribution of each to changes in histopathologic lesion from cocaine. It is also unclear to what extent strain-related differences may exist. The apparent ability of the hepatic enzyme-inducing agent phenobarbital, and perhaps duration of cocaine exposure, to produce a fundamental change in intralobular site of hepatic necrosis no doubt provides important clues as to the mechanism of cocaine-induced liver injury, but at present remains unexplained.

Biochemical and Functional Evidence of Hepatotoxicity

Cocaine-induced necrosis results in a massive release of hepatic aminotransferases, and serum alanine aminotransferase (ALT) or aspartate aminotransferase (AST) activities are often used to characterize the extent of hepatic injury. The time of peak ALT or AST activity following an hepatotoxic dose of cocaine may vary, but is usually between 8 and 24 h after administration (Shuster *et al.*, 1977; Evans and Harbison, 1978; James *et al.*, 1987).

Studies of cocaine hepatotoxicity in mice have found peroxidation of microsomal lipids. In DBA/2Ha mice, conjugated diene absorption indicative of lipid peroxidation was elevated within 1 h following a single dose of cocaine (Kloss *et al.*, 1983). Lipid peroxidation reached a maximum between 2 and 4 h post-administration, and decreased back to control levels at 8 h. Peroxidation of microsomal lipids has also been observed in rat and mouse isolated hepatocytes exposed *in vitro* to cocaine, and in hepatic microsomal suspensions following *in vitro* metabolism of cocaine (Kloss *et al.*, 1983; Teaf *et al.*, 1984; Donnelly *et al.*, 1988).

Cocaine hepatotoxicity in its early stages is accompanied by a depression in hepatic reduced glutathione content (Evans and Harbison, 1978; Thompson *et al.*, 1979; Evans, 1983; Thompson *et al.*, 1984; Suarez *et al.*, 1986; James *et al.*, 1987). Significant depression of hepatic-reduced glutathione may occur within 30 min following a hepatotoxic dose of cocaine, with the nadir typically 1–2 h post-administration. The maximal extent of depletion of hepatic reduced glutathione, depending upon the study, has been between 20 and 50%. Studies using isolated hepatocytes from the mouse and rat have found that depletion of reduced glutathione following exposure to cocaine preceded cytotoxicity, suggesting that these glutatione changes may be an integral part of events leading to cocaine-induced cell death (Donnelly *et al.*, 1988). In both mouse and rat hepatocytes, the loss of reduced glutathione was accompanied by an increase in concentrations of oxidized glutathione. Though oxidation may account for most of the decrease in reduced glutathione in hepatocytes, incomplete stoichiometric conversion of reduced to oxidized glutathione in rat hepatocytes suggested that factors other than oxidation might contribute to reduced glutathione loss.

A number of functional enzymatic changes occur as a consequence of liver damage from cocaine administration. Hepatic cytochrome P-450 content in mice can be diminished markedly by single or multiple doses of cocaine in a dose-dependent manner (Freeman *et al.*, 1981; Kloss *et al.*, 1982). This loss of cytochrome P450 can be seen in decreased hepatic microsomal drug metabolizing activities, as well as in prolonged duration of action of compounds such as pentobarbital or hexobarbital (Shuster *et al.*, 1977; Freeman and Harbison, 1981). In rats, changes in cytochrome P450 activities are more difficult to generalize, and cocaine administration has been observed to result in increases, decreases, or no change in enzyme activities related to cytochrome P450 activity (e.g., ethylmorphine *N*-demethylase and cocaine *N*-demethylase), depending upon the sex and strain of rat (Watanabe *et al.*, 1988). Such variability of hepatic enzyme changes with cocaine between rat strains and sexes may reflect or contribute to the overall variability in hepatotoxic response to cocaine in rats.Other important hepatic drug metabolizing or detoxification enzyme activities have been reported to be diminished by cocaine treatment. These include FAD-containing monooxygenase, superoxide dismutase, isocitric dehydrogenase, and glucuronyl transferase activities (Kloss *et al.*, 1982; Padilla *et al.*, 1984; Watanabe *et al.*, 1987).

ROLE OF METABOLISM IN COCAINE HEPATOTOXICITY

There is persuasive evidence from a variety of studies that the hepatic oxidative metabolism of cocaine plays a crucial role in its hepatotoxicity. Such evidence comes from comparisons of hepatotoxic potency between cocaine and its various metabolites, correlations between metabolic activity and susceptibility to cocaine hepatotoxicity among strains and species of laboratory animals, and the effects of inducers and inhibitors of cocaine metabolism on the extent of cocaine-induced hepatic injury.

Metabolism of Cocaine

The important pathways of cocaine metabolism are depicted in Figure 2. Cocaine is extensively metabolized in man, and only about 1–5% of a dose of cocaine is eliminated unchanged in the urine (Fish and Wilson, 1969; Wallace *et al.*, 1976). The predominant metabolic pathways for cocaine are hydrolytic. In humans, about 75–90% of a cocaine dose appears in the urine as benzoylecgonine and ecgonine methyl ester (Fish and Wilson, 1969; Wallace *et al.*, 1976; Jatlow, 1988). Plasma and liver esterases are responsible for the formation of the ecgonine methyl ester metabolite (Taylor *et al.*, 1976; Stewart *et al.*, 1977; Stewart *et al.*, 1979). No enzyme has yet been identified for the hydrolysis of cocaine to benzoylecgonine, and this reaction may occur nonenzymatically (Stewart *et al.*, 1979). The small amounts of ecgonine found in human urine may be formed from esteratic hydrolysis of benzoylecgonine and perhaps from nonenzymatic hydrolysis of ecgonine methyl ester (Stewart *et al.*, 1979).

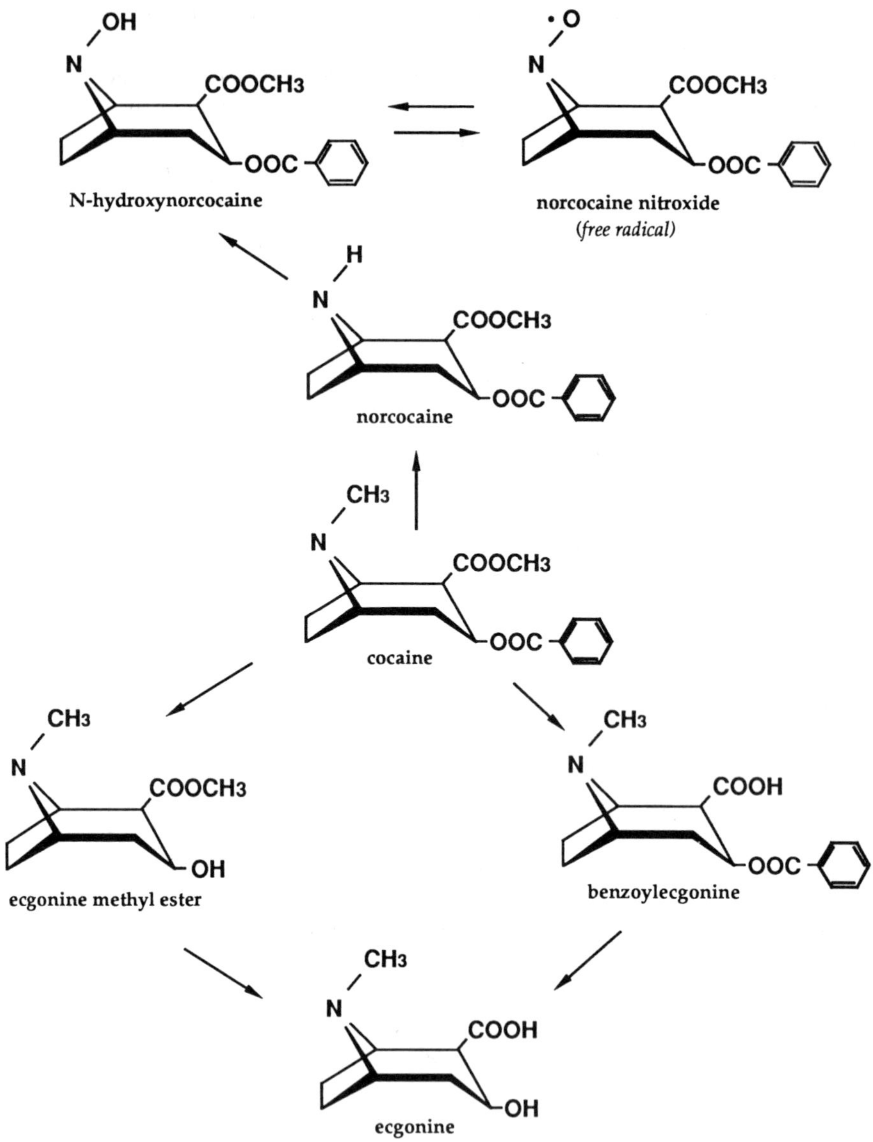

Figure 2. Esteratic and oxidative metabolism of cocaine.

Cocaine can also undergo stepwise oxidation of the tropane nitrogen through the activity of hepatic enzymes. The first step in this metabolic pathway, the N-demethylation of cocaine to norcocaine, has been observed in a variety of species (Leighty and Fentiman, 1974; Hawks *et al.*, 1974; Misra *et al.*, 1976; Inaba *et al.*, 1978; Stewart *et al.*, 1978). Norcocaine may be formed directly from cocaine via cytochrome P450 activity or through FAD-containing monooxygenase formation of a cocaine N-oxide intermediate, with subsequent cytochrome P450 oxidation to the norcocaine metabolite (Kloss *et al.*, 1983). The extent of N-demethylation of cocaine appears to vary widely among species. It has been estimated that the percentage of cocaine dose N-demethylated is 45% in the mouse, 17–20% in the rat, and 2–6% in humans (Werner, 1961; Englert *et al.*, 1976; Inaba *et al.*, 1978; Stewart *et al.*, 1978).

The next oxidation step is the N-hydroxylation of norcocaine to *N*-hydroxynorcocaine. This reaction appears to be mediated primarily by cytochrome P450, although FAD-containing monooxygenase may also play a role (Kloss *et al.*, 1982b; Shuster *et al.*, 1983). An unusual one-electron oxidation of *N*-hydroxynorcocaine to norcocaine nitroxide by cytochrome P450 has been observed in hepatic microsomal suspensions from mice, rats, and hamsters (Rauckman *et al.*, 1982). The nitroxide radical of norcocaine is relatively stable chemically, though it was found to be rapidly reduced back to the N-hydroxy derivative in the presence of hepatic microsomes and either NADPH or NADH.

It is uncertain whether there is further oxidative metabolism of cocaine beyond the nitroxide metabolite. Charkoudian and Shuster have demonstrated the electrochemical oxidation of synthesized norcocaine nitroxide to a reactive nitrosonium ion, and have postulated that a similar one-electron oxidation reaction may occur *in vivo* (Charkoudian and Shuster, 1985). However, such a reaction for norcocaine nitroxide has not yet been confirmed in a biological system. There is also uncertainty regarding the extent to which oxidative metabolism beyond the formation of norcocaine occurs in humans. Unfortunately, studies which have elucidated the oxidative metabolism of cocaine in rodent liver tissues have not been extended to include human tissues.

Hydrolysis Pathways and Hepatotoxicity

None of the esteratic metabolites of cocaine (i.e., benzoylecgonine, ecgonine methyl ester, and ecgonine) are hepatotoxic when administered to mice in high doses, indicating that the hydrolytic metabolism of cocaine does not lead to a toxic metabolite (Thompson *et al.*, 1979; Evans, 1983). A number of studies have found that inhibition of the hydrolytic metabolism of cocaine through the use of esterase inhibitors such as diazinon and tri-*ortho* tolylphosphate results in a substantial increase in the hepatotoxicity of cocaine in mice (Thompson *et al.*, 1979; Freeman and Harbison, 1981; Evans, 1983), confirming that esteratic metabolism is a detoxification pathway for cocaine.

Because the esterases compete with oxidative enzymes (i.e., cytochrome P450 and FAD-containing monooxygenase) for cocaine metabolism, inhibition of esterases results not only in increased levels of cocaine, but also increased formation of oxidative metabolites (Benuck *et al.*, 1988). For example, mice administered diazinon were found to have approximately 60-fold higher hepatic levels of *N*-hydroxynorcocaine 1 h after cocaine administration than mice without diazinon cotreatment (Shuster *et al.*, 1983). Thus, esteratic metabolism of cocaine, while not producing a toxic derivative, nonetheless exerts an important influence on cocaine hepatotoxicity through its effects on cocaine concentrations and those of its oxidative metabolites.

Oxidative Metabolism of Cocaine and Hepatotoxicity

In contrast to the enhancement of cocaine hepatotoxicity by inhibitors of esteratic metabolism, pretreatment of mice with inhibitors of cytochrome P450-mediated metabolism such as SKF 525A, chloramphenicol, iproniazid, metyrapone, or cimetidine can effectively prevent liver injury following cocaine administration (Thompson *et al.*, 1979; Kloss et al., 1982; Evans, 1983; Peterson *et al.*, 1983). This indicates that cocaine is not producing its effects on the liver directly, but rather must be converted to a toxic metabolite through one or more steps involving cytochrome P450.

The first step in the oxidative metabolism of cocaine is N-demethylation to norcocaine. Norcocaine has similar or greater potency as a hepatotoxicant as cocaine (Thompson *et al.*, 1979; Freeman and Harbison, 1981; Evans, 1983). However, pretreatment with inhibitors of cytochrome P450 blocks the development of liver damage from norcocaine, indicating that further oxidation of norcocaine is required to produce the hepatotoxic species (Thompson *et al.*, 1979). The next step in the sequential oxidation of cocaine is the N-hydroxylation of norcocaine to *N*-hydroxynorcocaine. *N*-hydroxynorcocaine is hepatotoxic when administered to mice, with a potency greater than either cocaine or norcocaine (Thompson *et al.*, 1979; Evans, 1983). Its hepatotoxicity is also effectively blocked by treatment with a cytochrome P450 inhibitor (SKF 525A), again indicating that at least one additional oxidative step must occur to derive the ultimate hepatotoxicant.

The metabolite formed from oxidation of *N*-hydroxynorcocaine is norcocaine nitroxide. Its hepatic effects have not been evaluated *in vivo*, though observations *in vitro* support the concept that oxidation at least through formation of the nitroxide is involved in cocaine hepatotoxicity. Using spin-trapping techniques, it was demonstrated that both *N*-hydroxynorcocaine and norcocaine nitroxide stimulate lipid peroxidation (Rosen *et al.*, 1982). In subsequent studies with microsomal lipid peroxidation measured by conjugated diene absorption as the toxic endpoint, it has been found that cocaine and norcocaine are approximately equally toxic, *N*-hydroxynorcocaine has greater toxicity, and norcocaine nitroxide is the most effective of the compounds at initiating lipid peroxidation (Kloss *et al.*, 1983).

Another means to examine the potential role of oxidative metabolism in cocaine hepatotoxicity is through the use of agents which increase the activity of cytochrome P450. However, unlike the experiments with inhibitors of oxidative metabolism and the direct testing of toxicity of oxidative metabolites, outcomes of cocaine toxicity studies involving enzyme inducing agents are more difficult to interpret. Studies using compounds which increase the cytochrome P450 metabolism of cocaine generally support the contention that toxicity occurs through oxidative metabolism, though the results are not entirely consistent. Several studies have found that pretreatment with the cytochrome P450-inducing agent phenobarbital increases the toxicity of cocaine in mice (Evans and Harbison, 1978; Kloss *et al.*, 1982; Evans, 1983); in fact, some studies observed significant liver injury only when the mice were phenobarbital pretreated. Hepatic enzyme induction from housing mice on pine bedding or through the administration of the essential oil β-ionone also have been reported to increase liver damage resulting from cocaine administration in mice (Thompson *et al.*, 1984). However, not all strains of mice respond to phenobarbital induction with increased susceptibility to cocaine hepatotoxicity (Boyer *et al.*, 1988), and differing effects of 3-methylcholanthrene induction on cocaine liver injury have been observed in two investigations using different mouse strains (Evans and Harrison, 1978; Thompson *et al.*, 1984). In other species such as the rat and hamster, phenobarbital was found to have no effect on cocaine hepatoxicity (Shuster et al., 1988).

Perhaps some of the inconsistency in the apparent effects of phenobarbital pretreatment on cocaine hepatotoxicity might arise from the fact that phenobarbital, while increasing cytochrome P450 activity, may also increase the activity of competing esterases (Shuster *et al.*, 1983; Benuck *et al.*, 1989). Thus, the effect of phenobarbital pretreatment on cocaine toxicity observed in a particular species, strain, and sex of laboratory animal may depend upon the relative effects of induction on these two competing metabolic pathways.

SPECIES, SEX, AND STRAIN VARIABILITY IN COCAINE HEPATOTOXICITY

A number of studies have found marked differences among species in susceptibility to cocaine hepatotoxicity. Typically, the greatest toxicity has been observed in mice, and often little or no discernable hepatotoxicity has been seen when equivalent cocaine doses were administered to other species such as rat, hamster, rabbit, or guinea pig (Evans and Harbison, 1978; Freeman and Harbison, 1981; Evans *et al.*, 1983). There are sufficient reports of cocaine-induced liver injury in rats, hamsters, and even monkeys that it would be incorrect to conclude that cocaine hepatotoxicity is somehow unique to the mouse (Watanabe *et al.*, 1987; Shuster *et al.*, 1988). However, it is difficult to make species generalizations beyond that the mouse appears to be the most sensitive laboratory species studied thus far.

One great difficulty in making generalizations about species differences in susceptibility to cocaine hepatotoxicity is that strains within species do not respond in a consistent fashion. This is most clearly demonstrated by the study of Boyer and colleagues, who examined cocaine hepatotoxicity in four inbred mouse strains (Boyer *et al.*, 1988). Hepatic response to a range of cocaine doses and to phenobarbital induction was determined for each strain. While naive males from each of the four strains — C57BL/6, DBA/2, C3H, and Balb/c — experienced substantial liver damage from a single cocaine dose, the differences in potency among strains were such that a midlevel cocaine dose of 50 mg/kg produced serum ALT activities ranging from less than 100 (C57BL/6) to approximately 10,000 Karmen units/ml (DBA/2). Males from the Balb/c and C57BL/6 strains responded to phenobarbital induction with increased cocaine hepatotoxicity, while phenobarbital induction had no effect on cocaine-induced liver injury in C3H and DBA mice. For cocaine doses above 50 mg/kg, ALT responses were higher (5- to 15-fold) in males than females in all strains. Thompson and co-workers also found sex and strain differences in cocaine hepatotoxicity in Balb/c and B6AF1 mice (Thompson *et al.*, 1984) and sex and strain differences in hepatic response to cocaine have been detailed in studies of rats (Watanabe *et al.*, 1987; Watanabe *et al.*, 1988).

The basis for these differences in susceptibility to cocaine hepatotoxicity cannot be stated with certainty at present. In view of the role of metabolism in cocaine hepatotoxicity, and the well-documented differences among species, strains, and sexes of laboratory animals in the metabolism of drugs and chemicals, species-, strain-, and sex-related differences in the bioactivation of cocaine provide an attractive explanation. Though the data are incomplete, there do not appear to be qualitative differences in the oxidative metabolism of cocaine between mice, as the most responsive species, and less sensitive species such as the rat. Both species appear to possess the complete oxidative pathway for cocaine metabolism, including the oxidation of *N*-hydroxynorcocaine to norcocaine nitroxide (Rauckman *et al.*, 1982). Cocaine is cytotoxic to hepatocytes from both the mouse and rat (Donnelly *et al.*, 1988), and the cytotoxicity of cocaine in rat hepatocytes is increased by cytochrome P450 induction and diminished by inhibition of cytochrome P450 in the same way that cocaine hepatotoxicity is influenced by these in mice *in vivo* (Roberts *et al.*, in press). It is more likely, therefore, that species differences in susceptibility to cocaine hepatotoxicity are a function of quantitative differences in metabolic pathways. Attempts have been made to correlate differences in the content or activity of enzymes such as cytochrome P450 with cocaine hepatotoxicity among strains and sexes with varying degrees of success (Thompson *et al.*, 1984; Boyer *et al.*, 1988; Watanabe *et al.*, 1988). The relationships are not simple, which is perhaps not surprising since the production of the ultimate hepatotoxicant from cocaine is dependent upon multiple, interrelated, enzymatic activities.

MECHANISM OF COCAINE HEPATOTOXICITY

Two fundamental mechanisms have been proposed for the toxic effects of cocaine on the liver. One proposed mechanism involves the formation of a highly reactive metabolite of cocaine which binds to critical cellular macromolecules, disrupting normal cellular function leading to cell death (Evans and Harbison, 1978; Evans, 1983). In the second proposed mechanism, a futile redox cycle between two oxidative metabolites of cocaine results in lipid peroxidation leading to cellular injury and death (Kloss *et al.*, 1984). These two possible mechanisms are discussed below, along with evidence for a role of the sympathetic nervous system in modulating cocaine hepatotoxicity.

Covalent Binding Mechanism

The evidence for a role of covalent (or irreversible) binding of cocaine in its hepatotoxicity comes primarily from a correlation between extent of covalent binding and tissue injury. Evans (1983) administered cocaine to naive and phenobarbital-induced mice, rats, and hamsters. Only phenobarbital-induced mice experienced liver damage, and these were the only animals in which administration of radiolabeled cocaine resulted in elevated levels of binding to liver proteins. By varying the position of the radiolabel on the cocaine molecule, Evans was able to provide evidence that both ester groups of the bound species were intact, but that N-demethylation preceded binding. The role of oxidative metabolism in creating the reactive form of cocaine was further supported by *in vitro* experiments in which the microsomal binding of *N*-hydroxynorcocaine was more extensive than either cocaine or norcocaine. Treatment of mice with inhibitors of esterase activity (tri-*ortho* tolylphosphate or *p*-nitrophenyl phosphate) increased both covalent binding and necrosis, while inhibition of cytochrome P450 with SKF 525A decreased covalent binding and necrosis. Interestingly, factors which increased or decreased cellular sulfhydryls (cysteine or diethyl maleate administration) significantly affected necrosis without altering the extent of covalent binding.

Though oxidative metabolism is clearly important in generating the form of cocaine which reacts with proteins, the identity of this metabolite is unknown. While oxidation of cocaine at least through the formation of norcocaine nitroxide appears to be associated with toxicity (see Role of Metabolism in Cocaine Hepatotoxicity), norcocaine nitroxide is a relatively stable radical and is unlikely to possess the reactivity to be the irreversibly binding species (Rauckman *et al.*, 1982). The nitrosonium ion derivative of cocaine proposed by Charkoudian and Shuster (1985) is probably sufficiently reactive, based upon observations with other nitrosonium ions (Shuster *et al.*, 1988), but has not yet been demonstrated to be formed during cocaine metabolism.

It is difficult to provide direct evidence which would indicate whether the binding of cocaine through its metabolite(s) to cellular proteins is responsible for its liver toxicity or merely a coordinated, independent event. A similar problem is shared by other compounds for which covalent binding has been proposed as a mechanism, and can only be resolved when it is determined precisely where and how such binding might lead to the death of the hepatocyte.

Redox Cycling and Lipid Peroxidation Mechanism

Studies of the metabolism of cocaine have provided evidence that the norcocaine nitroxide metabolite formed from *N*-hydroxynorcocaine may be reduced back to the N-hydroxy form through the activity of a flavoprotein (Rauckman *et al.*, 1982). Thus, the oxidative metabolism of cocaine may result in a futile redox cycle between *N*-hydroxynorcocaine and norcocaine nitroxide, generating superoxide anion during the reductive step and consuming NADPH during both oxidative and reductive reactions (Figure 3) (Kloss *et al.*, 1984). Dismutation of the superoxide anion would lead to peroxide formation with the potential for lipid peroxidation. Depletion of cellular NADPH by the futile redox cycle would diminish the ability of the cell to sustain concentrations of reduced glutathione (GSH) through glutathione reductase activity, and the peroxidative protection afforded the hepatocyte by glutathione peroxidase would therefore be compromised. Consistent with this proposed mechanism, hepatocellular GSH concentrations are diminished in cocaine-induced liver injury, and the hepatic injury has been observed to include lipid peroxidation (see Manifestations of Cocaine Hepatotoxicity in Laboratory Animals).

The role of a futile redox cycle involving oxidative metabolites in cocaine hepatotoxicity is far from certain. For example, it is unclear to what extent this futile redox cycle would be self-limiting; that is, the NAPDH depletion proposed to

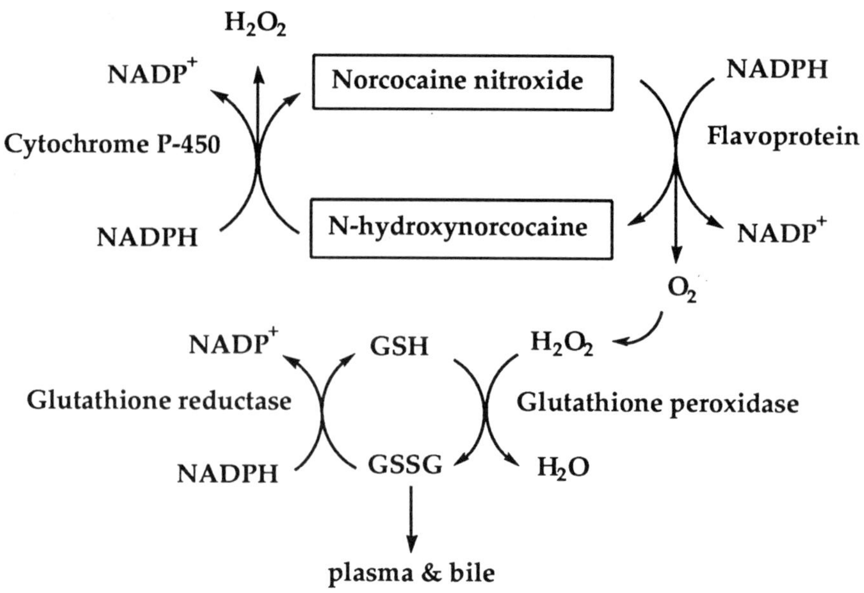

Figure 3. Proposed futile redox cycle for cocaine metabolites. (From Kloss *et al.*, 1984.)

ultimately impair the effectiveness of glutathione peroxidase would also presumably inhibit the activity of the redox cycle itself. Administration of a relatively high dose (100 mg/kg) of 2,2,6,6-tetramethylpiperidinoxyl (TEMPO), a compound which undergoes a redox cycle such as that proposed for cocaine, does not produce liver injury (Shuster *et al.*, 1988). Also, it is unclear whether the less than 50% depletion of hepatic GSH concentrations seen in cocaine hepatotoxicity is sufficient to truly impair the activity of glutathione peroxidase. Finally, recent studies by Donnelly and co-workers question the importance of lipid peroxidation in the hepatotoxicity of cocaine (Donnelly *et al.*, 1988). Preventing lipid peroxidation in isolated hepatocytes through the use of an antioxidant, *N,N'*-diphenylphenylene diamine, had no effect on the cytotoxicity of cocaine.

Role of the Sympathetic Nervous System

Whether cocaine produces its liver injury through covalent binding, a futile redox cycle, or another mechanism, it is clear that the sympathetic nervous system has an important influence. Cocaine causes increased sympathetic nervous system activity by both central and peripheral mechanisms (Van Dyke and Byck, 1977; Chiueh and Kopin, 1978). Pretreatment of mice with the sympathetic blocking drugs phentolamine, yohimbine, reserpine, prazosin, or dibenzylene produces a striking, and sometimes near-complete, antagonism of cocaine-induced liver injury (James *et al.*, 1987; Shuster, 1988). Studies of yohimbine antagonism of cocaine hepatotoxicity verified that this effect was not due to an inhibition of the metabolism and bioactivation of cocaine (James *et al.*, 1987), and presumably all of these structurally diverse compounds decrease the hepatotoxicity of cocaine through their common effect to block adrenergic receptor stimulation.

The administration of high doses of epinephrine has been observed to produce hepatic effects in some ways similar to those of cocaine, i.e., fatty liver changes and transaminase release. Cocaine has been found in some studies to increase circulating catecholamine levels, though it is unlikely that cocaine-induced catecholamine release alone is directly responsible for the liver injury resulting from cocaine. The doses of epinephrine required to elicit serum ALT activity increases of the magnitude seen with cocaine intoxication are quite large (i.e., greater than 2 mg/kg) (Shuster *et al.*, 1988), and probably result in serum epinephrine levels that would be difficult to achieve physiologically. Further, we have observed severe liver injury from cocaine in ICR mice that can be significantly diminished by phentolamine, yet occurs in the absence of increases in serum epinephrine or norepinephrine (unpublished observations). This indicates that the sympathetic influence on cocaine hepatotoxicity may, at least in some cases, be mediated predominantly through direct sympathetic innervation of the liver.

Rather than contributing directly to liver damage from cocaine, sympathetic activity may instead produce changes in the liver which increase the toxic response to cocaine. One possible mechanism involves effects on hepatic glutathione. Although the role of glutathione in cocaine hepatotoxicity is not well understood, cocaine-induced liver injury is diminished by agents which support hepatic glutathione (e.g., cysteine, cysteamine, and *N*-acetylcysteine) and increased by agents which deplete glutathione (e.g., diethyl maleate) (Evans and Harbison, 1978; Thompson *et al.*, 1979; Evans, 1983; Suarez *et al.*, 1986). Catecholamines and sympathomimetic drugs can depress hepatic glutathione concentrations as a consequence of adrenergic receptor stimulation (James *et al.*, 1983). An apparent sympathetic component to the hepatic glutathione loss induced by cocaine is prevented by adrenergic antagonist administration (James *et al.*, 1987). Whether this effect by adrenergic antagonists to minimize the decline in hepatic glutathione during cocaine intoxication is sufficient to account for their hepatoprotective effects is unclear, however.

INTERACTION OF COCAINE WITH OTHER SUBSTANCES OF ABUSE

At least three major types of substances of abuse have the potential to greatly enhance the hepatotoxicity of cocaine. One of these classes of compounds is the barbiturates. Numerous studies have found cocaine hepatotoxicity to be greatly enhanced by pre-exposure to phenobarbital, though, as discussed above, this effect may be somewhat strain and species dependent. The mechanism underlying this effect of phenobarbital is presumably through increased formation of oxidative metabolites of cocaine via cytochrome P450 induction. While phenobarbital is not typically encountered as a substance of abuse, other, more common barbiturates also have the ability to induce cytochrome P450 and thereby increase the hepatotoxic potential of cocaine.

A second, important substance of abuse with effects on cocaine hepatotoxicity is alcohol. A 5-d ethanol pretreatment regimen that itself produced no hepatotoxic effects has been observed to result in substantially increased cocaine mortality attributable to liver damage (Smith *et al.*, 1981). The ethanol enhancement of cocaine-induced liver injury was confirmed by serum aminotransferase activity measurements and direct histopathologic examination of the liver. A twofold increase in cytochrome P450 by the alcohol pretreatment was noted by these investigators, suggesting that the effect of alcohol to enhance cocaine toxicity in the liver may be through an increase in the oxidative metabolism of cocaine.

The third class of abused compounds shown to increase cocaine hepatotoxicity are the opioids, particularly morphine. Morphine itself is mildly hepatotoxic, producing fatty liver changes and small increases in serum alanine aminotransferase activities when administered to mice (Chang and Ho, 1979; Needham *et al.*, 1981). The combination of morphine and cocaine, however, results in a synergistic increase in toxicity (James *et al.*, 1982). Interestingly, this effect on cocaine hepatotoxicity occurs even when morphine is administered in a very small dose directly into the brain (Skoulis *et al.*, 1989). The mechanism for this interaction has not as yet been established, although morphine and other opioids cause a depression of hepatic glutathione (James *et al.*, 1982; Correia *et al.*, 1984). Like the effect to increase cocaine hepatotoxicity, the effect of opioids to diminish hepatic GSH can be elicited through intracerebroventricular administration of small doses (Roberts *et al.*, 1987; Skoulis *et al.*, 1989), indicating that the opioid effect is of central origin.

CLINICAL EVIDENCE FOR COCAINE HEPATOTOXICITY

An association between chronic coca use and serious liver disturbances dates to the previous century, and hepatomegaly may be the toxic effect most directly related to chronic coca chewing (Gutierrez-Noriega and Von Hagen, 1950). There are also more recent reports which suggest that cocaine abuse in a contemporary drug abuse setting may lead to serious hepatotoxicity. Marks and Chapple (1967) studied a population of cocaine and heroin users and found abnormalities in liver function tests that were unusual not only in their high incidence (80%), but also in their severity. These abnormalities disappeared upon cessation of drug use and reappeared when cocaine and heroin abuse was reinitiated.

Perino *et al.* (1987) have reported a case of fatal hepatic necrosis in a patient who used cocaine, apparently in combination with alcohol and barbiturates. Coagulative necrosis was found in hepatocytes in periportal and midzonal regions — centrilobular hepatocytes were undamaged. Mixed microvesicular and macrovesicular fatty changes were also noted. Hypoxia, shock, hyperthermia, hepatitis, and phenobarbital sensitivity syndrome were all ruled out in concluding that the hepatic injury resulted from cocaine.

Kanel *et al.* (1990) have recently reported an additional case of apparent cocaine hepatotoxicity. The patient stated on admission that he had been using cocaine, but denied recent use of other drugs. Results of a toxicological screen were consistent with this claim. Serum aminotransferase activities were initially only modestly elevated, but 48 h after admission rose to greater than 10,000 U/l. With clinical signs of fulminant hepatic failure and acute rhabdomyolysis, the patient's condition deteriorated rapidly and he died on the fourth day. Examination of the liver at autopsy revealed coagulative necrosis in the centrilobular and midzonal regions. Fatty changes, predominantly of the microvesicular type, were observed in the periportal region. As in the report by Perino *et al.*, (1987) these investigators were careful to rule out other potential causes of the hepatic injury before attributing the extensive hepatic necrosis to cocaine.

It is interesting to note that, in these two recent cases of apparent cocaine-induced liver injury, the zonal patterns of the lesions were different. Normally, hepatotoxins produce a consistent lesion type. However, studies of cocaine hepatotoxicity in animals have clearly shown that the zonal location of the necrosis can be influenced by agents which induce hepatic drug-metabolizing enzymes including barbiturates and alcohol (see Manifestations of Cocaine Hepatotoxicity in Laboratory Animals). Though it is impossible to draw conclusions from only two cases, the histopathology in these cases is, at least in general, consistent with animal observations. There was no evidence of exposure to other drugs in the patient examined by Kanel and colleagues, and the centrilobular hepatic lesion in this patient is similar to the hepatic injury typically seen in naive, or uninduced mice. The patient in the case report by Perino and co-workers had evidence of exposure to barbiturates along with cocaine. If this individual had used barbiturates recently before the cocaine intoxication episode, the opportunity would exist for hepatic enzyme induction. The periportal to midzonal necrosis in this patient is consistent with the periportal necrosis found in phenobarbital-pretreated mice administered cocaine.

IMPLICATIONS OF COCAINE HEPATOXICITY OF DRUG ABUSE

It is presently unclear as to the extent to which potential liver injury from cocaine abuse is a problem in humans. The number of clinical reports in which liver damage can be positively attributed to cocaine are few. This should not be surprising, however, in view of what we have learned about cocaine hepatotoxicity from laboratory animals and the inherent difficulties in attributing liver injury, in a clinical setting, to a single etiologic entity.

There are a number of practical problems in detecting cocaine hepatotoxicity clinically. Cocaine-induced liver injury may go undiagnosed for several reasons. First, abnormalities in liver function are quite common among drug abusers, and are usually attributed to hepatitis, possibly exacerbated by poor nutrition. The presence of liver function abnormalities in an individual with a history of drug abuse may therefore not be regarded as particularly remarkable by the attending clinician. Second, the appearance of clinical symptoms from the hepatotoxic insult would probably be delayed and may not appear for days after exposure. This would be consistent with observations in animals and, for example, in humans with an overdose of the hepatotoxic drug acetaminophen. Because of this delay, the patient and/or clinician may not associate the symptoms with cocaine use. Third, the patient may be reluctant to volunteer information about recent drug use, and any drug screen performed at the appearance of symptoms may be negative for cocaine, assuming no other cocaine use in the intervening period. Fourth, if clinicians do not regard cocaine as hepatotoxic, which is currently the case, they are unlikely to consider it as a potential diagnosis. There is no routine diagnostic test which will tell the physician that liver injury in a particular patient is the result of drug or chemical exposure, and the association may be particularly difficult if cocaine hepatotoxicity only appears under certain circumstances.

Even if cocaine hepatotoxicity is suspected, it is difficult to demonstrate. Ruling out other potential causes of hepatic abnormalities is a time-consuming endeavor that the clinician may be either unwilling or unable to pursue. Further, the number of cases in which clinically evident cocaine hepatotoxicity occurs in the absence of other factors which could cause or contribute to liver injury (e.g., other drugs, hypotension, or pyrexia) are probably few. Collectively, all of these factors make the detection and evaluation of cocaine hepatotoxicity in humans very difficult and create uncertainty as to whether the few clinical reports of cocaine hepatotoxicity in the clinical literature truly reflect the incidence of this phenomenon.

In studies of laboratory animals, cocaine hepatotoxicity is generally produced in naive animals only at relatively high doses. If these animal studies serve as reasonable models for man, significant liver damage from a single dose of cocaine in humans in the absence of other factors is probably a concern only for situations of cocaine overdose. For perspective, the size of the dose would be such that morbidity and mortality from acute cocaine effects would also be a significant concern. Animal studies indicate that hepatotoxicity can also be produced by repeated exposures to lesser doses, and it is reasonable to suspect that this may also be the case in humans.

An observation of particular concern from animal studies is that cocaine hepatotoxicity can be greatly enhanced by pre-or co-exposure to other drugs, especially those that increase oxidative metabolism or decrease esteratic metabolism. Thus, factors which delay the esteratic metabolism of cocaine, such as the congenital deficiency in cholinesterase which exists in a small percentage of the population (Jatlow *et al.*, 1979) or exposure to esterase inhibitors such as pesticides, might significantly increase the risk of liver injury from cocaine use. Interestingly, most of the drugs found to potentiate cocaine hepatotoxicity in animal studies are other prominent substances of abuse, i.e., alcohol, barbiturates, and opiates. These observations suggest that the abuse of cocaine, in combination with certain other drugs, might markedly enhance its hepatotoxic potential.

Considering the size and scope of cocaine abuse in the U.S. and other countries today, our understanding of the potential for cocaine-induced liver injury in users of cocaine is clearly inadequate. A better appraisal of the hazards of liver toxicity in cocaine abuse can only come from a combination of careful clinical observation and further research, particularly studies examining cocaine metabolism and effects in human liver. Hopefully, through these efforts, the puzzle of cocaine hepatotoxicity can be solved and an understanding achieved of doses and circumstances under which cocaine abuse poses a threat of injury to the liver.

REFERENCES

M. Benuck, M.E.A. Reith, and A. Lajtha. Presence of the toxic metabolite *N*-hydroxy-norcocaine in brain and liver of the mouse. *Biochem. Pharmacol.* **37**: 1169 (1988).

M. Benuck, M.E.A. Reith, H. Sershen, H.L. Wiener, and A. Lajtha. Oxidative metabolism of cocaine: comparison of brain and liver. *Proc. Soc. Exp. Biol. Med.* **190**: 7 (1989).

C.S. Boyer, D. Ross, and D.R. Petersen. Sex and strain differences in the hepatotoxic response to acute cocaine administration in the mouse. *J. Biochem. Toxicol.* **3**: 295 (1988).

Y.-Y.H. Chang and I.K. Ho. Effects of acute and continuous morphine administration on serum glutamate oxalacetate transaminase and glutamate pyruvate transaminase activities in the mouse. *Biochem. Pharmacol.* **28**: 1373 (1979).

J.C. Charkoudian and L. Shuster. Electrochemistry of norcocaine nitroxide and related compounds: implications for cocaine hepatotoxicity, *Biochem. Biophys. Res. Commun.*, **130**: 1044 (1985).

C.C. Chiueh and I.J. Kopin. Centrally mediated release by cocaine of endogenous epinephrine and norepinephrine from the sympathoadrenal medullary system of unanesthetized rats. *J. Pharmacol. Exp. Ther.* **205**: 148 (1978).

M.A. Correia, J.S. Wong, and E. Soliven. Morphine metabolism revisited: I. Metabolic activation of morphine to a reactive species in rats. *Chem. Biol. Interact.* **49**: 255 (1984).

D.A. Donnelly, C.S. Boyer, D.R. Petersen, and D. Ross. Cocaine-induced biochemical changes and cytotoxicity in hepatocytes isolated from both mice and rats. *Chem. Biol. Interact.* **67**: 95 (1988).

P. Ehrlich. Studien in der cocainreihe. *Deutsch Med. Wochenschr.* **16**: 717 (1890).

L.F. Englert, V.S. Estevez, and B.T. Ho. Demethylation in rats chronically treated with cocaine. *Res. Commun. Chem. Pathol. Pharmacol.* **13**: 555 (1976).

M.A. Evans and R.D. Harbison. Cocaine-induced hepatotoxicity in mice. *Toxicol. Appl. Pharmacol.* **45**: 739 (1978).

M.A. Evans. Role of protein binding in cocaine-induced hepatic necrosis. *J. Pharmacol. Exp. Ther.* **224**: 73 (1983).

F. Fish and W.D.C. Wilson. Excretion of cocaine and its metabolites in man. *J. Pharm. Pharmacol.* **21** (Suppl): 135 (1969).

R.W. Freeman and R.D. Harbison. The role of benzoylmethylecgonine in cocaine-induced hepatotoxicity. *J. Pharmacol. Exp. Ther.* **218**: 558 (1981).

R.W. Freeman and R.D. Harbison. Hepatic periportal necrosis induced by chronic administration of cocaine. *Biochem. Pharmacol.* **30**: 777 (1981).

M.R. Gottfried, M.W. Kloss, D. Graham, E.J. Rauckman, and G.M. Rosen. Ultrastructure of experimental cocaine hepatotoxicity. *Hepatology* **6**: 299 (1986).

C. Gutierrez-Noriega and V.W. Von Hagen. The strange case of the coca leaf. *Sci. Monthly* **70**: 81 (1950).

R.L. Hawks, K. J. Kopin, R.W. Colburn, and N.B. Thoa. Norcocaine: a pharmacological active metabolite of cocaine found in brain. *Life Sci.* **15**: 2189 (1974).

T. Inaba, D.J. Stewart, and W. Kalow. Metabolism of cocaine in man. *Clin. Pharmacol. Ther.* **23**: 547 (1978).

R.C. James, D.R. Goodman, and R.D. Harbison. Hepatic glutathione and hepatotoxicity: changes induced by selected narcotics. *J. Pharmacol. Exp. Ther.* **221**: 708 (1982).

R.C. James, S.M. Roberts, and R.D. Harbison. The perturbation of hepatic glutathione by α_2-adrenergic agonists. *Fund. Appl. Toxicol.* **3**: 303 (1983).

R.C. James, M.A. Schiefer, S.M. Roberts, and R.D. Harbison. Antagonism of cocaine-induced hepatotoxicity by the *alpha* adrenergic antagonists phentolamine and yohimbine. *J. Pharmacol. Exp. Ther.* **242**: 726 (1987).

P. Jatlow, P.G. Barash, C. Van Dyke, J. Radding, and R. Byck. Cocaine and succinylcholine sensitivity: a new caution. *Anesth. Analg.* **58**: 235 (1979).

P. Jatlow. Cocaine: analysis, pharmacokinetics and metabolic disposition. *Yale J. Biol. Med.* **61**: 105 (1988).

G.C. Kanel, W. Cassidy, L. Shuster, and T.B. Reynolds. Cocaine-induced liver cell injury. Comparison of morphologic features in man and in experimental models. *Hepatology* **11:** 646 (1990).

M.W. Kloss, G.M. Rosen, and E.J. Rauckman. Acute cocaine-induced hepatotoxicity in DBA/2Ha male mice. *Toxicol. Appl. Pharmacol* **65**: 75 (1982).

M.W. Kloss, J. Cavagnaro, G.M. Rosen, and E.J. Rauckman. Involvement of FAD-containing monooxygenase in cocaine-induced hepatotoxicity. *Toxicol. Appl. Pharmacol.* **64**: 88 (1982).

M.W. Kloss, G.M. Rosen, and E.J. Rauckman. Evidence of enhanced *in vivo* lipid peroxidation after acute cocaine administration. *Toxicol. Lett.* **15**: 65 (1983).

M.W. Kloss, G.M. Rosen, and E.J. Rauckman. *N*-Demethylation of cocaine to norcocaine. Evidence for participation by cytochrome P-450 and FAD-containing monooxygenase. *Mol. Pharmacol.* **23**: 482 (1983).

M.W. Kloss, G.M. Rosen, and E.J. Rauckman. Cocaine-mediated hepatotoxicity. A critical review. *Biochem. Pharmacol.* **33**: 169 (1984).

E.G. Leighty and A.F. Fentiman. Metabolism of cocaine to norcocaine and benzoyl ecgonine by an *in vitro* microsomal enzyme system. *Res. Commun. Chem. Pathol. Pharmacol* **8**: 65 (1974).

V. Marks and P.A.L. Chapple. Hepatic dysfunction in heroin and cocaine users. *Br. J. Addict.* **62**: 189 (1967).

P.L. MacLachlan and H.C. Hodge. The influence of cocaine feeding on the liver lipids of the white mouse. *J. Biol. Chem.* **127**: 721 (1939).

A.L. Misra, M.N. Patel, V.R. Alluri, S.J. Mulé, and P.K. Nayak. Disposition and metabolism of [^{3}H] cocaine in acutely and chronically treated dogs. *Xenobiotica* **6**: 537 (1976).

W.P. Needham, L. Shuster, G.C. Kanel, and M.L. Thompson. Liver damage from narcotics in mice. *Toxicol. Appl. Pharmacol.* **58**: 157 (1981).

G.M. Padilla, M.W. Kloss, E.J. Rauckman, and G.M. Rosen. Proliferative response and renewal of hepatic function following cocaine administration in mice. *Cell Tissue Kinet.* **17**: 153 (1984).

L.E. Perino, G.H. Warren, and J.S. Levine. Cocaine-induced hepatotoxicity in humans. *Gastroenterology* **93**: 176 (1987).

F.J. Peterson, R.G. Knodell, N.J. Lindeman, and N.M. Steele. Prevention of acetaminophen and cocaine hepatotoxicity in mice by cimetidine treatment. *Gastroenterology* **85**: 122 (1983).

E.J. Rauckman, G.M. Rosen, and J. Cavagnaro. Norcocaine nitroxide. A potential hepatotoxic metabolite of cocaine. *Mol. Pharmacol.* **21**: 458 (1982).

S.M. Roberts, N.P. Skoulis, and R.C. James. A centrally-mediated effect of morphine to diminish hepatocellular glutathione. *Biochem. Pharmacol.* **36**: 3001 (1987).

S.M. Roberts, J.G. Pounds, and R.C. James. Cocaine toxicity in cultured rat hepatocytes. *Toxicol. Lett.* (in press)

G.M. Rosen, M.W. Kloss, and E.J. Rauckman. Initiation of *in vitro* lipid peroxidation by *N*-hydroxynorcocaine and norcocaine nitroxide. *Mol. Pharmacol.* **22**: 529 (1982).

L. Shuster, F. Quimby, A. Bates, and M.L. Thompson. Liver damage from cocaine in mice. *Life Sci.* **20**: 1035 (1977).

L. Shuster, E. Casey, and S.S. Welankiwar. Metabolism of cocaine and norcocaine to *N*-hydroxynorcocaine. *Biochem. Pharmacol.* **32**: 3045 (1983).

L. Shuster, C.A. Garhart, J. Powers, Y. Grunfeld, and G. Kanel. Hepatotoxicity of cocaine, in: Mechanisms of Cocaine Abuse and Toxicity, NIDA Research Monograph Series 88, D. Clouet, K. Ashghar, and R. Brown, eds., U.S. Government Printing Office, Washington, D.C., p. 250 (1988).

N.P. Skoulis, R.C. James, R.D. Harbison, and S.M. Roberts. Depression of hepatic glutathione by opioid analgesic drugs in mice. *Toxicol. Appl. Pharmacol.* **99**: 139 (1989).

A.C. Smith, R.W. Freeman, and R.D. Harbison. Ethanol enhancement of cocaine-induced hepatotoxicity. *Biochem. Pharmacol.* **30**: 453 (1981).

D.J. Stewart, T. Inaba, B.K. Tang, and W. Kalow. Hydrolysis of cocaine in human plasma by cholinesterase. *Life Sci.* **20**: 1557 (1977).

D.J. Stewart, T. Inaba, and W. Kalow. *N*-Demethylation of cocaine in the rat and in isolated rat hepatocytes: comparison with aminopyrine demethylation. *J. Pharmacol. Exp. Ther.* **207**: 171 (1978).

D.J. Stewart, T. Inaba, M. Lucassen, and W. Kalow. Cocaine metabolism: cocaine and norcocaine hydrolysis by liver and serum esterases. *Clin. Pharmacol. Ther.* **25**: 464 (1979).

K.A. Suarez, P. Bhonsle, and D.L. Richardson. Protective effect of *N*-acetylcysteine pretreatment against cocaine induced hepatotoxicity and lipid peroxidation in the mouse. *Res. Commun. Substances Abuse* **7**: 7 (1986).

D. Taylor, V.S. Estevez, L.F. Englert, and B.T. Ho. Hydrolysis of carbon-labeled cocaine in human serum. *Res. Commun. Chem. Pathol. Pharmacol.* **14**: 249 (1976).

C.M. Teaf, R.W. Freeman, and R.D. Harbison. Cocaine-induced hepatotoxicity: lipid peroxidation as a possible mechanism. *Drug Chem. Toxicol.* **7**: 383 (1984).

M.L. Thompson, L. Shuster, and K. Shaw. Cocaine-induced hepatic necrosis in mice — the role of cocaine metabolism. *Biochem. Pharmacol.* **28**: 2389 (1979).

M.L. Thompson, L. Shuster, and E. Casey. Sex and strain differences in response to cocaine. *Biochem. Pharmacol.* **33**: 1299 (1984).

C. Van Dyke and R. Byck. Cocaine and other stimulants, in: *Advances in Behavioral Biology*. E.H. Ellinwood and M.M. Kelby, eds., Plenum Press, New York, (1977).

J.E. Wallace, H.E. Hamilton, D.E. King, D.J. Bason, H.A. Schwertner, and S.C. Harris. Gas-liquid chromatographic determination of cocaine and benzoylecgonine in urine. *Anal. Chem.* **48**: 34 (1976).

H.K. Watanabe, B. Hoskins, and I.K. Ho. Sensitivity difference to hepatotoxicity of cocaine in spontaneously hypertensive and Wistar Kyoto rats. *Alcohol Drug Res.* **7**: 363 (1987).

H.K. Watanabe, B. Hoskins, and I.K. Ho. Effects of subacute treatment with cocaine on activities of *N*-demethylase, UDP-glucuronyltransferase and sulfotransferase in WKY and SHR rat liver — sex and strain differences. *Life Sci.* **42**: 79 (1988).

V. G. Werner. Untersuchungen zum Stoffwechsel von Tropanalkaloiden bei einigen Säugetieren. *Planta Med.* **9**: 293 (1961).

3

Metabolic Mapping of the Effects of Cocaine in the Brain

LINDA J. PORRINO
Department of Physiology and Pharmacology
Bowman Gray School of Medicine
Wake Forest University
Winston-Salem, North Carolina

INTRODUCTION

Cocaine is a widely abused drug with particularly powerful reinforcing properties. It has complex actions in the central nervous system (CNS), but the primary neurochemical action of cocaine in the brain is the blockade of the re-uptake of dopamine (Heikkila *et al.*, 1979; Moore *et al.*, 1977) and norepinephrine (Hertting *et al.*, 1961: Moore *et al.*, 1977). Cocaine also releases catecholamines, but this effect occurs only at high doses (Dembiec, 1980). Serotonergic function is also altered by cocaine administration through re-uptake blockade (Ross and Renyi, 1967) and decreased turnover rates (Pradhan *et al.*, 1978). In addition, cocaine is known to have potent local anesthetic properties (Ritchie and Greene, 1980). Behaviorally, cocaine has a variety of actions, including reinforcing properties (Goldberg and Kelleher, 1976; Pickens and Thompson, 1968), discriminative stimulus properties (Colpaert *et al.*, 1978; Kilbey and Ellinwood, 1979), and locomotor activating properties (Post and Contel, 1983; Post and Rose, 1976; Scheel-Kruger *et al.*, 1977). Cocaine also has actions on the cardiovascular system (Fischman *et al.*, 1976) and on the liver (Shuster, 1977).

The response of an organism to cocaine administration depends on many factors, including the dose of cocaine, the route of administration, the physiological status of the animal, prior drug history, and the environmental setting in which the cocaine is administered. The overall response is unlikely to be the result of a single neurochemical action in a single central or peripheral site, but is more likely to be the product of multiple processes at a number of anatomical sites. In order to determine the neural substrates of the actions of cocaine, it is necessary to identify neural events in circuits and pathways throughout the brain, not just at one location. This requires either that we investigate small portions of the brain one at a time, as with electrophysiological methods, or that we use methods capable of surveying the entire brain simultaneously. The development of the 2-[^{14}C]deoxyglucose (2-DG) method by Louis Sokoloff and colleagues (1977) has provided such a method. The 2-DG method is a potent means by which neuroscientists can investigate the functional events in the brain related to various physiological, pharmacological, and behavioral states. It has been used extensively in neuropharmacology to identify the neural circuits that mediate the effects of a wide variety of pharmacological agents (for review, see McCulloch, 1982). In this chapter, we will describe the application of the 2-DG method to the identification of the neural substrates of the actions of cocaine.

The quantitative 2-DG autoradiographic method measures local rates of cerebral glucose utilization in all regions of the CNS simultaneously. The rate at which glucose is consumed in the brain is a measure of the amount of energy used because under normal physiological conditions, glucose is virtually the exclusive substrate for energy metabolism in the brain. In turn, the amount of energy used is a direct reflection of the amount of work done. In the brain, work is in the form of electrical or synaptic activity — functional activity (Sokoloff, 1982). The 2-DG method then does not measure functional activity directly, but measures glucose utilization, and changes in functional activity are inferred from the changes in glucose utilization. The validity of this assumption has been demonstrated in numerous experiments in which evoked changes in the functional activity of a variety of sensory and motor pathways have been shown by the 2-DG method to be associated with corresponding changes in rates of glucose utilization in the components of the activated or inhibited system (Sokoloff, 1982).

Although a number of energy-requiring processes (e.g., synthesis of neurotransmitters and amino acids) contribute to the basal rates of glucose utilization, the alterations in glucose utilization evoked by a pharmacological, physiological, or behavioral manipulation are thought to result principally from increases or decreases in electrical activity in the CNS. Furthermore, these changes occur mainly at nerve terminals rather than in cell bodies. This means that alterations in energy metabolism in a specific brain area are due to changes in the afferent inputs to that structure and not in the cell bodies contained within the structure.

The most significant advantage of the 2-DG method is that it affords a means to quantitatively measure a dynamic biochemical process in conscious, freely moving animals. Furthermore, the 2-DG method makes use of autoradiography which permits both high spatial resolution so that small anatomical regions and subregions can be identified, and the examination of the entire CNS simultaneously. There are, however, several important limitations of the method. First, it is not possible to identify the specific neurotransmitters, neuromodulators, or receptor types that are responsible for changes in brain electrical activity. The 2-DG method measures only glucose flux. Second, it is not possible to distinguish between inhibitory and excitatory processes, because both are energy consuming. Third, the 2-DG method is unable to differentiate between direct and indirect effects of a given stimulus. An entire pathway or circuit may be metabolically activated even though the direct action of the stimulus may occur only at the origin of or at some point along the pathway. Although this may be a limitation in some instances, as in the case of determining the primary site of action of a drug, it is an advantage when the goal is to identify the neural circuitry that mediates a specific behavioral response.

To measure rates of local cerebral glucose utilization, a radioactively labeled analogue of glucose, 2-[^{14}C]deoxyglucose, is used which, like glucose, is transported into cerebral tissue and phosphorylated by hexokinase but, unlike glucose, is trapped within cells. It is this trapping that allows quantitative autoradiography to be used to measure actual rates of glucose utilization in individual brain regions. Calculations are made from the levels of glucose and 2-[^{14}C]deoxygluxose in plasma measured during the experimental period and the concentration of radioactivity in the tissue measured autoradiographically. The details of the experimental procedures and an extensive discussion of the theoretical basis and mathematical derivation of the 2-DG method is beyond the scope of this chapter. For further discussion, see Sokoloff *et al.*, 1977; Sokoloff and Porrino, 1986.

FUNCTIONAL CONSEQUENCES OF ACUTE COCAINE ADMINISTRATION

The first report characterizing the functional consequences of cocaine administration in the brain was that of London and colleagues (1986). In this experiment, three doses of cocaine (1, 10, 30 mg/kg injected intraperitoneally) were tested in partially restrained rats. A limited number of changes in rates of brain energy metabolism were observed, even at the highest dose studied. Effects occurred mainly in the extrapyramidal motor system, generally considered to be important for the production of stereotyped behavior which was elicited at this dose. An interesting finding in this study was the decreased glucose utilization in the lateral habenula at the 10- and 30-mg/kg doses. Energy metabolism in the lateral habenula is decreased following the administration of other psychostimulants such as amphetamine (Wechsler *et al.*, 1979; Porrino *et al.*, 1984) and dopaminergic agonists such as apomorphine (McCulloch *et al.*, 1982). In addition to the habenula, the pattern of metabolic changes in other brain areas which resulted from cocaine administration in this study was generally similar to those patterns observed with stereotypy-inducing doses of other direct and indirect dopaminergic agonists.

We took a somewhat different experimental approach in our studies of the effects of cocaine on cerebral metabolic activity (Porrino *et al.*, 1988). In designing our experiments, two specific factors were considered. First, the intravenous (i.v.) route of administration was chosen because the bioavailability that results from the i.v. administration best approximates the bioavailability that results from the routes of administration most often used in humans (Jones, 1984). Rates of local cerebral glucose utilization were measured in rats following i.v. administration of three doses of cocaine (0.5, 1.0, and 5.0 mg/kg). The second factor considered was the behavioral response to i.v. cocaine administration. We wanted to relate the alterations in functional activity inferred from changes in glucose utilization to increases in locomotor activity elicited by cocaine. Measurements of locomotor activity were made in freely moving rats during the period of incorporation of the radiolabeled tracer, 2-[^{14}C]DG. The time of maximal behavioral response to cocaine was matched to the time of maximal incorporation to insure the greatest sensitivity of the method.

We found a highly selective pattern of distribution of alterations in cerebral energy metabolism at the lowest doses of cocaine tested, 0.5 and 1.0 mg/kg (Figure 1). Significant increments in glucose utilization were seen only in the medial prefrontal cortex and the nucleus accumbens (Figure 1). At 1.0 mg/kg, glucose utilization was also increased in the substantia nigra reticulata and decreased in the lateral habenula. Differences in mean rates of glucose utilization in 36 other structures were not statistically significant. Behaviorally, these doses produced small, non-significant increases in locomotor activity as compared to vehicle-injected controls.

The selectivity of effects at low doses was in striking contrast to the effects at the highest dose tested. Following the i.v. administration of 5.0 mg/kg, widespread elevations in glucose utilization were present throughout the brain, and these were most prominent in the extrapyramidal motor system (Figure 1). Additionally, glucose use was increased in a number of limbic structures, including the hippocampus and olfactory tubercle. In all, significant alterations in cerebral energy metabolism were found in 20 of the 38 structures examined. Behaviorally, animals were significantly more active following the administration of the 5.0-mg/kg dose of cocaine than after either vehicle or low-dose cocaine administration.

The significant correlations between locomotor activity and rates of glucose utilization that were found in portions of the extrapyramidal motor system, including the globus pallidus and substantia nigra reticulata, suggest that the changes in functional activity in these areas are a reflection of the alterations in motor function elicited by cocaine administration. In contrast, the pattern of alterations in local cerebral energy metabolism in the medial prefrontal cortex and nucleus accumbens is, however, more in concert with the role that these portions of the mesocorticolimbic dopaminergic system are thought to have in the mediation of cocaine's reinforcing effects rather than its effects on locomotion. The significance of the medial prefrontal cortex and nucleus accumbens has been demonstrated in a number of studies. Rats, for example, will self-administer cocaine directly into the prefrontal cortex (Goeders and Smith, 1983; Goeders *et al.*, 1986), and lesions of the nucleus accumbens can disrupt responding for the i.v. self-administration of cocaine (Roberts *et al.*, 1977; Roberts *et al.*, 1980; Koob, 1987). That the selective metabolic effects of cocaine in these areas occurred only at doses that are known to be reliably self-administered intravenously by rats is further support for the nucleus accumbens and medial prefrontal cortex as neural substrates underlying cocaine reinforcement.

Several considerations may account for the differences between the findings reported by London and colleagues (1986) and those of this study. A number of factors, including rat strain, use of restrained vs. freely moving preparations, and the route of administration, may account for the discrepancies. We tested one of these possibilities, route of administration, by comparing the effects on local cerebral glucose utilization of an intraperitoneal (i.p.) dose (10 mg/kg) of cocaine to those of an i.v. dose (1.0 mg/kg), both of which produced similar increases in locomotor activity (Porrino, 1988). As in previous studies, rates of glucose utilization did not

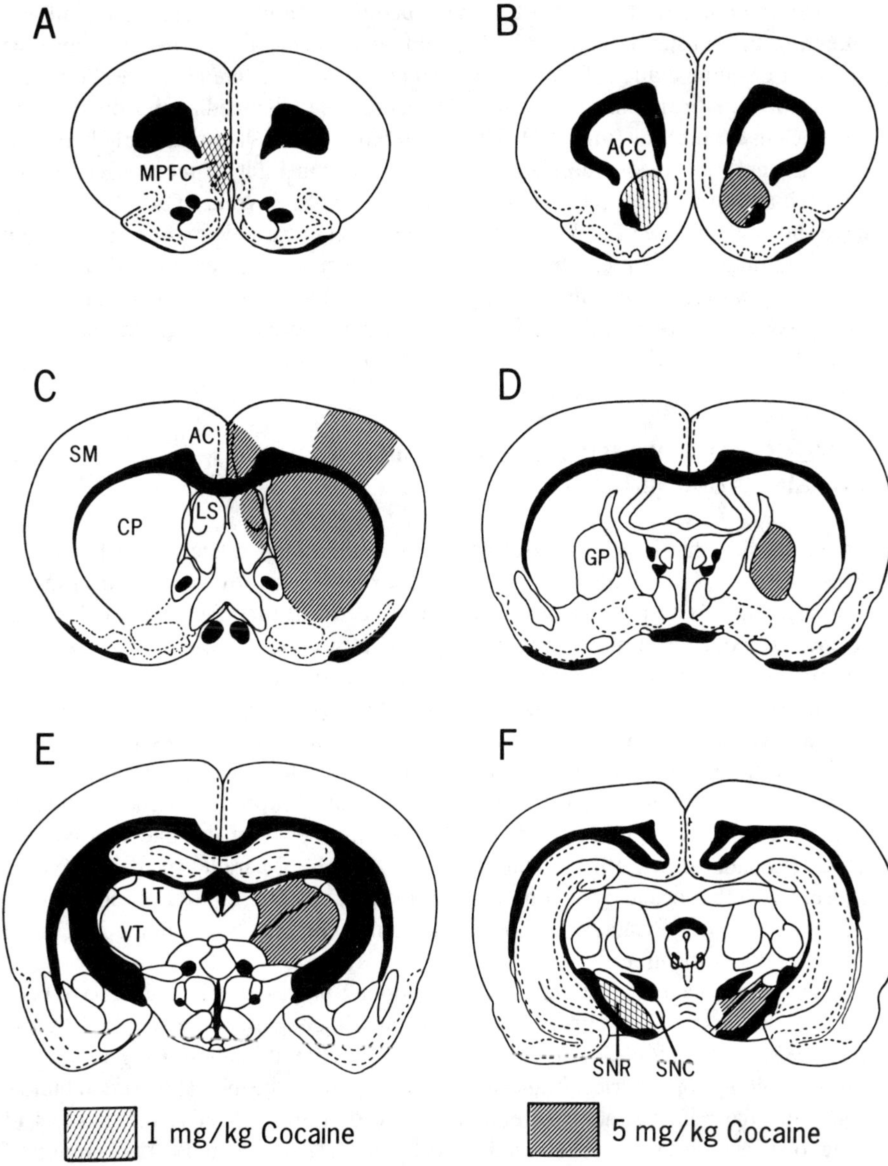

Figure 1. Diagrams illustrating the distribution of alterations in local cerebral glucose utilization associated with the administration of 1.0 and 5.0 mg/kg, i.v., as depicted in six frontal planes (A through F). Diagrams are adapted from Paxinos and Watson (1982). AC, Anterior cingulate cortex; ACC, nucleus accumbens; CP, caudate-putamen; GP, globus pallidus; LS, lateral septum; LT, lateral thalamus; MPFC, medial prefrontal cortex; SM, somatomotor cortex; SNC, substantia nigra pars compacta; SNR, substantia nigra pars reticulata; and VT, ventral thalamus.

differ from control values in most of the structures examined. Rates were similarly altered in the substantia nigra reticulata and lateral habenula following either type of cocaine administration (Figure 2). Intravenously administered cocaine, however, resulted in significant changes in rates of glucose use in the medial prefrontal cortex and nucleus accumbens (Figure 2). These differences in the distribution of alterations in local cerebral glucose utilization are interesting in light of recent findings by Spyraki and co-workers (Nomikos and Spyraki, 1988; Spyraki *et al.*, 1987), who have shown that route of administration is a significant variable in the establishment of a conditioned place preference in rats. Brain concentrations and the rate at which these concentrations are achieved vary with the route of administration and, therefore, may in turn be significant factors in determining the local metabolic effects as well as the behavioral consequences of cocaine administration.

FUNCTIONAL CONSEQUENCES OF CHRONIC COCAINE ADMINISTRATION

The studies that have been described up to this point relate to the functional consequences of the acute administration of a single dose of cocaine to drug-naive animals. The 2-DG method has also been used to identify the functional consequences of cocaine administration in more complex conditions. A recent study has examined the effects of chronic cocaine administration (Hammer *et al.*, in press). In this study, cocaine (10 mg/kg) was injected i.p. daily for 14 d and glucose utilization measured 10 min after the last dose. In addition to increases in metabolism in the substantia nigra and globus pallidus, metabolic rates in the nucleus accumbens were also elevated following this regimen. Repeated exposure to cocaine altered glucose utilization in this region, whereas a single dose of cocaine was without effect (London *et al.*, 1986). This difference may be a parallel of the behavioral sensitization that occurs with repeated cocaine administration (see Post and Contel, 1983) and provide evidence for a role for the nucleus accumbens in the process of behavioral sensitization.

The specific neural substrates of the reinforcing properties of cocaine have been investigated using brain stimulation reward as a model. In these studies (Bain *et al.*, 1987; Porrino *et al.*, 1989), local cerebral glucose utilization was measured in rats responding for electrical brain stimulation reward to the medial forebrain bundle following the administration of either cocaine or saline. Briefly, the patterns of functional activation were similar in most brain regions in these two groups of animals except in the olfactory tubercle. In this portion of the ventral striatum, significantly higher rates of energy metabolism were found in the cocaine-treated animals. These data, discussed in more detail by Lyons and Kornetsky (see Chapter 4), indicate that the olfactory tubercle may be important in mediating the reinforcing properties of cocaine.

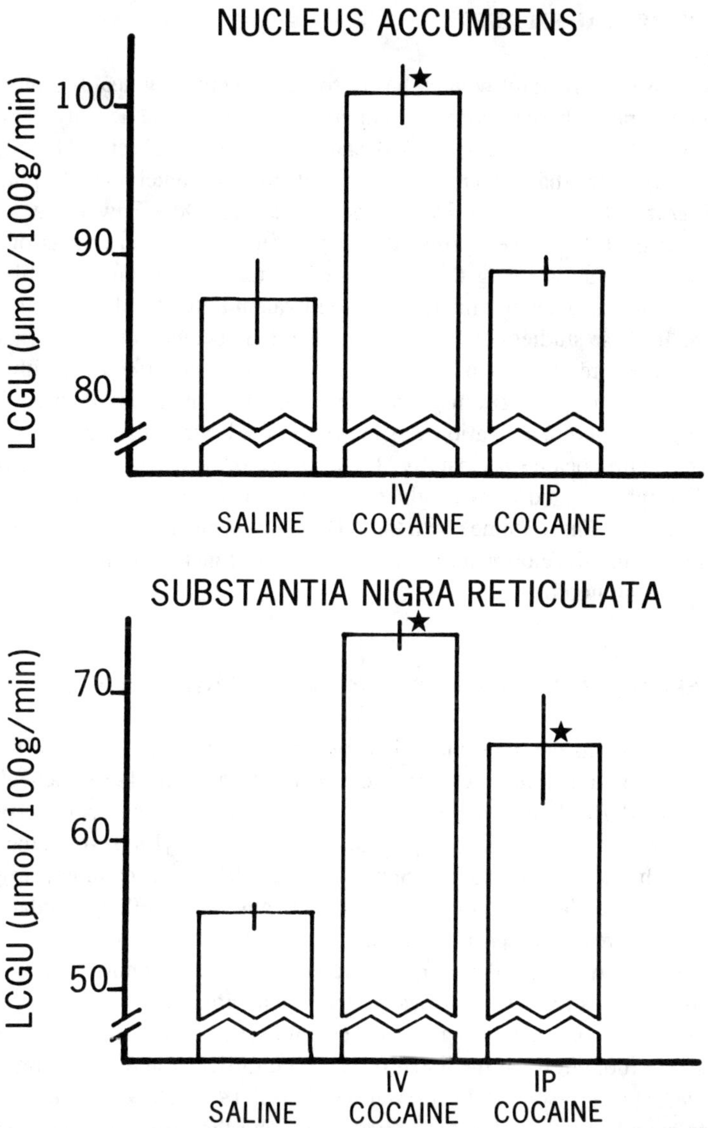

Figure 2. A comparison of the effects of saline vehicle, and acute intravenous (1 mg/kg) and intraperitoneal (10 mg/kg) administration of cocaine on local cerebral glucose utilization (LCGU) in the nucleus accumbens and substantia nigra pars reticulata in rats. Values represent means and standard errors of rates of LCGU (μmol/100 g/min) in groups of 6,4, and 4 rats, respectively. $p < 0.05$, significantly different from saline vehicle control, Bonferroni t-statistic.

DEVELOPMENTAL STUDIES

Cocaine use by pregnant women has become an increasing health problem in the U.S. Behaviorally, babies born to mothers who abuse cocaine during pregnancy tend to be jittery and have increased startle responses (Chasnoff *et al.*, 1985). Teratological effects have been seen in animal studies (Fantel and Macphail, 1982; Mahalik *et al.*, 1980). Dow-Edwards and colleagues (Dow-Edwards *et al.*, 1988; Dow-Edwards, 1988) have examined the long-term functional effects of neonatal cocaine administration using the 2-DG method to measure glucose utilization in adult rats treated at a developmental stage equivalent to the third trimester of human gestation. In these studies, pups were administered cocaine during the first 10 d of life, a treatment regimen which did not have significant effects on either weight gain or reflex ontogeny. Brain glucose metabolism was assessed in adulthood at 60 d of age. Metabolic activation was present in the brains of female, but not male, rats treated with cocaine perinatally. Increases were most evident in limbic structures, although elevated rates of glucose utilization were also seen in sensory and motor structures. These studies with the 2-DG method clearly illustrate that exposure to cocaine during development can have long-term functional consequences which persist into adulthood.

COMPARISON WITH OTHER PSYCHOMOTOR STIMULANTS

Psychostimulants can be divided into two separated drug classes on the basis of their actions at dopaminergic nerve terminals (McMillen, 1983; Scheel-Kruger, 1971; Shore, 1976). Drugs in the amphetamine class of stimulants act via stimulation of release of newly synthesized dopamine from an α-methyl-*para*-tyrosine sensitive pool (Chiueh and Moore, 1975b; Scheel-Kruger, 1971). In contrast, drugs in the nonamphetamine class of stimulants (e.g., cocaine and methylphenidate) release dopamine from reserpine-sensitive vesicular stores (Chiueh and Moore, 1975; Scheel-Kruger, 1971). Although there are differences in the biochemical actions of these drugs, there are strong similarities in their behavioral effects (e.g., reinforcing properties, locomotor activating properties, and stereotypy eliciting properties), all of which are thought to be mediated through their effects on dopaminergic systems (see Creese and Iverson, 1975; DeWit and Wise, 1977; Roberts *et al.*, 1975).

There have been a number of studies examining the effects of psychostimulants other than cocaine on brain glucose utilization. The dose-dependent nature of cocaine's effects on local cerebral energy metabolism is clearly analogous to the effects of other psychostimulants on functional activity. Low doses of either amphetamine (Porrino *et al.*, 1984) or methylphenidate (Porrino and Lucignani, 1987), for example, result in selective increases in glucose utilization in the nucleus accumbens (Figure 3). Again, as with cocaine, there appears to be a greater sensitivity of anatomical portions of the mesocorticolimbic dopaminergic system, as compared to components of other dopaminergic systems, to the actions of psychostimulants.

At higher doses, parallel effects on glucose utilization and behavior also occur.

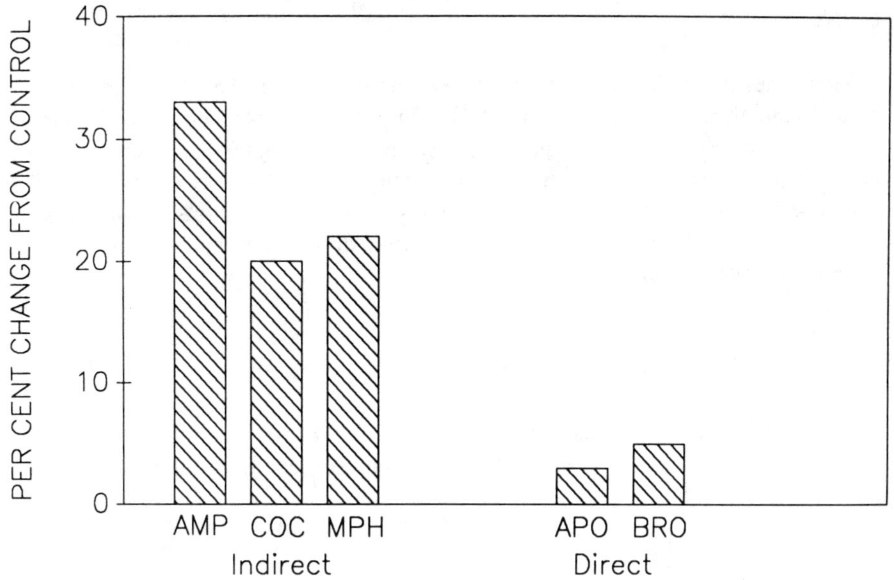

Figure 3. Comparison of the alterations in local cerebral glucose utilization accompanying the administration of selected dopaminergic agonists, indirect agonists, psychostimulants, and direct dopaminergic receptor agonist drugs. Values are percent change from control. AMPH, amphetamine (Porrino *et al.*, 1984); COC, cocaine (Porrino *et al.*); MPH, methylphenidate (Porrino and Lucignani, 1987); APO, apomorphine (McCulloch *et al.*, 1982; Porrino, unpublished observations); and BRO, bromocriptine (Pizzolato *et al.*, 1985).

Both amphetamine (Wechsler *et al.*, 1979; Porrino *et al.*, 1984) and methylphenidate (Bell *et al.*, 1982; Porrino and Lucignani, 1987) produce functional changes in many of the components of the extrapyramidal system similar to those produced by cocaine (Porrino *et al.*, 1988; London *et al.*, 1986). At these doses, all three psychostimulants elicit stereotypic behavior patterns which may include repetitive sniffing, grooming, rearing, circling, and/or gnawing (Fog, 1969).

Although this dose-dependent pattern of effects within the mesocorticolimbic and nigrostriatal dopaminergic systems seems to be typical of psychostimulants, such a pattern is not characteristic of other dopaminergic drugs. Apomorphine, a direct dopamine receptor agonist, for example, does not significantly alter glucose utilization in the mesocorticolimbic system. It is without effect on such areas as the nucleus accumbens (Figure 3) and prefrontal cortex (McCulloch *et al.*, 1982). This pattern is similar to that obtained following the administration of bromocriptine, another dopamine receptor agonist (Pizzolato *et al.*, 1985) (Figure 3). It appears that psychostimulants have a unique metabolic action in the mesocorticolimbic dopaminergic system.

CONCLUSIONS

This series of studies serves to emphasize the diversity of the neural circuits that mediate the effects of cocaine in the CNS. In particular, they show how variables such as dose, route of administration, and behavioral context can determine the neurochemical response to cocaine administration. The 2-DG method has provided a comprehensive view of the functional involvement of various brain areas in response to cocaine administration and may furnish information about the specific neural substrates responsible for cocaine abuse.

ACKNOWLEDGMENTS

Thanks to Alison Crane and Charles Kennedy, M.D., for their helpful comments on this manuscript and to Dory Feldman for her excellent editorial assistance.

REFERENCES

G.T. Bain, L.J. Porrino, C. Caplan, and C. Kornetsky. Cocaine effects on rewarding brain stimulation as assessed by the quantitative 2-deoxyglucose method. *Soc. Neurosci. Abstr.* **13**: 1322 (1987).

R.D. Bell, G.M. Alexander, R.J. Schwartzman, and J. Yu. The methylphenidate-induced stereotypy in the awake rat: local cerebral metabolism. *Neurology,* **32**: 377 (1982).

I.J. Chasnoff, W.J. Burns, S.H. Scholl, and K.A. Burns. Cocaine use in pregnancy. *N. Engl. J. Med.* **313**: 666 (1985).

C.C. Chiueh and K.E. Moore. Blockade by reserpine of methylphenidate-induced release of brain dopamine. *J. Pharmacol. Exp. Ther.* **193**: 559 (1975).

C.C. Chiueh and K.E. Moore. D-Amphetamine-induced release of newly synthesized and stored dopamine from the caudate nucleus *in vivo*. *J. Pharmacol. Exp. Ther.* **192**: 642 (1975).

F.C. Colpaert, J.E. Niemegeers, and P.A.J. Janssen. Neuroleptic interference with cocaine cue. Internal stimulus control of behavior and psychosis. *Psychopharmacology* **58**: 347 (1978).

I. Creese and S.D. Iverson, The pharmacological and anatomical substrates of the amphetamine response in the rat. *Brain Res.* **83**: 419 (1975).

D. Dembiec. Inhibition of potassium stimulated release of [^{3}H]-dopamine from rat striata as a result of prior exposure to cocaine, nomifensine, or mazindol. *Neurochem. Res.* **5**: 345 (1980).

H. DeWit and R.A. Wise. Blockade of cocaine reinforcement in rats with the dopamine blocker, pimozide, but not with the noradrenergic blockers phentolamine or phenoxybenzamine. *Can. J. Psychol.,* **31**: 195 (1977).

D.L. Dow-Edwards. Development effects of cocaine, in: Mechanisms of Cocaine Abuse and Toxicity. NIDA Monograph Series 88, D. Clouet, K. Ashghar, and R. Brown, eds., U.S. Government Printing Office, Washington, DC, p. 290 (1988).

D.L. Dow-Edwards, L.A. Freed, and T.H. Milhorat. Stimulation of brain metabolism by perinatal cocaine exposure. *Dev. Brain Res.* **42**: 137 (1988).

A.G. Fantel and T. Macphail. The teratogenicity of cocaine. *Teratology* **26**: 17 (1982).

M.W. Fischman, C.R. Schuster, L. Rosnekov, J.F.E. Shick, N.A. Krasnegor, W. Fennel, and D.X. Freedman. Cardiovascular and subjective effects of intravenous cocaine administration in humans, *Arch. Gen. Psychiat.* **33**: 983 (1976).

R. Fog. Stereotyped and non-stereotyped behavior in rats induced by various stimulant drugs. *Psychopharmacology,* **24**: 299 (1969).

N.E. Goeders, S.I. Dworkin, and J.E. Smith. Neuropharmacological assessment of cocaine self-administration into the medial prefrontal cortex. *Pharmacol. Biochem. Behav.* **24**: 1429 (1986).

N.E. Goeders and J.E. Smith. Cortical dopaminergic involvement in cocaine reinforcement. *Science* **221**: 773 (1983).

S.R. Goldberg and R.T. Kelleher. Behavior controlled by scheduled injections of cocaine in squirrel and rhesus monkeys. *J. Exp. Anal. Behav.* **25**: 93 (1976).

R.P. Hammer, Jr., S.J. Lee, D.W. Clow, and L.J. Porrino. Chronic cocaine increases glucose utilization in dopaminergic reward regions. *Am. Assoc. Anat. Abstr.* in press.

R.E. Heikkila, F.S. Cabbot, L. Manzino, and R.C. Duvoisin. Rotational behavior induced by cocaine analogs in rats with unilateral 6-hydroxydopamine lesions of the substantia nigra: dependence upon dopamine uptake inhibition. *J. Pharmacol. Exp. Ther.* **211**: 189 (1979).

G. Hertting, J. Axelrod, and L.G. Whitby. Effect of drugs on the uptake and metabolism of ^{3}H-norepinephrine. *J. Pharmacol. Exp. Ther.* **134**: 146 (1961).

R. Jones. The pharmacology of cocaine, in: Cocaine: Pharmacology, Effects and Treatment of Abuse. National Institute on Drug Abuse Research Monograph Series 50, J. Grabowski, ed., U.S. Government Printing Office, Washington, DC, p. 34 (1984).

M.M. Kilbey and E.H. Ellinwood. Discriminative stimulus properties of psychomotor stimulants in cats. *Psychopharmacology* **63**: 151 (1979).

G.F. Koob. Separate neurochemical substrates for cocaine and heroin reinforcement, in: *Quantitative Analyses of Behavior: Biological Determinants of Behavior*. Vol. 7, R.M. Church, M.L. Commons, J. Stellar, and A.R. Wagner, eds., Lawrence Erlbaum Associates, Hillsdale, NJ, p. 139 (1987).

E.D. London, G. Wilkerson, S.R. Goldberg, and M.E. Risner. Effects of L-cocaine on local cerebral glucose utilization in the rat. *Neurosci. Lett.* **68**: 73 (1986).

M.P. Mahalik, R.F. Gautieri, and D.E. Mann. Teratogenic potential of cocaine hydrochloride in CF-1 mice. *J. Pharm. Sci.* **69**: 703 (1980).

J. McCulloch. Mapping functional alterations in the CNS with [^{14}C]deoxyglucose, in: *Handbook of Psychopharmacology*. Vol. 15, L. Iverson, S. Iverson, and S. Snyder, eds., Plenum Press, New York, p. 321 (1982).

J. McCulloch, H.E. Savaki, M.C. McCulloch, J. Jehle, and L. Sokoloff. The distribution of alterations in energy metabolism in the rat brain produced by apomorphine. *Brain Res.* **243**: 67 (1982).

B.A. McMillen. CNS stimulants: two distinct mechanisms of action for amphetamine-like drugs. *Trends Pharmacol. Sci.* **4**: 429 (1983).

K.E. Moore, C.C. Chiueh, and G. Zeldes. Release of neurotransmitters from the brain *in vivo* by amphetamine, methylphenidate and cocaine, in: *Cocaine and Other Stimulants,* E.H. Ellinwood and M.M. Kilbey, eds., Plenum Press, New York, p. 143 (1977).

G.G. Nomikos and C. Spyraki. Cocaine-induced place preference conditioning: importance of route of administration and other procedural variables. *Psychopharmacology* **94**: 119 (1988).

G. Paxinos and C. Watson. *The Rat Brain in Stereotaxic Coordinates.* Academic Press, New York (1982).

R. Pickens and T. Thompson. Cocaine reinforced behavior in rats. Effects of reinforcement magnitude and fixed ratio size. *J. Pharmacol. Exp. Ther.* **161**: 122 (1968).

G. Pizzolato, T.T. Soncrant, and S.I. Rapoport. Time-course and regional distribution of the metabolic effects of bromocriptine in the rat brain. *Brain Res.* **341**: 303 (1985).

L.J. Porrino, Effects of cocaine on local cerebral glucose utilization depend on rate of administration. *FASEB* **2**: A1138 (1988).

L.J. Porrino, F.R. Domer, A.M. Crane, and L. Sokoloff. Selective alterations in cerebral metabolism within the mesocorticolimbic dopaminergic system produced by acute cocaine administration in rats. *Neuropsychopharmacology* **1**: 109 (1988).

L.J. Porrino and G. Lucignani. Different patterns of brain energy metabolism associated with high and low doses of methylphenidate: implications for stimulant drug action in hyperactive children. *Biol. Psychiat.* **22**: 125 (1987).

L.J. Porrino, G. Lucignani, D. Dow-Edwards, and L. Sokoloff. Dose-dependent effect of acute administration on functional brain metabolism in rats. *Brain Res.* **307**: 311 (1984).

L.J. Porrino, D. Lyons, G. Bain, L. Sokoloff, and C. Kornetsky. Similarities in the cerebral metabolic effects of cocaine and morphine on rewarding brain stimulation. *FASEB* **3**: A419 (1989).

R.M. Post and H. Rose. Increasing effects of repetitive cocaine administration in the rat. *Nature* **260**: 731 (1976).
R.M. Post and N.R. Contel. Human and animal studies of cocaine: implications for the development of behavioral pathology, in: *Stimulants: Neurochemical, Behavioral and Clinical Perspectives*. Creese, I., ed., Raven Press, New York, p. 169 (1983).
S.N. Pradhan, A.K. Battacharyya, and S. Pradhan. Serotonergic manipulation of the behavioral effects of cocaine in rats. *Commun. Psychopharmacol.* **2**: 481 (1978).
J.M. Ritchie and N.M. Greene. Local anesthetics, in: *The Pharmacological Basis of Therapeutics*. L.S. Goodman and A. Gilman, eds., Macmillan, Toronto, p. 300 (1980).
D.C.S. Roberts, M.E. Corcoran, and H.C. Fibiger. On the role of ascending catecholaminergic systems in intravenous self-administration of cocaine. *Pharmacol. Biochem. Behav.* **6**: 615 (1977).
D.C.S. Roberts, G.F. Koob, P. Klonoff, and H.C. Fibiger. Extinction and recovery of cocaine self-administration following 6-hydroxydopamine lesions of the nucleus accumbens. *Pharmacol. Biochem. Behav.* **12**: 781 (1980).
D.C.S. Roberts, A.P. Zis, and H.C. Fibiger. Ascending catecholamine pathways and amphetamine-induced locomotor activity: importance of dopamine and apparent non-involvement of norepinephrine. *Brain Res.* **93**: 441 (1975).
S.F. Ross and A.L. Renyi. Inhibition of the uptake of tritiated catecholamines by antidepressants and related agents. *Eur. J. Pharmacol.* **2**: 181 (1967).
J. Scheel-Kruger. Comparative studies of various amphetamine analogues demonstrating different interactions with the metabolism of the catecholamines in the brain. *Eur. J. Pharmacol.* **14**: 47 (1971).
J. Scheel-Kruger, C. Braestrup, M. Nielson, K. Golembiowska, and E. Mogilnicka. Cocaine: discussion on the role of dopamine in the biochemical mechanism of action, in: *Cocaine and Other Stimulants*. E.H. Ellinwood and M.M. Kilbey, eds., Plenum Press, New York, p. 373 (1977).
P.A. Shore. Actions of amfonelic acid and other nonamphetamine stimulants on the dopamine neuron. *J. Pharm. Pharmacol.* **28**: 855 (1976)
L. Shuster, F. Quimby, A. Bates, and M.L. Thompson. Liver damage from cocaine in mice. *Life Sci.*, **20**: 1035 (1977).
L. Sokoloff. The radioactive deoxyglucose method: theory, procedure, and applications for the measurement of local glucose utilization in the central nervous system, in: *Advances in Neurochemistry*. B.W. Agranoff and M.H. Aprison, eds., Plenum Press, New York, p. 7 (1982).
L. Sokoloff and L.J. Porrino. Some fundamental considerations in the application of the deoxyglucose method to pharmacological studies, in: *Pharmacology of Cerebral Ischemia*, I.Kriegelstein, ed., Elsevier, Amsterdam, p. 65 (1986).
L. Sokoloff, M. Reivich, C. Kennedy, M. H. DesRosiers, C.S. Patlak, K.D. Pettigrew, O. Sakurada, and M. Shinohara. The [^{14}C]deoxyglucose method for the measurement of local cerebral glucose utilization: theory, procedure and normal values in the conscious and anesthetized albino rat. *J. Neurochem.* **28**: 897 (1977).
C. Spyraki, G.G. Nomikos, and D.D. Varonos. Intravenous cocaine-induced place preference: attenuation by haloperidol. *Behav. Brain Res.* **26**: 57 (1987).
L.R. Wechsler, H.E. Savaki, and L. Sokoloff. Effects of D-amphetamine and L-amphetamine on local cerebral glucose utilization in the conscious rat. *J. Neurochem.* **32**: 15 (1979).

4

Brain-Stimulation Reward as a Model of Drug-Induced Euphoria: Comparison of Cocaine and Opiates

DAVID HUSTON-LYONS and CONAN KORNETSKY
Laboratory of Behavioral Pharmacology
Boston, Massachusetts

INTRODUCTION

This chapter presents the thesis that cocaine as well as morphine (and other sychomotor stimulants and opiates that are abused) share a common underlying site of neuronal action that is, at least in part, responsible for their hedonic effects.

Although cocaine and morphine have many opposing effects on a variety of physiological functions [e.g., general central nervous system (CNS) activity, body temperature, and blood pressure (Ritchie and Greene, 1985; Jaffe and Martin, 1985)] both are abused and induce feelings of euphoria in humans (Jaffe, 1985), are self-administered by animals (Deneau *et al.*, 1965; Pickens and Thompson, 1968; Weeks, 1962), increase the sensitivity of animals to rewarding electrical intracranial stimulation (Esposito *et al.*, 1978; Marcus and Kornetsky, 1974), a model of drug-induced euphoria, and increase dopamine turnover (Iverson *et al.*, 1973; Wood and Altar, 1988). These opposing and similar effects are illustrated in Figure 1. The effects shared by these two classes of drugs, represented by cocaine and morphine, provide a window into the neuronal bases for their hedonic actions, the necessary ingredient for their positive reinforcing effects. Despite the hedonic quality of abuse substances, this aspect of their action has often been conspicuously absent from concepts of addiction and abuse. A common theme often found in theories of drug abuse is that abused substances restore normal psychological or neurophysiological function (Lettieri *et al.*, 1980). Although abused substances may be restorative (e.g., a depressed existence may become more tolerable by the use of a drug that elevates mood), reinstatement of normal or near-normal function would not occur

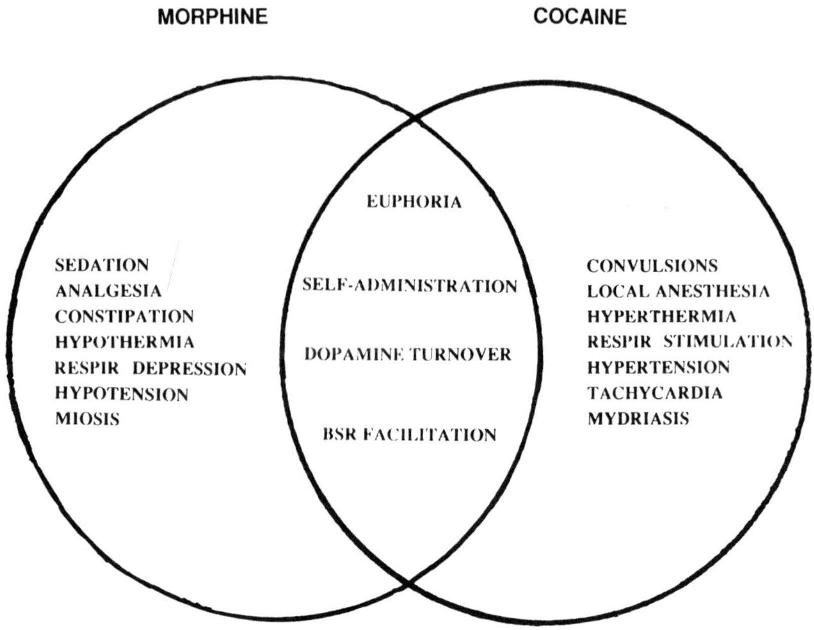

Figure 1. The effects of morphine are shown on the left and those of cocaine on the right. The area of overlap represents the common effects produced by both drugs.

if the abused substance did not, in some way, cause some feeling of well-being, euphoria, or what is often described as a "high". This hedonic effect, although not sufficient, is necessary for abuse and addiction to occur (Kornetsky, 1985).

In this chapter we will review the effects that cocaine and morphine, as well as related drugs, share on brain-stimulation reward (BSR), dopamine metabolism, and local cerebral glucose metabolism as evidence that common neural substrates mediate the reinforcing effects of many drugs of abuse.

BRAIN-STIMULATION REWARD (BSR)

Soon after the discovery that aversive intracranial electrical brain stimulation could be a negative reinforcer in learning experiments (Delgado *et al.*, 1954; Hess, 1957), Olds and Milner (1954) found that stimulation to sites in the septum and hypothalamus of rats could act as a positive reinforcer. Later studies confirmed that positive reinforcement could be obtained from intracranial electrical stimulation of human subjects as well (Heath *et al.*, 1960). In 1956, Olds and co-workers described the first study of the effects of the drugs chlorpromazine and reserpine on self-stimulation, in which a decreased rate of lever pressing for stimulation was found. In subsequent years, pentobarbital (Killam *et al.*, 1957), amphetamine (Killam *et al.*, 1957; Stein and Ray, 1960), morphine (Olds and Travis, 1960), and finally cocaine (Stein, 1964) were each found to increase an animal's sensitivity to rewarding brain stimulation.

Maps of brain sites that support self-stimulation indicate that this behavior may be elicited from discrete sites throughout the brain, from the telencephelon to the cerebellum (Phillips, 1984). However, the most reliable and robust self-stimulation behavior may be obtained from mesocorticolimbic sites such as the ventral tegmental area and the medial forebrain bundle (Crow, 1972). Furthermore, behavior motivated by natural rewards, e.g., food (Anand and Brobeck, 1951) and sex (Singer, 1968; Hitt *et al.*, 1973), are affected by discrete hypothalamic lesions within or adjacent to areas that support BSR. It is unlikely to be only a coincidence that a variety of drugs of abuse increase an animal's sensitivity to stimulation to these areas. Data from studies such as these demonstrate the existence of a discrete brain reward system that supports not only natural rewards but also the reinforcing effects of drugs.

Several methods for assessing drug effects on BSR are commonly employed (Kornetsky and Bain, 1990). These methods may be differentiated by the dependent measure used, which falls into two categories: (1) rate of responding, usually lever pressing for stimulation or (2) the threshold level of stimulation required to illicit a response, e.g., the current intensity of stimulation for which an animal will work 50% of the time. Increased sensitivity to rewarding brain stimulation is manifest as an increase in rate of responding or a lowering of the threshold.

The major hurdle for any measure of BSR is the separation of reward-specific drug effects from performance deficits and locomotor activation (Liebman, 1983). Several methods may adequately measure reward-specific changes if the proper controls for drug effects on baseline responding and locomotor activity are performed (Kornetsky and Bain, 1990; Liebman, 1983). The value of studies of BSR is twofold. Increased sensitivity to stimulation is a predictor of abuse liability, that is, drugs that increase sensitivity to BSR are euphorigenic and abused by man. This method has the additional advantage, which studies of intravenous self-administration and drug discrimination do not, that sites of stimulation that show drug-induced increased sensitivity are implicated as mediators of the rewarding effects of drugs.

COCAINE, MORPHINE, AND BRAIN-STIMULATION REWARD

Stein (1964) described the first investigation of the effects of cocaine on BSR that showed cocaine facilitated rate of lever pressing for BSR and potentiated the rate facilitating effects of amphetamine. As described in Table 1, subsequent investigators have consistently found that cocaine increased the rate of responding and lowered the threshold for stimulation. As an example of cocaine's ability to increase an animal's sensitivity to rewarding brain stimulation, Figure 2 depicts the threshold-lowering effects of 2.5–20 mg/kg of cocaine in animals. This figure illustrates the dissociation of an animal's sensitivity to electrical stimulation of

Table 1
Effects of Cocaine on Brain-Stimulation Reward

Site of stimulation	Method	Dose range tested (mg/kg)	Result	Ref.
VTA	CRF	4–5	Facilitated rate; potentiated amphetamine effects	Stein, 1964
VTA	CRF	5	Facilitated rate	Crow, 1970
MFB-LH	CRF	0.63–10	Facilitated rate	Wauquier and Niemegeers, 1974
LH	CRF	5	Facilitated rate; adrenergic antagonists failed to block effect	Bailey and Pradhan, 1975
LH	CRF	10	Facilitated rate; effect blocked by pimozide	Wauquier and Niemegeers, 1976
MFB	Discrete trial threshold	1–60	Lowered reward threshold	Esposito *et al.*, 1978
PH, VTA	CRF	2–60	Facilitated rate to both sites	Aulakh *et al.*, 1979
PH, VTA	CRF	10	Facilitated rate	Bhattacharyya and Pradhan, 1979
MFB	Discrete trial threshold, detection	2.5–20	Lowered reward threshold at doses with no effect on detection	Kornetsky and Esposito, 1981
MFB-LH	Rate-intensity function	5–7.5	Facilitated rate; cross tolerance to amphetamine	Leith and Barrett, 1981
MFB, NACC, OT, SN	CRF in rat pups	5–30	Facilitated rate	Barr and Lithgow, 1986
LH	Rate-intensity function	5–20	Facilitated rate; lowered threshold; potentiated by imipramine	Glick and Badalamenti, 1986
MFB, VTA	Discrete trial threshold	10–15	Naloxone blocked cocaine's threshold-lowering effect	Bain and Kornetsky, 1987
VTA	Rate-train duration function	10–30	Facilitated rate; lowered threshold	Frank *et al.*, 1988
VTA	Rate-train duration function	10–15 daily for 18d	Facilitated rate; lowered threshold; no tolerance or sensitization	Frank *et al.*, 1988
VTA	Two lever reset threshold	5–30	Facilitated rate; lowered threshold; naloxone did not block cocaine's effect	van Wolfswinkel *et al.*, 1988

Note: NACC, nucleus accumbens; OT, olfactory tubercle; MFB, medial forebrain bundle; LH, lateral hypothalamus; PH, posterior hypothalamus; VTA, ventral tegmental area; SN, substantia nigra.

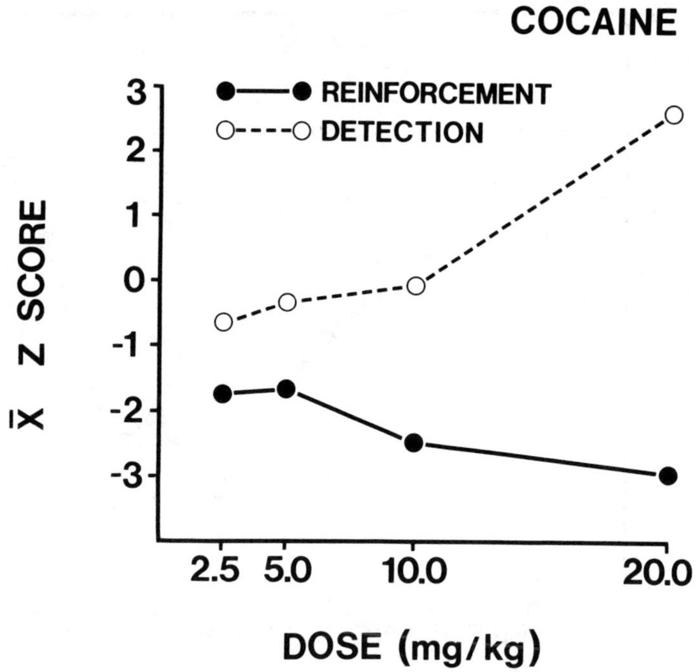

Figure 2. Mean z-score (standard score) changes in the reinforcement and detection thresholds from pre- to post-drug administration as a function of dose of intraperitoneally injected cocaine (n = 4). The absolute difference between post- minus predrug thresholds measured in microamperes has been converted to z-scores based on the mean and standard deviation of the difference found after administration of saline. A z-score of ±2 defines the 95% confidence limits for individual animals; thus, a z-score less than −2 indicates a lowering of the threshold and an increase in sensitivity, a z-score greater than +2 indicates a decrease in sensitivity.

sufficient current intensity to be rewarding and sensitivity to a just-detectable intensity of stimulation to the same site. Cocaine lowers the reinforcement threshold, that is, lowers the current intensity of stimulation for which an animal will work, while simultaneously impairing the animal's ability to discriminate low-level stimulation. Additional work has demonstrated that cocaine and neuroleptics play antagonistic roles with respect to self-stimulation and reverse each others effects (Aulakh *et al.*, 1979; Wauquier and Niemegeers, 1976), that is neuroleptics decrease and cocaine increases the rate of responding. Despite these well-established findings, further investigation is necessary to account for additional conflicting results. For example, differences in methodology may explain why tolerance to the effects of cocaine on BSR was found in some studies (Leith and Barrett, 1981) but not others (Frank *et al.*, 1988), as well as whether naloxone is effective (Bain and Kornetsky, 1987) or ineffective (van Wolfswinkel *et al.*, 1988) in blocking the threshold-lowering effects of cocaine. Studies to date clearly demonstrate that cocaine increases an animal's sensitivity to rewarding stimulation to various sites in the mesolimbic system (see Table 1) and these effects are mediated by catecholaminergic activity (Aulakh *et al.*, 1979; Wauquier and Niemegeers, 1976).

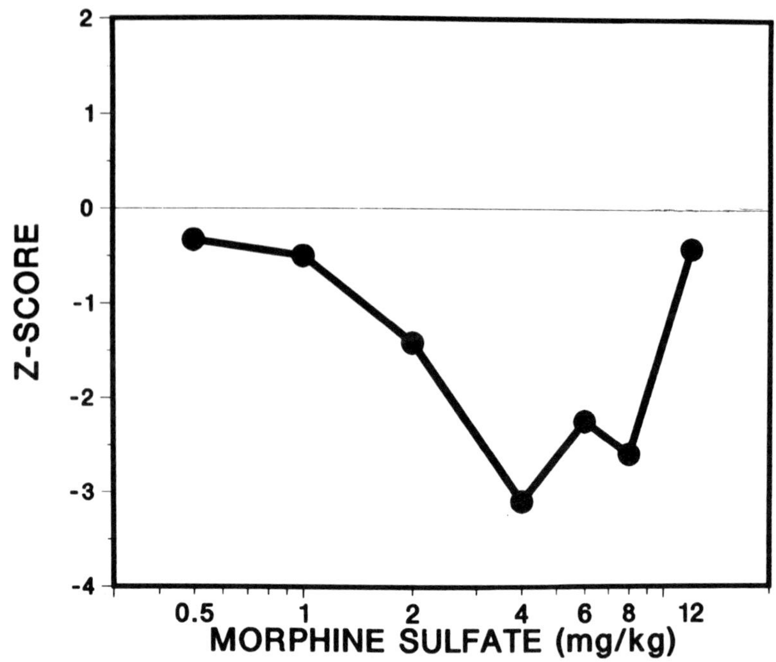

Figure 3. Mean z-score (standard score) changes in the reward threshold from pre- to post-drug administration as a function of dose of subcutaneously injected morphine ($n = 10$). The absolute difference between post- minus predrug thresholds measured in microamperes has been converted to z-scores based on the mean and standard deviation of the difference found after administration of saline. A z-score of ± 2 defines the 95% confidence limits for individual animals; thus, a z-score less than -2 indicates a lowering of the threshold and an increase in sensitivity, a z-score greater than $+2$ indicates a decrease in sensitivity.

A considerably larger volume of literature exists on the effects of other abused stimulants on BSR (especially amphetamine), discussion of which is beyond the scope of this chapter. However, both cocaine and amphetamine are indirect dopamine agonists (Groppeti *et al.*, 1973; Iverson, 1973; Weiner, 1985), abused, and possess similar subjective effects (Fischman and Schuster, 1982). For this reason, throughout the remainder of this chapter the stimulants cocaine and amphetamine are considered to have comparable effects.

The effects of opiates on BSR have also been described elsewhere (Esposito and Kornetsky, 1978). As shown in Figure 3, morphine, the prototypical μ-agonist opioid, increases sensitivity to rewarding stimulation at doses of 4–8 mg/kg (Esposito *et al.*, 1979). These effects are dose-dependently blocked by low doses (0.06 mg/kg or greater) of the opiate receptor antagonist naloxone (Moolten and Kornetsky, 1987). In contrast, similar increases in sensitivity to BSR are not found in animals treated with the κ-agonist opioids (Unterwald *et al.*, 1987).

BRAIN-STIMULATION REWARD SITE SPECIFICITY: STIMULANTS AND OPIATES

Although there are brain sites that support BSR in which dopamine does not appear to be directly involved (Phillips, 1984), robust responding is difficult to establish from these brain areas. Although other neurotransmitters may be implicated in reward processes, current evidence clearly argues for the prepotency of dopamine in the genesis of BSR. There is also considerable evidence that at least part of a common pathway for stimulant and opiate reward may be found in the dopaminergic mesocorticolimbic system. Perhaps the best support for this neural system as the common link for opiate and stimulant reward is simply the similar effect drugs of abuse have on the sensitivity of mesocorticolimbic neurons. However, additional support comes from studies of drug effects on BSR of sites outside the mesocorticolimbic system. If sites that show similar effects of stimulants and opioids are of a common path, then sites that show disparate effects may be excluded.

Of the long-length ascending dopaminergic systems (Cooper *et al.*, 1982) — the mesolimbic, the mesocortical, and the nigrostriatal systems — the mesolimbic and mesocortical systems originate in the ventral tegmental area and project to the nucleus accumbens, olfactory tubercle, septum, amygdala and piriform cortex, and the prefrontal, cingulate, and entorhinal areas, respectively. A clear delineation between the mesolimbic and the mesocortical systems is often not made, and the combined substrate is referred to as the mesocorticolimbic system. The nigrostriatal tract projects from the substantia nigra pars compacta to the caudate-putamen. Electrodes placed in the cell bodies or terminal areas of any of these systems will maintain BSR (Prado-Alcala *et al.*, 1984; Prado-Alcala and Wise, 1984); however, the effects of stimulants and opiates on BSR to these sites vary. A schematic representation of some of the sites in which there are increases in sensitivity for BSR after the administration of morphine or the stimulants amphetamine or cocaine is shown in Figure 4. Note that the greatest overlap between the stimulants and morphine is found in the mesolimbic system. Although there are no studies in which the effects of morphine were determined from stimulation sites in the olfactory tubercle or caudate, it is likely that similar facilitory effects would have been found in mesocorticolimbic sites such as the olfactory tubercle.

Cocaine increases an animal's sensitivity to rewarding brain stimulation to the ventral tegmental area, substantia nigra, medial forebrain bundle-lateral hypothalamus, septum, nucleus accumbens, and olfactory tubercle (see Table 1). Amphetamine is known to augment BSR to several dopaminergic sites tested: prefrontal cortex (Franklin and Robertson, 1980), nucleus accumbens (Phillips *et al.*, 1975), caudate (Spencer and Revzin, 1976), lateral hypothalamus-medial forebrain bundle (Cassens and Mills, 1973; Esposito *et al.*, 1980), and the substantia nigra (Franklin and Vaccarino, 1983). However, amphetamine facilitation of prefrontal cortex self-stimulation is more difficult to obtain than from the medial forebrain bundle-lateral hypothalamic area (Carey *et al.*, 1975; Goodall and Carey, 1975; Robertson *et al.*, 1981; Hand and Franklin, 1983), suggesting that the mesolimbic system is more sensitive to the effects of amphetamine than the mesocortical system. In addition,

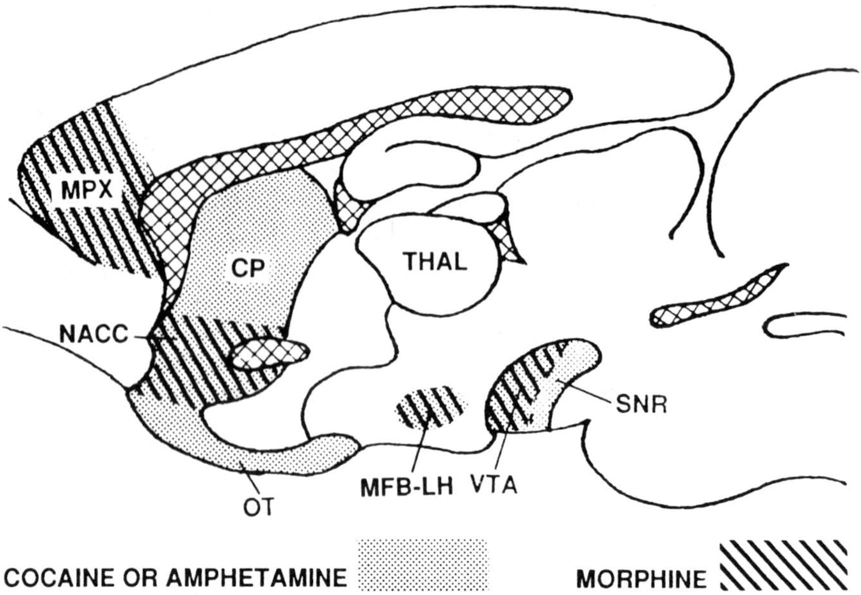

Figure 4. Schematic representation of brain stimulation sites which are part of the mesocorticolimbic and nigrostriatal dopaminergic systems that show increased sensitivity to rewarding brain stimulation in the rat after systemic injection of morphine or the stimulants cocaine or amphetamine. Areas of altered sensitivity are superimposed on an adaptation of a schematic saggital section from Heimer *et al.*, 1985. MPX, medial prefrontal cortex; CP, caudate-putamen; NACC, nucleus accumbens; OT, olfactory tubercle; MFB-LH, medial forebrain bundle-lateral hypothalamus; THAL, thalamus; VTA, ventral tegmental area; SNR, substantia nigra pars reticulata.

destruction of ascending dopamine pathways more effectively diminished self-stimulation of the ventral tegmental area than the substantia nigra (Clavier and Fibiger, 1977). Nigral self-stimulation was only temporarily disrupted, and return of this behavior was presumably maintained by a nondopaminergic substrate. These data suggest, as well, that mesolimbic self-stimulation is more sensitive and dependent on dopaminergic pathways.

The effects of morphine on BSR have been previously reviewed (Esposito and Kornetsky, 1978). Briefly, systemic injection of morphine has been shown to facilitate responding to stimulation of the ventral tegmental area (Esposito *et al.*, 1979), lateral hypothalamus-medial forebrain bundle (Marcus and Kornetsky, 1974), prefrontal cortex (Lorens, 1973; Lorens, 1976), and nucleus accumbens (van Wolfswinkel and van Ree, 1985), but not the substantia nigra (Liebman and Segal, 1977; Nazzaro *et al.*, 1981; Nelson *et al.*, 1981). Thus, the substantia nigra is relatively insensitive to opiate effects on BSR. These data, taken together, suggest that the common system, as a mediator of both stimulant and opiate reward, is found in mesocorticolimbic pathways with special emphasis on mesolimbic sites.

DRUG SITE SPECIFICITY: MICROINJECTION OF STIMULANTS AND OPIATES

A valuable technique that is compatible with BSR is the microinjection of drugs. Microinjection studies, which require administration of microliter amounts of drug solution through a cannula stereotaxically implanted in brain, are capable of implicating discrete brain sites as substrates for drug-induced behaviors. With respect to BSR, disparate sites appear to mediate the effects of stimulants and opiates. Microinjection of amphetamine into the nucleus accumbens facilitated responding to stimulation to the ventral tegmental area (Broekkamp *et al.*, 1975). In contrast, morphine injections into the caudate or nucleus accumbens decreased or had no effect on rates of responding for stimulation (Broekkamp and Van Rossum, 1975) and did not lower thresholds for rewarding brain stimulation (Van Wolfswinkel and Van Ree, 1985) to the ventral tegmental area. The microinjection of dopamine antagonists into the nucleus accumbens (Broekkamp and Van Rossum, 1975; Mogenson *et al.*, 1979; Stellar *et al.*, 1985), but not the caudate or prefrontal cortex, attenuated the rewarding effects of BSR with minimal performance disruption (Stellar *et al.*, 1985). These data suggest that dopaminergic activity in the nucleus accumbens is necessary to maintain the rewarding effects of BSR and the increased sensitivity to rewarding stimulation that results from administration of amphetamine. Studies of the intracranial self-administration of amphetamine (Lyness *et al.*, 1979) or cocaine (Roberts *et al.*, 1977; Roberts *et al.*, 1980) further support the notion that the nucleus accumbens is a sufficient substrate for stimulant reward.

The lack of effect of microinjections of morphine into the caudate or accumbens (Broekkamp and Van Rossum, 1975; Van Wolfswinkel and Van Ree, 1985) appears to rule out these sites as substrates for morphine-induced increased sensitivity for BSR, but it is important to note that other forebrain areas such as the olfactory tubercle have not been tested. It is known, however, that intraventral tegmental administration of morphine facilitates responding for stimulation to the lateral hypothalamus (Broekkamp *et al.*, 1979) and lowers the threshold for stimulation to the nucleus accumbens (Van Wolfswinkel and Van Ree, 1985), which strongly suggests that opiates activate the mesocorticolimbic system at the level of the ventral tegmental area. Studies of the intracranial self-administration of morphine corroborate this conclusion and have further demonstrated that the ventral tegmental area is a substrate for opiate reward that is both necessary (Britt and Wise, 1981) (i.e., i.v. opiate self-administration is altered by intraventral tegmental area administration of an opiate antagonist) and sufficient (Bozarth, 1983), that is, opiates are self-administered directly into the ventral tegmental area.

Recently, Wise (1987) has made an argument for forebrain-mediated opiate reward based on the evidence that methionine enkephalin is self-administered into the nucleus accumbens (Geoders *et al.*, 1984) and the quaternary opiate antagonist methyl naloxonium chloride more effectively altered i.v. heroin self-administration when it was injected into the nucleus accumbens than it did in the ventral tegmental area (Vaccarino *et al.*, 1985). It was also suggested that previous failures to establish intranucleus accumbens self-administration of morphine (Bozarth and Wise, 1982) may be due to the use of relatively small volumes that were unable to diffuse over a sufficient area within the bounds of the nucleus accumbens (Wise, 1987). We suggest that the disparate results may be due to the necessity for opiates to diffuse into the olfactory tubercle in order to be rewarding. The olfactory tubercle sits adjacent and below the nucleus accumbens, contains opiate receptors (Mansour *et al.*, 1987), is a major projection area of the mesocorticolimbic system, and has been implicated in both stimulant and opiate reward in studies of local cerebral glucose utilization (Huston-Lyons *et al.*, 1989) (see Glucose Utilization: Cocaine, Morphine, and BSR).

DRUG INTERACTION ON BSR: STIMULANTS AND OPIATES

Street use of drugs includes combining stimulants and opiates in order to increase the euphoria experienced over each drug taken alone (Chambers *et al.*, 1973; Ellinwood *et al.*, 1976; Greene and DuPont, 1973; Greene *et al.*, 1974). Studies from this laboratory of the interaction of drugs of abuse on BSR to the medial forebrain bundle are in agreement with these reports. Although moderate doses of amphetamine (Esposito *et al.*, 1980) or morphine (Marcus and Kornetsky, 1974) alone lowered the threshold for BSR to the medial forebrain bundle, the co-administration of an ineffective or minimally effective dose of amphetamine with a range of doses of morphine, or a low dose of morphine with a range of doses of amphetamine, shifted the dose-response curves to the left and down (Hubner *et al.*, 1987), that is, each drug potentiated the effects of the other. An example of the interactive effects of amphetamine and morphine on BSR may be found in Figure 5. The specificity for dopaminergic interaction is given by the experiments of Izenwasser and Kornetsky (1989) in which amfonelic acid, a selective indirect dopamine agonist, potentiated the threshold-lowering effect of morphine, but nisoxetine, a relatively specific noradrenergic uptake blocker, did not. Although these studies do not directly address the mechanism by which stimulants and opiates interact, they suggest that, when both substances are co-administered, they work in concert to increase the sensitivity of neurons of the medial forebrain bundle.

Naloxone's effects on BSR have been reviewed by Shaefer (1988), who notes that when naloxone is administered alone it consistently reduces responding for intermittent schedules of responding, but has no effect on thresholds for stimulation that are rate independent. Shaefer (1988) also concludes that the opiate antagonist, naloxone, modifies the effects of dopaminergic drugs on BSR (e.g., Seeger *et al.*, 1981). Studies from this laboratory are in agreement and have shown that naloxone blocks the threshold-lowering effects of several drugs with actions on dopaminergic systems on BSR to the medial forebrain bundle: cocaine (Bain and Kornetsky, 1987), amphetamine (Esposito *et al.*, 1980), amfonelic acid (Knapp and Kornetsky, 1989), and methylenedioxymethamphetamine (MDMA) (Hubner *et al.*, 1988). Figure 6 summarizes the effects of naloxone on the maximally effective threshold-lowering dose of various drugs; as shown, nicotine is the only drug we have tested thus far that was not blocked by naloxone (Lyons *et al.*, 1988). That opiate antag-

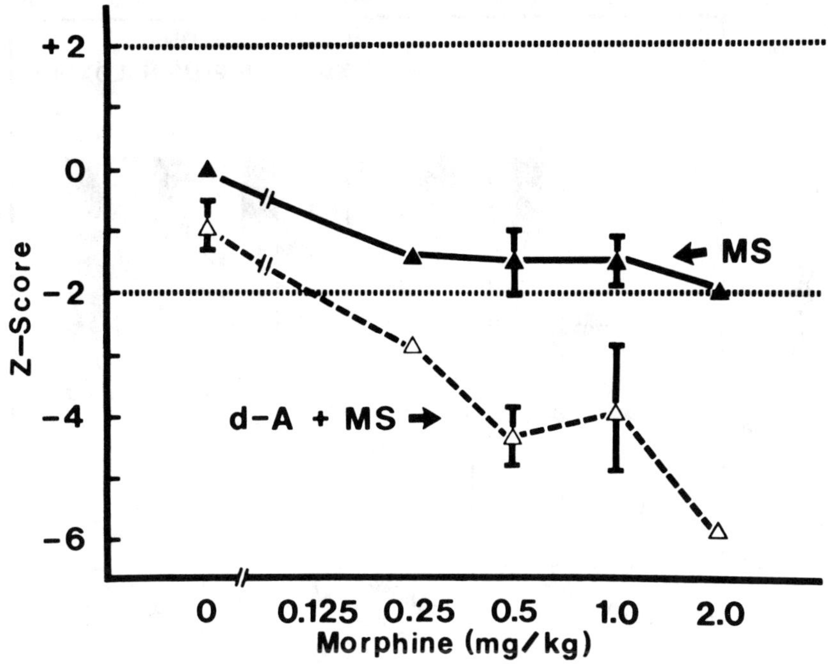

Figure 5. Mean z-score changes in the reward threshold from pre- to postdrug after administration of the ineffective dose of D-amphetamine with various doses of morphine. The absolute difference between post- minus predrug thresholds has been converted to z-scores (standard scores) based on the mean and standard deviation of the difference found after administration of saline. A z-score of ±2 defines the 95% confidence limits for individual animals; thus, a z-score less than −2 indicates a lowering of the threshold and an increase in sensitivity, and a z-score greater than +2 indicates a decrease in sensitivity. Error bars depict ±1 standard error.

onists block the effects of stimulants on BSR further demonstrates the interdependency of stimulant and opiate action on neurons in the medial forebrain bundle and specifically indicates that, when animals are tested under rate-independent conditions, endogenous opiate activity is necessary for many stimulants to manifest their rewarding effects.

DOPAMINE METABOLISM: COCAINE, MORPHINE, AND BSR

Cocaine

It has been well established that cocaine has local anesthetic properties and blocks the re-uptake of monoamines (Groppeti *et al.*, 1973; Iverson, 1973). It is generally agreed that the ability of cocaine to block the re-uptake of dopamine is responsible for its euphorigenic effects (Fibiger, 1978; Wise and Bozarth, 1982). In this regard, Kuhar and co-workers (Kuhar *et al.*, 1988; Ritz *et al.*, 1987) have recently demonstrated that the reinforcing effects of stimulants are significantly correlated with

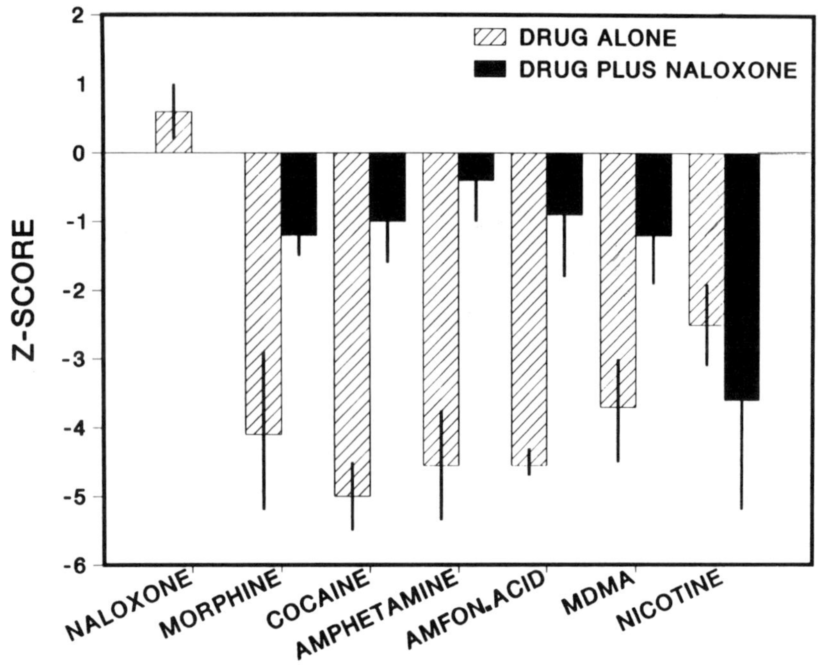

Figure 6. Maximum mean z-score changes in the reward threshold after administration of various abused drugs and challenge with naloxone. The absolute difference between post- minus predrug thresholds measured in microamperes has been converted to z-scores based on the mean and standard deviation of the difference found after administration of saline. A z-score of ±2 defines the 95% confidence limits for individual animals. Data from at least four animals make up each example. Error bars refer to ±1 standard error. The dose of naloxone was 4 or 5 mg/kg in all cases, except for morphine which was blocked at lower doses (naloxone 0.125 mg/kg). The remaining drug doses (in milligrams per kilogram) were as follows: morphine (4); cocaine (10–15); amphetamine (0.5–1.0); amfonelic acid (0.5–1); methylenedioxymethamphetamine (MDMA) (2–4); and nicotine (0.5–0.75).

the ability to bind to dopamine re-uptake sites. Moreover, it is dopaminergic antagonists, and not noradrenergic antagonists, that increase the rate of self-administration of amphetamine (Yokel and Wise, 1975; Yokel and Wise, 1976) or cocaine (DeWit and Wise, 1977; Risner and Jones, 1980). In fact, not only stimulants like cocaine and amphetamine, but other abused drugs (e.g., nicotine, opiates, and ethanol), increase dopamine concentrations in mesolimbic and nigrostriatal terminal fields (Di Chiara and Imperato, 1988). Studies of cocaine (Goeders *et al.*, 1986; Wise, 1984; Roberts and Koob, 1982; Roberts *et al.*, 1980), as well as amfonelic acid (Porrino *et al.*, 1988), self-administration further support that dopaminergic neurons of the mesocorticocolimbic systems maintain cocaine reward. Figure 7 illustrates the brain areas that show increased dopamine turnover after either systemic injection of morphine or BSR of the midbrain or medial forebrain bundle. Although not indicated, cocaine increases dopamine turnover at all projection sites of the mesocorticolimbic, as well as the nigrostriatal, system.

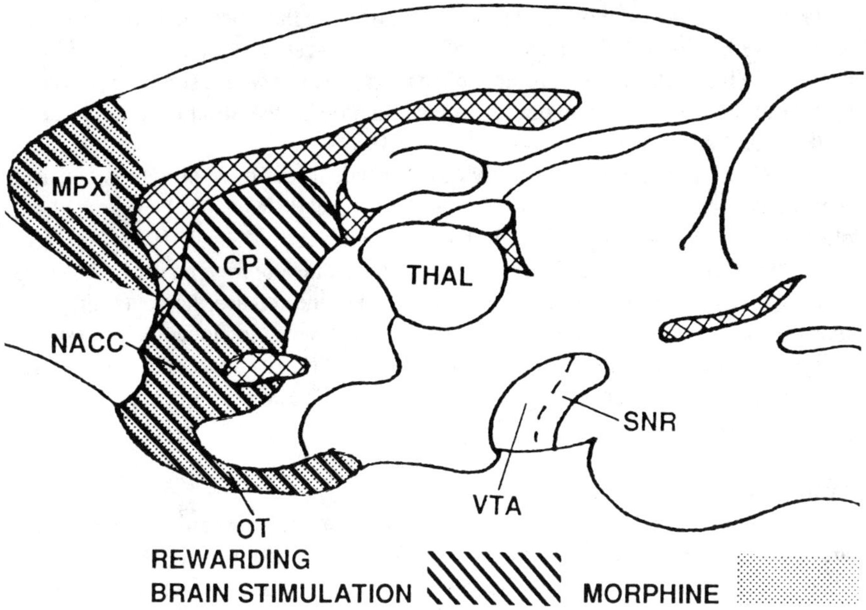

Figure 7. Schematic representation of the areas of the rat brain that most often demonstrate increased dopamine turnover after rewarding electrical stimulation to the midbrain or systemic administration of morphine. Areas of increased turnover are superimposed on an adaption of a schematic saggital section from Heimer *et al.,* 1985. MPX, medial prefrontal cortex; CP, caudate-putamen; NACC, nucleus accumbens; OT, olfactory tubercle; THAL, thalamus; VTA, ventral tegmental area; and SNR, substantia nigra pars reticulata.

Morphine

Morphine preferentially binds to μ opiate receptors, which can be found throughout the brain and in proximity to virtually all dopaminergic structures (Mansour *et al.,* 1987). Furthermore, μ opiate receptors have been demonstrated to exist on presynaptic dopaminergic elements in the septum and nucleus accumbens (Pollard *et al.,* 1977). The effects of morphine on dopamine systems have been reviewed by Wood and Altar (1988). Acute administration of morphine increased dopamine levels in nucleus accumbens dialysates with only small increases in striatal dialysates (DiChiara and Imperato, 1988). Increased DOPAC and 3-methoxytryamine (3-MT) levels were found in morphine-treated rats in the prefrontal, cingulate, and pyriform cortex, the olfactory tubercle, and nucleus accumbens (Kim *et al.,* 1986; Wood and Altar, 1988). As a metabolite of dopamine that only arises outside of the neuron, 3-MT is arguably a more accurate measure of dopamine release than other metabolites (Wood and Altar, 1988). Thus, although rat striatal DOPAC concentrations are consistently increased after administration of morphine, similar increases in striatal 3-MT levels are not found in the rat (Westerink and Spaan, 1982; Wood *et al.,* 1980) or strains of mice that do not show morphine-induced increased locomotor activity. Because morphine apparently acts at the level of the substantia nigra to increase locomotor activity in strains of mice that show activation (Racagni *et al.*, 1979; Wood and Richard, 1982), some investigators have concluded that the relative insensitivity to morphine in the rat (and some mouse strains) nigrostriatal system is due to a "presynaptic clamping" action in the striatum (Biggio *et al.*, 1978; Wood *et al.*, 1980; Wood and Altar, 1988).

The selective morphine sensitivity of the mesocorticolimbic system is consistent with evidence that morphine-induced facilitation of BSR cannot be obtained from electrodes in the substantia nigra in the rat (Liebman and Segal, 1977; Nazzaro *et al.*, 1981; Nelson *et al.*, 1981). At present, it remains controversial whether opiate reward is dependent on mesocorticolimbic dopamine release; however, it is clear that opiate reward is influenced by various treatments to dopaminergic systems (Bain and Kornetsky, 1987; Bozarth and Wise, 1981; Esposito *et al.*, 1980; Glick and Cox, 1975; Hanson and Cimini-Venema, 1972; Hubner *et al.*, 1987; Spyraki *et al.*, 1983).

Brain-Stimulation Reward

Dopamine neurons of the ventral tegmental area and substantia nigra have been attributed with long spike durations (>3 ms), slow random firing rates (one to six spikes per second) and slow conduction velocities (0.5 m/s) (Yim and Mogenson, 1980; Guyenet and Aghajanian, 1978). Experimenter-delivered stimulation at about 100 Hz or less (Wood *et al.*, 1982) with a pulse width of 0.5 m or greater (Millar *et al.*, 1985) is required for release of dopamine in forebrain areas from neurons with axons in the medial forebrain bundle presumably with these characteristics. However, dopamine turnover has been observed in several projection areas of the mesocorticolimbic system in self-stimulating animals (Phillips and Fibiger, 1989), that is, prefrontal cortex, nucleus accumbens, olfactory tubercle, as well as the striatum, regardless of whether the parameters of stimulation were consistent with direct stimulation of dopamine neurons (Fibiger *et al.*, 1987; Garriques and Cazala, 1983; Phillips *et al.*, 1987) or not (Mitchell *et al.*, 1982; Simon *et al.*, 1979). Although a recent study by McCown and colleagues (1986) reported DOPAC concentrations in forebrain areas did not increase after hypothalamic stimulation, Phillips and Fibiger (1989) point out that the DOPAC/DA ratio, which is considered a more sensitive measure of dopamine metabolism, was increased in the nucleus accumbens and olfactory tubercle. Earlier studies that failed to find changes in dopamine metabolism after rewarding brain stimulation (Saint Laurent *et al.*, 1975; Miliaressis *et al.*, 1975; Myers and Mora, 1977) have also been criticized elsewhere (Simon *et al.*, 1979). Apparently, whether the directly stimulated neuron, or "primary reward" neuron, is an ascending dopamine neuron or a nondopaminergic descending neuron, the ascending neuron is eventually activated. A reasonable hypothesis has emerged which suggests that, in the case of direct stimulation of only descending neurons, the subsequent impulse activates dopaminergic midbrain neurons which, in turn, release dopamine in the forebrain (Wise, 1980; Yeomans, 1982).

Additional evidence for the mediation of BSR by dopamine comes from studies which showed that neuroleptics reduce the rewarding impact of brain stimulation

(Liebman, 1983; Wise, 1978). Attenuation of the reinforcing effects of BSR has been noted when neuroleptics were administered systemically (Lynch and Wise, 1985; Nakajima and McKenzie, 1986) or microinjected into the nucleus accumbens (but not the prefrontal cortex) (Mogensen *et al.*, 1979), which further supports the contention that dopaminergic activity not only covaries with BSR but is necessary to maintain it.

Taken together, these data indicate that cocaine, morphine, and rewarding brain stimulation either block uptake or cause dopamine release. In either case, increased amounts of dopamine are available for post-synaptic activity. Furthermore, this increase in available dopamine is necessary to maintain the reinforcing properties of cocaine and brain stimulation, and, at the least, modulates the reinforcing effects of morphine.

GLUCOSE UTILIZATION: COCAINE, MORPHINE, AND BSR

Stimulants

The development by Sokoloff and colleagues (1977) of a quantitative autoradiographic method for determining differential rates of glucose metabolism (LCMRglu) in tissue has provided an effective tool to examine the neural substrate of drug-induced increases in sensitivity of reward pathways. A detailed description of this method and discussion of the effects of cocaine may be found in Chapter 3. In brief, studies to date that examined changes in glucose metabolism after administration of stimulants have established that abused stimulants [e.g., cocaine (Porrino *et al.*, 1988), amphetamine (Porrino *et al.*, 1984; Weschler, 1979), and methylphenidate (Porrino and Lucignani, 1987)] increased metabolism in the nucleus accumbens and the medial prefrontal cortex at doses that are self-administered by animals (Porrino *et al.*, 1988). We have substantially replicated the findings by Porrino and co-workers (1988) of the effects of cocaine on brain metabolic activity. These effects are described below (see Brain-Stimulation Reward). Higher doses which yield increased locomotion and incidence of stereotypy increased metabolism in the extrapyramidal system as well (London *et al.*, 1986; Porrino, *et al.*, 1988). Other dopamine agonists with limited abuse liability [e.g., apomorphine (McCulloch *et al.*, 1982) and bromocriptine (Pizzolato *et al.*, 1985)] showed increased metabolism in the extrapyramidal system only.

Morphine

Previous studies of the effects of morphine on brain metabolic activity have indicated that doses (>8 mg/kg) that are above those that lower the threshold for rewarding brain stimulation (2–8 mg/kg) (Marcus and Kornetsky, 1974; Esposito and Kornetsky, 1977) found only decreased rates of brain glucose utilization, which is probably the result of morphine-induced hypercapnia resulting from respiratory depression (Sakurada *et al.*, 1976). Similar results were obtained by other investigators (Hiesinger *et al.*, 1983; Fanelli *et al.*, 1987), who reported only decreased rates of metabolism after systemic injection of 5 or 8 mg/kg of morphine when metabolism was assessed approximately 15 min after drug administration. Ito and colleagues (1983), in assessing the effects of 5 mg/kg of morphine approximately 60 min after drug administration, found increased rates of metabolism in the substantia nigra pars reticulata and the mamillary bodies and a depressed rate in the medial part of the lateral habenula. Except for the report of Ito and colleagues, the general effect of morphine on glucose metabolism was to decrease metabolic activity or have no effect in the structures studied. The disparate results of studies in which approximately the same dose was given may have been due to differences in time after morphine administration of the determination of metabolic activity. Our experiments (see below), however, clearly show that at doses of 4 mg/kg, that morphine will increase metabolic activity in the mesocorticolimbic system.

Brain-Stimulation Reward

Self-stimulation to the ventral tegmental area (Esposito *et al.*, 1990) or the medial forebrain bundle (Porrino *et al.*, 1989) increased metabolic activity throughout the mesocorticolimbic system, including projection areas such as the nucleus accumbens, lateral septum, and the medial prefrontal cortex. A significant increase in metabolism was evident in the olfactory tubercles from stimulation in the medial forebrain bundle; a trend was noted from stimulation to the ventral tegmental area. Good evidence that the metabolic effects of self-stimulation were reward specific comes from Porrino and colleagues (1984) who found that self-stimulation, in contrast to experimenter-delivered stimulation, to the ventral tegmental area resulted in increased metabolism in mesocorticolimbic structures as well (a trend toward increased metabolism was noted in the olfactory tubercle in self-stimulating animals only).

Not all studies of the metabolic effects of self-stimulation are in agreement with these conclusions. Gallistel and colleagues, in a series of experiments (Gallistel *et al.*, 1985; Gomita, *et al.*, 1982; Yadin *et al.*, 1983), concluded that BSR does not activate projection areas of the mesocorticolimbic system. Although there are significant differences in stimulation parameters, more important differences in the method of determining glucose metabolism exist between these studies and those of others (Porrino *et al.*, 1984; Porrino *et al.*, 1985; Esposito *et al.*, 1984). The method employed by Gallistel and co-workers makes use of a relative optical density method of analysis which, in this case, appears to lack the sensitivity of the quantitative method established earlier by Sokoloff and colleagues (1977) and severely diminishes the accuracy of the estimation of the accumulation of tracer (Porrino *et al.*, 1990; Porrino and Crane, 1990). Furthermore, the extensive literature on the pharmacology of BSR clearly indicates a role for forebrain dopaminergic systems in this form of reward (Fibiger, 1978; Wise and Bozarth, 1982).

We have recently completed an examination of the cerebral metabolic effects of cocaine and morphine, and the interaction of these drugs on BSR. Our findings on the effects of intraperitoneal cocaine (10 mg/kg) (Huston-Lyons *et al.*, 1989) were, in general, consistent with previous studies of cocaine (Porrino *et al.*, 1988) and other abused stimulants (Porrino *et al.*, 1984, Porrino and Lucignani, 1987; Weschler *et al.*, 1979). Areas showing increases in metabolism were the medial prefrontal cortex, olfactory tubercle, the medial dorsal thalamus, and, to a lesser extent, the ventral tegmental area. The effects of morphine (4 mg/kg) were more circumscribed than cocaine, with significant increases found only in the olfactory tubercle (Huston-Lyons *et al.*, 1989). That cocaine and morphine, at doses which are rewarding (Esposito *et al.*, 1978; Marcus and Kornetsky, 1974), activate the olfactory tubercle suggests that this structure plays a unique role in the mediation of drug-induced reward. A schematic representation of the sites of increased glucose metabolism induced by cocaine (10 mg/kg) and morphine (4 mg/kg) in our laboratory may be found in Figure 8. At present, we have no data to compare to the increased metabolism in the mamillary bodies noted by Ito and colleagues (1983), but it is interesting that amphetamine (Porrino *et al.*, 1984; Weschler *et al.*, 1979) and cocaine (Porrino *et al.*, 1988) induced increased metabolism in the mamillary bodies as well.

Additional support for the notion that the olfactory tubercle plays a pivotal role in drug-induced reward comes from studies of the interaction of cocaine or morphine in self-stimulating animals. Cocaine administered to self-stimulating animals increased metabolsim in projection areas of the mesocorticolimbic system (e.g., the medial prefrontal cortex, nucleus accumbens, and olfactory tubercle) above levels seen in animals that self-stimulated only or were administered cocaine only. Furthermore, morphine administered to self-stimulating animals increased metabolism above either treatment separately in the olfactory tubercle alone. Figure 8 also illustrates the effects of cocaine (10 mg/kg) and morphine (4 mg/kg) on animals actively self-stimulating. These data indicate the olfactory tubercle is sensitive to cocaine, morphine, and BSR. Furthermore, the interaction of self-stimulation and cocaine or morphine on the metabolism of the olfactory tubercle parallels the ability of these drugs to increase behavioral sensitivity to BSR.

SUMMARY

Several lines of evidence implicate the mesocorticolimbic system as mediator of the rewarding effects of cocaine and other drugs of abuse. The rewarding impact of BSR to sites within the mesocorticolimbic system is increased after administration of stimulants or opiates; moreover, these drugs work in concert when co-administered. Cocaine, morphine, and BSR each cause increased dopamine turnover. Even with stimulation parameters that reportedly do not directly stimulate dopamine neurons, dopamine systems are activated as long as the animal is working for stimulation. Furthermore, determination of drug effects on glucose metabolism has directly implicated the olfactory tubercle, a primary projection area of the mesocorticolimbic system, as a possible mediator of the reinforcing actions of both stimulants and opiates.

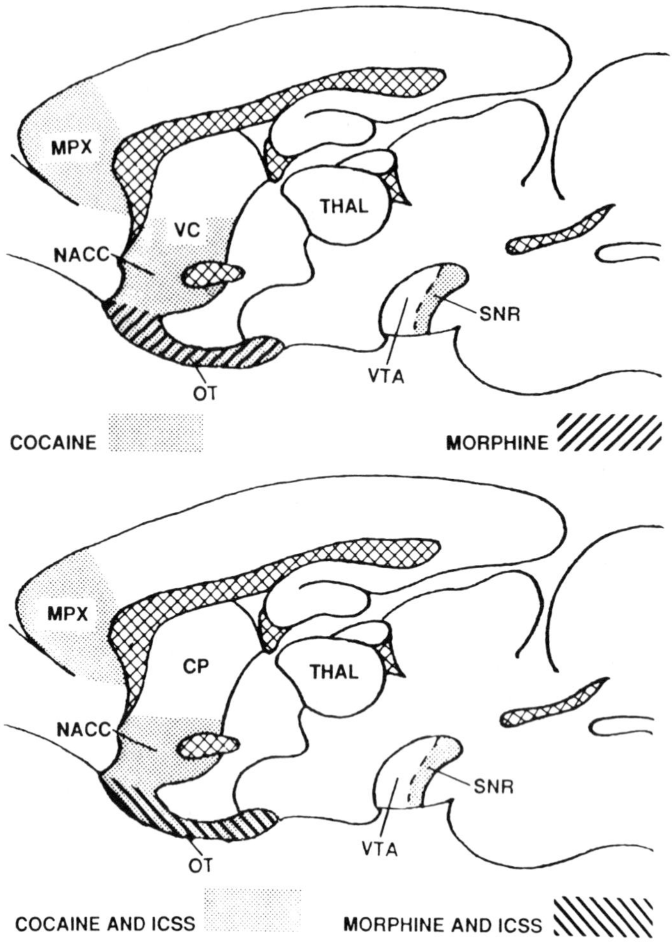

Figure 8. Upper panel: Schematic representation of the regions in the rat brain that show increased LCMRglu after systemic administration of cocaine (10 mg/kg) or morphine (4 mg/kg) compared to saline. Lower panel: Representation of the effects of cocaine (10 mg/kg) or morphine (4 mg/kg) in animals self-stimulating compared to animals self-stimulating after administration of saline (lower panel). Areas of increased metabolism are superimposed on an adaptation of a schematic saggital section from Heimer *et al.*, 1985. MPX, medial prefrontal cortex; CP, caudate-putamen; VC, ventral caudate; NACC, nucleus accumbens; OT, olfactory tubercle; THAL, thalamus; VTA, ventral tegmental area; SNR, substantia nigra pars reticulata.

REFERENCES

B.K. Anand and J.R. Brobeck. Hypothalamic control of food intake in rats and cats. *Yale J. Biol. Med.* **24**: 123 (1951).

C.S. Aulakh, B. Ghosh, and S.N. Pradhan. Actions and interactions of cocaine on self-stimulation in rats. *Psychopharmacology* **63**: 75 (1979).

P.T. Bailey and S.N. Pradhan. Interactions of adrenergic stimulants and blockers on self-stimulation behavior in rats. *Res. Comm. Chem. Path. Pharmacol.* **11**: 543 (1975).

G. Bain and C. Kornetsky. Naloxone attenuation of the effect of cocaine on rewarding brain stimulation. *Life Sci.* **40**: 1119 (1987).

G.A. Barr and T. Lithgow. Pharmaco-ontogeny of reward: enhancement of self-stimulation by D-amphetamine and cocaine in 3- and 10-day-old rats. *Brain Res.* **389**: 193 (1986).

A.K. Bhattacharyya and S.N. Pradhan. Behavioral interactions in cocaine-treated rats. *Life Sci.* **24**: 1855 (1979).

G. Biggio, M. Casa, M.G. Corda, C. DiBello, and G.L. Gessa. Stimulation of dopamine synthesis in caudate nucleus by intrastriatal enkephalins and antagonism by naloxone. *Science* **200**: 552 (1978).

M.A. Bozarth. Opiate reward mechanisms mapped by intracranial self-administration, in: *The Neurobiology of Opiate Reward Processes.* J.E. Smith and J.D. Lane, eds., Elsevier Biochemical Press, New York, p. 331 (1983).

M.A. Bozarth and R.A. Wise. Heroin reward is dependent on a dopaminergic substrate. *Life Sci.* **29**: 1881 (1981).

M.A. Bozarth and R.A. Wise. Localization of the reward relevant opiate receptors, in: Problems of Drug Dependence. NIDA Research Monograph, pp. 158-164. L.S. Harris, ed., U.S. Government Printing Office, Washington, DC, p. 158 (1982).

M.D. Britt and R.A. Wise. Attenuation of intravenous heroin reward by ventral tegmental opiate receptor blockade. *Absts. Soc. Neurosci.* **7**: 607 (1981).

C.L. Broekkamp, A.G. Phillips, and A.R. Cools. Facilitation of self-stimulation behavior following intracerebral microinjections of opioids into the ventral tegmental area. *Pharmacol. Biochem. Behav.*, **11**: 289 (1979).

C.L.E. Broekkamp, A.J.J. Pijnenburg, A.R. Cools, and J.M. Van Rossum. The effect of microinjections of amphetamine into the neostriatum and the nucleus accumbens on self-stimulation behavior. *Arch. Int. Pharmacocyn. Ther.* **217**: 110 (1975).

C.L.E. Broekkamp and J.M. Van Rossum. The effect of microinjections of morphine and haloperidol into the neostriatum and the nucleus accumbens on self-stimulation behavior. *Arch. Int. Pharmacodyn.* **217**: 110 (1975).

R.J. Carey, E. Goodall, and S.A. Lorens. Differential effects of amphetamine and food deprivation on self-stimulation of the lateral hypothalamus and medial frontal cortex. *J. Comp. Physiol. Psychol.* **88**: 224 (1975).

G.P. Cassens and A.W. Mills. Lithium and amphetamine: opposite effects on threshold of intracranial reinforcement. *Psychopharmacologia* **30**: 283 (1973).

C.D. Chambers, W.J.R. Taylor, and P.V. Walter. Drug abuse during ambulatory detoxification, in: *Methadone: Experiences and Issues.* C.O. Chambers and L. Brill, eds., Behavioral Publications, New York, p. 203 (1973).

R.M. Clavier and H.C. Fibiger. On the role of ascending catecholaminergic projections in intracranial self-stimulation of the substantia nigra. *Brain Res.* **131**: 271 (1977).

J.R. Cooper, F.E. Bloom, and R.H. Roth. *The Biochemical Basis of Neuropharmacology.* Oxford University Press, New York (1982).

T.J. Crow. Enhancement by cocaine of intracranial self-stimulation in the rat. *Life Sci.* **9**: 375 (1970).

T.J. Crow. Catecholamine-containing neurons and electrical self-stimulation. I. A review of some data. *Psychol. Med.* **2**: 414 (1972).

G.A. Deneau, T. Yanagita, and H.M. Seevers. Psychic dependence studies of self-administration techniques in the rhesus monkey. Minutes of the meeting of the Committee on Problems of Drug Dependence, NAS-NRC, Houston, TX, (1965).

J.M.R. Delgado, W.W. Roberts, and N.E. Miller. Learning motivated by electrical stimulation of the brain. *Am. J. Physiol.* **179**: 587 (1954).

H. DeWit and R.A. Wise. Blockade of cocaine reinforcement in rats with the dopamine receptor blocker pimozide, but not with the noradrenergic blockers phentolamine or phenoxybenzamine. *Can. J. Psychol.* **31**: 195 (1977).

G. DiChiara and A. Imperato. Opposite effects of mu and kappa opiate agonists on dopamine release in the nucleus accumbens and the dorsal caudate of freely moving rats. *J. Pharmacol. Exp. Ther.* **244**: 1067 (1988).

G. DiChiara and A. Imperato. Drugs abused by humans preferentially increase synaptic dopamine concentrations in the mesolimbic system of freely moving rats. *Proc. Natl. Acad. Sci. U.S.A.* **85**: 5274 (1988).

E.H. Ellinwood, R.D. Eibergen, and M.M. Kilbey. Stimulants: interaction with clinically relevant drugs. *Ann. N.Y. Acad. Sci.* **281**: 393, (1976).

R.U. Esposito and C. Kornetsky. Morphine lowering of self-stimulation thresholds: lack of tolerance with long-term administration. *Science* **195**: 189 (1977).

R.U. Esposito and C. Kornetsky. Opioids and rewarding brain stimulation. *Neurosci. Biobehav. Rev.* **2**: 115 (1978).

R.U. Esposito, S. McLean, and C. Kornetsky. Effects of morphine on intracranial self-stimulation to various brain stem loci. *Brain Res.* **168**: 425 (1979).

R.U. Esposito, A.H. Motola, and C. Kornetsky. Cocaine: acute effects on reinforcement thresholds for self-stimulation behavior to the medial forebrain bundle. *Pharmacol. Biochem. Behav.* **8**: 437 (1978).

R.U. Esposito, W. Perry, and C. Kornetsky. Effects of D-amphetamine and naloxone on brain stimulation reward. *Psychopharmacology* **69**: 187 (1980).

R.U. Esposito, L.J. Porrino, T.F. Seeger, A.M. Crane, H.D. Everist, and A. Pert. Changes in local cerebral glucose utilization during rewarding brain stimulation. *Proc. Natl. Acad. Sci. U.S.A.* **81**: 635 (1984).

R.J. Fanelli, M. Szikszay, D.R. Jasinski, and E.D. London. Differential effects of μ and κ opioid analgesics on cerebral glucose utilization in the rat. *Brain Res.* **422**: 257 (1987).

H.C. Fibiger. Drugs and reinforcement mechanisms: a critical review of the catecholamine theory. *Annu. Rev. Pharmacol. Toxicol.* **18**: 37 (1978).

H.C. Fibiger, F.G. Le Piane, A. Jakubovic, and A.G. Phillips. The role of dopamine in intracranial self-stimulation of the ventral tegmental area. *J. Neurosci.* **7**: 3888 (1987).

M.W. Fischman and C.R. Schuster. Cocaine self-administration in humans, *Fed. Proc.* **41**: 241 (1982).

F.A. Frank, S. Martz, and T. Pommering. The effect of chronic cocaine on self-stimulation train-duration thresholds. *Pharmacol. Biochem. Behav.* **29**: 755 (1988).

R.A. Frank, T. Pommering, and D. Nitz. The interactive effects of cocaine and imipramine on self-stimulation train-duration thresholds. *Pharmacol. Biochem. Behav.* **30**: 1 (1988).

K.B. Franklin and A. Robertson. 5HT blockade and the stimulant effects of D- and L-amphetamine: no interaction in self-stimulation of prefrontal cortex, hypothalamus, or dorsal tegmentum. Unexpected lethality in hippocampal sites. *Pharmacol. Biochem. Behav.* **13**: 365 (1980).

K.B.J. Franklin and F.J. Vaccarino, Differential effects of amphetamine isomers on SN self-stimulation: evidence for DA neuron subtypes. *Pharmacol. Biochem. Behav.* **18**: 747 (1983).

C.R. Gallistel, Y. Gomita, E. Yadin, and K.A. Campbell. Forebrain origins and terminations of the medial forebrain bundle metabolically activated by rewarding stimulation or by reward-blocking doses of pimozide. *J. Neurosci.* **5**: 1246 (1985).

A. Garriques and P. Cazala. Central catecholamine metabolism and hypothalamic self-stimulation behavior in two inbred strains of mice. *Brain Res.* **265**: 265 (1983).

S.D. Glick and J.I. Badalamenti. Sex difference in reward asymmetry and effects of cocaine. *Neuropharmacology* **25**: S33 (1986).

S.D. Glick and R.D. Cox. Dopaminergic and cholinergic influences on morphine self-administration in rats. *Res. Comm. Chem. Path. Pharm.* **12**: 17 (1975).

N.E. Goeders, S.L. Dworkin, and J.E. Smith. Neuropharmacological assessment of cocaine self-administration into the medial prefrontal cortex. *Pharmacol. Biochem. Behav.* **24**: 1429 (1986).

N.E. Goeders, J.D. Lane, and J.E. Smith. Self-administration of methionine enkephalin into the nucleus accumbens. *Pharmacol. Biochem. Behav.* **20**: 451 (1984).

Y. Gomita and C.R. Gallistel. Effect of reinforcement blocking doses of pimozide on neural systems driven by rewarding stimulation of the MFB: a 14C-2-deoxyglucose analysis. *Pharmacol. Biochem. Behav.* **17**: 841 (1982).

E.B. Goodall and R.J. Carey. Effects of D- versus L-amphetamine, food deprivation, and current intensity on self-stimulation of the lateral hypothalamus, substantia nigra, and medial frontal cortex of the rat. *J. Comp. Physiol. Psychol.* **89**: 1029 (1975).

M.H. Greene and R.L. DuPont. An outbreak of intravenous amphetamine abuse in heroin addicts, in: *Fifth National Conference on Methadone Treatment.* R.L. Dupont and R.S. Freeman, eds., NAPAN, New York, p. 776 (1973).

M.H. Greene, N. Turner, and R.L. Dupont. Amphetamines in the District of Columbia. *Int. J. Addict.* **9**: 653 (1974).

A. Groppetti, F. Zambotti, A. Biazzi, and P. Mantegazza. Amphetamine and cocaine on amine turnover, in: *Frontiers in Catecholamine Research.* E. Usdin and S.H. Synder, eds., Pergamon Press, Oxford, p. 917 (1973).

P.G. Guyenet and G.K. Aghajanian. Antidromic identification of dopaminergic and other output neurons of the substantia nigra. *Brain Res.* **150**: 69 (1978).

T.H. Hand and T.H. Franklin. The influence of amphetamine on preference for lateral hypothalamic versus prefrontal cortex of ventral tegmental area stimulation. *Pharmacol. Biochem. Behav.* **18**: 695 (1983).

H.M. Hanson and C.A. Cimini-Venema. Effects of haloperidol on self-administration of morphine in rats. *Fed. Proc.* **31**: 503 (1972).

R.G. Heath, W.A. Mickle, C.W. Sem-Jacobsen, and A. Torkildsen. Seven years experience with depth electrode studies. In: *Electrical Studies on the Unanesthetized Brain.* E.R. Ramey and D.S. O'Doherty, eds., Paul B. Hoeber, New York, p. 214 (1960).

L. Heimer, G.F. Alheid, and L. Zaborsky. Basal ganglia, in: *The Rat Nervous System.* Vol. 1 G. Paxinos, ed., Academic Press, New York, p. 37 (1985).

W.R. Hess. *The Functional Organization of the Diencephalon.* Grune & Stratton, New York (1957).

E.M. Hiesiger, R.M. Voorhies, L. Lipshutz, G. Baster, W.R. Shapiro, and G.W. Pasternak. The effects of morphine on glucose metabolism (^{14}C-2-deoxyglucose) in rat brain as measured by quantitative autoradiography. *Abstr. Soc. Neurosci.* **9**: 138 (1983).

J.C. Hitt, D.M. Bryon, and D.T. Modianos. Effects of rostral medial forebrain bundle and olfactory tubercle lesions upon sexual behavior of male rats. *J. Comp. Physiol. Psychol.* **82**: 30 (1973).

C.B. Hubner, G.T. Bain, and C. Kornetsky. The combined effect of morphine and D-amphetamine on the threshold for brain-stimulation reward. *Pharmacol. Biochem. Behav.* **28**: 311 (1987).

C.B. Hubner, M. Bird, S. Rassnick, and C. Kornetsky. The threshold lowering effects of MDMA (ecstasy) on brain-stimulation reward. *Psychopharmacology* **95**: 49 (1988).

D. Huston-Lyons, L.J. Porrino, G.T. Bain, and C. Kornetsky. Both morphine and cocaine increase glucose metabolism in the olfactory tubercle in freely moving rats. *Abst. Soc. Neurosci.* (1989).

M. Ito, S. Sumio, H. Namba, L. Sokoloff, and C. Kennedy. Effects of acute morphine administration on local cerebral glucose utilization in the rat *J. Cereb. Blood Flow Meta.* **3**: S574 (1983).

L.L. Iverson. Catecholamines. *Br. Med. Bull.* **29**: 91 (1973).

S. Izenwasser and C. Kornetsky. Pharmacological effects of morphine on brain-stimulation reward. *Psychopharmacology* **93**: 136 (1987).

S. Izenwasser and C. Kornetsky. The effect of amfonelic acid or nisoxetine in combination with morphine on brain-stimulation reward. *Pharmacol. Biochem. Behav.* **32**: 983 (1989).

J.H. Jaffe. Drug addiction and drug abuse, in: *The Pharmacological Basis of Therapeutics.* A.G. Gilman, L.S. Goodman, T.W. Rall, and F. Murad, eds., Macmillan, New York, p. 532 (1985).

J.H. Jaffe and W.R. Martin. Opioid analgesics and antagonists, in: *The Pharmacological Basis of Therapeutics.* A.G. Gilman, L.S. Goodman, T.W. Rall, and F. Murad, eds., Macmillan, New York, p. 498 (1985).

K.F. Killam, J. Olds, and J. Sinclair. Further studies on the effects of centrally acting drugs on self-stimulation. *J. Pharmacol. Exp. Ther.* **19**: 157 (1957).

H.S. Kim, S. Iyengar, and P.L. Wood. Opiate actions on mesocortical dopamine metabolism in the rat. *Life Sci.* **39**: 2033 (1986).

C.M. Knapp and C. Kornetsky. The effects of amfonelic acid alone and in combination with naloxone on brain-stimulation reward. *Pharmacol. Biochem. Behav.* **32**: 977 (1989).

C. Kornetsky. Brain-stimulation reward: a model for the neuronal bases for drug induced euphoria, in: *Neuroscience Method in Drug Abuse Research.* R. Brown, D.P. Friedman and Y. Nimit, eds., NIDA Research Monograph Series 62, U.S. Government Printing Office, Washington, DC, p. 30 (1985).

C. Kornetsky and G. Bain. Brain-stimulation reward: a model of drug-induced euphoria, in: *Modern Methods in Pharmacology: Testing and Evaluation of Drugs of Abuse.* M.W. Adler and A. Cowen, eds., Alan R. Liss, New York, p. 211 (1990).

C. Kornetsky and R.U. Esposito. Reward and detection thresholds for brain stimulation: dissociative effects of cocaine. *Brain Res.* **209**: 496 (1981).

M.J. Kuhar, M.C. Ritz, and J. Sharkey. Cocaine receptors on dopamine transporters mediate cocaine-reinforced behavior, in: *Mechanisms of Cocaine Abuse and Toxicity.* D. Clouet, K. Asghar, and R. Brown, eds., NIDA Research Monograph Series 88, U.S. Government Printing Office, Washington, DC, p. 14 (1988).

N.J. Leith and R.J. Barrett. Self-stimulation and amphetamine: tolerance to D and L isomers and cross tolerance to cocaine and methylphenidate. *Psychopharmacology (Berlin),* **74**: 23 (1981).

D.J. Lettieri, M. Sayer, and H.W. Pearson. *Theories on Drug Abuse: Selected Contemporary Perspectives.* NIDA Research Monograph, U.S. Government Printing Office, Washington, DC, (1980).

J.M. Liebman. Discriminating between reward and performance: a critical review of intracranial self-stimulation methodology. *Neurosci. Biobehav. Rev.* **7**: 45 (1983).

J.M. Liebman and S.D. Segal. Differential effects of morphine and D-amphetamine on self-stimulation from closely adjacent regions in rat midbrain. *Brain Res.* **136**: 103 (1977).

E.D. London, G. Wilkerson, S.R. Goldberg, and M.E. Risner. Effects of L-cocaine on local cerebral glucose utilization in the rat. *Neurosci. Lett.* **68**: 73 (1986).

S.A. Lorens. Influence of morphine on lateral hypothalamic self-stimulation in the rat. *Psychopharmacologia* **32**: 271 (1973).

S.A. Lorens. Comparison of the effects of morphine on hypothalamic and medial frontal cortex self-stimulation in the rat. *Psychopharmacologia* **48**: 217 (1976).

M.R. Lynch and R.A. Wise. Relative effectiveness of pimozide, haloperidol and trifluoperazine on self-stimulation rate-intensity functions. *Pharmacol. Biochem. Behav.* **23**: 777 (1985).

W.H. Lyness, N.H. Friedle, and K.E. Moore. Destruction of dopaminergic nerve terminals in nucleus accumbens: effect on D-amphetamine self-administration. *Pharmacol. Biochem. Behav.* **11**: 553 (1979).

D.J. Lyons, G.T. Bain, and C. Kornetsky. The effects of nicotine on brain-stimulation reward. *Abst. Soc. Neurosci.* **14**: 765 (1988).

A. Mansour, H. Khachaturian, M.E. Lewis, H. Akil, and S.J. Watson. Autoradiographic differentiation of mu, delta, and kappa opioid receptors in the rat forebrain and midbrain. *J. Neurosci.* **7**: 2445 (1987).

R. Marcus and C. Kornetsky. Negative and positive intracranial reinforcement thresholds: effects of morphine. *Psychopharmacologia* **38**: 1 (1974).

T.J. McCown, T.C. Napier, and G.R. Breese. Effects of chronic electrode implantation in dopaminergic neurons *in vivo. Pharmacol. Biochem. Behav.* **25**: 63 (1986).

J. McCulloch, H.E. Savaki, M.C. McCulloch, J. Jehle, and L. Sokoloff. The distribution of alterations in energy metabolism in the rat brain produced by apomorphine. *Brain Res.* **243**: 67 (1982).

E. Miliaressis, N.B. Thoa, Y. Tizabi, and D.M. Jacobowitz. Catecholamine concentrations in discrete brain areas following self-stimulation of the ventromedial tegmentum of the rat. *Brain Res.* **100**: 192 (1975).

J. Millar, J.A. Stamford, Z.L. Kruk, and R.M. Wightman. Electrochemical, pharmacological and electrophysiological evidence of rapid dopamine release and removal in the rat caudate nucleus following electrical stimulation of the medial forebrain bundle. *Eur. J. Pharmacol.* **109**: 341 (1985).

M.J. Mitchell, N.M. Nicolaou, G.W. Arbuthnott, and G. Yates. Increases in dopamine metabolism are not a general feature of intracranial self-stimulation. *Life Sci.* **30**: 1081 (1982).

G.R. Mogenson, M. Takigawa, A. Robertson, and M. Wu. Self-stimulation of the nucleus accumbens and ventral tegmental area of tsai attenuated by microinjections of spiroperidol into the nucleus accumbens. *Brain Res.* **171**: 247 (1979).

M. Moolten and C. Kornetsky. Reversal of morphine induced facilitation on rewarding electrical brain-stimulation by naloxone: evidence for direct pharmacological activation of opiate receptors. *Abstr. Soc. Neurosci.* **13**: 1322 (1987).

R.D. Myers and F. Mora. *In vivo* neurochemical analysis, by push-pull perfusion, of the mesocortical dopaminergic system of the rat during self-stimulation. *Brain Res. Bull.* **2**: 105 (1977).

J.M. Nazzaro, T.F. Seeger, and E.L. Gardner. Morphine differentially affects ventral tegmental and substantia nigra brain reward thresholds. *Pharmacol. Biochem. Behav.* **14**: 325 (1981).

S. Nakajima and G.M. McKenzie. Reduction of the rewarding effect of brain stimulation by a blockade of dopamine D1 receptor with SCH 23390. *Pharmacol. Biochem. Behav.* **24**: 919 (1986).

W.T. Nelson, Jr., S.S. Steiner, M. Brutus, R. Farrell, and S.J. Ellman. Brain site variation in effects of morphine on electrical self-stimulation. *Psychopharmacology* **74**: 58 (1981).

J. Olds, K.F. Killam, and P. Bach-y-Rita. Self-stimulation of the brain used as a screening method for tranquilizing drugs. *Science* **124**: 265 (1956).

J. Olds and P. Milner. Positive reinforcement produced by electrical stimulation of septal area and other regions of rat brain. *J. Comp. Physiol. Psych.* **47**: 419 (1954).

J. Olds and R.P. Travis. Effects of chlorpromazine, meprobamate, pentobarbital and morphine on self-stimulation. *J. Pharmacol. Exp. Ther.* **128**: 397 1960.

M.E. Olds. Reinforcing effects of morphine in the nucleus accumbens. *Brain Res.* **237**: 429 (1982).

A.G. Phillips. Brain reward circuitry: a case for separate systems. *Brain Res. Bull.* **12**: 195 (1984).

A.G. Phillips, S.M. Brooke, and H.C. Fibiger. Effects of amphetamine isomers and neuroleptics on self-stimulation from the nucleus accumbens and dorsal noradrenergic bundle. *Brain Res.* **85**: 13 (1975).

A.G. Phillips and H.C. Fibiger. *The Neuropharmacological Basis of Reward.* J.M. Liebman and S.J. Cooper, eds., Clarendon Press, Oxford, p. 66 (1989).

A.G. Phillips, A. Jakubovic, and H.C. Fibiger. Increased *in vivo* tyrosine hydroxylase activity in rat telencephalon produced by self-stimulation of the ventral tegmental area. *Brain Res.* **402**: 109 (1987).

R. Pickens and T. Thompson. Cocaine-reinforced behavior in rats: effects of reinforcement magnitude and fixed-ratio size. *J. Pharmacol. Exp. Ther.* **161**: 122 (1968).

G. Pizzolato, T.T. Sonerant, and S.I. Rappoport. Time-course and regional distribution of the metabolic effects of bromocriptine in the rat brain. *Brain Res.* **341**: 303 (1985).

H. Pollard, C. Llorens, J. J. Bonnet, J. Costentin, and J.C. Schwartz. Opiate receptors on mesolimbic dopaminergic neurons. *Neurosci. Lett.* **7**: 295 (1977).

L.J. Porrino and A.M. Crane. Metabolic mapping of the effects of drugs of abuse, in: *Modern Methods in Pharmacology: Testing and Evaluation of Drugs of Abuse.* M.W. Adler and A. Cowen, eds., Alan R. Liss, New York, p. 147 (1990).

L.J. Porrino, F.R. Domer, A.M. Crane, and L. Sokoloff. Selective alterations in cerebral metabolism within the mesocorticolimbic dopaminergic system produced by acute cocaine administration in rats. *Neuropsychopharmacology* **1**: 109 (1988).

L.J. Porrino, R.U. Esposito, T.F. Seeger, and A.M. Crane. Patterns of brain energy metabolism associated with rewarding brain stimulation of the substantia nigra, *J. Cereb. Blood Flow Metab.* **5**: S211 (1985).

L.J. Porrino, R.U. Esposito, T.F. Seeger, A.M. Crane, A. Pert, and L. Sokoloff. Metabolic mapping of the brain during rewarding self-stimulation. *Science* **224**: 306 (1984).

L.J. Porrino, N.L. Goodman, and L.G. Sharpe. Intravenous self-administration of the indirect dopamine agonist amfonelic acid by rats. *Pharmacol. Biochem. Behav.* **31**: 623 (1988).

L.J. Porrino, D. Huston-Lyons, G. Bain, L. Sokoloff, and C. Kornetsky. The distribution of changes in local cerebral energy metabolism associated with brain stimulation reward to the medial forebrain bundle of the rat. *Brain Res.* **511**: 1 (1990).

L.J. Porrino and G. Lucignani. Different patterns of brain energy metabolism associated with high and low doses of methylphenidate. *Biol. Psychiat.* **22**: 125 (1987).

L.J. Porrino, G. Lucignani, D. Dow-Edwards, and L. Sokoloff. Dose-dependent effect of acute amphetamine administration on functional brain metabolism in rats. *Brain Res.* **307**: 311 (1984).

R. Prado-Alcala and R.A. Wise. Brain stimulation reward and dopamine terminal fields. I. Caudate-putamen, nucleus accumbens and amygdala. *Brain Res.* **297**: 265, (1984).

R. Prado-Alcala, A. Streather, and R.A. Wise. Brain stimulation reward and dopamine terminal fields. II. Septal and cortical projections. *Brain Res.* **301**: 209 (1984).

C. Racagni, F. Bruno, E. Iuliano, and R. Paoletti. Differential sensitivity to morphine-induced analgesia and motor activity in two inbred strains of mice: behavioral and biochemical correlations. *J. Pharmacol. Exp. Ther.* **209**: 111 (1979).

M.E. Risner and B.E. Jones. Intravenous self-administration of cocaine and norcocaine by dogs. *Psychopharmacology* **71**: 83 (1980).

M.J. Ritchie and N.M. Greene. Local anesthetics, in: *The Pharmacological Basis of Therapeutics*. A.G. Gilman, L.S. Goodman, T.W. Rall, and F. Murad, eds., MacMillan, New York, p. 302 (1985).

M.C. Ritz, R.J. Lamb, S.R. Goldberg, and M.J. Kuhar. Cocaine receptors on dopamine transporters are related to self-administration of cocaine. *Science* **237**: 1219 (1987).

D.C.S. Roberts, M.E. Corcoran, and H.C. Fibiger. On the role of the ascending catecholaminergic systems in intravenous self-administration of cocaine. *Pharmacol. Biochem. Behav.* **6**: 615 (1977).

D.C.S. Roberts and G.F. Koob. Disruption of cocaine self-administration following 6-hydroxydopamine lesions of the ventral tegmental area of rats. *Pharmacol. Biochem. Behav.* **17**: 901 (1982).

D.C.S. Roberts, G.F. Koob, P. Klonoff, and H.C. Fibiger. Extinction and recovery of cocaine self-administration following 6-hydroxydopamine lesions of the nucleus accumbens. *Pharmacol. Biochem. Behav.* **12**: 781 (1980).

A. Robertson, A. LaFerrier, and K.B.J. Franklin. Amphetamine and increases in current intensity modulate reward in the hypothalamus and substantia nigra but not the prefrontal cortex. *Physiol. Behav.* **26**: 809 (1981).

J. Saint-Laurent, M.F. Roizen, E. Miliaressis, and D.M. Jakobowitz. The effects of self-stimulation on the catecholamine concentration of discrete areas of the rat brain. *Brain Res.* **99**: 194 (1975).

O. Sakurada, M. Shinohara, W.A. Klee, C. Kennedy, and L. Sokoloff. Local cerebral glucose utilization following acute and chronic morphine administration and withdrawal. *Abst. Soc. Neurosci.* **2**: 613 (1976).

G. J. Schaefer. Opiate antagonists and rewarding brain stimulation. *Neurosci. Biobehav. Rev.* **12**: 1 (1988).

T.F. Seeger and K.R. Carlson. Amphetamine and morphine: additive effects on ICSS threshold. *Abst. Soc. Neurosci.* **7**: 924 (1981).

H. Simon, L. Stinus, J. P. Tassin, S. Lavielle, G. Blanc, A.M. Thierry, J. Glowinski, and M. LeMoal. Is the dopaminergic mesocorticolimbic system necessary for intracranial self-stimulation? *Behav. Neural. Biol.* **27**: 125 (1979).

J.J. Singer. Hypothalamic control of male and female sexual behavior in female rats. *J. Comp. Physiol. Psychol.* **84**: 38 (1968).

L. Sokoloff, M. Reivich, C. Kennedy, M.H. DesRosiers, C.S. Patlack, K.D. Pettigrew, O. Sakurada, and M. Shinohara. The [^{14}C]deoxyglucose method for the measurement of local cerebral glucose utilization: theory, procedure, and normal values in the conscious and anesthetized albino rat. *J. Neurochem.* **28**: 897 (1977).

J. Spencer and A. Revzin. Amphetamines, chlorpromazine and clonidine effects on self-stimulation in caudate and hypothalamus of the squirrel monkey. *Pharmacol. Biochem. Behav.* **5**: 149 (1976).

C. Spyraki, H.C. Fibiger, and A.G. Phillips. Attenuation of heroin reward in rats by disruption of the mesolimbic dopamine system. *Psychopharmacology (Berlin)* **79**: 278 (1983).

L. Stein. Self-stimulation of the brain and the central stimulant action of amphetamine. *Fed. Proc.* **23**: 836 (1964).

L. Stein and O.S. Ray. Brain stimulation reward "thresholds" self determined in the rat. *Psychopharmacologia.* **1**: 251 (1960).

J.R. Stellar, D. Corbett, and A.L. Hamilton. A forebrain map of dopamine's relevance to lateral hypothalamic stimulation reward as based on intercranial neuroleptic injection. *Abst. Soc. Neurosci.* **11**: 48 (1985).

E. Unterwald, S. Sasson, and C. Kornetsky. Evaluation of the supraspinal analgesic activity and abuse liability of ethylketocyclazocine. *Eur. J. Pharmacol.* **133**: 275 (1987).

F.J. Vaccarino, F.E. Bloom, and G.F. Koob. Blockade of nucleus accumbens opiate receptors attenuates intravenous heroin reward in rats. *Psychopharmacology* **85**: 37 (1985).

L. Van Wolfswinkel, W.F. Seifert, and J.M. Van Ree. Catecholamines and endogenous opioids in ventral tegmental self-stimulation reward. *Pharmacol. Biochem. Behav.* **30**: 589 (1988).

L. Van Wolfswinkel and J.M. Van Ree. Site of rewarding action of morphine in the mesolimbic system determined by intracranial electrical self-stimulation. *Brain Res.* **358**: 349 (1985).

A. Wauquier and C.J.E. Niemegeers. Intracranial self-stimulation in rats as a function of various stimulus parameters. V. Influence of cocaine on medial forebrain bundle stimulation with monopolar electrodes. *Psychopharmacology* **38**: 201 (1974).

A. Wauquier and C.J.E. Niemegeers. Restoration of self-stimulation inhibited by neuroleptics. *Eur. J. Pharmacol.* **40**: 191 (1976).

L.R. Wechsler, H.E. Savaki, and L. Sokoloff. Effects of D-amphetamine and *l*-amphetamine on local cerebral glucose utilization in the conscious rat. *J. Neurochem.* **32**: 15 (1979).

J.R. Weeks. Experimental morphine addiction: method for automatic intravenous injections in unrestrained rats. *Science* **138**: 143 (1962).

N. Weiner. Norepinephrine, epinephrine and the sympathomimetic amines, in: *The Pharmacological Basis of Therapeutics*. A.G. Gilman, L.S. Goodman, T.W. Rall, and F. Murad, eds., MacMillan, New York, p. 145 (1985).

B.H.C. Westerink and S.J. Spaan. Estimation of the turnover of 3-methoxytyramine in the rat striatum by HPLC with electrochemical detection: implications for the sequence in the cerebral metabolism of dopamine. *J. Neurochem.* **38**: 342 (1982).

R.A. Wise. Catecholamine theories of reward: a critical review. *Brain Res.* **152**: 215 (1978).

R.A. Wise. Action of drugs of abuse on brain reward systems. *Pharmacol. Biochem. Behav.* **13**: 213 (1980).

R.A. Wise. Neural mechanisms of the reinforcing actions of cocaine, in: *Cocaine: Pharmacology, Effects and Treatment of Abuse*. J. Grabowski, ed., NIDA Research Monograph Series 50, DHHS Pub. No. (ADM) 84-1326, U.S. Government Printing Office, Washington, DC, p. 15 (1984).

R.A. Wise. The role of reward pathways in the development of drug dependence. *Pharmacol. Ther.* **35**: 227 (1987).

R.A. Wise and M.A. Bozarth. Action of drugs of abuse on brain reward systems: an update with specific attention to opiates. *Pharmacol. Biochem. Behav.* **17**: 239 (1982).

R.A. Wise and M.A. Bozarth. Brain reward circuitry: four circuit elements ''wired'' in apparent series, *Brain Res. Bull.* **12**: 203 (1984).

P.L. Wood and C.A. Altar. Dopamine release *in vivo* from nigrostriatal, mesolimbic and mesocortical neurons: utility of 3-methoxytryramine measurements. *Pharmacolog. Rev.* **40**: 163 (1988).

P.L. Wood, N.P.V. Nair, and M. Bozarth. Striatal 3-methoxytyramine as an index of dopamine release: effects of electrical stimulation. *Neurosci. Lett.* **32**: 291 (1982).

P.L Wood and J.W. Richard. Morphine and nigrostriatal function in the rat and mouse: the role of nigral and striatal opiate receptors. *Neuropharmacology* **21**: 1305 (1982).

P.L. Wood, M. Sotland, J.W. Richard, and A. Rackham. Actions of mu, kappa, sigma, delta, and agonist/antagonist opiates on striatal dopaminergic function. *J. Pharmacol. Exp. Ther.* **215**: 697 (1980).

E. Yadin, V. Guarini, and C.R. Gallistel. Unilaterally activated systems in rats self-stimulating at sites in the medial forebrain bundle, medial prefrontal cortex, or locus coeruleus. *Brain Res.* **266**: 39 (1983).

J. Yeomans. The cells and axons mediating medial forebrain bundle reward, in: *The Neuronal Basis of Feeding and Reward*. R.G. Hoebel and D. Novin, eds., Haer Institute, Brunswick, ME, p. 405 (1982).

C.Y. Yim and G.J. Mogenson. Electrophysiological studies of neurons in the ventral tegmental area of tsai. *Brain Res.* **181**: 301 (1980).

R.A. Yokel and R.A. Wise. Increased lever-pressing for amphetamine after pimozide in rats: implications for a dopamine theory of reward. *Science* **187**: 547 (1975).

R.A. Yokel and R.A.C. Wise. Attenuation of intravenous amphetamine reinforcement by central dopamine blockade in rats. *Psychopharmacology* **48**: 311 (1976).

5

Neural Substrates Mediating Cocaine Reinforcement: The Role of Monoamine Systems

DAVID C.S. ROBERTS
Department of Psychology
Carleton University
Ottawa, Canada

INTRODUCTION

Literally thousands of pharmaceuticals exist which gain access to the brain, yet relatively few of them interact with the central nervous system (CNS) in such a way as to produce a rewarding effect. The main groups which have this special property are (1) stimulant drugs such as cocaine and the amphetamines, (2) opiates, (3) anxiolytis such as benzodiazepines and alcohol, and (4) hallucinogens. Where do these drugs gain access to the critical neural circuitry and how do they act? It is apparent that the majority of these drugs bind to, and interact with, specific receptors which are related to neural transmission. The term "receptor" is used here in a general sense to mean a molecular site to which a drug could specifically bind. This might be a post-synaptic receptor for a particular neurotransmitter, but it could also include enzymes or other recognition sites important in transmitter function (uptake, metabolism, etc.).

This chapter will review data concerning the identity of the receptors involved in the reinforcing effects of psychomotor stimulants, with a special emphasis on cocaine. Perhaps more is known about the specific receptor mechanisms which mediate the rewarding action of psychomotor stimulants than for any other class of rewarding compounds. Electrophysiological and biochemical experiments have identified specific sites in the brain where cocaine produces physiological responses, and the present volume attests to this progress. Obviously, drugs could have numerous actions throughout the nervous sytem, and only a few anatomical sites may ultimately prove to be important for the rewarding effects; effects in other brain regions may be relevant to other physiological responses which are unrelated to reinforcement. This chapter deals with the literature which attempts to distinguish the anatomical and physiological effects which are important for cocaine reward from those which are not.

The principal mode of action of cocaine appears to be at norepinephrine (NA), dopamine (DA), and serotonin (5-HT) neuronal synapses. By binding to the transporter of these monoamines and blocking their re-uptake into the terminal, cocaine interrupts the most important processes responsible for inactivating these neurotransmitters. Thus, the duration of monoamine stimulation of the receptor is prolonged. Because cocaine increases the synaptic efficacy of NE, DA, and 5-HT without directly stimulating the receptor, cocaine is said to be an ''indirect'' monoamine agonist.

NE, DA, and 5-HT each possess a characteristic distribution within the CNS. Each of these transmitters is generally localized to cell clusters in the pons/medulla and mesencephalon which send projections in varying densities throughout the neuro-axis. If every monoamine synapse is affected by cocaine, and if every region of the CNS receives a significant innervation from one or more of the monoamine transmitters, it follows that all areas of the brain are probably influenced. This being the case, the research question is not to discover the brain regions in which cocaine has a pharmacological effect, but those systems or areas essential for the rewarding actions of cocaine.

Evidence will be reviewed which demonstrates that the reinforcing effects of cocaine are dependent on a specific dopaminergic projection from a group of cells in the ventral midbrain. This DA group lies in the ventral tegmental region and projects to the nucleus accumbens and other forebrain areas. Data will also be presented which suggest that forebrain serotonergic systems may also play an important role — one which is antagonistic to cocaine reinforcement, that is, the reinforcing effects of cocaine have been demonstrated to be enhanced following treatments which interfere with 5-HT function, although the exact projection areas important for this effect have not yet been determined.

The prototypical drug for studying psychomotor stimulant reward in animals has been dextroamphetamine. Its duration of action is much longer than that of cocaine, but many of the behavioral effects are similar. Substantial *in vitro* data indicate that the net effect of amphetamine and cocaine is similar in that they both produce an increase in monoamine content at the synapse. As noted above, cocaine blocks the re-uptake of monoamines, whereas amphetamine has the additional property of releasing newly synthesized stores of monoamine transmitters. While this chapter will focus on the substrates of cocaine reinforcement, many parallels between cocaine and amphetamine will be made with the understanding that the behavioral pharmacology of amphetamine and cocaine can differ.

ASSESSMENT OF COCAINE SELF-ADMINISTRATION BEHAVIOR

Much of the evidence linking the reinforcing efficacy of cocaine to monoamine systems comes from self-administration experiments. In these studies, laboratory

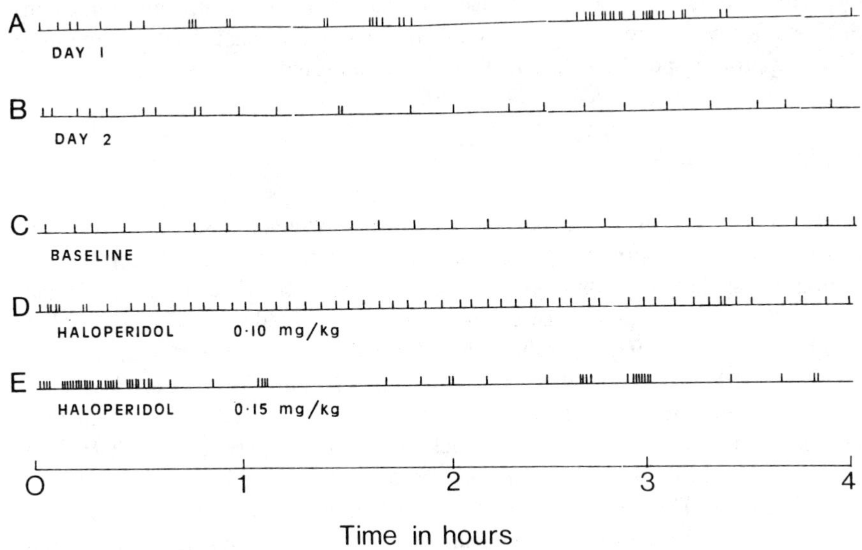

Figure 1. Event records showing pattern of i.v. cocaine self-administration (0.6 milligrams per injection). Line A is an example of the response pattern typical of the acquisition phase (during the first self-administration session) before a regular response pattern has emerged. A consistent interinfusion interval develops quickly, in this case on day 2 (line B). After several days, the baseline pattern is extremely regular (line C). Pretreatment with the DA antagonist, haloperidol (1.1 mg/kg) produces a pronounced increase in self-administration rate (line D) over baseline (line C) or at higher doses (0.15 mg/kg), an extinction-like response pattern (line D).

animals, usually rats or monkeys, are given the opportunity to perform some task to obtain an i.v. injection of cocaine. Animals quickly learn to depress a lever when each response produces an injection (i.e., a continuous reinforcement schedule or CRF) and they settle into a pattern of self-administration with regularly spaced infusions (see Figure 1). Occasionally, animals will overdose during their first self-administration session, suggesting that the surviving animals have learned two things. First, the lever produces a reinforcing drug injection and, second, too many injections can produce a toxic reaction. The animals also learn to pause for some length of time after each infusion. There is no doubt that animals are entirely capable of responding during these pauses, for they will work for other reinforcers (Wise *et al.*, 1977). This implies that the injection frequency taken reflects a preferred rate of drug intake rather than an imposed rate limited by drug incapacitation. This preferred rate of drug intake is exquisitely sensitive to changes in dose. It is also an inverse relationship, such that higher injection doses are self-administered more slowly, with lower doses being self-administered at a relatively high rate (Pickens and Thompson, 1968). Since the rate of self-administration is so sensitive to alterations in dose and, as it seems intuitive that any treatment which would influence a drug's rewarding action should change the rate of drug intake, most of the studies to date have used rate of self-administration as the dependent measure to study the effects of various pharmacological pretreatments and lesions. Such studies will be reviewed in this chapter, with emphasis placed on the pitfalls of this approach and alternative methodologies described.

DOPAMINE

Pharmacological manipulations yielded the first evidence to support the idea that DA systems are important for the manifestation of cocaine reward. Self-administration of psychomotor stimulants has been shown to be affected in a predictable way by DA receptor antagonist drugs. Low dosages of the DA antagonists haloperidol (Gill *et al.*, 1975), perphenazine (Johanson *et al.*, 1976), pimozide (Yokel and Wise, 1975; Risner and Jones, 1976), butaclamol (Yokel and Wise, 1975), chlorpromazine (Wilson and Schuster, 1972), α-flupenthixol, sulpiride, metoclopramide, thioridazine (Roberts and Vickers, 1984), spiperone, and SCH 23390 (Koob *et al.*, 1987) all produce an increased intake of amphetamine and cocaine (see Figure 1). This increase in self-administration rate is replicable and robust. While there is some debate whether drugs selective for dopamine D_1 or D_2 receptors mediate this effect (see Koob *et al.*, 1987; Woolverton, 1986), it would appear that at appropriate doses and pretreatment intervals both types of DA antagonists are effective. However, since the factors which govern drug intake are not completely understood (see below), the interpretation of the results has been open to some debate.

Perhaps the most powerful argument has been presented by Yokel and Wise (1975, 1976), who have drawn an analogy between the increase in drug intake following DA receptor blockade and the increase in injection rate seen in animals after a reduction in the unit dose. Recalling that rate of cocaine self-administration is dependent on injection dose, and that a reduction in injection dose produces a compensatory increase in injection rate, Yokel and Wise suggest that DA antagonist-induced increases in self-administration also represent a compensatory response. They argue that partial antagonism of dopamine receptors produces a partial block of the stimulant-reinforcing effects and that the increase in drug intake represents an attempt by the animal to overcome this blockade. By this logic, if DA receptor blockade is the same as a reduction in dose, then it follows that DA receptor stimulation is critical for cocaine reward.

The analogy goes further. Lowering the cocaine injection dose will increase responding only to a point, past which the self-administration behavior will extinguish. This is normally characterized by bursts of responding followed by response cessation. A similar phenomenon occurs with DA antagonists, such that pretreatment of animals self-administering cocaine with DA receptor blockers causes a dose-dependent increase in stimulant intake until, at some point, the behavior degenerates into a pattern reminiscent of behavioral "extinction" (DeWit and Wise, 1977; Yokel and Wise, 1975; Yokel and Wise, 1976). Thus, there are striking parallels between DA blockade and reductions in unit drug dose. These data bolster the hypothesis that DA receptor stimulation is the essential site of action for cocaine reward.

Although the effects of DA antagonists appear convincing, caution must be exercised when interpreting results involving changes in self-administration rate. All of the factors which control the rate of drug intake are simply not understood, and alternate explanations exist. Since animals adjust their injection rate so that the total amount of drug remains relatively constant (Pickens and Thompson, 1968), it would appear that there is a preferred level of drug intake which falls within a surprisingly small range. Logically, one might ask why the animals do not exceed this level. Clearly, one factor which must determine the upper limit of this preferred range (and this is presumably the case for all self-administered compounds) is the degree of toxicity associated with high blood or brain concentrations. All drugs are toxic at some dose level, and animals presumably learn to restrict their intake accordingly to avoid such adverse effects. Therefore, it is likely that the self-administration rate reflects an integrated function which both maximizes the positive effects and minimizes the adverse effects. Treatments which influence either the positive or the adverse effects might be expected to alter this delicate balance.

With the above assumptions, increases in the rate of self-administration might reflect either (1) a compensatory response to a decrease in reinforcement strength as suggested by Yokel and Wise (1975) or (2) an increase in the preferred drug level due to a decrease in toxicity (see Wilson and Schuster, 1973). In one case, the conclusion is that the rewarding effect is reduced and, in the other, the conclusion is that the net rewarding effect is increased! Hence, it appears that the same data could be taken as evidence in support of either of the two opposing positions. When one considers that rate of drug intake could also be affected by treatments which alter the distribution, metabolism, and clearance of cocaine, it appears that a change in drug intake has limited value as an indicator of a change in drug reward.

Most pharmacologists have long recognized the difficulties in interpreting changes in rate of self-administration. Unfortunately, alternative methods to examine changes in drug reinforcement have been lacking. For example, fixed ratio (FR) schedules, on which an animal must respond a predetermined number of times in order to earn a reinforcement, offer no more information than a continuous reinforcement schedule (i.e., FR 1); other more complex schedules which yield more information than rate of drug intake are obviously necessary. Indeed, in primates, procedures involving concurrent (Iglauer and Woods, 1974) and second-order schedules (Kelleher and Goldberg, 1977; Johanson, 1982), choice procedures (Johanson and Aigner, 1981), and discrete trials (Stretch, 1977) have contributed elegantly to our understanding of self-administration as an operant. However, most of the lesion and pharmacological challenge experiments which address the neurochemical aspects of the behavior have used rats as subjects. In fear of lengthy training protocols and potential problems of interpreting disruptions of self-administration behavior, researchers have shied away from these more complicated tasks when rodents are used as subjects.

Recently, we have proposed a procedure which can be used feasibly in rats and permits a rate-free interpretation of the data (Roberts, 1989; Roberts *et al.*, 1989).

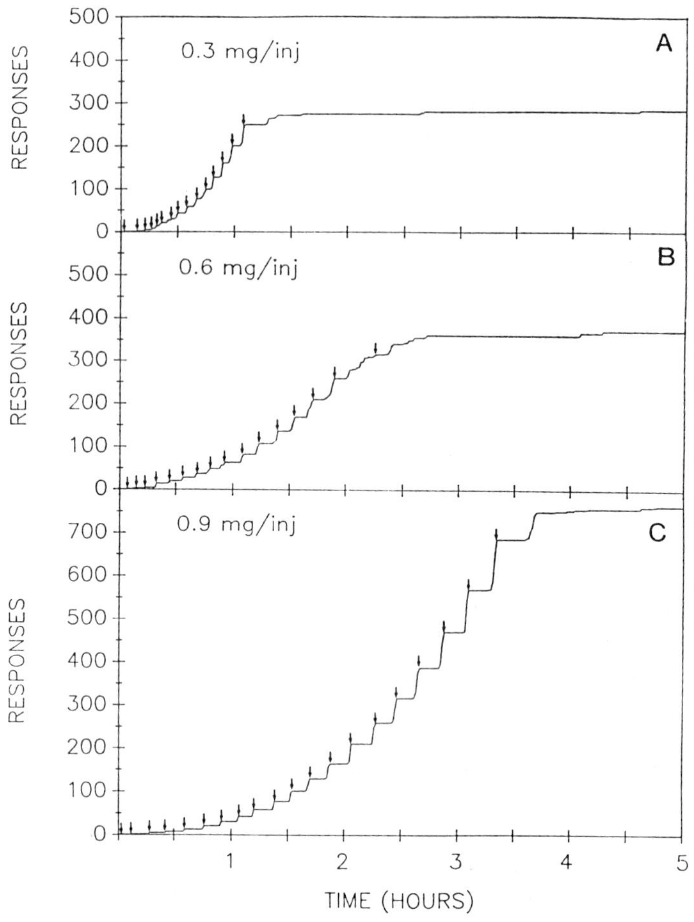

Figure 2. Computer-reconstructed cumulative records of self-administration behavior of a rat responding under a progressive ratio (PR) schedule for three doses of cocaine reward. Vertical increments indicate lever responses. Infusions are marked by arrows. Break points are defined as the last ratio completed prior to a 1-h period of nonreward. In this instance, the break points were 57, 70, and 138 for the 0.3, 0.6, and 0.9 milligram per injection doses, respectively. (From Roberts, D.C.S., Loh, E., and Vickers, G. *Psychopharmacology* **97**: 535 (1989). With permission.)

This procedure employs a progressive ratio (PR) schedule of reinforcement. With this schedule, the response requirements to earn a drug injection escalate with each reinforcement. For example, the first injection of each test session requires only one response on the lever; however, the number of responses required to earn subsequent injections follows through an approximately logarithmic series such as 2, 4, 6, 9, 12, 15, 20, 25, 32, 40 . . . until the self-administration behavior extinguishes.

Figure 2 shows the pattern of responding of one animal on a PR schedule reinforced on different days by three different doses of cocaine. During the early part

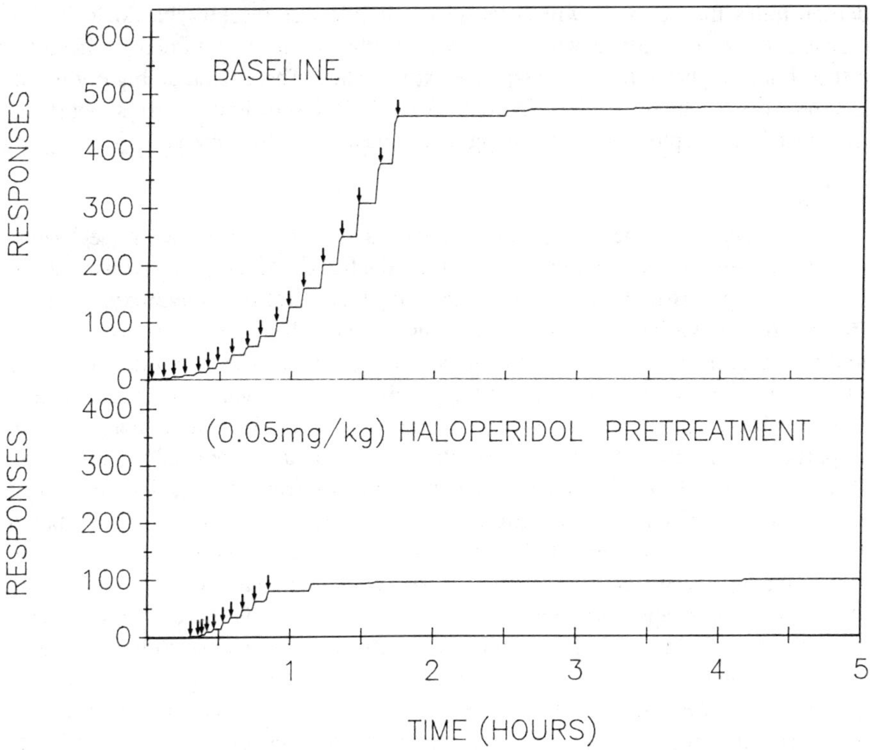

Figure 3. Computer-reconstructed cumulative record showing the effect of haloperidol (0.05 mg/kg), a DA antagonist, on cocaine self-administration reinforced on a progressive ratio schedule. Vertical increments indicate lever responses. Infusions are marked by arrows. The break point after haloperidol treatment was 32 compared to the baseline break point of 219.

of each test session, infusions of cocaine are regularly spaced. Each infusion is followed by a remarkably consistent post-infusion pause, after which the animal quickly responds until the requirements for the next infusion are met. At some point in the session, the response requirements exceed the reinforcing efficacy of the drug dose and the lever-pressing behavior extinguishes. The last ratio completed before extinction is defined as the ''break-point'', which serves as the dependent measure. In general, higher doses of cocaine produce higher breaking points, although it should be emphasized that this is not necessarily the case for all animals. Consistent with the research in primates by Bedford *et al.* (1978), we have found that some rats show an inverted U-shaped dose/response function with a midrange dose producing the highest breaking points (Roberts *et al.*, 1989).

Figure 3 shows the effect of 0.05 mg/kg haloperidol pretreatment on cocaine self-administration on a PR schedule. This dose, which we have shown previously to cause an increase in cocaine intake (Roberts and Vickers, 1984), was found to lower the breaking point on the PR schedule. It is interesting to note that, even on the PR schedule, the rate of intake of the first few injections was increased. This demonstrates that animals will compensate for the DA receptor blockade by increasing their drug intake when the drug injections can be ''earned'' with little effort, but at higher ratios, the response extinguishes. There is no ambiguity in the interpretation of these results. The data from the PR schedule clearly support the idea that DA receptor blockade antagonizes cocaine reinforcement.

Lesion Studies

Initial studies designed to explore the effects of discrete catecholamine lesions on cocaine self-administration were first conducted by Fibiger and co-workers at the University of British Columbia. By injecting the neurotoxin 6-hydroxydopamine (6-OHDA) into various terminal regions and fiber bundles, the involvement of DA and NE systems could be explored. It was during these studies that the importance of the nucleus accumbens septi (NAS) in cocaine reinforcement was first identified.

Following on the observations of Pijnenberg *et al.* (1975), who demonstrated an important role for the DA innervation of the NAS in stimulant-induced hyperactivity, Roberts *et al.* (1977) showed that depletion of DA from this nucleus abolished cocaine self-administration on a FR 1 schedule of reinforcement; rats simply failed to respond for cocaine reward after intra-accumbens infusions of 6-OHDA. These same animals that would not respond for cocaine would continue to self-administer the direct agonist apomorphine. Thus, cocaine injections failed to reinforce self-administration behavior in DA-lesioned animals that were shown to be capable of responding and sensitive to the rewarding effect of another drug. In a subsequent report (Roberts *et al.*, 1980), we demonstrated that self-administration behavior recovered after the lesion and the time course of this recovery was correlated with the degree of DA depletion in the accumbens. No recovery was observed if substantial (i.e., greater than 90%) depletions of DA were achieved. Figure 4 shows the event records obtained from animals with severe depletions of DA which failed to self-administer cocaine, yet would readily self-administer apomorphine.

Pettit *et al.* (1984) have provided further evidence of the pharmacological specificity of the effect by showing that, while 6-OHDA accumbens lesions disrupt the rate of cocaine self-administration, they apparently leave heroin self-administration unaffected. In addition, Koob *et al.* (1987) have reinvestigated this phenomenon with the PR schedule and confirmed that, indeed, the motivation to self-administer cocaine is reduced after 6-OHDA lesions of the NAS.

6-OHDA lesions of the DA cells within the ventral tegmental area, which innervate the nucleus accumbens (see Figure 5), also produce a disruption of cocaine self-administration, although many animals continued to respond at a reduced rate (Roberts and Koob, 1982). Curiously, we found that the reductions in rate did not correlate with DA levels in the nucleus accumbens. However, it must be reemphasized that rate does not measure the motivation to self-administer a drug (see also Roberts, 1989); for this reason, these data are not necessarily inconsistent with the hypothesis that the nucleus accumbens is critically involved.

Other studies have shown similar effects of 6-OHDA lesions on amphetamine self-administration. Lyness *et al.* (1979) tested the effects of 6-OHDA injections into the nucleus accumbens in animals trained to self-administer amphetamine. The

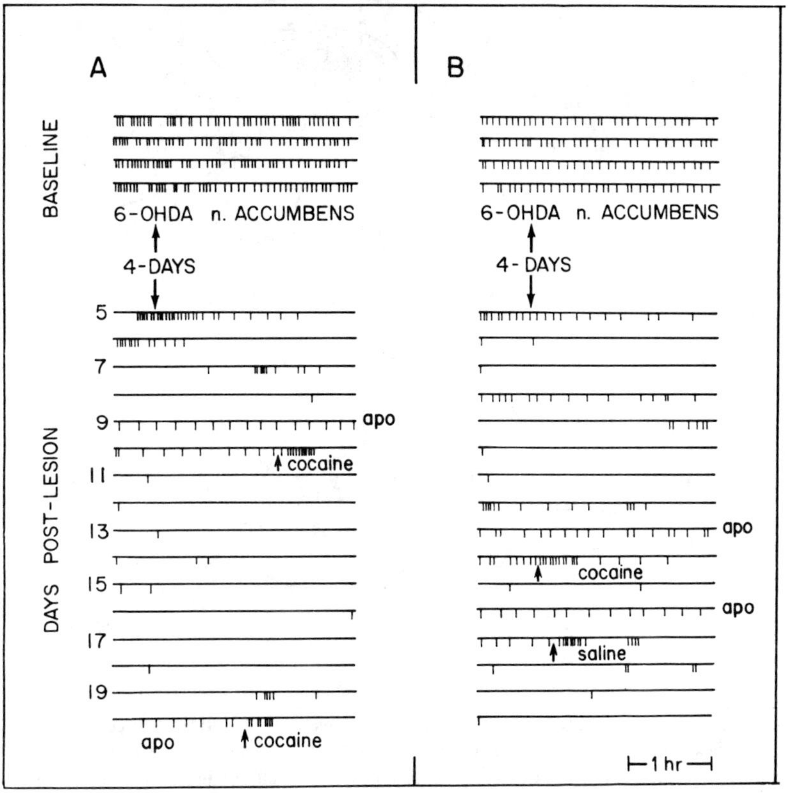

Figure 4. Event records showing the effect in two animals with 6-OHDA-induced lesions of the nucleus accumbens on cocaine self-administration on a CRF schedule of reinforcement. Each pen deflection indicates an i.v. cocaine infusion (0.75 milligrams per injection). The top four lines represent baseline responding on daily 3-h sessions. On the 5th day following the 6-OHDA treatment, responding declined to minimal levels. Regularly spaced self-administration behavior was reinstated when apomorphine (0.06 milligram per injection) was substituted for cocaine; however, when cocaine was again offered (indicated by arrows), self-administration behavior again declined. (From Roberts, D.C.S., Koob, G.F., Klonoff, P., and Fibiger, H.C. *Pharmacol. Biochem. Behav.* **17**: 781 (1980). With permission.)

treatment was found to abolish the behavior in trained rats and also prevented the acquisition of amphetamine self-administration in untrained, naive animals.

It is important to note that stereotaxically placed injections of 6-OHDA do not disrupt i.v. cocaine self-administration unless they are placed within the mesolimbic system. Injections of the neurotoxin into a number of other areas have been found to have no effect. Figure 5 illustrates the 6-OHDA injection sites which have produced positive and negative effects. Intracerebral infusions of 6-OHDA into the caudate nucleus (Roberts and Zito, 1987), the dorsal tegmentum (Roberts *et al.*, 1977), or the prefrontal cortex (Martin-Iverson *et al.*, 1986) produce no change in the subsequent rate of cocaine self-administration. Interestingly, Deminiere *et al.* (1988) have reported that 6-OHDA lesions of the amygdala actually facilitate ac-

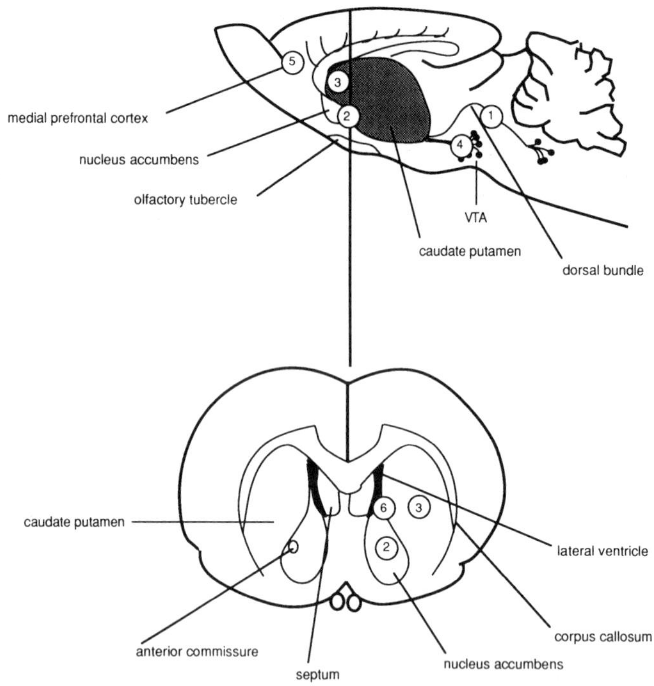

Figure 5. Locations of 6-OHDA injection sites which have been investigated in cocaine self-administration studies. Stereotaxically placed injections of 6-OHDA into the nucleus accumbens (2) or the ventral tegmental area (4) have been found to disrupt cocaine self-administration. Injections into the dorsal NE bundle (1), the caudate nucleus (3,6), or the medial prefrontal cortex (5) are without effect. The upper panel represents a rat brain in sagittal plane. The lower panel represents a coronal section taken at the level indicated.

quisition of amphetamine self-administration. This observation may be viewed as complementary with the nucleus accumbens hypothesis, since they also showed that their amygdala lesions also produced an increase in the turnover rate of DA in the nucleus accumbens.

Inconsistencies within the Dopamine Hypothesis

It would appear that the above lesion data offer substantial support for the hypothesis that the nucleus accumbens is the critical site for cocaine reward and that other forebrain projections are relatively unimportant or modulatory in nature. However, conflicting evidence for this position comes from studies which have examined *intracerebral* self-administration of cocaine. Goeders has convincingly shown that rats will self-administer cocaine directly into the prefrontal cortex, but not into the nucleus accumbens or ventral tegmental area (Goeders and Smith, 1983). The behavior was blocked by infusing the DA antagonist sulpiride concurrently with cocaine. Furthermore, Goeders and Smith (1986) showed that the presynaptic DA terminal site was necessary since animals would no longer self-administer cocaine into the prefrontal cortex after 6-OHDA infusions into this area, although their rats would respond for prefrontal DA injections.

The data of Goeders and Smith (1986) clearly show that the prefrontal DA innervation is critical for intracranial self-administration of cocaine. However, the fact that Martin-Iverson *et al.* (1986) found that prefrontal 6-OHDA lesions had no effect on i.v. cocaine self-administration would seem contradictory. How is it that DA in the prefrontal area is important for intracranial but not i.v. self-administration? One possibility is that, since Martin-Iverson *et al.* (1986) examined only the rate of self-administration after frontal lesions, a change in motivation to self-administer cocaine may have gone undetected. Changes in motivation to self-administer are not necessarily reflected in rate, and we have documented several instances where large changes in break-point on the PR schedule were not paralleled by changes in injection frequency (Roberts, 1989; Roberts *et al.*, 1989). The discrepancy will remain until the effect of prefrontal 6-OHDA lesions on cocaine self-administration is reexamined with the PR schedule.

Also inconsistent with a predominant role of the DA innervation of the accumbens in cocaine reinforcement is the finding by Goeders and Smith (1983) that this nucleus does not support intracranial self-administration of cocaine. Since the accumbens does support intracranial self-administration of both amphetamine (Hoebel *et al.*, 1983) and DA (Guerin *et al.*, 1984), it is possible that the local anesthetic effects of cocaine interfere with the manifestation of the rewarding properties. However, it is not clear why such an effect would interfere with cocaine self-administration into the nucleus accumbens but not with self-administration into the prefrontal area.

Two additional studies are also inconsistent with the DA hypothesis. It has been reported that rats acquired a self-administration response *more* quickly after radiofrequency (Le Moal *et al.*, 1979) or 6-OHDA lesions (Deminiere *et al.*, 1984) of the ventral tegmental area. It is not immediately apparent how these data can be reconciled with the idea that the mesolimbic system is a critical substrate, although procedural differences may be important. Deminiere *et al.* (1984) used very small injection doses of amphetamine (7.5 micrograms per injection) which promote self-injection rates upwards of 80 injections per hour. By comparison, other labs typically use dosages which lead to rates of four to ten injections per h (e.g., 100 micrograms per injection amphetamine). It may be worth considering whether self-administration behavior reinforced by a large bolus of drug that is easily discriminated might involve different mechanisms than the response facilitating effects of extremely low doses. For example, one might speculate that a large drug bolus has an immediate suppressive (or satiating) effect, whereas lower doses are relatively more effective in enhancing specific responses. Indeed, the neural mechanisms responsible for a drug cue may be different from those that control response facilitation.

To summarize, not withstanding a few inconsistencies, there is a substantial literature which supports the hypothesis that the DA projection from the ventral midbrain to the nucleus accumbens is essential for cocaine self-administration. What remains to be understood is the relative contributions of other forebrain sites, including the amygdala and prefrontal cortex.

SEROTONIN

There have been no studies published to date that have examined cocaine self-administration behavior following direct manipulation of serotonin systems, although a number of reports by Lyness and co-workers have examined the role of serotonin in mediating amphetamine reinforcement. In their initial study, in which intraventricular injections of the neurotoxin 5,7-dihydroxytryptamine (5,7-DHT) were used to deplete forebrain 5-HT, they found that lesioned animals consistently self-administered more amphetamine than control animals (Lyness *et al.*, 1980). Leccesse and Lyness (1984) subsequently showed that 5,7-DHT lesions of the median forebrain bundle (MFB) also produced substantial increases in amphetamine intake, indicating that an ascending system was involved, although curiously, 5,7-DHT infusions into the nucleus accumbens had no effect (Lyness *et al.*, 1980). Consistent with these data are the results of Simon *et al.* (1980) who showed that radiofrequency lesions of the dorsal or median raphe, 5-HT nuclei which project to the forebrain, accelerate the acquisition of amphetamine self-administration. The findings of both labs support the idea that the action of amphetamine on 5-HT systems is aversive and, by removing the 5-HT action, the drug becomes more reinforcing.

If an action at 5-HT terminals were in fact aversive, it might be predicted that augmentation of 5-HT function would decrease amphetamine reinforcement. Some support for this idea comes from studies using pharmacological pretreatments, although this literature is somewhat inconsistent as well. Leccesse and Lyness (1984) presented some intriguing data showing that amphetamine self-administration was decreased following either L-tryptophan administration, which increases brain 5-HT content (see also Smith *et al.*, 1986) or fluoxetine administration, which acts as an indirect 5-HT agonist. Compatible with the idea that these manipulations require intact 5-HT systems, L-tryptophan or fluoxetine did not alter amphetamine self-administration in 5,7-DHT lesioned animals. Administration of 5-HT antagonists have yielded confusing results. Whereas metergoline pretreatment caused an increase in amphetamine intake (Lyness and Moore, 1983), methysergide and cyproheptadine produced a decrease (Leccesse and Lyness, 1984). Nonspecific actions of the various drugs may account for these discrepant findings, but the important point is that 5-HT manipulations seem to have an important influence on amphetamine self-administration.

Again, it should be reemphasized that the above results are difficult to interpret due to the reliance on rate of self-administration as the dependent measure. In order to better assess the role of 5-HT systems in cocaine reward, we have recently used the PR schedule to investigate the effects of 5,7-DHT lesions on cocaine self-administration. In preliminary studies, we found that some animals responded to very high ratios in order to obtain a single injection of cocaine after intraventricular injections of 5-HT neurotoxin (unpublished observations). Intracerebral infusions of 5,7-DHT into the MFB or the amygdala also produced significant, although less spectacular, increases in break point (Loh and Roberts, 1990; Figure 6). On average, the MFB lesioned group responded to break points that were a third higher on the escalating scale than their previous baseline. These data provide clear evidence that

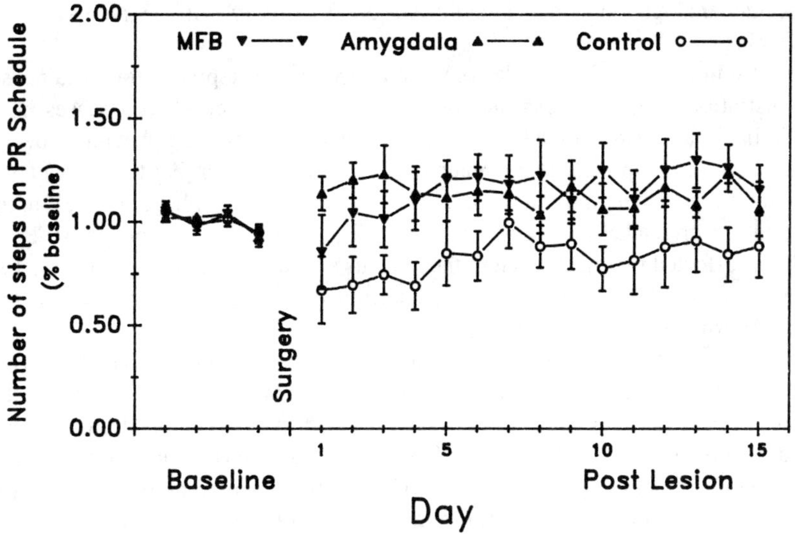

Figure 6. The effect of 5,7-DHT lesions of the median forebrain bundle (MFB) on cocaine self-administration on a PR schedule of reinforcement. Daily break points were converted to a percentage score of the 4-d baseline (prelesion). Points represent the mean (± S.E.M.) of groups of animals which received stereotaxic injections of 5,7-DHT (MFB) or vehicle injections (control).

5-HT systems are critically involved in cocaine self-administration, and support the idea that 5-HT systems are antagonistic to the reinforcing effects of psychomotor stimulants.

The data are consistent with the idea that there is an aversive or toxic reaction from cocaine produced by its indirect agonist properties at 5-HT synapses. By removing the 5-HT terminal, this particular action is attenuated and the relative reinforcing effects of the drug are enhanced. However, before this hypothesis can be seriously entertained, it must be shown that the 5,7-DHT lesions do not increase break points on a PR schedule for any reward, (e.g., food or apomorphine). If it were to be found that 5,7-DHT lesions also affect the break points of other reinforcers which do not interact directly with 5-HT systems, the data would suggest a more general phenomenon, such as behavioral preservation or resistance to extinction. Assuming for the moment that this is not the case, it can be tentatively concluded that the reinforcing property of cocaine is directly related to its action at DA terminals and negatively related to its action at 5-HT terminals. This is exactly the relationship reported recently by Ritz *et al.* (1987). These authors studied a variety of cocaine analogues for their ability to compete for binding sites with agents with specific affinities for DA, NE, and 5-HT transporters. They found that there was a positive correlation between the potencies of the various cocaine analogues for the DA transporter and their reinforcing potencies as measured in self-administration studies. Of particular interest to the present discussion was their report that there was a *negative* partial correlation between the reinforcing effects of the drugs and their potencies at 5-HT transporter sites. Consistent with our lesion effects, it seems that a cocaine analogue that has less intrisic 5-HT activity will be relatively more reinforcing.

Our finding that 5,7-DHT lesions increase the break points for cocaine self-administration permits an explanation of the previously reported increases in amphetamine intake. Animals self-administer greater amounts of amphetamine because some aversive component is attenuated. Our results also parallel the amphetamine observations in other important ways. Both cocaine and amphetamine self-administration are affected by injection of 5,7-DHT into the ventricle or the MFB, and neither is affected by injections into the nucleus accumbens (Loh and Roberts, 1990; Lyness *et al.*, 1980). It remains to be determined whether these or other forebrain areas are critical for this effect.

It is important to note, however, that there are important differences between amphetamine and cocaine self-administration behavior. In spite of the fact that we found a significant increase in break point for cocaine self-administration on a PR schedule, no changes in rate of drug intake were observed (Loh and Roberts, 1990). Recalling that 5,7-DHT lesions increase rate of amphetamine intake (Lyness *et al.*, 1980; Leccese and Lyness, 1984), it remains unclear why 5,7-DHT lesions produce an increase in the intake of amphetamine and not cocaine. The effects of the 5-HT antagonist metergoline are also different. While this drug has drastic effects on amphetamine intake (Lyness and Moore, 1983), we have been unable to demonstrate a significant effect of metergoline pretreatment on the rate of cocaine self-administration (unpublished observations). It would appear that the rate of cocaine self-administration is not easily modified, even in cases when it can be shown that breaking points on a PR schedule are substantially altered.

NOREPINEPHRINE

Several of the early, self-administration studies showed that a number of drug pretreatments which are known to affect both NE and DA function exerted a significant influence on the rate of cocaine self-administration. For example, α-methyl-*p*-tyrosine, which blocks the synthesis of DA and NE, was shown to increase cocaine (Wilson and Schuster, 1974) and methamphetamine self-administration (Pickens *et al.*, 1968), whereas a decrease in intake was shown after pretreatment of a monoamine oxidase inhibitor (Wilson and Schuster, 1974). Further support for NE involvement was provided by Davis *et al.* (1975) using an acquisition paradigm. The procedure they used was to allow rats to acquire a lever response for small i.v. doses of amphetamine and then to extinguish this behavior. They reported that pretreatment with the dopamine-β-hydroxylase inhibitors (DBH), U-14,624 and DDC, which block the conversion of DA to NE, disrupts re-acquisition of the amphetamine self-administration behavior. However, subsequent investigations using specific NE antagonists have led to the conclusion that NE is not of primary importance. Thus, it may be that the DBH inhibitors used by Davis *et al.* (1975) caused a nonspecific or toxic reaction. As we have found that DDC can act as an aversive stimulus (Roberts and Fibiger, 1976) and, if this experience were paired with an amphetamine injection, such toxicity could be expected to attenuate the reinforcing experience.

In contrast to the powerful influence of DA antagonists, pretreatment with selective NE antagonists generally has been found to exert no consistent effect on rate of cocaine self-administration. Neither α-adrenergic antagonists, such as phentolamine and phenoxybenzamine (Wilson and Schuster, 1974; Risner and Jones, 1976; DeWit and Wise, 1977), nor the β-adrenergic antagonist propranolol (DeWit and Wise, 1977) caused a change in amphetamine or cocaine intake.

Lesions of NE fibers systems have also been consistently without effect. The major innervation of the forebrain arises from a brain stem cell group known as the locus coeruleus which projects through the dorsal midbrain in a relatively constrained bundle of fibers. Injections of 6-OHDA into this tegmental tract causes a near complete loss of telencephalic NE. Despite this NE denervation, animals continue to respond for i.v. cocaine at rates not significantly different from prelesion values (Roberts *et al.*, 1977). Again, it is possible that NE treatments may affect the motivational effects of cocaine and not change the rate of intake; however, until the effects of NE lesions are evaluated on the PR schedule, a firm conclusion cannot be reached. However, the preponderance of evidence to date suggests that NE does not play an important role in either cocaine or amphetamine reinforcement.

FUTURE DIRECTIONS

A great deal of progress has been made in the search for the "site of action" of the reinforcing effects of cocaine. However, identifying the locus where a drug gains entry into the neural circuitry is only the beginning; the more important questions remain. Presentation of natural reinforcers (e.g., food) influence a variety of behavioral effects, including hyperactivity, response perseveritivity, memory formation, and affective state. It is likely that different neural mechanisms or anatomical regions are involved in each of these. How cocaine, or any other reinforcing drug, taps into these different processes is not known, but it is possible that each class of drug (opiate, stimulant, anxiolytic, etc.) taps into these processes in a unique manner. It is certainly clear that self-administration of stimulants is qualitatively different from self-administration of opiates. For example, our experience with the PR schedule has shown us that the motivation to respond for cocaine increases after the first few injections, that is, ingestion of cocaine increases the desire to take it, a phenomenon that recently has been reported in the human population (Jaffe *et al.*, 1989). The motivation to self-administer heroin, on the other hand, is most powerful before the initial injection of the session. (Roberts and Bennett, submitted). Future directions will probably emphasize the differences between the different classes of "reinforcing" drugs (see Chapter 4) rather than focusing on their similarities.

ACKNOWLEDGMENTS

This work was supported by grants from the Medical Research Council and the Natural Sciences and Engineering Research Council of Canada.

REFERENCES

R.L. Balster and C.R. Schuster. Fixed interval schedule of cocaine self-administration: effect of dose and infusion duration. *J. Exp. Anal. Behav.* **20**: 119 (1973).

J.A. Bedford, L.P. Bailey, and M.C. Wilson. Cocaine reinforced progressive ratio performance in the rhesus monkey. *Parmacol. Biochem. Behav.* **9**: 631 (1978).

W.M. Davis and S.G. Smith. Central cholinergic influence on self-administration of morphine and amphetamine. *Life Sci.* **16**: 237 (1985).

W.M. Davis, S.G. Smith, and J.H. Khalsa. Noradrenergic role in the self-administration of morphine or amphetamine. *Pharmacol. Biochem. Behav.* **3**: 477 (1974).

J.M. Deminiere, H. Simon, J.P. Herman, and M. Le Moal. 6-Hydroxydopamine lesions of the dopamine mesocortico-limbic cell bodies increases (+)-amphetamine self-administration. *Psychopharmacology* **83**: 281 (1984).

J.M. Deminiere, K. Taghzouti, J.P. Tassin, M. Le Moal, and H. Simon. Increased sensitivity to amphetamine and facilitation of amphetamine self-administration after 6-hydroxydopamine lesions of the amygdala. *Psycholpharmacology* **94**: 232 (1988).

H. DeWit and R.A. Wise. Blockade of cocaine reinforcement in rats with the dopamine receptor blocker pimozide, but not the noradrenergic blockers phentolamine and phenoxybenzamine. *Can. J. Psychol.* **31**: 195 (1977).

C.A. Gill, W.C. Holz, and C. Zirkel. The role of dopamine in self-administration of psychomotor stimulants. Reported to the NAS-SCR Committee on Problems of Drug Dependence. (1975)

N.E. Goeders and J.E. Smith. Cortical dopaminergic involvement in cocaine reinforcement. *Science* **221**: 733 (1983).

N.E. Goeders and J.E. Smith. Reinforcing properties of cocaine in the medial prefrontal cortex: primary action on presynaptic dopaminergic terminals. *Pharmacol. Biochem. Behav.* **25**: 191 (1986).

G.F. Guerin, N.E. Goeders, S.I. Dworkin, and J.E. Smith. Intracranial self-administration of dopamine into the nucleus accumbens. *Neurosci. Abst.* **10**: 1072 (1984).

B.G. Hoebel, A.P. Monaco, L. Hernandez, E.F. Aulisi, B.G. Stanley, and L. Lenard. Self-injection of amphetamine directly into the brain. *Psychopharmacology* **81**: 158 (1983).

C. Iglauer and J.H. Woods. Concurrent performances: reinforcement by different doses of intravenous cocaine in rhesus monkeys. *J. Exp. Anal. Behav.* **22**: 179 (1974).

J.H. Jaffe, N.G. Cascella, K.M. Kumor, and M.A. Sherer. Cocaine-induced cocaine craving. *Psychopharmacology* **97**: 59 (1989).

C.E. Johanson and T. Aigner. Comparison of the reinforcing properties of cocaine and procaine in rhesus monkeys. *Pharmacol. Biochem. Behav.* **15**: 49 (1981).

C.E. Johanson, D.A. Kandel, and K. Bonese. The effects of perphenazine on self-administration behavior. *Pharmacol. Biochem. Behav.* **4**: 427 (1976).

C.E. Johanson. Behavior maintained under fixed-interval and second-order schedules of cocaine or pentobarbitol in rhesus monkeys. *J. Pharm. Exp. Ther.* **221**: 384 (1982).

R.T. Kelleher and S.R. Goldberg. Fixed-interval responding under second-order schedules of food presentation and cocain injection. *J. Exp. Anal. Behav.* **28**: 221 (1977).

G.F. Koob, F. Vaccarino, F. Amalric, and F.E. Bloom. Positive reinforcement properties of drugs: search for neural substrates, in: *Brain Reward Systems and Abuse*. J. Engel and L. Oreland, eds., Raven Press, New York, p. 35 (1987).

G.F. Koob, H.T. Le, and I. Creese. The D1 dopamine receptor antagonist SCH 23390 increases cocaine self-administration in the rat. *Neurosci. Lett.* **79**: 315 (1987).

M. Le Moal, L. Stinus, and H. Simon. Increased sensitivity to (+)-amphetamine self-administration by rats following meso-cortico-limbic dopamine neurone destruction. *Nature* **280**: 156 (1979).

A.P. Lecesse and W.H. Lyness. The effects of putative 5-hydroxytryptamine receptor active agents on D-amphetamine self-administration in controls and rats with 5,7-dihydroxytryptamine median forebrain bundle lesions. *Brain Res.* **303**: 153 (1984).

E. A. Loh and D. C. S. Roberts: Break-points on a progressive ratio schedule reinforced by intravenous cocaine increase following depletion of forebrain serotonin. *Psychopharmacology* **101**: 262-266 (1990).

W.H. Lyness, N.M. Friedle, and K.E. Moore. Destruction of dopaminergic nerve terminals in nucleus accumbens: effect on D-amphetamine self-administration. *Pharmacol. Biochem. Behav.* **11**: 553 (1979).

W.H. Lyness, N.M. Friedle, and K.E. Moore. Increased self-administration of D-amphetamine after destruction of 5-hydroxytryptamine neurons. *Pharmacol. Biochem. Behav.* **12**: 937 (1980).

W.H. Lyness and K.E. Moore. Increased self-administration of D-amphetamine by rats pretreated with metergoline. *Pharmacol. Biochem. Behav.* **18**: 721 (1983).

M.T. Martin-Iverson, C. Szostak, and H.C. Fibiger. 6-Hydroxydopamine lesions of the medial prefontal cortex fail to influence intravenous self-administration of cocaine . *Psychopharmacology* **88**: 310 (1986).

H.O. Pettit, A. Ettenberg, F.E. Bloom, and G.F. Koob. Destruction of dopamine in the nucleus accumbens selectively attenuates cocaine but not heroin self-administration in rats. *Psychopharmacology* **84**: 167 (1984).

R. Pickens and T. Thompson. Cocaine reinforced behavior in rats: effect of reinforcement magnitude and fixed ratio size. *J. Pharm. Exp. Ther.* **161**: 122 (1968).

R. Pickens, R. Meisch, and J. Dougherty. Chemical interactions with methamphetamine reinforcement. *Psychol. Rep.* **23**: 1267 (1968)

A.J.J. Pijnenberg, W.M.M. Honig, and J.M. van Rossum. Inhibition of D-amphetamine-induced locomotor activity by injection of haloperidol into the nucleus accumbens. *Psychopharmacologia (Berlin)* **41**: 87 (1975).

M.E. Risner and B.E. Jones. Role of noradrenergic and dopaminergic processes in amphetamine self-administration. *Pharmacol. Biochem. Behav.* **5**: 477 (1976).

M.C. Ritz, R.J. Lamb, S.R. Goldberg, and M.J. Kuhar. Cocaine receptors on dopamine transporters are related to self-administration of cocaine. *Science* **237**: 1219 (1987).

D.C.S. Roberts. Breaking points on a progressive ratio schedule reinforced by intravenous apomorphine increase daily following 6-hydroxydopamine lesions of the nucleus accumbens. *Pharmacol. Biochem. Behav.* (1989) in press.

D. C. S. Roberts and S. A. L. Bennett: Self-adminisration of heroin under progressive ratio schedules of reinforcement in rats. Submitted.

D.C.S. Roberts, S.A.L. Bennett, and G.J. Vickers. The estrous cycle affects cocaine self-administration on a progressive ration schedule. *Psychopharmacology,* in press.

D.C.S. Roberts, M.E. Corcoran, and H.C. Fibiger. On the role of ascending catechamine systems in intravenous self-administration of cocaine. *Pharmacol. Biochem. Behav.* **6**: 615 (1977).

D.C.S. Roberts and H.C. Fibiger. Conditioned taste aversion induced by diethyldithiocarbamate (DDC). *Neurosci. Lett.* **2**: 339 (1976).

D.C.S. Roberts, G.F. Koob, P. Klonoff, and H.C. Fibiger. Extinction and recovery of cocaine self-administration following 6-hydroxydopamine lesions of the nucleus accumbens. *Pharmacol. Biochem. Behav.* **17**: 781 (1980).

D.C.S. Roberts and G.F. Koob. Disruption of cocaine self-administration following 6-hydroxydopamine lesions of the ventral tegmental area in rats. *Pharmacol. Biochem. Behav.* **17**: 901 (1982).

D.C.S. Roberts, E. Loh, and G. Vickers. Self-administration of cocaine on a progressive ratio schedule in rats: dose-response relationship and effect of haloperidol pretreatment. *Psychopharmacology* **97**: 535 (1989).

D.C.S. Roberts and G. Vickers. Atypical neuroleptics increase self-administration of cocaine: an evaluation of a behavioral screen for antipsychotic activity. *Psychopharmacology* **82**: 135 (1984).

D.C.S. Roberts and G. Vickers. The effect of haloperidol on cocaine self-administration is augmented with repeated administrations. *Psychopharmacology* **94**: 526 (1987).

D.C.S. Roberts and K.A. Zito. Interpretation of lesion effects on stimulant self-administration, in: *Methods of Assessing the Reinforcing Properties of Abused Drugs*. M. A. Bozarth, ed., Springer-Verlag, New York, p. 87 (1987).

H. Simon, L. Stinus, and M. Le Moal. Effets de la lesion des noyaux du raphe sur l'autoadministration de D-amphetamine chez le rat: augmentation considerable de l'appetence au toxique. *C. R. Acad. Sci. Paris* **290**: 225 (1980).

F.L. Smith, D.S.L. Yu, D.G. Smith, A.P. Leccese, and W.H. Lyness. Dietary tryptophan supplements attenuate amphetamine self-administration in the rat. *Pharmacol. Biochem. Behav.* **25**: 849 (1986).

R. Stretch. Discrete trial control of self-injection behavior in squirrel monkeys: effects of morphine, naloxone, and chlorpromazine. *Can. J. Physiol. Pharm.* **54**: 778 (1977).

M.C. Wilson and C.R. Schuster. The effects of chlorpromazine on psychomotor stimulant self-administration in the rhesus monkey. *Psychopharmacology* **26**: 115 (1972).

M.C. Wilson and C.R. Schuster. Cholinergic influence on intravenous cocaine self-administration by rhesus monkeys. *Pharmacol. Biochem. Behav.* **1**: 649 (1973).

M.C. Wilson and C.R. Schuster. Aminergic influences on intravenous cocaine self-administration by rhesus monkeys. *Pharmacol. Biochem. Behav.* **2**: 563 (1974).

R.A. Wise, P.A. Hansson, and G.J. Gerber. Concurrent intracranial self-stimulation and amphetamine self-administration in rats. *Pharmacol. Biochem. Behav.* **7**: 459 (1977).

W.L. Woolverton. Effects of a D1 and a D2 dopamine antagonist on self-administration of cocaine and piribedil by rhesus monkeys. *Pharmacol. Biochem. Behav.* **24**: 531 (1986).

R.A. Yokel and R.A. Wise. Increased lever pressing for amphetamine after pimozide in rats: implications for a dopamine theory of reward. *Science* **187**: 547 (1975).

R.A. Yokel and R.A. Wise. Attenuation of intravenous amphetamine reinforcement by central dopamine blockade in rats. *Psychopharmacology* **48**: 311 (1976).

6

Neurobiology of Cocaine: Reinforcing and Stimulus Effects

KATHRYN A. CUNNINGHAM
Department of Pharmacology and Toxicology
University of Texas Medical Branch
Galveston, Texas

STEVEN I. DWORKIN AND JAMES E. SMITH
Department of Physiology and Pharmacology
Bowman Gray School of Medicine
Wake Forest University
Winston-Salem, North Carolina

INTRODUCTION

Numerous behavioral assays have been employed to investigate the neurobiological mechanisms of action of abused drugs. Among others, these procedures include electrical brain stimulation, conditioned place preference, drug self-administration, and drug discrimination. Each of these behavioral procedures has its own strengths and weaknesses, a fact that necessitates the use of several procedures to thoroughly investigate the behavioral effects of any drug. All of these procedures have been, or are currently used, to investigate the abuse liability of cocaine and the neurobiological mechanisms involved in cocaine abuse. This chapter will review the results obtained to date of investigations of the reinforcing and stimulus properties of cocaine.

REINFORCING PROPERTIES OF COCAINE

The search for the neuronal substrates of reinforcement began shortly after the initial demonstration that animals could be conditioned to emit responses that resulted in electrical stimulation of a brain region (Olds and Milner, 1954). The results of this study suggest that discrete and, perhaps, even dedicated neuronal circuits may mediate the processes collectively known as reinforcement. This hypothesis is supported by intracranial self-stimulation (ICSS) studies that have shown stimulation of specific brain sites to be reinforcing, while stimulation of other sites is not or is sometimes even aversive (see Stellar and Stellar, 1985). Studies using procedures other than ICSS have shown that lesions of some brain regions or discrete innervations or neuronal projections modulate the response-contingent presentation of a reinforcer without altering the ability of the animal to emit the required response. These data suggest that neuronal circuits may be dedicated to the central processes of reinforcement. If the supposition that discrete neuronal networks are responsible for reinforcement is correct, drugs that are consistently self-administered should produce reinforcing effects by activating these circuits. Activation of the neuronal networks that mediate reinforcement may be the underlying process that maintains the self-administration of the substance. Since it is improbable that unique and separate neuronal systems evolved for the exclusive mediation of drug reinforcement, abused substances probably activate systems that mediate these processes for other nondrug reinforcers (food, water, etc.). The chemically induced activation by a drug is probably somewhat exaggerated (like ICSS) in contrast to the initiation of similar, endogenous processes resulting from exposure to reinforcing stimuli received through peripheral sense organs (i.e., visual, auditory, or tactile).

This hypothesis concerning the general mechanisms of the reinforcing action of abused drugs is supported by recent demonstrations of binding sites for some of these compounds. Brain binding sites for opiates (Pasternak, 1988), benzodiazepines (Young *et al.*, 1981), and a number of other drugs have been extensively characterized, while such sites for stimulants have been more difficult to identify. Recent evidence suggests that cocaine binds at sites that modulate the uptake of dopamine (DA; Kennedy and Hanbauer, 1983; Ritz *et al.*, 1987), serotonin (5-HT; Reith *et al.*, 1980; Sershen *et al.*, 1982), and norepinephrine (NE; Koe, 1976). Thus, cocaine may activate neuronal circuits responsible for reinforcement similar to ICSS, but through specific receptors by chemical, rather than electrical, mechanisms. The following section presents the current understanding of the neurobiological mechanisms of reinforcement which are operationally defined as the neuronal events that ultimately result in increased propensity of the organism to self-administer cocaine. Several animal laboratory models have been used to investigate such actions of cocaine. These include intracranial electrical stimulation (ICSS), conditioned place preference (CPP), and i.v. self-administration.

Effects of Cocaine on ICSS Threshold

The ICSS procedure uses the delivery of a weak electrical stimulus to a discrete brain region to engender and maintain operant responding (i.e., lever press or nose poke). If responding is maintained, the electrical stimulus is thought to directly initiate reinforcing neuronal activity. Abuse potential of some drugs has been assessed by evaluating the action of drugs on such reinforcing brain stimulation (Olds and Travis, 1960). Cocaine potentiates ICSS (Crow, 1970) which likely results from a lowering of the threshold. These initial experiments monitored ICSS response rates and were subject to the rate effects of the drug, independent of any direct effects on the neuronal events maintaining the behavior. To circumvent this potential problem of interpretation, methodologies were developed to measure the sensitivity of the neuronal processes that are responsible for ICSS. One useful method is to decrement the current intensity for stimulation some moderate amount after each contingent presentation (Stein and Ray, 1959). A response on a second operandum would reset the current intensity with the average reset current considered to be the threshold for reinforcing brain stimulation. Another method is a discrete trial procedure in which a brief response-independent electrical stimulus is delivered to the brain region with the opportunity to obtain an additional response-dependent stimulation at that site available on a limited hold (Marcus and Kornetsky, 1974). Ascending and descending sequences are used with a method of limits to determine the stimulus intensity that maintains responding at 50% of the available opportunities. Cocaine lowers the ICSS threshold for stimulation in the lateral hypothalamus (Esposito *et al.*, 1978). However, the dose that lowered the ICSS thresholds appeared to increase the detection threshold for the discriminative effects of the stimulation at that same site. This was evident since the detection of the sampling stimulus was decremented, but those stimuli that were detected resulted in the same percentage of self-stimulation responses (Kornetsky and Esposito, 1981). Cocaine effects on ICSS thresholds have also been studied with the stimulus train duration technique (Frank *et al.*, 1988). The stimulation that maintains half the maximal response rate is considered the ICSS threshold with the train duration procedure. Thresholds for ICSS in the ventral tegmental area determined with this technique also show that cocaine lowers these indices, which is interpreted as an increase in the reinforcing efficacy of the stimulation (Frank *et al.*, 1988).

The ICSS threshold-lowering actions of cocaine likely result from the actions of the drug that ultimately increase the efficacy of transmission in the neuronal circuits or networks that mediate the reinforcing effects of the ICSS. These effects are not likely to result from a direct reinforcing effect of the response-independent presentation of cocaine which is additive with the reinforcing actions of the stimulation since response-independent presentation cannot be assumed to have reinforcing consequences for the following reasons: (1) noncontingent presentation of identical patterns of previously presented contingent brain stimulation (ICSS) may be aversive since rats will respond to terminate such response-independent delivery (Steiner *et al.*, 1969) and (2) noncontingent intracranial electrical stimulation at intensities and sites where stimulation was contingently delivered alters glucose utilization determined with 2-deoxyglucose autoradiography (below) in different brain regions than those activated in animals self-administering the electrical stimulation (Porrino *et al.*, 1984). The areas showing increased glucose utilization with ICSS were generally in limbic regions that have been identified to have increased neurotransmitter turnover rates in rats self-administering i.v. morphine compared to yoked, morphine-infused controls (Smith *et al.*, 1980; 1982; 1984 a and b). In these studies, the turnover rates of acetylcholine, the biogenic amines, and amino acid neurotransmitters were measured in small brain regions of rats, either intravenously self-administering morphine or receiving yoked infusions of morphine or the vehicle. The yoked, noncontingent infusions of morphine generally affected different neurons than those modulated in self-administering littermates. Recent data from studies using similar treatment groups indicate that response-independent, yoked i.v. infusions of cocaine are a greater physiological stressor than identical contingent infusions to self-administering littermates (Dworkin, 1990 submitted). Greater mortality is seen in rats receiving noncontingent administration of cocaine than in rats self-administering identical doses of the drug. These data collectively indicate that response-independent administration of a drug is physiologically distinct from response-contingent administration of this psychostimulant.

Conditioned Place Preference to Cocaine

Conditioned place preference (CPP) is a procedure in which noncontingent administration of either saline or drug is immediately followed by confinement in one side of a two-sided chamber, each side with characteristics that are differentially discriminable and, thus, specific to each treatment. The drug-free animal is subsequently placed in the environment and allowed to move freely between the two sides, with the time spent on each side recorded. Because the environment associated with the noncontingent administration of the drug serves as a conditioned reinforcer, the time spent on that side of the chamber increases. This procedure assumes that the administration of a substance has intrinsic hedonic value in the absence of any behavior on the part of the organism, and that noncontingent presentation is thought to have the same positive effect as reinforcement. Thus, ''reward'' neurons are activated independent of the control of the organism for the delivery of the putative reinforcer or response strengthener. To conclude that response-independent presentation of any potential reinforcer is pleasing may be in error. For these reasons, CPP, which may be an effective screen for potential drugs of abuse, may not be an adequate behavioral model for the investigation of the underlying neurobiological mechanisms of reinforcement.

Cocaine Self-Administration

Self-administration is currently the most useful, nonhuman laboratory model that can directly address the reinforcing effects of a drug. However, the self-administration environment is complex because all of the behavioral effects of a drug are present. This includes response-contingent presentation as well as the stimulus properties, rate effects, effects on ingestive behaviors, and any of the other behavioral actions of the drug. Thus, the self-administered drug has stimulus properties (see below) that can set the occasion for the emission of other responses. It can increase or decrease the rates of behaviors, including subsequent self-infusions. It can modulate the intake of food or water which, in turn, can further increase or decrease drug intake; or it can modulate other unspecified physiological functions in either a direct or indirect manner that may also alter drug intake. In spite of these complexities, experiments utilizing self-administration methodologies have provided significant information with regard to the neurobiological basis of cocaine reinforcement.

Table 1
Effects of Systemically-Administered Neurotransmitter Antagonists on the ICSS Threshold-Lowering Actions of Cocaine, Place Preference Conditioning to Cocaine and the Reinforcing Efficacy of Intravenous Cocaine Self-Administration

	Neurotransmitter receptor antagonized				
Procedure	**ACh**	**DA**	**NE**	**5HT**	**Opioid**
ICSS threshold lowering	?	?	?	?	D
CPP	?	D[a] and NE	?	?	?
i.v. self-administration	D	D	I	D	D[b]

Note: D = decreased; I = increased; NE = no effect; ? = not known.

[a] Decreased with intravenous cocaine. No effect with intraperitoneal cocaine.
[b] In food-satiated animals.

NEURONAL CIRCUITS OF COCAINE REINFORCEMENT

Pharmacological Antagonism Studies

One of the first experimental approaches used to assess underlying brain mechanisms of the reinforcing actions of cocaine was the assessment of the effects of systemic administration of receptor antagonists on the behavioral actions of cocaine (see Table 1). The ICSS threshold-lowering effects of cocaine were antagonized with naloxone (Bain and Kornetsky, 1987), suggesting involvement of opioid neurons in the mediation pathways. Clearly influenced by the DA hypothesis which states that reinforcement is mediated by DA circuitry, attempts to pharmacologically block cocaine CPP have focused on DA antagonists which either do (Spyraki *et al.*, 1987) or do not blunt cocaine CPP (Spyraki *et al.*, 1982). Furthermore, CPP induced by i.v., but not i.p., administration of cocaine is reduced by DA blockade (Nomikos and Spyraki, 1988), which may be related to the differential effects of i.p. and i.v. cocaine on brain glucose metabolism (see below).

Intravenous self-administration of cocaine was antagonized by blockade of DA receptors with haloperidol (de Wit and Wise, 1977; Ettenberg *et al.*, 1982). As far as the specific DA receptor subtypes involved, some evidence suggests that the D_2, but not D_1, receptors are necessary (Woolverton, 1986). However, more recent data suggest that *both* D_1 and D_2 receptors are potentially important in these reinforcing processes (Koob *et al.*, 1987.) In addition, cholinergic muscarinic receptor antagonists decreased the reinforcing efficacy of i.v. cocaine (Wilson and Schuster, 1973), while substances that antagonized NE neuronal activity may increase these effects (Goldberg and Gonzales, 1976). However, this latter demonstration may be context dependent since opposite effects have been reported (De La Garza and Johanson, 1982). Serotonergic agonists may decrease the reinforcing efficacy of cocaine (Mattia *et al.*, 1986), while opiate receptor blockade attenuates the self-administration of cocaine in food-satiated animals (Carroll *et al.*, 1986).

In general, antagonist experiments provide evidence for the potential involvement of particular neurotransmitters in the neural processes responsible for the reinforcing

Table 2
Effects of 6-Hydroxydopamine (6-OHDA), 5,7-Dihydroxytryptamine (5,7-DHT) or Kainic Acid Lesions of Various Brain Regions on Intravenous Cocaine Self-Administration

	Lesion effect		
Brain region	6-OHDA	5,7-DHT	Kainic acid
Nucleus accumbens	D	D	D
Ventral tegmentum	D	?	?
Ventral pallidum-substantia innominata	?	?	D
Medial thalamus	?	?	D
Medial prefrontal cortex	NE		?

Note: D = decreased; NE = no effect; ? = not known.

actions of cocaine. However, these effects may not result from specific alterations in reinforcing efficacy since disruption of either sensory input or response output processes could also be responsible for the observed modifications. Changes could also result from alterations in the rate effects of cocaine as well as the stimulus properties (the ability of the organism to detect the quantity of drug present). These experiments seldom evaluated more than one dose of cocaine or the antagonist, which obviates the possibility of determining the direction of potential changes in reinforcing efficacy. In general, with the above qualifications, these data suggest the involvement of acetylcholine, DA, 5-HT, and opioid neurons in the global processes responsible for the reinforcing effects of cocaine. Antagonism of self-administration behavior suggests that acetylcholine, DA, and 5-HT neurons are important in the expression of these behaviors, that DA systems appear necessary for CPP, and that opioid systems are essential to the ICSS threshold-lowering effects of cocaine. There is a significant concordance for the involvement of DA in the processes maintaining the behaviors by these three paradigms, but this results more from the limited scope of evaluation rather than from evidence for a unique role. DA is obviously important to the cocaine reinforcement process; however, proposing *only* these neuronal systems to mediate these central events may be premature.

Specific Brain Regions Involved

The brain sites involved in the reinforcing actions of cocaine have been investigated with 2-deoxyglucose autoradiography and neurotoxin lesions. Neurotoxins are used to destroy discrete neuronal systems so that the role of these neurons in the behavior being evaluated can be determined.

Neurotoxin Lesions

Intracranial administration of neurotoxins has been utilized to identify brain systems involved in cocaine CPP and i.v. cocaine self-administration, but these techniques have not been used to assess neurons implicated in the ICSS threshold-lowering effects of the drug (see Tables 2 and 3). 6-Hydroxydopamine (6-OHDA) lesions of DA innervation of the nucleus accumbens attenuated CPP conditioned

Table 3
Effects of 6-Hydroxydopamine Lesions of the Nucleus Accumbens on Various Behaviors

Behaviors	Effects
i.v. cocaine self-administration	D
i.v. opiate self-administration	D
Conditioned place preference to cocaine	D
Conditioned place preference to opiates	D
Locomotor activity effects of cocaine	I
Licking	I
Stereotypy produced by cocaine	D

Note: D = decreased; I = increased.

to i.v. cocaine (Spyraki *et al.*, 1987), but not to intraperitoneal cocaine (Spyraki *et al.*, 1982). The potential importance of these differential effects is discussed in the next section. Similar 6-OHDA lesions attenuated i.v. cocaine self-administration (Pettit *et al.*, 1984; Roberts *et al.*, 1977; 1982), while kainic acid lesions of the nucleus accumbens also decreased the reinforcing efficacy of cocaine (Zito *et al.*, 1985), suggesting that not only are DA innervations important to the processes maintaining this behavior, but also interneurons and/or projections from this region. Neurotoxic 6-OHDA lesions of mesolimbic DA cell bodies in the ventral tegmental area decreased cocaine self-administration (Roberts and Koob, 1982), further supporting the involvement of these neurons in the maintenance of i.v. self-administration. Recently, kainic acid lesions of the ventral pallidum-substantia innominata (Hubner and Koob, 1987) and medial thalamus (Hubner and Koob, 1988) were shown to decrease the self-administration of i.v. cocaine, indicating that other brain regions and neuronal systems are also important to the processes maintaining this behavior. The effects of lesions on progressive ratio schedules of cocaine presentation have also recently been assessed. Lesions of 5-HT systems produced by 5,7-dihydroxytryptamine appear to increase the break point, suggesting *increases* in the reinforcing efficacy of cocaine (Loh and Roberts, 1990).

In general, these lesion effects have been evaluated on single doses of cocaine and have not been shown to be independent of alterations in more global brain processes (i.e., the ability of the animal to respond, the ability of the animal to detect the presence of the drug). However, recent studies in animals on concurrent schedules of i.v. cocaine, food, and water presentation showed that 6-OHDA lesion of the nucleus accumbens specifically decreased cocaine intake without altering food- or water-maintained responding (Dworkin and Smith, 1988). Rats were trained to lever press on separate levers for either i.v. cocaine, food pellets, or water presentation. Stable baselines of responding, dose-effect determinations and food/water extinction probes were established. Subsequently, the animals received either bilateral injections of 6-OHDA or sham vehicle treatment. The lesion of DA innervation of the nucleus accumbens specifically decreased i.v. cocaine intake without altering food- or water-maintained responding. These intracranial neurotoxin lesion studies suggest that mesolimbic DA innervation of the nucleus accumbens originating in the ventral tegmental area as well as interneurons and projections of the nucleus accumbens, ventral pallidum-substantia innominata, and the medial thalamus are involved in the central neural events subservient to the reinforcing properties of cocaine.

Neurotoxic 6-OHDA lesions of the nucleus accumbens also attenuate the discriminative stimulus properties of cocaine (Dworkin and Bimle, 1989) and modify the rate effects of cocaine (Robbins *et al.*, 1983) and the effects of cocaine on motor activity (Roberts *et al.*, 1975). Any of these actions could be partially or wholly responsible for the effects of this lesion on cocaine self-administration. Specific changes resulting from decreases in reinforcing efficacy are yet to be demonstrated *independent* from any of these potential contributing factors. Of course, this hypothesis assumes that the neuronal networks mediating reinforcement can be differentiated from those that mediate these other behavioral actions of cocaine.

Glucose Utilization

2-Deoxyglucose (2-DG) autoradiography is used to determine glucose utilization in discrete brain regions during exposure to a behavioral paradigm. The rate of glucose utilization is thought to represent the rate of neuronal activity in that region since neurons do not store glucose, which is the lone energy source for support of such activity. Transport of 2-DG into neurons is via the same uptake processes as glucose with the exception that 2-DG is not broken down through glycolysis. If radioactive 2-DG is used, then the location and amount of this substance in discrete brain regions can be determined as a measure of neuronal activity with autoradiographic procedures.

The effects of systemic, noncontingent cocaine presentation to restrained rats (London *et al.*, 1986) and noncontingent i.v. cocaine to nonrestrained rats (Porrino *et al.*, 1988) have recently been evaluated on cerebral glucose utilization using 2-DG autoradiography. Intraperitoneal injections (30 mg/kg) into restrained rats and high i.v. doses (5 mg/kg) into unrestrained rats resulted in increased glucose utilization in extrapyramidal systems which, in the latter case, may be correlated with the activation of motor systems that is often reflected in increased locomotion. The brain areas showing increased glucose utilization included both dopaminergic and nondopaminergic regions. In contrast, noncontingent i.v. doses that are readily self-administered by rats (1.0 mg/kg) resulted in specific activation of mesolimbic-mesocortical DA projection areas (medial prefrontal cortex and nucleus accumbens). These data are consistent with the contrasting effects of neuroleptics and neurotoxin lesions on CPP after systemic and i.v. cocaine (Nomikos and Spyraki, 1988; Spyraki *et al.*, 1987). Neuroleptic treatment or 6-OHDA lesions of the nucleus accumbens did not alter CPP to i.p. cocaine, but did attenuate these behaviors conditioned to i.v. cocaine. Although there are numerous differences between systemic and i.v. administration (i.e., rate of rise in brain levels, peak brain levels, etc.), these differences in CPP clearly underline the concerns for this procedure as a model for study of brain reinforcement mechanisms. If CPP is the result of the reinforcing actions of the drug, then the differential effect of neuroleptic and neurotoxin lesions dependent upon route of administration would indicate that *either* the reinforcing properties resulting from the two routes of administration are produced through different neuronal mechanisms *or* that the procedure itself may not accurately define reinforcement. Thus, the CPP procedure has several inherent weaknesses for use in the study of the neurobiology of drug reinforcement. Furthermore, as previously stated, neither of these studies used contingent administration so that no assumptions of the reinforcing effects of such administration can be made. In its most innocuous form, these data suggest that the activation of extrapyramidal DA cues may be dominant with i.p. cocaine, while mesolimbic-mesocortical DA systems may be responsible for these behaviors upon i.v. administration. Thus, the lowest dose of i.v. cocaine selectively alters glucose utilization in the medial prefrontal cortex and nucleus accumbens, which is more consistent with regions important to intracranial self-administration (Goeders and Smith, 1983) and i.v. self-administration (Roberts *et al.*, 1977, 1982; Roberts and Koob, 1982; Zito *et al.*, 1985), respectively.

Microdialysis, In Vivo Voltammetry and Single Unit Activity

Experiments using either microdialysis or voltammetry to monitor DA release in the nucleus accumbens of rats self-administering i.v. cocaine have reported increased DA release (Finlay *et al.*, 1988; Hurd *et al.*, 1989; Pettit and Justice, 1989; Weiss *et al.*, 1988). However, these data are not totally in agreement. Two of these studies have reported sustained, increased release across the session with repeated self-infusions (Finlay *et al.*, 1988; Pettit and Justice, 1989), while the other two studies (Hurd *et al.*, 1989; Weiss *et al.*, 1988) showed increased release only in animals exposed to the initial self-administration session which was not present in experienced self-administering animals. It currently appears that by day 8 or 9 of exposure to self-administration, there may be a change in the characteristics of DA involvement in cocaine actions, at least in the nucleus accumbens. In rats orally self-administering cocaine, single unit recordings from microwires implanted in the nucleus accumbens indicated that an initial excitation of firing rates is associated with cocaine self-administration; greater activation of neurons in the nucleus accumbens was also associated with locomotor movement after, but not prior to, cocaine consumption (Wolske and West, 1988). Collectively, these data generally suggest that cocaine self-administration is accompanied by increased activity of dopaminergic innervations of the nucleus accumbens from the ventral tegmental area.

In conclusion, discrete and perhaps dedicated neuronal networks appear to mediate reinforcement for a number of reinforcers and may be artificially activated by direct electrical stimulation (ICSS) or by chemical stimulation with drugs (self-administration). Drugs may initiate or modulate activity in these circuits by altering the firing rate of component neurons and/or of neurons that innervate these circuits. Proposing that the neurons mediating reinforcement are organized into systems or networks has distinct advantages over proposing such complex processes to be the result of changes in one or two neurotransmitter systems. First, this hypothesis is more parsimonious with published data indicated that a variety of pathways and neurotransmitters is involved. Second, it is more consistent with what is known about complex brain function. It is unlikely that a central neuronal system is independent of influences from other neurons or that intricate behavioral processes can be conveyed by one or two neuronal pathways. Reinforcement involves the separation of stimulus inputs from the broad stream of external or internal events occurring at any moment so that a reinforcing stimulus is detected and, once it is, responses must be initiated to complete the necessary consummatory behavior. Reinforcers are not immutable or constant. The state of the organism (deprivation, satiation) modulates the saliency of a putative reinforcer. Many reinforcers are not primary (necessary to the existence of the organism), but develop reinforcing efficacy through conditioning. All of these factors are important to the processes that comprise reinforcement. Rather than one or two neuronal pathways, neuronal circuits or networks involving a number of pathways and neurotransmitters are probably necessary to control and integrate these complex phenomena. A theorist may conclude from ICSS data that the pleasing or rewarding effect of the direct electrical stimulation of a brain region results from the activation of a specific pathway or group of pathways and be correct, since this is exactly what this procedure involves. However, theories of brain mechanisms based on such phenomena may not account for the more global complex processes of reinforcement. Issues that an effective general theory of reinforcement must address include the dynamic nature of reinforcers (satiation, deprivation, context), conditioned reinforcement, positive and negative reinforcement, and the response-strengthening effects of reinforcement. When these broader issues are addressed, a hypothesis with greater utility and applicability to a broader spectrum of behaviors can be proposed.

DISCRIMINATIVE STIMULUS PROPERTIES OF COCAINE

Investigations of the neurobiological aspects of the discriminative stimulus effects of cocaine have resulted in a better understanding of the complex profile of physiological and psychological effects elicited by the drug. A full appreciation of the specific brain mechanisms and loci of action underlying the stimulus effects of the drug is confounded by the fact that cocaine has at least two prominent actions in neural tissue: local anesthesia and potentiation of monoamine neurotransmission. Investigations of the neurochemical basis for monoamine potentiation by cocaine have illustrated that it inhibits the re-uptake mechanisms for DA, NE (Heikkila *et al.*, 1975; Taylor and Ho, 1978), and 5-HT (Ross and Renyi, 1969; Taylor and Ho, 1978). While neurochemical findings suggest that the actions of cocaine on monoaminergic systems are related primarily to indirect mechanisms (reuptake inhibition), a role for direct membrane actions cannot be ruled out, as suggested by both its local anesthetic actions (Ritchie and Greengard, 1966) and its affinity for a 5-HT (5-HT_3) receptor (Kilpatrick *et al.*, 1989).

The potent reinforcing effects of cocaine are one of the most dramatic behavioral effects of the drug (above). Human self-administer cocaine to experience euphoria and a sense of well-being and confidence; cocaine will also support self-administration in animals (above). The subjective aspects of the reinforcing effects of cocaine can be inferred from the study of the interoceptive stimulus properties of the drug in animals using drug discrimination procedures; the neural mechanisms underlying this *in vivo* effect of cocaine are accessible to pharmacological analysis using this behavioral assay. In the typical drug discrimination task, subjects are trained to make one of two operant responses (press the left lever) following injection of the ''training'' drug and to make the other response (press the right lever) in the absence of the drug to obtain reinforcement. Thus, drug-induced internal states become biologically meaningful and function as interoceptive stimuli (''cues'') which signal the availability of reinforcement (Appel *et al.*, 1982). These subjective ''cues'' associated with a drug probably play an important role in establishing and maintaining drug dependence (Childress *et al.*, 1988). Drug discriminations with psychoactive drugs are easily trained and, in most cases, result in good stimulus control, as measured by orderly generalization gradients with steep slopes (Appel *et al.*, 1982). After acquiring such a discrimination, various pharmacological manipulations can be conducted in an effort to discern the mechanism(s) which underlie the ability of a drug to act as a cue (Appel *et al.*, 1982; Cunningham and Appel, 1988) and, thus, gain insight into the mechanisms which underlie substance dependence.

Drug discrimination is one of the few *in vivo* assays which is reliable, robust, sensitive, and pharmacologically specific (Appel *et al.*, 1982; Cunningham and Appel, 1988). Its primary value for neuropharmacology is derived from its specificity within a given drug class; that is, (1) only functionally similar drugs or drugs which act through similar neuronal mechanisms have effects which substitute for that of the training drug and (2) only specific receptor antagonists block agonist-induced cues (Appel *et al.*, 1982). Thus, compounds will not substitute *solely* on the basis of psychoactive effects, nor will randomly selected drugs mask the stimulus effects of a drug (Appel *et al.*, 1982). The study of the discriminative stimulus properties of cocaine, predominantly in rodents, has provided significant information concerning the neurobiologic mechanisms underlying the intense behavioral effects of cocaine.

The action of cocaine on central monoamine neurotransmitter function is indirect, that is, cocaine inhibits the re-uptake of DA, NE, and 5-HT into the presynaptic terminal which potentiates the action of the endogenous neurotransmitter in the synapse. Thus, cocaine might be expected to share stimulus effects with those drugs which indirectly amplify monoamine activity. This is the case with transmitter releasers (e.g., amphetamine derivatives; D'Mello and Stolerman, 1977), monoamine oxidase inhibitors (MAOIs; e.g., deprenyl; Colpaert *et al.*, 1980), and re-uptake inhibitors (e.g., GBR 12909; Cunningham and Callahan, in press) which act preferentially on DA systems and have been shown to completely mimic the cocaine state (see Table 4). In contrast, 5-HT and NE re-uptake inhibitors (Colpaert *et al.*, 1979; Cunningham and Callahan, in press; McKenna and Ho, 1980; Silverman and Schultz, 1989), MAOIs not selective for the DA substrate (Colpaert *et al.*, 1980), and the 5-HT releaser fenfluramine (McKenna and Ho, 1980; Wood and Emmett-Oglesby, 1988) engender partial or no cocaine-like responding. Thus, augmentation of DA synaptic transmission by a variety of ''indirect'' mechanisms induce discriminative stimuli which are perceived as similar to those of cocaine;

Table 4
Results of Substitution Tests in Animals Trained to Discriminate Cocaine from Saline

Complete substitution >80% drug-lever responses	Partial substitution 60–79% drug-lever responses	No substitution <59% drug-lever responses
	Cocaine Analogs	
n-Allyl-norcocaine (2)		Benzoylecgonine (27)
Intraventricular (35)		Benzoylnorecgonine (27)
Intra-accumbens (38)		Cocaine methiodide (28)
Norcocaine (2,23,27)		Ecgonine methyl ester (27)
Procaine (24)		Lidocaine (15,16,30)
Win 35065-2 (23)		Procaine (9,23,27,30)
Win 35428 (23)		Pseudococaine (28)
		Intra-prefrontal cortex (38)
		Intra-striatal (38)
	DA Agonists: "Indirect"	
d-Amphetamine (1,4,7,9, 14,17,20,23,28,31)	*l*-MDA (3)	Hydroxyamphetamine (9,23,31)
l-Amphetamine (9)	*d*-MDMA (3)	
dl-Cathionine (21)	*l*-MDMA (3)	
Diethylpropion (37)	Methamphetamine (33)	
d-MDA (3)	Methylphenidate (28)	
Methylamphetamine (9)	Phenylethylamine (33)	
Methylphenidate (9,17,20,30,37)		
Phenmetrazine (37)		
Phentermine (37)		
	DA Agonists: "Direct"	
Apomorphine (6,28,31)	Amantadine (9)	Apomorphine (15,16,23,24)
Quinpirole (1,4)	Apomorphine (9)	Bromocriptine (9)
	SKF 38393 (4,36)	Piribedil (9)
		SKF 38393 (1)
	Uptake Inhibitors and MAOIs	
Deprenyl (10)	Desipramine (13)	Clorgyline (10)
GBR 12909 (13)	Nisoxetine (30)	Desipramine (9)
Nomifensine (9)		Fluoxetine (13)
Pargyline (10)		Imipramine (9,28)
Pheniprazine (10)		Nialamide (10)
Tranylcypromine (10)		Tranylcypromine (9)
	5-HT Agonists	
Quipazine (32)	Mescaline (28)	Fenfluramine (28,37)
		LSD (9,23)
		Mescaline (9)

Table 4 (continued)
Results of Substitution Tests in Animals Trained to Discriminate Cocaine from Saline

Complete substitution >80% drug-lever responses	Partial substitution 60–79% drug-lever responses	No substitution <59% drug-lever responses
	NE Agonists or Antagonists	
Propranolol (20)	Clonidine (34)	Isoproterenol (9)
	Propranolol (9,30)	Phentolamine (20)
		Practolol (9)
		Salbutamol (9)
	Others	
Benztropine (9)	Caffeine (18,30)	Atropine (9)
Dexetimide (9)	Caffeine/epinephrine/phenyl-propanolamine (18)	Bay K 8644 (5)
Phencyclidine (9)		Catapresan (28)
		Epinephrine/phenyl-propanolamine (18)
		Fentanyl (9)
		Isopropamide (9)
		Morphine (23,24)
		Pentobarbital (24)
		Pentylenetetrazol (17)
		Phencyclidine (12)
		Strychnine (28)
		Δ9-THC (23,24)

Note: Numbers refer to citations below; complete citations are found in the Reference Section.

1. Barrett and Appel, 1989; 2. Bedford *et al.*, 1981; 3. Broadbent *et al.*, 1989; 4. Callahan *et al.*, 1991; 5. Callahan and Cunningham, 1990; 6. Colpaert *et al.*, 1976; 7. Colpaert *et al.*, 1978a; 8. Colpaert *et al.*, 1978b; 9. Colpaert *et al.*, 1979; 10. Colpaert *et al.*, 1980; 11. Colpaert, 1986; 12. Cunningham and Appel, 1982; 13. Cunningham and Callahan, in press; 14. D'Mello and Silverman, 1977; 15. De La Garza and Johanson, 1983; 16. De La Garza and Johanson, 1985; 17. Emmett-Oglesby *et al.*, 1983; 18. Gauvin *et al.*, 1989; 19. Goudie and Reid, 1988; 20. Ho and McKenna, 1978; 21. Huang and Wilson, 1986; 22. Jarbe, 1978; 23. Jarbe, 1981; 24. Jarbe, 1984; 25. Kleven *et al.*, 1988; 26. McKenna and Ho, 1977; 27. McKenna *et al.*, 1979; 28. McKenna and Ho, 1980; 29. Silverman and Ho, 1977; 30. Silverman and Schultz, 1989; 31. Stolerman and D'Mello, 1981; 32. Winter and Slifer, 1989; 33. Wood *et al.*, 1984; 34. Wood *et al.*, 1985; 35. Wood *et al.*, 1987; 36. Wood and Emmett-Oglesby, 1987; 37. Wood and Emmett-Oglesby, 1988; 38. Wood and Emmett-Ogelsby, 1989.

simply enhancing NE and 5-HT transmission does *not* appear to be sufficient to produce a cocaine-like stimulus.

Although re-uptake inhibitors for NE and 5-HT elicit no more than a partial substitution for cocaine (Cunningham and Callahan, in press; Silverman and Schultz, 1989), the low doses of desipramine and fluoxetine, as well as GBR 12909, enhance the cocaine cue produced by low doses of cocaine (Cunningham and Callahan, in press). The DA re-uptake inhibitor mazindol is also reported to potentiate the lethal and convulsant effects of cocaine in rodents (Jaffe *et al.*, 1989). While the cellular

Table 5
Results of Combination Tests in Animals Trained to Discriminate Cocaine from Saline

Complete antagonism <30% drug-lever responses	Partial antagonism 31–60% drug-lever responses	No antagonism >61% drug-lever responses
	DA Antagonists	
SCH 23390 (4,25)	Chlorpromazine (22)	α-Methyl-*p*-tyrosine (22)
	Haloperidol (4,7,11, 22,24,28,30)	Haloperidol (1,6,8,21,31)
	Pimozide (20,22,27)	Pimozide (6,8)
	SCH 23390 (1)	Pipamperone (6)
	Spiperone (1)	Reserpine (27)
		Spiperone (8)
	NE Antagonists	
		Dibenzamine (6)
		Nisoxetine (30)
		Phenoxybenzamine (22,28)
		Phentolamine (30)
		Propranolol (6,22)
		Sotalol (28)
		Yohimbine (34)
	5-HT Antagonists	
		Cinanserin (28)
		Cyproheptadine (9)
		Methysergide (9)
	Others	
		Atropine (28)
		Nimodipine (5)
		Physostigmine (20)

Note: Numbers refer to citations listed in Table 4.

mechanisms by which re-uptake inhibitors potentiate the interoceptive and toxic effects of cocaine are unknown, pharmacokinetic variables such as inhibition of cocaine metabolism (Misra *et al.*, 1986) could contribute to the potentiated discriminability of cocaine observed with these monoamine re-uptake inhibitors. Alternatively, NE and 5-HT systems may modulate the interoceptive effects of cocaine.

The results of substitution tests with compounds that act at presynaptic terminals suggest that DA is the primary neurotransmitter involved in the interoceptive cocaine state. Substitution and combination tests with receptor agonists and antagonists (Tables 4 and 5) also bolster this conclusion, although some aspects about *which* DA systems or *what* changes in DA neurons are responsible for the cocaine cue remain to be clearly established. As reflected in Tables 4 and 5, there is considerable confusion within the literature as to the conditions under which DA agonists can be shown to mimic, and DA antagonists to block, the cocaine cue. Thus, the prototypical DA receptor agonist apomorphine is reported to mimic fully (Colpaert *et al.*, 1976; McKenna and Ho, 1980; Stolerman and D'Mello, 1981) or partially in rats (Colpaert *et al.*, 1979), but not substitute for cocaine in pigeons or monkeys (De La Garza and Johanson, 1983, 1985; Jarbe, 1981, 1984). Two additional DA agonists, bromocriptine and piribedil, elicited a maximum of 50–70% cocaine-lever responding in rats (Colpaert *et al.*, 1979). Furthermore, DA antagonists do not *reliably* and *consistently* antagonize the cocaine cue. For example, while haloperidol and other neuroleptics are reported to block its stimulus effects, the cocaine cue is often resistant to antagonism by these same DA blockers (Table 5; Barrett and Appel, 1989; Callahan *et al.*, 1991; Colpaert *et al.*, 1976, 1978a, 1978b; Cunningham and Appel, 1982; Silverman and Schultz, 1989). Although pharmacokinetic and pharmacodynamic factors have not been ruled out, the lack of selectivity and specificity of both cocaine and the neuroleptics in their neural actions may account for the inconsistencies seen in this neuroleptic antagonism of cocaine.

Dopamine D_1 and D_2 receptors can be differentiated based upon their interaction with adenylate cyclase (Stoof and Kebabian, 1984). Utilizing current DA neuropharmacological tools specific for D_1 and D_2 receptors, several laboratories have recently assessed the role of these DA receptors in the cocaine cue. The D_2 agonist quinpirole mimics, while the D_1 agonist SKF 38393 elicits moderate (50–75%) or no stimulant-lever responding in cocaine-trained animals (Barrett and Appel, 1989; Callahan *et al.*, 1991; Wood and Emmett-Oglesby, 1987); similar findings are reported in D-amphetamine-trained animals (Callahan *et al.*, 1991; Nielsen and Jepsen, 1985; Nielsen *et al.*, 1989; Smith *et al.*, 1989; Woolverton, 1984). Thus, substitution tests suggest that D_2 receptors are more critically involved in the stimulus effects of cocaine than are D_1 receptors. Interestingly, at low doses, the D_1 antagonist SCH 23390 elicits a partial to full antagonism of the cocaine (and D-amphetamine) cue(s), as well as the substitution of cocaine in D-amphetamine-trained animals (Callahan *et al.*, 1991; Barrett and Appel, 1989; Kleven *et al.*, 1988; Nielsen *et al.*, 1989; Smith *et al.*, 1989). SCH 23390 has no affinity for D_2 receptors (Iorio *et al.*, 1983), but has been shown to block other DA-mediated behaviors classically thought to involve D_2 receptor stimulation (Arnt *et al.*, 1987; Braun *et al.*, 1986; White *et al.*, 1988). A synergistic coupling between D_1 and D_2 receptors has been suggested to account for these findings (see Clark and White, 1987). The inability of the D_1 agonist SKF 38393 to completely substitute while a D_1 antagonist (SCH 23390) blocks the cocaine cue might be explained if stimulation of D_1 receptors with a D_1 agonist (which is not *sufficient* to mimic a stimulant cue) might be *necessary* for "enabling" the interoceptive state. Thus, SCH 23390 would block D_2-mediated stimulant cues by attenuating the "enabling" actions produced by D_1 stimulation (Clark and White, 1987). Support for this hypothesis is found in a recent study by Smith and colleagues (1989), in which the discriminability of a low dose of D-amphetamine (0.3 mg/kg) was potentiated by increasing doses of SKF 38393 (0.5–10 mg/kg) in animals trained to discriminate D-amphetamine (1 mg/kg) from saline. However, in another recent study, SKF 38393 (10 mg/kg), in combination with cocaine, did *not* shift the cocaine dose-response relationship to the left, as would be predicted by the "enabling" hypothesis (Callahan *et al.*, 1991). Similarly, *combinations* of selective D_1 and D_2 agonists may (Smith *et al.*, 1989) or may not (Nielsen *et al.*, 1989) be perceived as similar to the training drug in D-amphetamine-trained rats.

The efficacious and potent SCH 23390 antagonism of the cocaine cue observed in some laboratories (Callahan *et al.*, 1991; Kleven *et al.*, 1988), as has been observed with other behavioral effects of stimulants (see Clark and White, 1987), might alternatively be explained by the high affinity of this benzazepine, not only for D_1 (Iorio *et al.*, 1983) but also 5-HT_2 receptors (Bischoff *et al.*, 1986; McQuade *et al.*, 1988). The possibility that 5-HT might be involved in mediating some aspects of the *in vivo* effects of cocaine is not unfounded. Cocaine inhibits (1) the uptake of 5-HT and its precursor tryptophan (Ross and Renyi, 1969), (2) the activity of the 5-HT synthetic enzyme tryptophan hydroxylase (Knapp and Mandell, 1972), and (3) the impulse activity of 5-HT neurons (Cunningham and Lakoski, 1988; 1990). While cocaine mimics the cue produced by the 5-HT releaser fenfluramine (White and Appel, 1981), single doses of fenfluramine (McKenna and Ho, 1980), the 5-HT re-uptake inhibitor imipramine (Colpaert *et al.*, 1979), or the hallucinogens D-lysergic acid diethylamide (LSD; Colpaert *et al.*, 1979) or mescaline (McKenna and Ho, 1980) do *not* mimic cocaine. Fluoxetine also does not mimic, although this 5-HT re-uptake inhibitor does potentiate the cocaine cue (Cunningham and Callahan, in press). Single doses of the 5-HT antagonists cinanserin (McKenna and Ho, 1980), cyproheptadine, or methysergide (Colpaert *et al.*, 1979) also did not antagonize the cocaine cue. Despite these findings, Winter and Slifer (1989) have recently reported that the 5-HT agonist-antagonist quipazine substituted for cocaine.

Although the NE re-uptake inhibitor desipramine did not mimic (Colpaert *et al.*, 1979; Cunningham and Callahan, in press), the β-NE blocker propranolol reportedly mimics or partially mimics cocaine, although other β-blockers (practolol, isoproterenol, and salbutamol) are ineffective in this regard (Colpaert *et al.*, 1979; Ho and McKenna, 1978; Silverman and Schultz, 1989). The α_2-NE agonist clonidine also partially substituted for cocaine (Wood *et al.*, 1985). In light of the hypothesis that some aspects of cocaine withdrawal may be related to central 5-HT depletion and the attempts to combat cocaine euphoria and withdrawal with 5-HT and NE uptake inhibitors and precursors (see Nunes and Rosecan, 1987), a systematic study of the role of 5-HT and NE in the stimulus effects of cocaine is warranted and will provide insight into the complexity of this interoceptive stimulus.

Attempts to determine whether the subjective effects of cocaine are mediated by its local anesthetic actions have met with mixed results. In general, procaine and lidocaine elicit a maximum of 50 to 60% drug-lever responding when tested for substitution (Colpaert *et al.*, 1979; De La Garza and Johanson, 1983, 1985; Jarbe, 1981; McKenna *et al.*, 1979; Silverman and Schultz, 1989), although there is one report of procaine substitution in pigeons (Jarbe, 1984). A lack of involvement of the local anesthetic actions in the subjective effects might not be unexpected since large doses of cocaine are required for systemic local anesthesia. Comparative studies of cocaine metabolites and analogs (such as the phenyltropanes), which differ with respect to potency as monoamine re-uptake inhibitors and local anesthetics, could provide a means to study the relative contribution of these actions to the cocaine cue. To date, the centrally active metabolites norcocaine (Bedford *et al.*, 1981; Jarbe, 1981; McKenna *et al.*, 1979) and *n*-allyl-norcocaine (Bedford *et al.*, 1981), as well as the phenyltropane analogues Win 35,428 and 35,065-2 (Jarbe, 1981), have been shown to mimic cocaine in a potency order similar to that shown for their DA re-uptake properties (not their local anesthetic potency).

The loci of action and circuitry involved in the stimulus effects of cocaine have been investigated with the aid of intracerebral microinjection techniques. Injection of cocaine into the lateral ventricles (Wood *et al.*, 1987) or nucleus accumbens, but not the striatum or prefrontal cortex, completely substitutes for cocaine (Wood and Emmett-Oglesby, 1989). Similarly, amphetamine injected into the nucleus accumbens, but not the striatum, mimics a systemically trained amphetamine cue (Nielsen and Scheel-Kruger, 1986). Since 6-OHDA lesions of the nucleus accumbens also attenuate the discriminability of amphetamine (Dworkin and Bimle, 1989), DA mesolimbic, but not nigrostriatal, pathways seem to be critical in the behavioral effects of these stimulants. As argued by Nielsen and Scheel-Kruger (1986), however, the nucleus accumbens might instead be an important nucleus in the *performance* of a drug discrimination task in general, since the cue produced by systemic injection of the hallucinogen LSD can also be elicited by the intra-accumbens LSD administration. Nevertheless, the exact circuitry involved in mediating stimulant discriminative properties does not overlap exactly with that which underlies the reinforcing effects of cocaine since microinjections of cocaine into the prefrontal cortex, but *not* the nucleus accumbens, maintains self-administration behavior in rats (Goeders and Smith, 1983; above).

The primary conclusion to be drawn with any conviction from the study of the discriminative stimulus properties of cocaine to date is that mesolimbic DA systems play a primary role in mediating these interoceptive aspects of cocaine's behavioral effects. An understanding of the exact modulatory importance of other neurotransmitters and neuronal circuitry, as well as which DA receptor systems are most critically involved, is requisite groundwork in the quest to understand drug addiction processes and to identify pharmacological compounds potentially useful in the therapeutic treatment of cocaine dependence.

ACKNOWLEDGMENTS

Supported in part by USPHS grants DA-05708, DA-06511, DA-1999, DA-3628, DA-3631, DA-3832, DA-0114, and NIDA Contract # 271-87-8118. We extend our thanks to Dr. Blaire McPherson for her editorial assistance and Ms. Rosalba Ortiz for her secretarial assistance.

REFERENCES

J.B. Appel, F.J. White, and A.M. Holohean. Analysing mechanisms of hallucinogenic drug action with drug discrimination procedures. *Neurosci. Biohav. Rev.* **6**: 529 (1982).

J. Arnt, J. Hyttel, and J. Perregaard. Dopamine D_1 receptor agonists combined with the selective D_2 agonist quinpirole facilitate the expression of oral stereotyped behavior in rats. *Eur. J. Pharmacol.* **133**: 137 (1987).

G.T. Bain and C. Kornetsky. Naloxone attenuation of the effect of cocaine of rewarding brain stimulation. *Life Sci.* **40**: 1119 (1987).

R.L. Barrett and J.B. Appel. Effects of stimulation and blockade of dopamine receptor subtypes on the discriminative stimulus properties of cocaine. *Psychopharmacology* **99**: 13 (1989).

J.A. Bedford, G.L. Nail, R.F. Borne, and M.C. Wilson. Discriminative stimulus properties of cocaine, norcocaine, and *n*-allylnorcocaine. *Pharmacol. Biochem. Behav.* **14**: 81 (1981).

S. Bischoff, M. Heinrich, J.M. Sonntag, and J. Krauss. The D-1 dopamine receptor antagonist SCH 23390 also interacts potently with brain serotonin ($5\text{-}HT_2$) receptors. *Eur. J. Pharmacol.* **129**: 367 (1986).

A.R. Braun, P. Barone, and T.N. Chase. Interaction of D_1 and D_2 dopamine receptors in the expression of dopamine agonist induced behaviors, in: *Neurobiology of Central D_1 Dopamine Receptors.* Breese, G.R. and Creese, I. eds., Plenum Press, New York, p. 151 (1986).

J. Broadbent, E.K. Michael, and J.B. Appel. Generalization of cocaine to the isomers of 3,4-methylenedioxyamphetamine and 3,4-methylenedioxymethamphetamine: effects of training dose. *Drug Dev. Res.* **16**: 443 (1989).

P.M. Callahan, J.B. Appel, and K.A. Cunningham. Dopamine D_1 and D_2 mediation of the discriminative stimulus properties of D-amphetamine and cocaine. *Psychopharmacology* **103**: 50 (1991).

P.M. Callahan and K.A. Cunningham. The discriminative stimulus properties of cocaine: effects of BAY K 8644 and nimodipine. *Eur. J. Pharmacol.* **86**: 143 (1990).

M.E. Carroll, S.T. Lac, M.J. Walker, R. Kragh, and T. Newman. Effects of naltrexone on intravenous cocaine self-administration in rats during food satiation and deprivation. *J. Pharmacol. Exp. Ther.* **238**: 1 (1986).

A. Childress, R. Ehrman, A. McLellan, and C. O'Brien. Update on behavioral treatments for substance abuse. In: *Problems of Drug Dependence.* L. S. Harris, ed., National Institute on Drug Abuse, Rockville, MD, p. 183 (1988).

D. Clark and F.J. White. D_1 dopamine receptor: the search for a function. *Synapse* **1**: 347 (1987).

F.C. Colpaert. Interactions of haloperidol with discriminative responding controlled by 10 mg/kg of cocaine. *Drug Dev. Res.* **9**: 125 (1986).

F.C. Colpaert, C.J.E. Niemegeers, and P.A.J. Janssen. Cocaine cue in rats as it relates to subjective drug effects: a preliminary report. *Eur. J. Pharmacol.* **40**: 195 (1976).

F.C. Colpaert, C.J.E. Niemegeers, and P.A.J. Janssen. Neuroleptic interference with the cocaine cue: internal stimulus control of behavior and psychosis. *Psychopharmacology* **58**: 247 (1978a).

F.C. Colpaert, C.J.E. Niemegeers, and P.A. J. Janseen. Discriminative stimulus properties of cocaine and D-amphetamine, and antagonism by haloperidol: a comparative study. *Neuropharmacology.* **17**: 937 (1978b).

F.C. Colpaert, C.J.E. Niemegeers, and P.A.J. Janssen. Discriminative stimulus properties of cocaine: neuropharmacological characteristics as derived from stimulus generalization experiments. *Pharmacol. Biochem. Behav.* **10**: 535 (1979).

F.C. Colpaert, C.J.E. Niemegeers, and P.A.J. Janssen. Evidence that a preferred substrate for type B monoamine oxidase mediates stimulus properties of MAO inhibitors: a possible role for beta-phenylethylamine in the cocaine cue. *Pharmacol. Biochem. Behav.* **13**: 513 (1980).

T.J. Crow. Enhancement by cocaine of intracranial self-stimulation in the rat. *Life Sci.* **9**: 375 (1970).

K.A. Cunningham and J.B. Appel. Discriminative stimulus properties of cocaine and phencyclidine: similarities in mechanism of action, in: *Drug Discrimination: Applications in CNS Pharmacology.* F.C. Colpaert and J.L. Slangen. eds., Elsevier/North-Holland Biomedical Press, Amsterdam, p. 181 (1982).

K.A. Cunningham and J.B. Appel. Hallucinogenic and non-hallucinogenic 5-HT agonists: differences in subjective effects parallel differences in receptor dynamics, in: *5-HT Agonists as Psychoactive Drugs*. R.H. Rech and G.A. Gudelsky. eds., NPP Press, Ann Arbor, MI, p. 217 (1988).

K.A. Cunningham and P.M. Callahan. Monoamine reuptake inhibitors potentiate the discriminative state induced by cocaine in the rat. *Psychopharmacology* (in press).

K.A. Cunningham and J.M. Lakoski. Electrophysiological effects of cocaine and procaine on dorsal raphe serotonin neurons. *Eur. J. Pharmacol.* **148**: 457 (1988).

K.A. Cunningham and J.M. Lakoski. The interactions of cocaine with serotonin dorsal raphe neurons: single unit extracellular recording studies. *Neuropsychopharmacology*. **3**: 41 (1990).

G.D. D'Mello and I.P. Stolerman. Comparison of the discriminative stimulus properties of cocaine and amphetamine in rats. *Br. J. Pharmacol.* **61**: 415 (1977).

R. De La Garza and C.E. Johanson. Effects of haloperidol and physostigmine on self-administration of local anesthetics. *Pharmacol. Biochem. Behav.* **17**: 1295 (1982).

R. De La Garza and C.E. Johanson. The discriminative stimulus properties of cocaine in the rhesus monkeys. *Pharmacol. Biochem. Behav.* **19**: 145 (1983).

R. De La Garza and C.E. Johanson. Discriminative stimulus properties of cocaine in pigeons. *Psychopharmacology* **85**: 23 (1985).

H. de Wit, and R.A. Wise. Blockade of cocaine reinforcement in rats with the dopamine receptor blocker pimozide, but not with the noradrenergic blockers phentolamine or phenoxybenzamine. *Can. J. Psychol.* **31**: 195 (1977).

S.I. Dworkin. Lethal effects of cocaine altered by behavioral contingencies-noncontingent vs. contingent administration. *Science* (submitted, 1990).

S.I. Dworkin and C. Bimle. 6-Hydroxydopamine lesions of the nucleus accumbens attenuate the discriminative stimulus effects of D-amphetamine. *Drug Dev. Res.* **16**: 435 (1989).

S.I. Dworkin and J.E. Smith. Neurobehavioral pharmacology of cocaine. In: *Mechanisms of Cocaine Abuse and Toxicity*. D. Clouet, K. Ashgar, and R. Brown, eds., Research Monograph 88, National Institute on Drug Abuse, Rockville, MD, p. 185 (1988).

M.W. Emmett-Oglesby, M. Wurst, and H. Lal. Discriminative stimulus properties of a small dose of cocaine. *Neuropharmacology* **22**: 97 (1983).

R.U. Esposito, A.H.D. Motola, and C. Kornetsky. Cocaine: acute effects on reinforcement thresholds for self-stimulation behavior to the medial forebrain bundle. *Pharmacol. Biochem. Behav.* **8**: 437 (1978).

A. Ettenberg, H.O. Pettit, F.E. Bloor, and G.F. Koob. Heroin and cocaine intravenous self-administration rats: mediation by separate neural systems. *Psychopharmacology* **78**: 204 (1982).

J.M. Finlay, H.C. Fibiger, C.D. Blaha, and A.G. Phillips. *In vivo* electrochemical detection of dopamine within the nucleus accumbens of rats self-administering cocaine. *Neurosci. Abstr.* **14**: 742 (1988).

R.A. Frank, S. Martz, and T. Pommering. The effects of chronic cocaine on self-stimulation train-duration thresholds. *Pharmacol. Biochem. Behav.* **29**: 755 (1988).

D.V. Gauvin, R.D. Harland, R.C. Michaelis, and F.A. Holoway. Caffeine-phenylethylamine combinations mimic the cocaine discriminative stimulus. *Life Sci.* **44**: 67 (1989).

N.E. Goeders and J.E. Smith. Cortical dopamine involvement in cocaine reinforcement. *Science* **221**: 773 (1983).

S.R. Goldberg and F.A. Gonzales. The effects of propranolol on behavior maintained under fixed ratio schedules of cocaine injection or food presentation in squirrel monkeys. *J. Pharmacol. Exp. Ther.* **198**: 626 (1976).

A. Goudie and D. Reid. Qualitative discrimination between cocaine and amphetamine in rats. *Eur. J. Pharmacol.* **151**: 471 (1988).

R.E. Heikkila, H. Orlansky, and G. Cohen. Studies on the distinction between uptake inhibition and release of ^{3}H-dopamine in rat brain tissue slices. *Biochem. Pharmacol.* **24**: 847 (1975).

B.T. Ho and M. McKenna. Discriminative stimulus properties of central stimulants, in: *Drug Discrimination and State-Dependent Learning*. B.T. Ho, D.W. Richards, and D.L. Chute, eds., Academic Press, New York, p. 67 (1978).

D. Huang and M.C. Wilson. Comparative discriminative stimulus properties of DL-cathinone, D-amphetamine and cocaine in rats. *Pharmacol. Biochem. Behav.* **24**: 205 (1986).

C.B. Hubner and G.F. Koob. Ventral pallidal lesions produce decreases in cocaine and heroin self-administration in the rat. *Neurosci. Abstr.* **13**: 1717 (1987).

C.B. Hubner and G.F. Koob. Lesions of the dorsomedial nucleus of the thalamus produce decreases in cocaine self-administration in the rat. *Neurosci. Abstr.* **14**: 660 (1988).

Y.L. Hurd, F. Weiss, G.F. Koob, N.-E. Anden, and U. Ungerstedt. Cocaine reinforcement and extracellular dopamine overflow in rat nucleus accumbens: an *in vivo* microdialysis study. *Brain Res.* **498**: 199 (1989).

L.C. Iorio, A. Barnett, F.H. Leitz, V.P. Houser, and C.A. Korduba. SCH 23390, a potential benzazepine antipsychotic with unique interactions on dopaminergic systems. *J. Pharmacol. Exp. Ther.* **226**: 462 (1983).

J.H. Jaffe, J.M. Witkin, S.R. Goldberg, and J.L. Katz. Potential toxic interactions of cocaine and mazindol. *Lancet II* **8654**: 111 (1989).

T.U.C. Jarbe. Cocaine as a discriminative cue in rats: interactions with neuroleptics and other drugs. *Psychopharmacology* **59**: 183 (1978).

T.U.C. Jarbe. Cocaine cue in pigeons: time course studies and generalization to structurally related compounds (norcocaine, WIN 35,428 and 35,065-2) and (+)-amphetamine. *Br. J. Pharmacol.* **73**: 843 (1981).

T.U.C. Jarbe. Discriminative stimulus properties of cocaine: effects of apomorphine, haloperidol, procaine and other drugs. *Neuropharmacology* **23**: 899 (1984).

L.T. Kennedy and I. Hanbauer. Sodium-sensitive cocaine binding to rat striatal membrane: possible relationship to dopamine uptake sites. *J. Neurochem.* **41**: 172 (1983).

G.J. Kilpatrick, B.J. Jones, and M.B. Tyers. Binding of the 5-HT_3 ligand GR65630, to rat area postrema vagus nerve and the brain of several species. *Eur. J. Pharmacol.* **159**: 157 (1989).

M.S. Kleven, E.W. Anthony, L.I. Goldberg, and W.L. Woolverton. Blockade of the discriminative stimulus effects of cocaine in rhesus monkeys with the D_1 dopamine antagonist SCH 23390. *Psychopharmacology* **95**: 427 (1988).

S. Knapp and A.J. Mandell. Narcotic drugs: effects on the serotonin biosynthetic systems of the brain. *Science* **177**: 1209 (1972).

B.K. Koe. Molecular geometry of inhibitors of the uptake of catecholamines and serotonin in synaptosomal preparations of rat brain. *J. Pharmacol. Exp. Ther.* **199**: 649 (1976).

G.F. Koob, H.T. Le, and I. Creese. The D_1 dopamine receptor antagonist SCH 23390 increases cocaine self-administration in the rat. *Neurosci. Lett.* **79**: 315 (1987).

C. Kornetsky and R.U. Esposito. Reward and detection thresholds for brain stimulation: dissociative effects of cocaine. *Brain Res.* **209**: 496 (1981).

E.A. Loh and D.C.S. Roberts. Break points on a progressive ratio schedule reinforced by intravenous cocaine increase following depletion of forebrain serotonin. *Psychopharmacology* **101**: 262 (1990).

E.D. London, G. Wilkerson, S.R. Goldberg, and M.E. Risner. Effects of L-cocaine on local cerebral glucose utilization in the rat. *Neurosci. Lett.* **68**: 73 (1986).

R. Marcus and C. Kornetsky. Negative and positive intracranial reinforcement thresholds: effects of morphine. *Psychopharmacology* **38**: 1 (1974).

A. Mattia, A.P. Leccesse, and J.E. Moreton. Effects of quipazine on cocaine self-administration in rats. *Pharmacologist* **28**: 151 (1986).

M. McKenna and B.T. Ho. Induced tolerance to the discriminative stimulus properties of cocaine. *Pharmacol. Biochem. Behav.* **7**: 273 (1977).

M.L. McKenna and B.T. Ho. The role of dopamine in the discriminative stimulus properties of cocaine. *Neuropharmacology* **19**: 297 (1980).

M.L. McKenna, B.T. Ho, and L.F. Englert. Generalization of norcocaine to the discriminative stimulus properties of cocaine. *Pharmacol. Biochem. Behav.* **10**: 273 (1979).

R.D. McQuade, D. Ford, R.A. Duffy, R.E. Chipkin, L.C. Iorio, and A. Barnett. Serotoninergic component of SCH 23390: *in vitro* and *in vivo* binding assays. *Life Sci.* **43**: 1861 (1988).

A.L. Misra, N.L. Vadlamani, and R.B. Pontani. Cocaine-desipramine interaction. *Res. Comm. Subs. Abuse* **7**: 85 (1986).

E.B. Nielsen and S.A. Jepsen. Antagonism of the amphetamine cue by both classical and atypical antipsychotic drugs. *Eur. J. Pharmacol.* **111**: 167 (1985).

E.B. Nielsen, K. Randrup, and P.H. Anderson. Amphetamine discrimination: effects of dopamine receptor agonists. *Eur. J. Pharmacol.* **160**: 253 (1989).

E.B. Nielsen and J. Scheel-Kruger. Cueing effects of amphetamine and LSD: elicitation by direct microinjection of the drugs into the nucleus accumbens. *Eur. J. Pharmacol.* **125**: 85 (1986).

G.G. Nomikos and C. Spyraki. Cocaine-induced place conditioning: importance of route of administration and other procedural variables. *Psychopharmacology* **94**: 119 (1988).

E.V. Nunes and J.S. Rosecan. Human neurobiology of cocaine, in: *Cocaine Abuse: New Directions in Treatment and Research.* H.I. Spitz and J.S. Rosecan, eds., Brunner/Mazel, New York, p. 48 (1987).

J. Olds and P. Milner. Positive reinforcement produced by electrical stimulation of septal area and other regions of rat brain. *J. Comp. Physiol. Psychol.* **47**: 419 (1954).

J. Olds and R.P. Travis. Effects of chlorpromazine, meprobamate, pentobarbital and morphine on self-stimulation. *J. Pharmacol. Exp. Ther.* **128**: 397 (1960).

G.W. Pasternak. ed., *Opiate Receptors.* Humana Press, Clifton, NJ, p. 499 (1988).

H.O. Pettit, A. Ettenberg, F.E. Bloom, and G.F. Koob. Destruction of dopamine in the nucleus accumbens selectively attenuates cocaine but not heroin self-administration in rats. *Psychopharmacology* **84**: 167 (1984).

H.O. Pettit and J.B. Justice, Jr. Dopamine in the nucleus accumbens during cocaine self-administration as studied by *in vivo* microdialysis. *Pharmacol. Biochem. Behav.* **34**: 899 (1989).

L.J. Porrino, F.R. Domer, A.M. Crane, and L. Sokoloff. Selective alterations in cerebral metabolism within the mesocorticolimbic dopaminergic system produced by acute cocaine administration in rats. *Neuropsychopharmacology* **1**: 109 (1988).

L.J. Porrino, R.U. Esposito, T.F. Seeger, A.M. Crane, A. Pert, and L. Sokoloff. Metabolic mapping of the brain during rewarding self-stimulation. *Science* **224**: 306 (1984).

M.E.A. Reith, H. Sershen, D.L. Allen, and A. Lajtha. Saturable [^{3}H]-cocaine binding in the central nervous system of mouse. *Life Sci.* **27**: 1055 (1980).

J.M. Ritchie and P. Greengard. On the mode of action of local anesthetics. *Annu. Rev. Pharmacol.* **6**: 405 (1966).

M.C. Ritz, R.J. Lamb, S.R. Goldberg, and M.J. Kuhar. Cocaine receptors on dopamine transporters are related to self-administration of cocaine. *Science* **237**: 1219 (1987).

T.W. Robbins, D.C.S. Roberts, and G.F. Koob. Effects of D-amphetamine and apomorphine upon operant behavior and schedule-induced licking in rats with 6-hydroxydopamine induced lesions of the nucleus accumbens. *J. Pharmacol. Exp. Ther.* **224**: 662 (1983).

D.C.S. Roberts, M.E. Corcoran, and H.C. Fibiger. On the role of ascending catecholaminergic systems in intravenous self-administration of cocaine. *Pharmacol. Biochem. Behav.* **6**: 615 (1977).

D.C.S. Roberts and G.F. Koob. Disruption of cocaine self-administration following 6-hydroxydopamine lesions of the ventral segmental area in rats. *Pharmacol. Biochem. Behav.* **17**: 901 (1982).

D.C.S. Roberts, G.F. Koob, P. Klonoff, and H.C. Fibiger. Extinction and recovery of cocaine self-administration following 6-hydroxydopamine lesions of the nucleus accumbens. *Pharmacol. Biochem. Behav.* **17**: 781 (1982).

D.C.S. Roberts, A.P. Zis, and H.C. Fibiger. Ascending catecholamine pathways and amphetamine-induced locomotor activity: importance of dopamine and apparent non-involvement of norepinephrine. *Brain Res.* **93**: 441 (1975).

S.B. Ross and A.L. Renyi. Inhibition of the uptake of tritiated 5-hydroxytryptamine in brain tissue. *Eur. J. Pharmacol.* **7**: 270 (1969).

H. Sershen, M.E.A. Reith, and A. Lajtha. Comparison of the properties of central and peripheral binding sites for cocaine. *Neuropharmacology* **21**: 469 (1982).

P.B. Silverman and B.T. Ho. Characterization of discriminative response control by psychomotor stimulants, in *Discriminative Stimulus Properties of Drugs.* H. Lal. ed., Plenum Press, New York, p. 107 (1977).

P.B. Silverman and K.A. Schultz. Comparison of cocaine and procaine discriminative stimuli. *Drug Dev. Res.* **16**: 427 (1989).

F.L. Smith, C. St. John, T.F.T. Yang, and W.H. Lyness. Rose of specific receptor subtypes in amphetamine discrimination. *Psychopharmacology* **97**: 501 (1989).

J.E. Smith, C. Co. M.E. Freemen, and J.D. Land. Brain neurotransmitter turnover correlated with morphine-seeking behavior in rats. *Pharmacol. Biochem. Behav.* **16**: 509 (1982).

J.E. Smith, C. Co, M.E. Freemen, M.P. Sands, and J.D. Land. Neurotransmitter turnover in the striatum of rat brains is correlated with morphine self-administration. *Nature* **287**: 152 (1980).

J.E. Smith, C. Co, and J.D. Lane. Limbic acetylcholine turnover rates correlated with rat morphine-seeking behaviors. *Pharmacol. Biochem. Behav.* **20**: 429 (1984a).

J.E. Smith, C. Co, and J.D. Lane. Limbic muscarinic cholinergic and benzodiazepine receptor changes with chronic intravenous morphine self-administration. *Pharmacol. Biochem. Behav.* **20**: 443 (1984b).

C. Spyraki, H.C. Fibiger, and A.G. Phillips. Cocaine-induced place preference conditioning: lack of effects of neuroleptics and 6-hydroxydopamine lesions. *Brain Res.* **253**: 195 (1982).

C. Spyraki, G.G. Nomikos, and D.D. Varonos. Intravenous cocaine-induced place preference: attenuation by haloperidol. *Behav. Brain Res.* **26**: 57 (1987).

J.R. Stellar and E. Stellar. *Neurobiology of Motivation and Reward.* Springer-Verlag, New York, (1985).

L. Stein and O. Ray. Self-regulation of brain stimulating current intensity in the rat. *Science* **130**: 570 (1959).

S.S. Steiner, B. Beer, and M.M. Shaffer. Escape from self-produced rates of brain stimulation. *Science* **163**: 90 (1969).

I.P. Stolerman and G.D. D'Mello. Role of training conditions in discrimination of central nervous system stimulants by rats. *Psychopharmacology* **73**: 295 (1981).

J.C. Stoof and J.W. Kebabian. Two dopamine receptors: biochemistry, physiology and pharmacology. *Life Sci.* **35**: 2281 (1984).

D. Taylor and B.T. Ho. Comparison of inhibition of monoamine uptake by cocaine, methylphenidate and amphetamine. *Res. Comm. Chem. Pathol. Pharmacol.* **21**: 67 (1978).

F. Weiss, Y.L. Hurd, G.F. Koob, N.-E. Anden, and U. Ungerstedt. *In vivo* microdialysis study of DA function in the rat nucleus accumbens during acute and repeated cocaine self-administration. *Neurosci. Abstr.* **14**: 658 (1988).

F.J. White and J.P. Appel. A neuropharmacological analysis of the discriminative stimulus properties of fenfluramine *Psychopharmacology* **73**: 110 (1981).

F.J. White, L.M. Bednarz, S.R. Wachtel, S. Hjorth, and R. Brooderson. Is stimulation of both D_1 and D_2 receptors necessary for the expression of dopamine-mediated behaviors? *Pharmacol. Biochem. Behav.* **30**: 189 (1988).

M.C. Wilson and C.R. Schuster. Cholinergic influence on intravenous cocaine self-administration by rhesus monkeys. *Pharmacol. Biochem. Behav.* **1**: 643 (1973).

T.A. Winter and B.L. Slifer. Discriminative stimulus properties of cocaine: is there mediation by serotonin? *FASEB J.* **3**: 295 (1989).

M. Wolske and M.O. West. Single unit activity in the nucleus accumbens of the freely moving rat during oral self-administration of cocaine. *Neurosci. Abstr.* **14**: 930 (1988).

D.M. Wood and M.W. Emmett-Oglesby. Evidence for D_2-receptor involvement in tolerance to the discriminative stimulus properties of cocaine. *Soc. Neurosci. Abstr.* **13**: 1717 (1987).

D.M. Wood and M.W. Emmett-Oglesby. Substitution and cross-tolerance profiles of anorectic drugs in rats trained to detect the discriminative stimulus properties of cocaine. *Psychopharmacology* **95**: 364 (1988).

D.M. Wood and M.W. Emmett-Oglesby. Mediation in the nucleus accumbens of the discriminative stimulus produced by cocaine. *Pharmacol. Biochem. Behav.* **33**: 453 (1989).

D.M. Wood, H. Lal, and M. Emmett-Oglesby. Acquisition and recovery of tolerance to the discriminative stimulus properties of cocaine. *Neuropharmacology* **23**: 1419 (1984).

D.M. Wood, H. Lal, S. Yaden, and M.W. Emmett-Oglesby. One-way generalization of clonidine to the discriminative stimulus produced by cocaine. *Pharmacol. Biochem. Behav.* **23**: 529 (1985).

D.M. Wood, K.C. Retz, and M.W. Emmett-Oglesby. Evidence of a central mechanism mediating tolerance to the discriminative stimulus properties of cocaine. *Pharmacol. Biochem. Behav.* **28**: 401 (1987).

W.L. Woolverton. Pharmacological analysis of the discriminative stimulus properties of D-amphetamine in rhesus monkeys. *Pharmacologist* **26**: 161 (1984).

W.L. Woolverton. Effects of a D_1 and D_2 dopamine antagonist on the self-administration of cocaine and piribedil by rhesus monkeys. *Pharmacol. Biochem. Behav.* **24**: 531 (1986).
W.S. Young, III, D. Niehoff, M.J. Kuhar, B. Beer, and A.S. Lippa. Multiple benzodiazepine receptor localization by light microscopic radiohistochemistry. *J. Pharmacol. Exp. Ther.* **216**: 425 (1981).
K.A. Zito, G. Vickers, and D.C.S. Roberts. Disruption of cocaine and heroin self-administration following kainic acid lesions of the nucleus accumbens. *Pharmacol. Biochem. Behav.* **23**: 1029 (1985).

7

Sensitization and Kindling Effects of Chronic Cocaine Administration

ROBERT M. POST, SUSAN R. B. WEISS, AND AGU PERT
Biological Psychiatry Branch,
National Institute of Mental Health
Bethesda, Maryland

INTRODUCTION

Repeated administration of the psychomotor stimulant-local anesthetic agent cocaine can result in two different phenomena which reflect an increased, rather than a decreased, drug effect over time. In behavioral sensitization, there is an increased motor activation or stereotypic response, whereas in kindling major motor seizures develop to a dose of cocaine that was previously subconvulsant. The mechanisms underlying these two phenomena appear quite different.

Behavioral sensitization is a common property of psychomotor stimulants and direct dopamine agonists (Post and Contel, 1981; Ellinwood and Kilbey, 1980), but not other catecholamine or indoleamine direct or indirect agonists, suggesting that a dopaminergic agonist effect is critical for its production. In this respect, cocaine shares many pharmacological properties with amphetamine and methylphenidate. These psychomotor stimulant effects of cocaine are thought to relate to its positive properties of arousal, motor activation, and intense euphoria (Van Rossum, 1970; Post *et al.*, 1987; Wise, 1984; Scheel-Kruger *et al.*, 1977; Gawin and Ellinwood, 1988) but, in excess, these effects of cocaine may also be associated with the production of increasing dysphoria, manic-like psychosis, and a paranoid psychosis that is difficult to distinguish from schizophrenia (Owens, 1912; Lewin, 1931; Gay *et al.*, 1973; Post, 1975, 1976; Jeri *et al.*, 1978; Byck *et al.*, 1980; Manschreck *et al.*, 1988; Washton and Gold, 1984; Sherer *et al.*, 1988).

Cocaine is also a local anesthetic of equal potency to lidocaine (Goodman and Gilman, 1965). This provides cocaine with an additional set of pharmacological properties distinct from other psychomotor stimulants. By comparing the action of cocaine to the pure local anesthetic lidocaine or to the relatively pure local anesthetic procaine, one may be able to differentiate components of cocaine's effects that are

Table 1
Electrical Kindling: Major Characteristics

1.	Repeated stimulations a. Initially subseizure threshold b. Intermittent
2.	Local afterdischarges and seizure activity a. Increases in amplitude, frequency b. Increase in duration c. Increase in complexity of wave form d. Increase in anatomical spread
3.	Replicable sequence of seizure stages Behavioral arrest, blinking and masticatory movements, head nodding, opsithotonis, contralateral then bilateral forelimb, clonus, rearing, and falling
4.	Discharges kindle in quantum jumps
5.	Limbic system kindled more readily than cortex
6.	In kindled animals the history of convulsion development is recapitulated as seizure builds
7.	Transfer effects to secondary sites; kindling facilitated in other sites even after primary site destroyed
8.	Interference: a secondary kindled site interferes with primary site kindling
9.	No toxic or neuropathological changes evident: kindling is a transsynaptic process
10.	Relatively permanent change in connectivity; a kindled animal will still have seizures after a 1-year seizure-free interval
11.	Seizures may develop spontaneously in chronically kindled animals
12.	Interictal spikes and spontaneous epileptiform potentials develop

Note: See Goddard *et al.* (1969) regarding 1, 4–10; Wada and Sato (1974) regarding 2–4, 6; Wada *et al.* (1974) regarding 10–12; Racine (1978) and Pinel and Rovner (1978a, b).

related to its local-anesthetic compared with its psychomotor-stimulant properties. While low doses of cocaine (5–10 mg/kg i.p.) are able to engage its psychomotor stimulant effects (Table 1), higher doses (50–65 mg/kg i.p.) are generally required to observe its local anesthetic properties (see Table 5). For example, kindling of seizures can occur with repeated administration of cocaine at relatively high doses (50–65 mg/kg) (Weiss *et al.*, 1989b), and, similarly, seizures can also be kindled with the local anesthetic lidocaine (Post and Kopanda, 1976; Post *et al.*, 1975, 1981, 1984b). Thus, it is likely that the kindling-like property of cocaine resides in its pharmacological properties as a local anesthetic. It is also possible that the local-anesthetic effects of cocaine are related to a variety of clinical phenomena that develop after repeated use, including panic attacks and seizures in man (Post *et al.*, 1987).

Kindling was first described by Goddard *et al.* in 1969 using repeated electrical stimulation of the amygdala. Electrical stimulation, which was initially below the threshold for eliciting seizure activity (afterdischarge), eventually produced afterdischarges. Moreover, these afterdischarges increased progressively in duration, spread throughout the brain, became more complex (Racine, 1978), eventually resulting in the production of a full-blown major motor seizure (Table 2). It is now clear that repeated, intermittent administration of electrical current is required for elicitation of the kindling phenomena and that, once it is elicited, it is essentially a permanent change in the animal's excitability (Goddard *et al.*, 1969; Racine,

Table 2
Characteristics of Behavioral Sensitization to Psychomotor Stimulants

1.	Shorter latency to maximum response following repeated exposure
2.	Increased magnitude of behavioral response to same dose
3.	Selective increase in some behaviors, not others
4.	Alterations in responsivity are long lasting (week to months)
5.	Intermittent administration produces greater sensitization, as compared with continuous administration
6.	Effects are dose-related
7.	Genetic component is evident in degree of sensitization
8.	Sensitization is not related to alterations in blood levels
9.	Sensitization shows environmental context dependency; i.e., is conditionable
10.	Cross-sensitization occurs among many stimulants and dopamine agonists
11.	Cross-sensitization can occur between stimulants and a variety of stresses (tail-pinch, shock, starvation), but may be stressor-, intensity- and time-dependent
12.	Neurotransmitter, receptor, and subreceptor changes postulated
13.	Hormones and peptides (i.e., vasopressin) modulate and may produce behavioral sensitization
14.	Dopaminergic substrates in nucleus accumbens (and amygdala for environmental context-dependent component) may be critical to sensitization

Note: See Post and Contel, 1981; Antelman and Chiodo, 1983; Ellinwood and Kilbey, 1977; Weiss *et al.*, 1989a for additional references.

1978). Furthermore, if the kindling stimulation is repeated enough, and hundreds of seizures are elicited, spontaneity can occur whereby the animal begins to demonstrate seizures in the absence of exogenous electrophysiological stimulation (Pinel and Rovner, 1978b; Pinel, 1981).

The response to cocaine can, therefore, increase with repetition by at least two different mechanisms. In behavioral sensitization, motor activity or stereotypy is the major end point measured, while in kindling changes in convulsive responsivity are measured. The behavioral and physiological consequences of repeated cocaine administration will thus be described in the context of these two phenomena in animals and their potential clinical implications for human use suggested.

It is important to note that, in addition to many of the other dangers of cocaine use described elsewhere in this volume, both behavioral sensitization and pharmacological kindling have their unique hidden liabilities; that is, the potential behavioral and physiological toxicities associated with these two processes are not immediately evident in the initial use of the drug (Table 6). Even without dose escalation, which is a common phenomenon with cocaine use, repeated administration of the same dose of the drug may be associated with increasingly severe behavioral and convulsive toxicities. The phenomenology, pharmacology, and possible neural substrates of sensitization and kindling are, therefore, of great clinical and basic interest.

BEHAVIORAL SENSITIZATION

Phenomenology

As illustrated in Figure 1, repeated administration of the same dose of cocaine

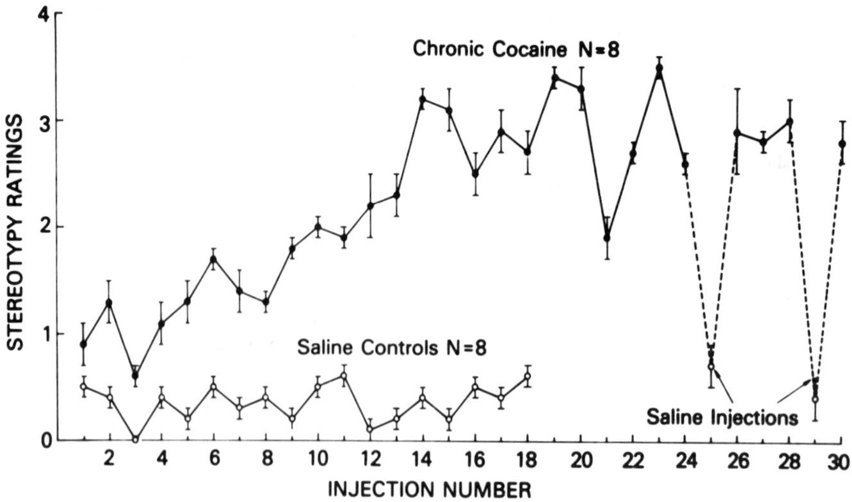

Figure 1. Rats treated with cocaine (10 mg/kg i.p.) develop progressively more intense and constricted behavioral stereotypic patterns including corner-to-corner motor sequences, vertical rearing and nose poking, and head bobbing.

(10 mg/kg, i.p.) is associated with increasing degrees of stereotypy. A similar phenomenon is seen in relationship to motor activity. As illustrated in Figure 2, there is relatively little motor activity following the first cocaine dose but progressively more is seen following successive doses, such that, by the fifth dose, substantial degrees of hyperactivity are evident. A major component of behavioral sensitization appears to be context dependent and thus reflects conditioning (Post *et al.*, 1981; Weiss *et al.*, 1989a); that is, animals repeatedly injected in the context of the same test cage show increases in locomotor activity, while animals injected with equal doses of cocaine in their home cage are not as activated when injected in the test cage, as illustrated in Figure 3.

This increase in motor activation upon repeated cocaine administration is relatively long lasting (Post and Contel, 1983; Post *et al.*, 1987). As illustrated in Figure 4, animals given chronic cocaine administration for 12 d show significantly more motor activity upon rechallenge, up to 87 d later. However, if animals are only given one large dose of cocaine, and rechallenged with a small dose the next day, the behavioral sensitization that is revealed appears to last approximately 5–10 d only. As illustrated in Figure 5, behavioral sensitization to a single large dose of cocaine is also environmental-context dependent (Weiss *et al.*, 1989a). On day 1, animals are injected with cocaine (40 mg/kg) either in their test chamber or in their home cage and compared with animals that received saline injections. On day 2, all animals are rechallenged with a small dose of cocaine (10 mg/kg i.p). Only those animals that received cocaine in the context of the test chamber on day 1 showed behavioral sensitization. Remarkably, animals that received a large dose of cocaine in a different chamber are no more active than saline-injected controls

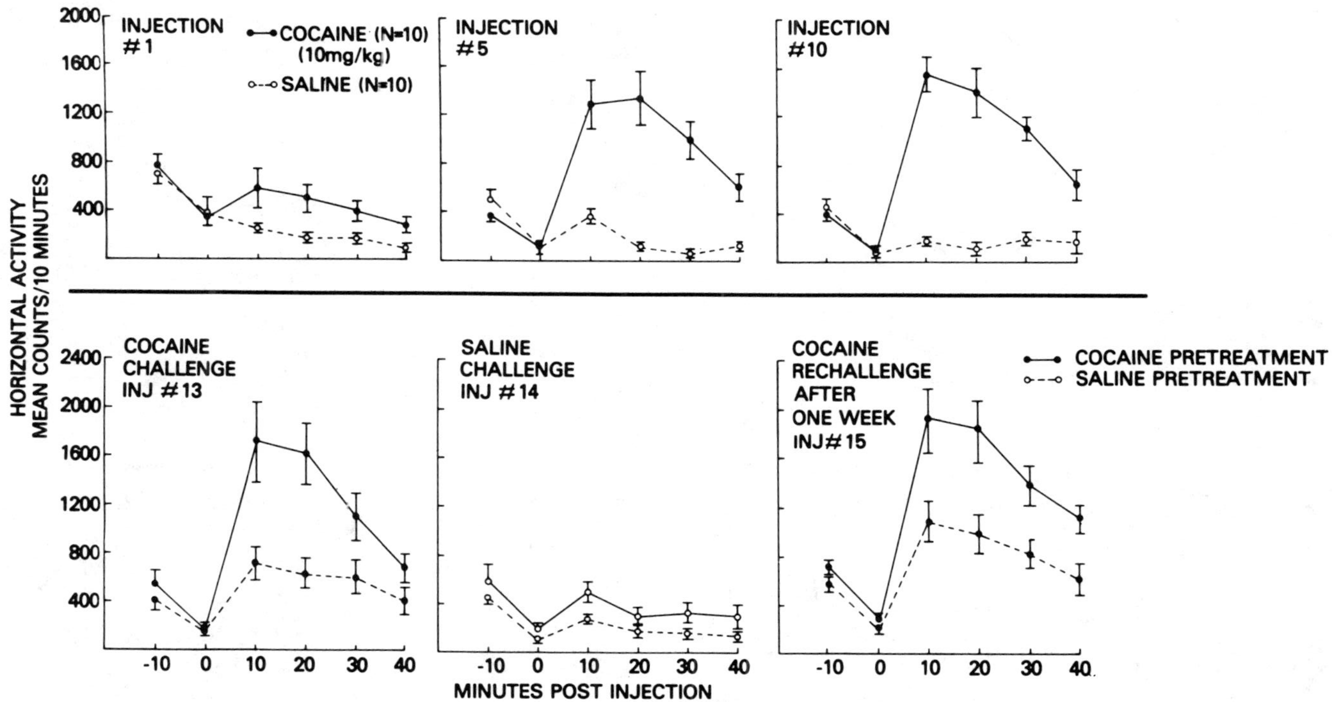

Figure 2. Development and persistence of behavioral sensitization to cocaine's effects on motor activity. Rats treated with cocaine (10 mg/kg i.p.) show increases in motor activity following repeated injections (#1–#10) and an increased response compared with saline controls (injections #13 and #17). A small conditioned response to the saline injection (#14) was apparent; chronic cocaine animals showed significantly greater activity after the saline injection compared with saline-pretreated animals (which had recieved only one prior cocaine injection).

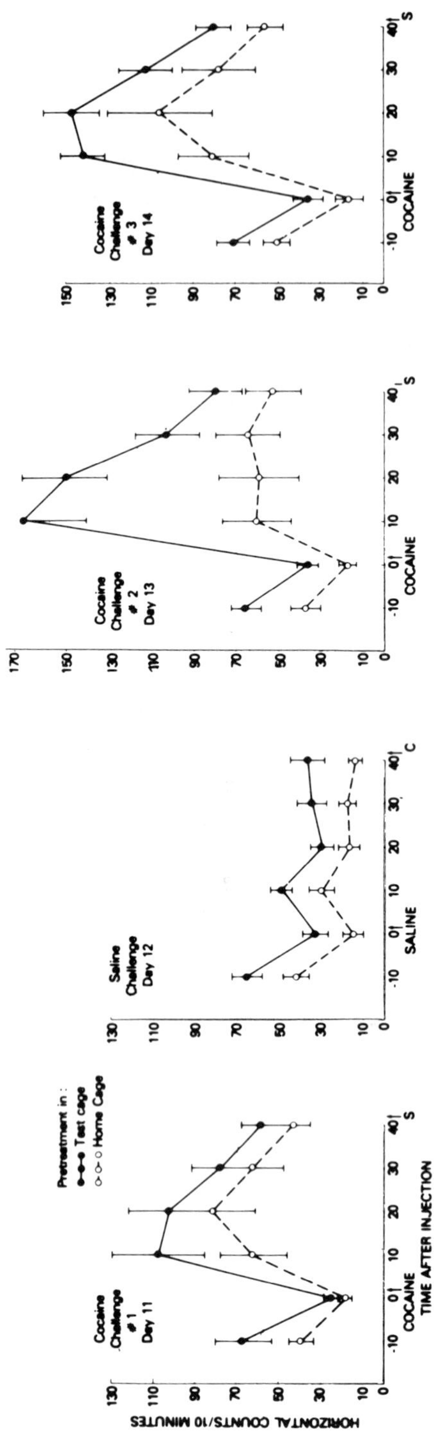
HORIZONTAL COUNTS/10 MINUTES
Cocaine Challenge # 1 Day 11
Pretreatment in :
Test cage
Home Cage
COCAINE
TIME AFTER INJECTION
Saline Challenge Day 12
SALINE
Cocaine Challenge # 2 Day 13
COCAINE
Cocaine Challenge # 3 Day 14
COCAINE

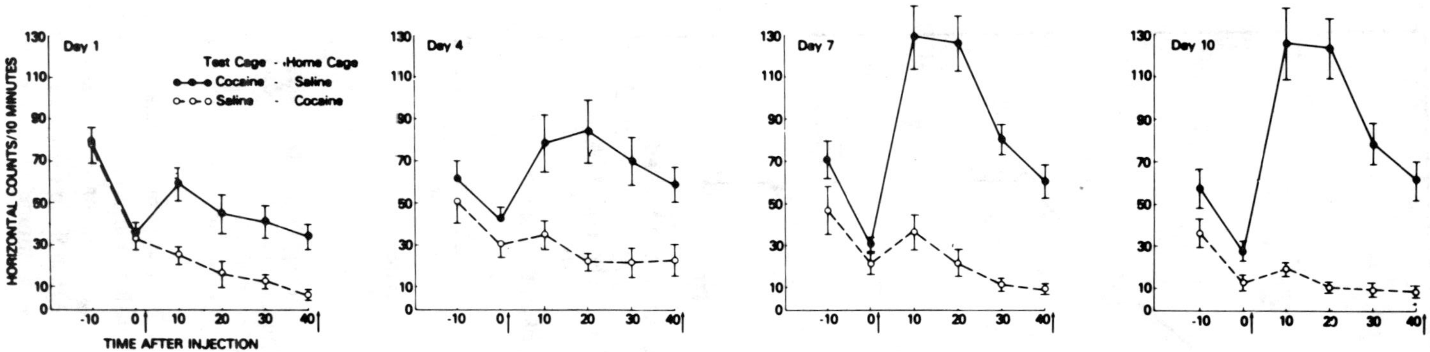

Figure 3. Top: Cocaine pretreatment in different environmental contexts: cocaine sensitization. Bottom: Effect of environmental context on cocaine sensitization. Increasing motor response to once-daily cocaine (10 mg/kg i.p.) is observed only in test cage-treated animals (days 1–10). Moreover, test cage-treated animals showed greater motor activity following cocaine challenge #2 ($p < 0.01$) and #3 ($p < 0.05$) than animals that received the same number of previous cocaine doses, but in their home cage. Test cage-treated animals were also more active before and after saline challenge (day 12) than those treated in their home cage ($p < 0.001$). These data support a role for conditioning variables in cocaine-induced behavioral sensitization.

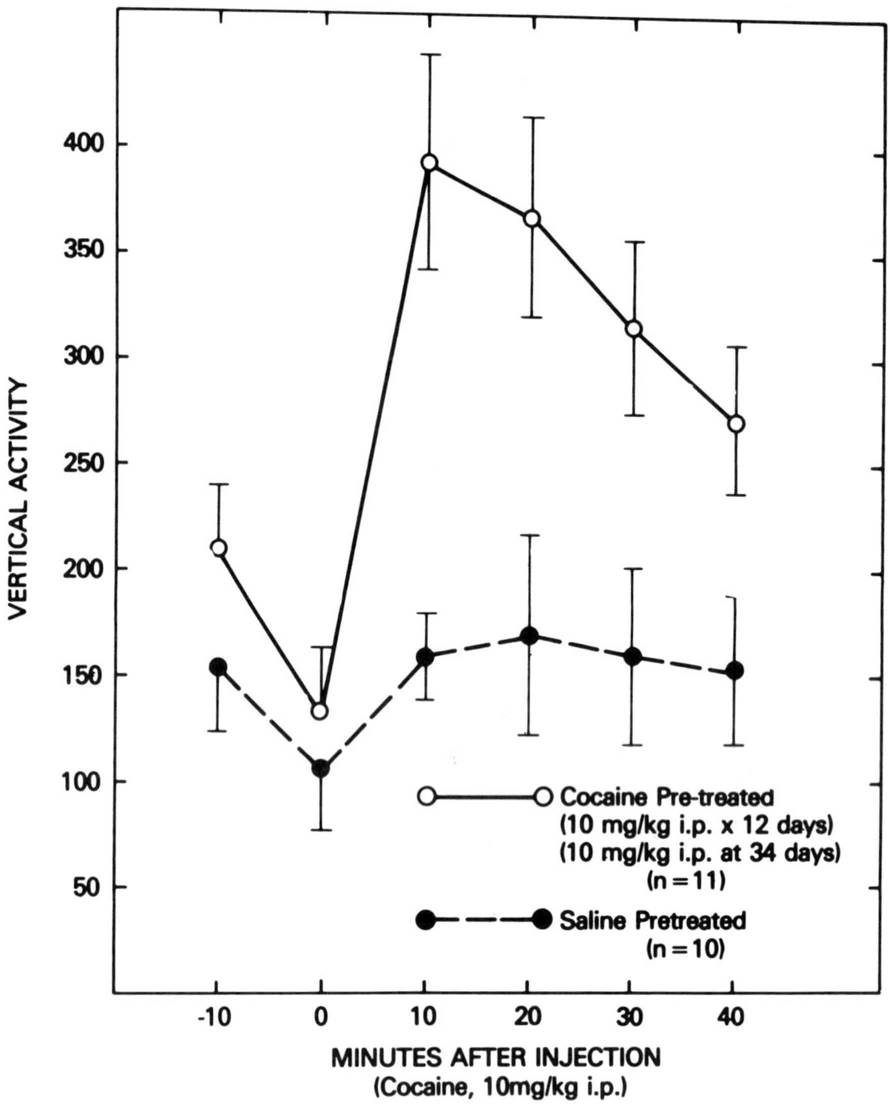

Figure 4. Time (pre- and post-injection) is plotted on the abscissa and vertical activity is plotted on the ordinate. Rats were pretreated for 12 d with either cocaine (10 mg/kg) or saline injections (i.p.). When challenged with an acute cocaine injection 87 d following the termination of chronic cocaine, the cocaine-pretreated animals continued to exhibit much greater activity than saline-pretreated controls.

(Figure 5). This demonstrates that in the one-day, high-dose paradigm, the conditioned component of cocaine-induced behavioral sensitization accounts entirely for the increased effect on repeated exposure; that is, sensitization is context dependent and not induced solely by prior cocaine. The degree of similarity between the test cage and challenge environments appears to determine the degree of

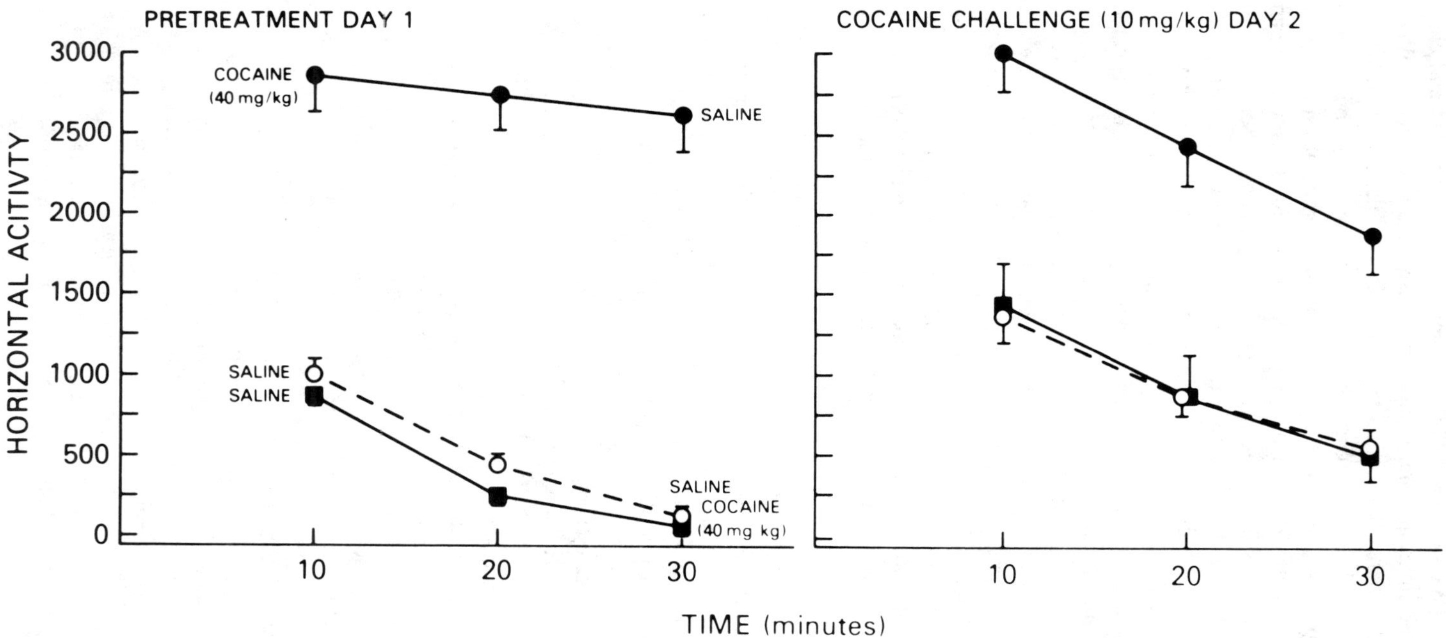

Figure 5. Left panel: Horizontal locomotor activity is illustrated on the ordinate and time on the abscissa for three groups of rats (n = ten per group) receiving the following day-1 pretreatments: cocaine in the test cage and saline in the home cage (●); saline in the test cage and cocaine in the home cage (■); saline in both cages (○). Right panel: Horizontal activity is illustrated for these three groups following a cocaine challenge (10 mg/kg). Only the rats that received cocaine in the test apparatus on day 1 showed sensitization to cocaine on day 2.

behavioral sensitization (Weiss *et al.*, 1989a). When animals are tested and challenged in highly similar environments, behavioral sensitization is evident. When they are tested in environments that are quite dissimilar from the initial challenge environment, sensitization does not occur. Thus, the induction of behavioral sensitization to cocaine appears to follow principles of stimulus generalization.

In other paradigms, a context-indendent component of cocaine sensitization can be demonstrated, however. As illustrated in Figure 6, if high doses of cocaine (40 mg/kg i.p.) are repeated for 3 d and animals then receive the low dose (10 mg/kg i.p.) cocaine challenge on the 4th d, increases in motor activity and stereotypy are observed in both groups of cocaine-pretreated animals compared with those injected with saline. In this instance, the high degrees of stereotypy induced by repeated high doses of cocaine appear to elicit both a context-dependent and -independent component of sensitization, that is, stereotypy is higher in animals repeatedly injected with cocaine in the test cage (context dependent) compared with those administered cocaine in the home cage (context independent). However, in this latter group, cocaine-induced stereotypies are still significantly greater than the saline-pretreated controls. These latter data are consistent with those of Gale and colleagues (1984), who demonstrated that very high doses of cocaine (160 mg/kg s.c.) produce a context-independent stereotypy that cannot be blocked with a neuroleptic pretreatment.

Taken together, these data suggest that there is an important interaction between dose and number of repetitions in determining the intensity, quality, and duration of cocaine-induced behavioral sensitization, as illustrated in the dose-repetition matrix in Figure 7, that is, a single dose of cocaine (10 mg/kg i.p.) is insufficient to produce behavioral sensitization. However, repetition of this dose for 7–10 d produces behavioral sensitization that is both robust (Figures 2 and 3) and long lasting (Figure 4). It is also associated with significant increases in motor activity elicited by a saline challenge (see Figure 3). A single dose of 40 mg/kg produces behavioral sensitization to a rechallenge with 10 mg/kg the next day, which lasts for approximately 1 week. In contrast to those after many repetitions, these animals are not more responsive to a saline challenge than controls. Thus, more repetititions and higher doses appear to produce both longer-lasting sensitization and one that progressively can become independent of environmental context and conditioning factors, i.e., is more "hard wired" (Figure 7). At the very highest doses of cocaine administration, kindling of seizures occurs in a fashion that is independent of conditioning, as described in a following section.

Pharmacology of Behavioral Sensitization

Using the 1-d high-dose sensitization paradigm, we have examined some of the pharmacological components of the development and expression of sensitization. As illustrated in Figure 8, the administration of the neuroleptic haloperidol on day 1 with cocaine, which is sufficient to block the initial manifestations of cocaine-induced hyperactivity, blocks the development of cocaine sensitization, as revealed by a low-dose challenge on day 2 (Weiss *et al.*, 1989a). In contrast, if cocaine-induced hyperactivity is expressed on day 1, moderate or high doses of haloperidol

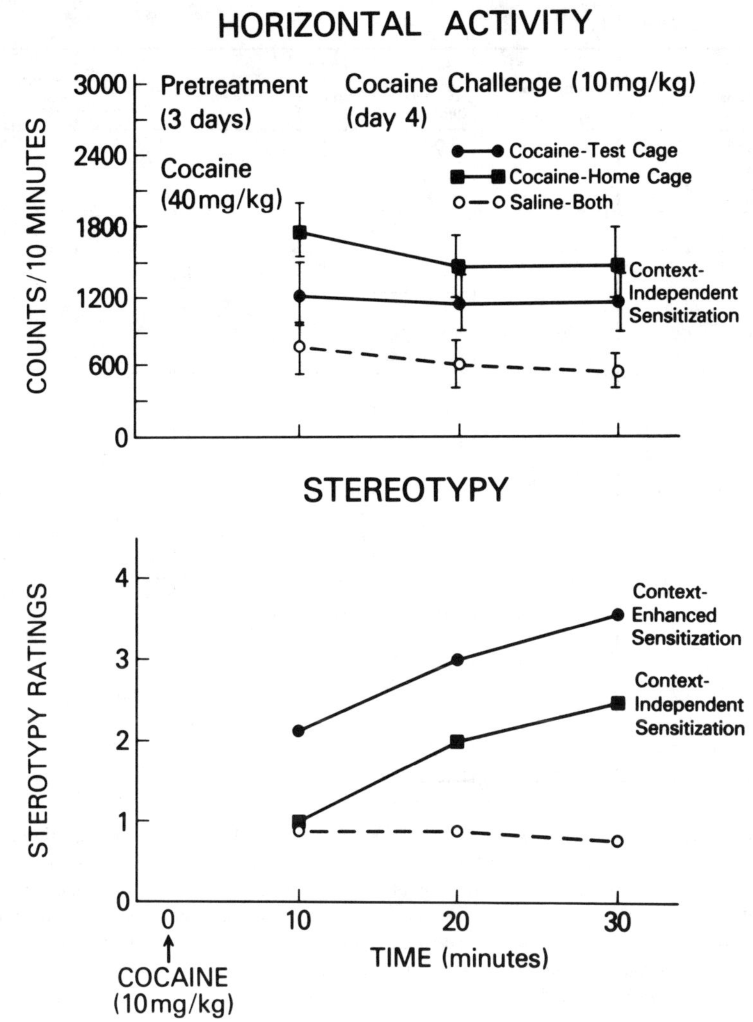

Figure 6. Rats received three once-daily injections of cocaine (40 mg/kg i.p.) in the plexiglas activity test cage, and saline in their home cage (●), saline in the test cage and cocaine in their wire home cage (■), or saline in both cages (○). On the fourth day, all three groups were challenged with low dose cocaine (10 mg/kg i.p.), and locomotor counts were recorded (top), and stereotypy rated every 10 min (bottom). Both groups of cocaine-pretreated animals showed greater stereotypy than saline controls to the low dose challenge on day 4. Those pretreated with cocaine in the home cage showed more stereotypy than saline controls, revealing a context-independent component to repeated high dose cocaine sensitization. As previously noted, those that received cocaine in the context of the test cage showed significantly greater stereotype than either of the other groups, demonstrating a context-dependent or conditioned component of the sensitization.

COCAINE		EFFECT									
Number of Injections	Dose mg/kg i.p.	Behavioral Sensitization	Duration	Activity Context Dep.	Activity Context Indep.	Stereotypy Context Dep.	Stereotypy Context Indep.	Saline Conditioning	Sensitization Neuroleptic Independent	Seizure Kindling	Death
↑↑↑	COC_{65}									++	++
x 10 days	COC_{160} subcut. (K. Gale)	++			++		++		++		
↑↑↑↑↑↑↑↑↑↑	COC_{10}	++	months	++	0			++			
↑↑↑	COC_{40}	+		0	++	++	++	±			
↑	COC_{40}	++	days	++	0	0	0	0	0		
↑ ↑ ↑	COC_{20}	0	0	0	0	0	+				
↑ ↑ ↑	COC_{10}	0	0	0	0	0	0				
↑	COC_{10}	0	0								

Figure 7. Size and number of arrows indicate dose of cocaine and a number of administrations. COC = cocaine and dose in milligrams per kilogram administered once daily i.p., except in the study of Gale, when doses were subcutaneous. 0 = no effect; ± = equivocal, + = moderate effect; + + = marked or definite effect. Effects of cocaine increase and are more persistent with dose and number of repetitions. They also become less dependent on environmental context and conditioning. The highest doses may be associated with seizures, kindling, and their associated lethality. Thus, effects shift from behavioral sensitization to kindling as the dose is increased

administered prior to the day 2 challenge are insufficient to block the expression of cocaine-induced behavioral sensitization, as illustrated in Figure 9. Similar data are illustrated in Figure 10 with a higher dose of haloperidol showing that neuroleptic blocks the development but not the expression of cocaine-induced behavioral sensitization.

These data suggest that dopaminergic mechanisms are critically involved in the induction or development phase of behavioral sensitization to the psychomotor stimulant cocaine, but once the behavior is induced the expression of sensitization does not depend on an unblocked dopamine system. Elsewhere, we have discussed the possibility that this animal model represents an interesting one for the development of neuroleptic-unresponsive behavioral and psychotic states (Post and Weiss, 1988; Weiss *et al.*, 1989a). To the extent that some components of human psy-

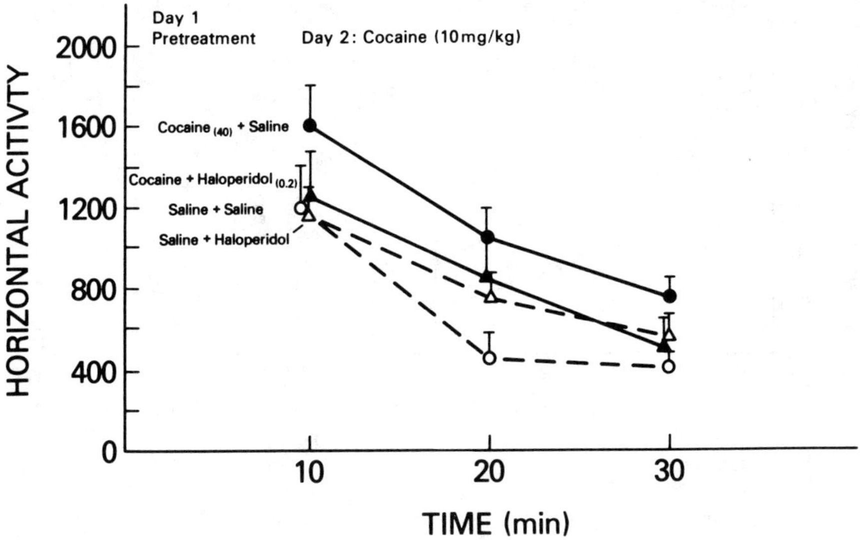

Figure 8. Haloperidol administered prior to day 1 cocaine (40 mg/kg i.p.) (▲) blocks the development of cocaine sensitization as measured by rechallenge with cocaine (10 mg/kg i.p.) on day 2; note these animals are no more active than saline-pretreated controls, but are less active than those treated with cocaine (40 mg/kg) plus saline. More marked effects are observed with ha' peridol (0.5 mg/kg; see Figure 10) which produced a more complete inhibition of cocaine-induced hypersensitivity on day one. Note, however, that the D2 antagonist sulpiride will block the development of sensitization even though it does *not* block the cocaine-induced hypersensitivity on day 1 (Fontana and Pert, unpublished observation, 1990).

chopathology such as mania and psychosis are modeled by stimulants and are related to context-dependent behavioral sensitization, the current data suggest the possibility that these may become neuroleptic refractory if the initial episodes are not pharmacologically or otherwise blocked. Wyatt and colleagues (1988) have reviewed clinical data convergent with this hypothesis indicating greater neuroleptic refractoriness in schizophrenic patients when initial episodes are untreated. While psychomotor stimulant-induced psychoses in man appear to be generally highly responsive to neuroleptics, anecdotal reports suggest that stimulant-induced psychotic behavior in some individuals may not be neuroleptic responsive.

It is of considerable interest that several manipulations which decrease the initial degrees of cocaine-induced hyperactivity are sufficient to block the development of behavioral sensitization. Thus, high doses of diazepam, which nonspecifically sedate the animal, are sufficient to block the cocaine-induced hyperactivity on day 2 (Figure 11, right panel). Moreover, it is noteworthy that, in contrast to neuroleptics, diazepam pretreatment on day 2 is also sufficient to block the expression of cocaine-induced behavioral sensitization (Figure 12). A similar phenomenon is observed with clonidine, the α_2 agonist. It, too, is capable of blocking both the development (Figure 13) and the expression (Figure 14) of cocaine-induced behavioral sensitization. However, when animals are sedated on day 1 with high doses

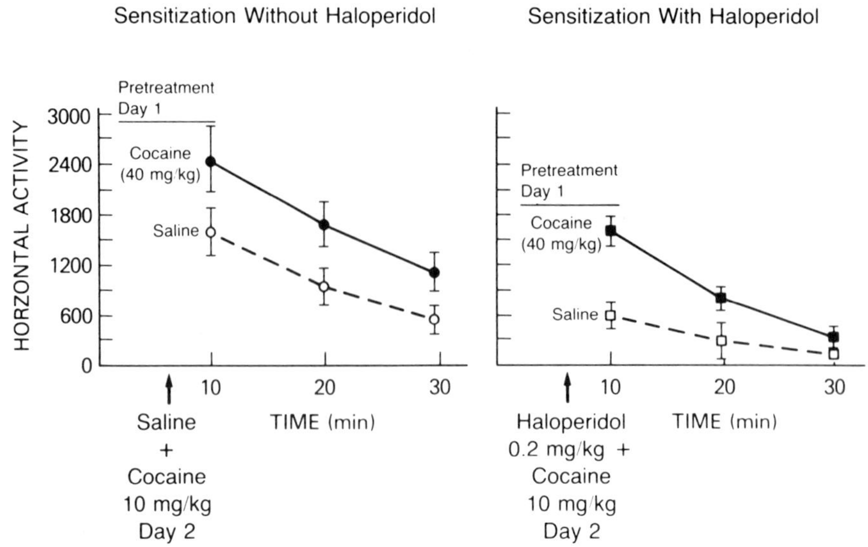

Figure 9. Haloperidol does not block the expression of sensitization to cocaine once it is induced. Group means and standard errors for horizontal activity are plotting over time in animals challenged with saline prior to cocaine (10 mg/kg) (left panel) or haloperidol (0.2 mg/kg) prior to cocaine (10 mg/kg) on day 2 (right panel). Animals that received cocaine (40 mg/kg) pretreatment on day 1 are indicated in the filled symbols; those given saline, open symbols. The left panel illustrates the magnitude of cocaine-induced sensitization without haloperidol; the right panel shows the persistent sensitization effects with haloperidol.

of the GABA agonist muscimol sufficient to block the cocaine-induced hyperactivity, behavioral sensitization is not blocked. This may, in part, be the case because GABA agonists given alone on day 1 increase cocaine-induced hyperactivity on day 2 in a fashion that is not related to the context-dependent or conditioned sensitization phenomenon.

In the 1-d paradigm, the anticonvulsant carbamazepine is neither able to block the development (when administered prior to day 1 [Figure 15]) nor the expression (when administered prior to day 2 challenge [Figure 16]) of cocaine-induced behavioral sensitization. However, as illustrated in Figure 17, while carbamazepine is unable to block the major components of behavioral sensitization related to locomotor activity, it does seem able to attenuate the most severe and long-lasting components of cocaine-induced stereotypic behavior measured in the last 30 min after injection.

Thus, pharmacological interventions in cocaine-induced behavioral sensitization appear highly similar to those involved in amphetamine-induced behavioral sensitization where neuroleptics appear capable of blocking the development but not the expression of sensitization (Tadokoro and Kuribara, 1986; Beninger and Hahn, 1983; Beninger and Herz, 1986). Diazepam and clonidine are capable of blocking both the development and the expression of behavioral sensitization, while the anticonvulsant carbamazepine blocks neither. This lack of effect on sensitization

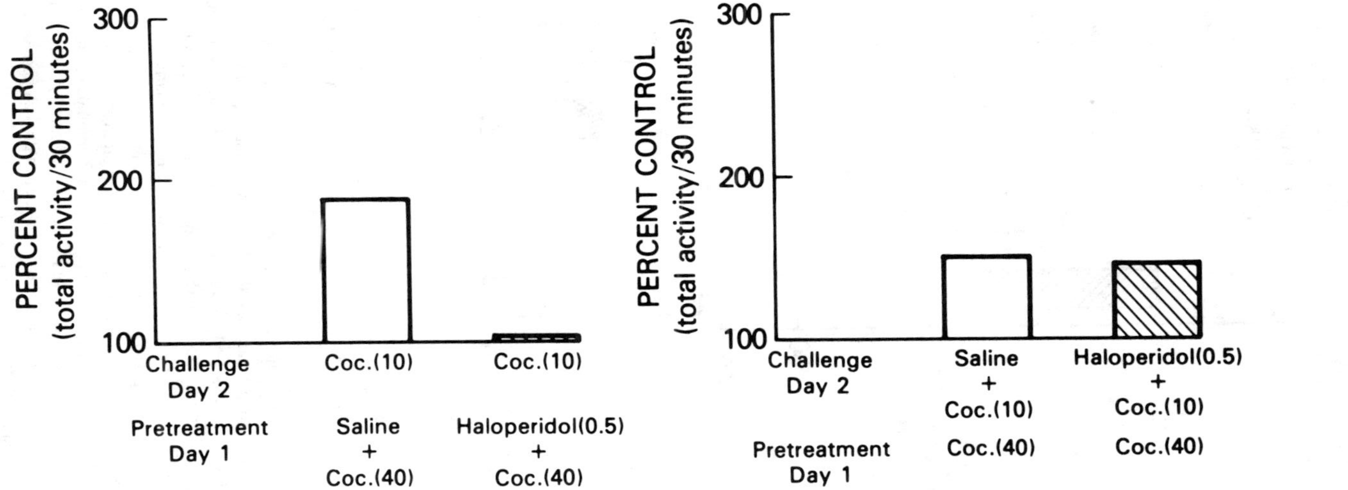

Figure 10. In the left panel, the control group (open bar) was pretreated with two saline injections on day 1 and was challenged with cocaine (10 mg/kg) on day two; the second control group (shaded bar) received haloperidol (0.5 mg/kg i.p.) and saline injections on day 1 and was challenged with cocaine (10 mg/kg) on day 2. In the right panel the control group (open bar) was pretreated with saline on day 1 and then challenged with cocaine (10 mg/kg) on day 2; the second control group (shaded bar) was treated with saline on day 1 but received haloperidol and cocaine (10 mg/kg) on day 2. These data show that even when higher doses of haloperidol (0.5 mg/kg) than those employed in Figure 9 are used, haloperidol is still unable to block the expression of sensitization (i.e., when given prior to day 2 challenge). This dose is very effective in blocking the development of sensitization when it is administered prior to the day 1 cocaine treatment, however.

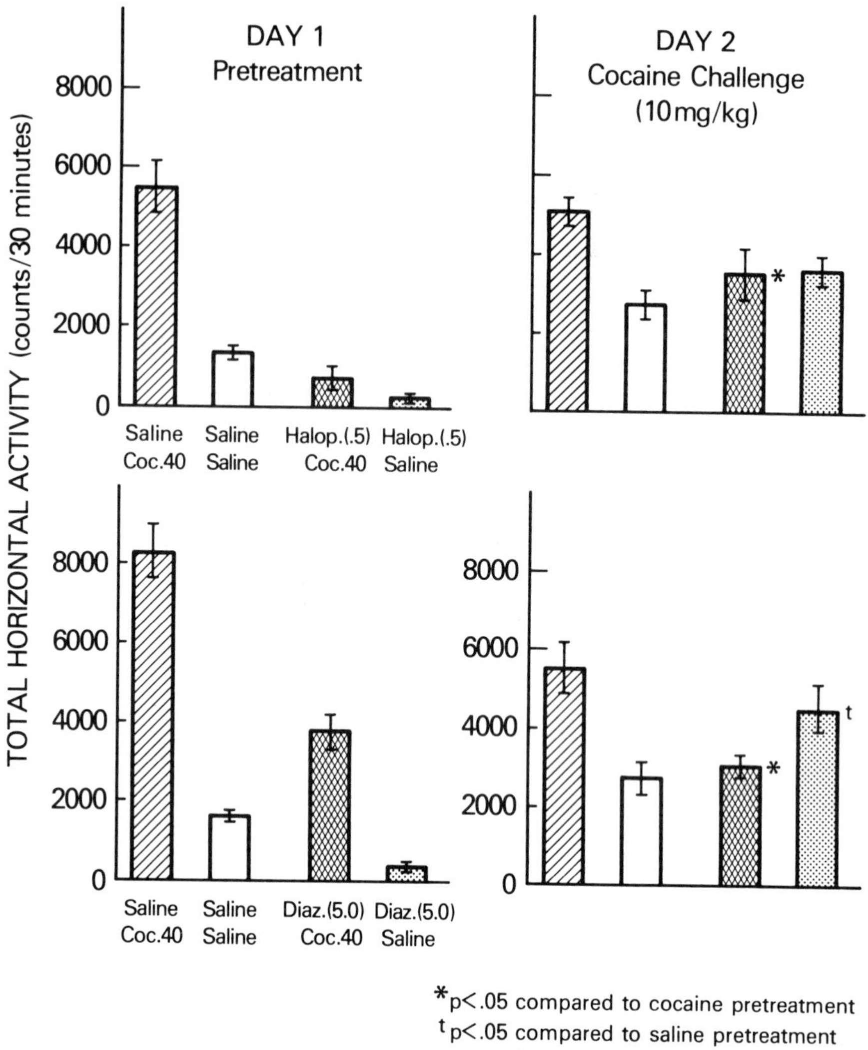

Figure 11. The effect of pharmacological agents that block cocaine-induced hyperactivity on the development of cocaine sensitization. Left side: Group mean and standard error values are illustrated for horizontal activity on the pretreatment day. Rats received cocaine (40 mg/kg) or saline following pretreatment with saline or active drug, haloperidol (top), diazepam (bottom). The haloperidol data from Figure 10 are re-presented and re-plotted for clarity and for ease in visualization of the comparison with diazepam. Right side: Total horizontal activity following a cocaine challenge (10 mg/kg) to all groups on day 2. Group means and standard errors are plotted. Haloperidol and diazepam pretreatment on day 1 blocked the development of cocaine-induced behavioral sensitization. In each case, animals pretreated with cocaine showed an increased response to the day 2 challenge dose of cocaine (column 1) compared with saline-injected controls (column 2). In contrast, those animals pretreated with haloperidol or diazepam plus cocaine on day 1 did not significantly differ from their controls following the cocaine challenge on day 2, i.e., there was no sensitization. Diazepam alone on day 1 potentiated the response to cocaine on the challenge day (last column).

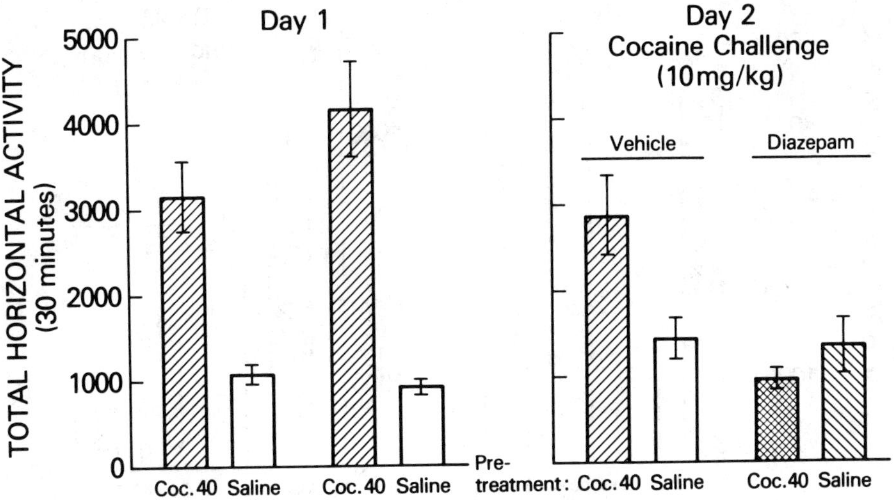

Figure 12. In contrast to the inability of haloperidol (Figure 9) as well as selective D1 and D2 antagonists (David Fontana and Agu Pert, unpublished observations, 1990) to block the expression of sensitization, diazepam (0.5 mg) pretreatment on day 2 is able to block the expression of cocaine-induced sensitization, i.e., contrast second to last column on right (cross-hatched) with first column on right (diagonal).

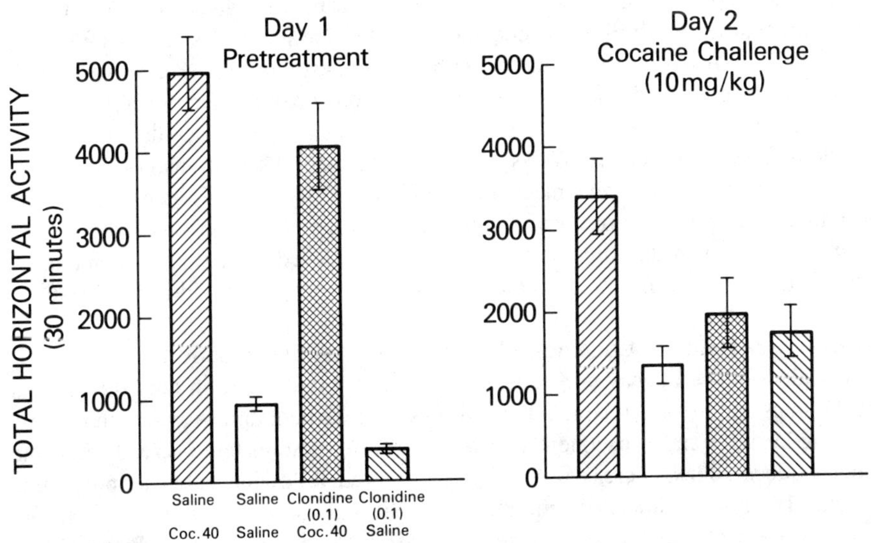

Figure 13. Pretreatment with the α_2 agonist clonidine (0.1 mg/kg) on day 1 prior to cocaine is able to block the development of cocaine-induced behavioral sensitization when all animals are challenged with a low dose of cocaine on day 2 (i.e., compare cross-hatched column on right with diagonal column).

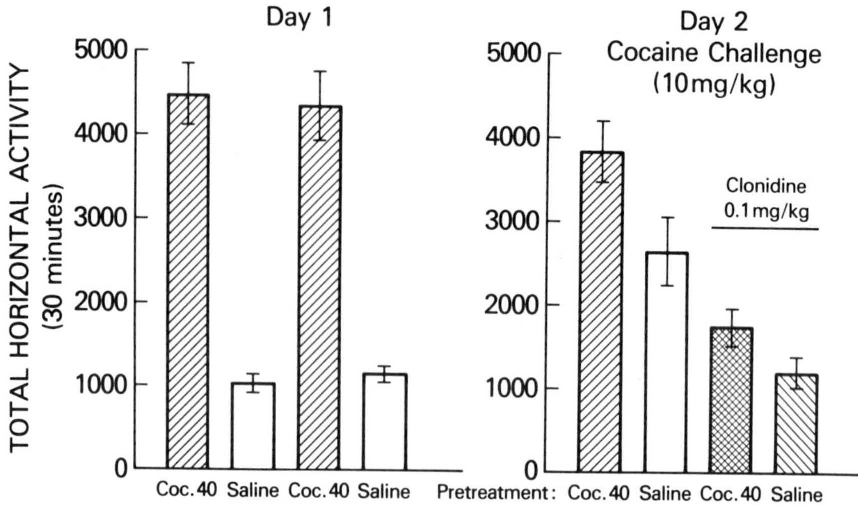

Figure 14. As in Figure 12 for diazepam, clonidine is able to block the expression of cocaine-induced behavioral sensitization when it is administered prior to the day 2 challenge dose of cocaine. Thus, clonidine (in contrast to neuroleptics) blocks both the development and expression of cocaine-induced behavioral sensitization.

is in interesting contrast to the ability of carbamazepine to block the development of pharmacological kindling of cocaine- or lidocaine-induced seizures (Weiss *et al.*, 1989b). The data are consistent with the suggestion that the development of cocaine-induced behavioral sensitization depends on dopaminergic substrates, which are blocked by neuroleptics but not carbamazepine. This phase can also be affected by clonidine and diazepam (Weiss *et al.*, unpublished observations, 1988), implicating α_2 and benzodiazepine mechanisms either directly or by virtue of their ability to inhibit initial degrees of cocaine-induced hyperactivity. Tadokoro and Kuribara (1986) have shown that blocking motor activity induced by methamphetamine by placing treated mice in a jar also blocks the development of behavioral sensitization.

Neural Substrates of Cocaine-Induced Behavioral Sensitization

Cocaine-induced behavioral sensitization appears to depend on both the nucleus accumbens and the amygdala. When animals are repeatedly pretreated with cocaine (10 mg/kg i.p.) and subsequently tested with drug administered directly into the nucleus accumbens, increased activity is observed following either amphetamine (Figure 18) or saline injection (not illustrated). In contrast, injections into the caudate nucleus are not differentially affected by prior cocaine pretreatment (Figure 19).

The effects of lesions of the nucleus accumbens also suggest the importance of this structure for the phenomena of behavioral sensitization. When animals are given 6-hydroxydopamine into the nucleus accumbens in order to produce dopamine terminal degeneration and subsequent depletion (approximately 65% of control values), initial degrees of cocaine-induced hyperactivity on day one are not atten-

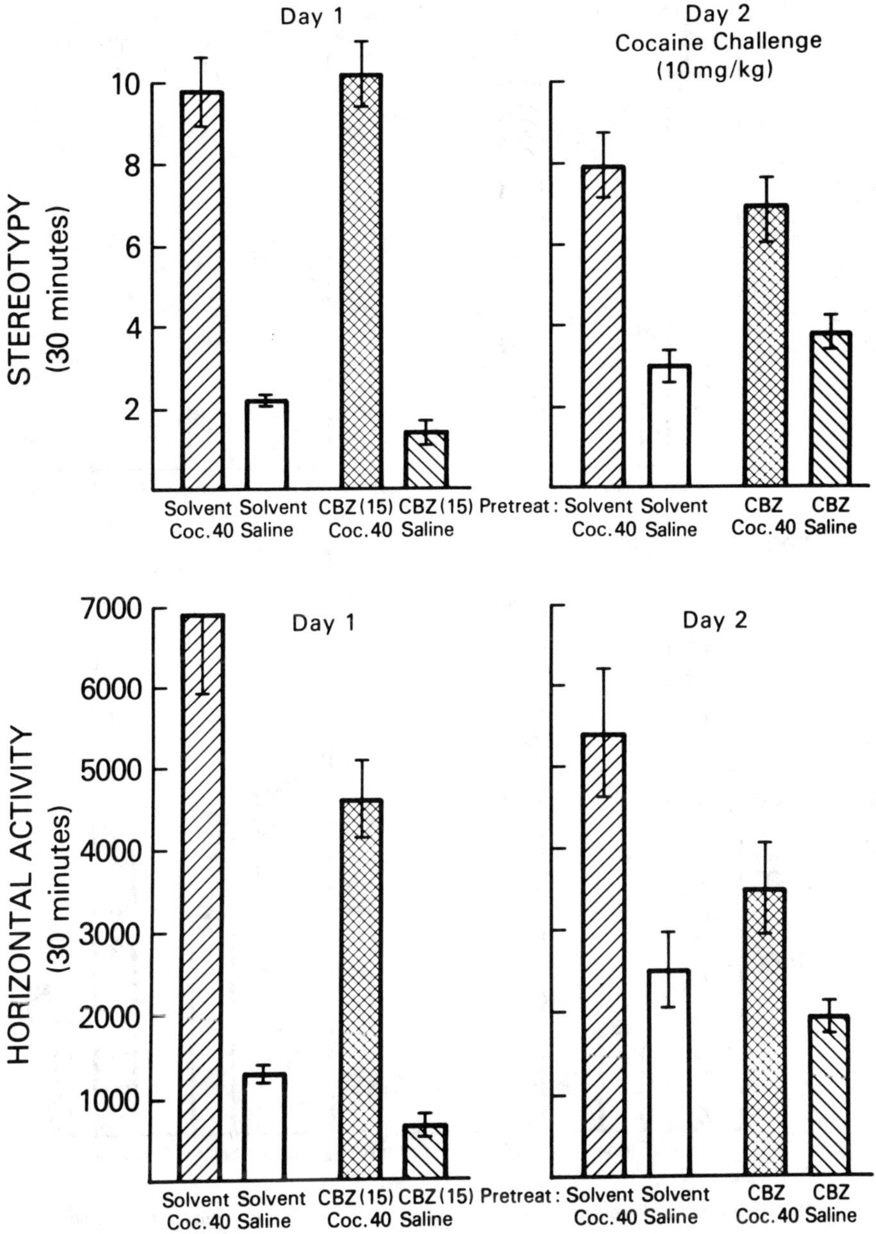

Figure 15. Carbamazepine (CBZ, 15 mg/kg i.p.), which is a highly effective anticonvulsant in blocking completed amygdala-kindled seizures at this dose, is unable to block the development of cocaine-induced behavioral sensitization. Carbamazepine was administered prior to the day 1 cocaine treatment and did not significantly inhibit either stereotypy (top, left) or horizontal motor activity (bottom, left). On the day 2 challenge with cocaine, animals pretreated with CBZ and cocaine on day 1 were more stereotypic and active than those pretreated with CBZ and saline.

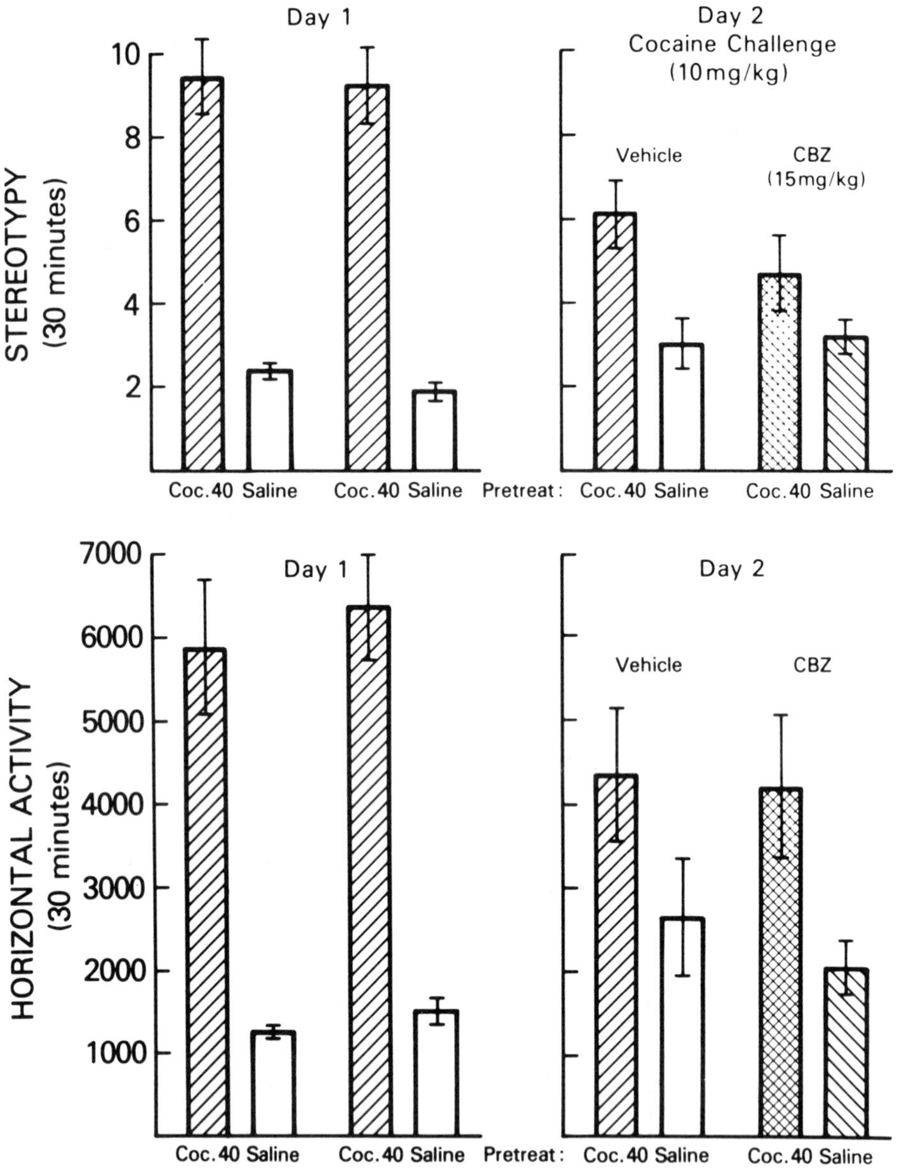

Figure 16. In contrast to diazepam (Figure 12) or clonidine (Figure 14), carbamazepine does not block the expression of cocaine-induced behavioral sensitization when administered prior to the day 2 challenge.

uated compared with sham lesion controls (Figure 20, top) Weiss *et al.*, unpublished). However, remarkably, cocaine-induced hyperactivity on day 2 in response to a low-dose cocaine challenge is attenuated (Figure 20, bottom), suggesting the importance of the nucleus accumbens in cocaine-induced behavioral sensitization.

Similarly, electrolytic lesions of the amygdala are unable to block cocaine-induced

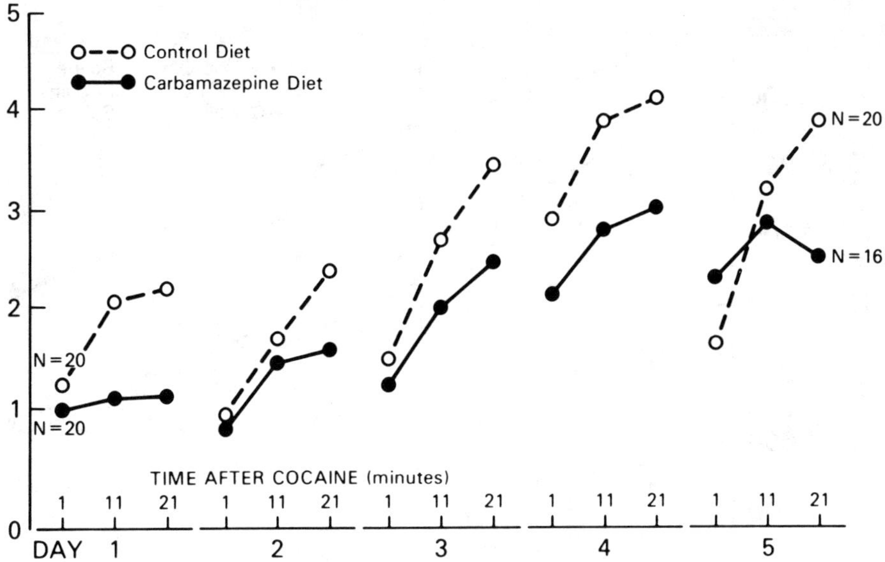

Figure 17. The group mean stereotypy response in animals receiving cocaine (40 mg/kg) is illustrated for the first five cocaine administrations. The ratings were conducted 1, 11, and 21 min after cocaine injections. Half the rats were fed a carbamazepine-containing diet and the other half were fed a control diet. Carbamazepine inhibited the maximum stereotypy response but did not inhibit the sensitization to stereotypy or its daily temporal pattern.

hyperactivity on day 1 (Figure 21, top), but also result in the blockade of cocaine-induced behavioral sensitization, as revealed by challenge on day 2 (Figure 21, bottom) (Weiss *et al.*, unpublished). In contrast, lesions of the dorsal and ventral hippocampus are unable to block either the induction of cocaine-induced hyperactivity on day 1 or the sensitization on day 2. Likewise, lesions of the cerebellum are unable to block this phenomenon. We have also replicated the effects of amygdala lesions and extended them to demonstrate that dopamine depletion in the amygdala achieved with 6-hydroxydopamine (and DMI to protect noradrenergic systems) was sufficient to block cocaine-induced behavioral sensitization.

Taken together, we believe that the nucleus accumbens substrate is critical in the mediation of the motor program associated with cocaine-induced hyperactivity and potentially in mediating its motivational and reward value. Without an intact nucleus accumbens, the animals do not show the increased responsivity upon rechallenge on day 2. The amygdala substrate also appears critical in mediating the environmental-context components of cocaine-induced behavioral sensitization. We hypothesize that the "knowledge" regarding previous exposure to cocaine in that same environment is dependent on an intact amygdala (see Gloor, 1978; Murray and Mishkin, 1985; Mishkin and Appenzeller, 1987) and the absence of this information produces responses similar to those seen in animals injected with cocaine in different environments, or in saline controls; that is, no sensitization is observed.

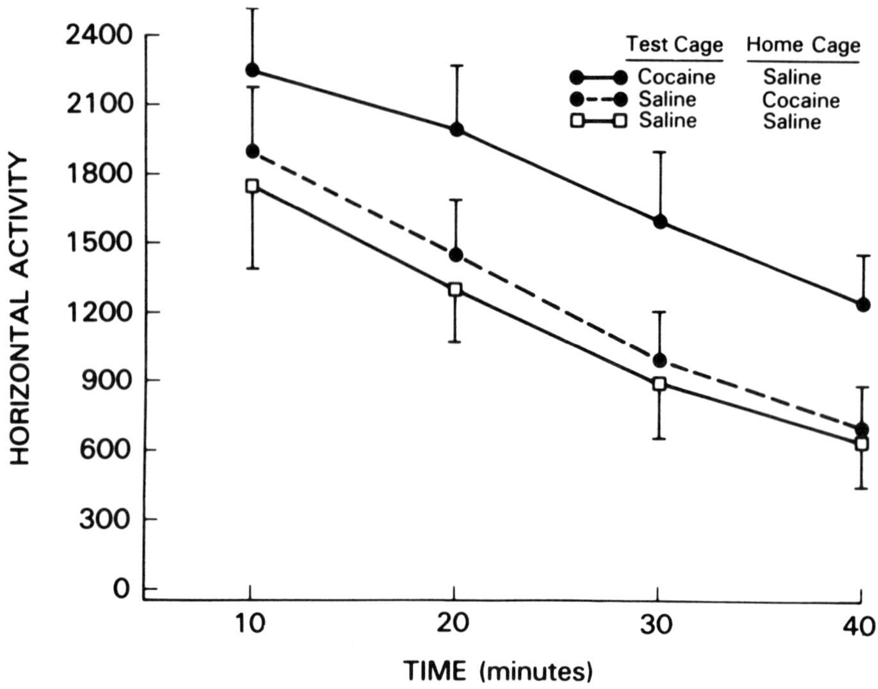

Figure 18. Three groups of rats received two daily i.p. injections of cocaine (10 mg/kg) or saline for 10 d. One group received cocaine in the test environment and saline in the home cage; a second group received saline in the test environment and cocaine in the home cage; and the third group received two saline injections. Following this chronic regimen, an acute challenge with amphetamine (10 μg) injected directly into the nucleus accumbens bilaterally revealed a significant difference in the locomotor activity exhibited by the rats that previously received cocaine in the test chamber and those that had never received cocaine. No difference between the rats that received cocaine in the home cage and those that had never received cocaine was seen.

PHARMACOLOGICAL KINDLING

Phenomenology

As illustrated in Figure 22, repeated administration of lidocaine (65 mg/kg i.p.) initially does not produce seizures. However, upon repetition, an increasing percentage of animals (usually some 40–50% in a given group) eventually develop lidocaine-induced seizures to the same dose. This induction of seizures appears to represent a true kindling phenomenon. If diazepam is co-administered with the local anesthetic, seizure induction does not occur and animals are not "sensitized" to a rechallenge with lidocaine in the drug-free state (Post *et al.*, 1984b, 1987). Moreover, there is no increase in lidocaine blood levels following repeated administration. In fact, animals showing lidocaine-induced seizures display lower blood levels of lidocaine and its active metabolite than those not seizing (Post *et al.*, 1987).

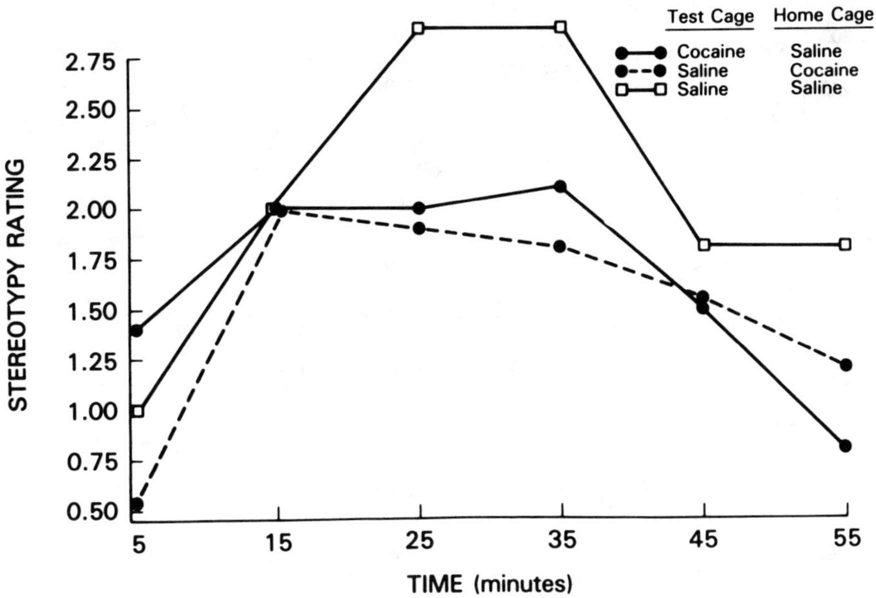

Figure 19. Prior cocaine treatment does not sensitize animal to amphetamine injected into the caudate nucleus. Three groups of rats received two daily i.p. injections of cocaine or saline for a total of 17 days. Stereotypy ratings were made using a modified version of the Creese and Iverson scale (1974) at 10-min intervals beginning 5 min after the injection and lasting for 55 min postinjection. The group mean ratings are plotted on the ordinate and time is plotted on the abscissa. One injection was given to the rats just prior to being placed in the test cage in which stereotypy was rated and a second injection occurred when the rats were returned to the home cage. One group received cocaine in the test environment and saline in the home cage; a second group received saline in the test environment and cocaine in the home cage; and the third group received two saline injections. Following this chronic regimen, an acute challenge with amphetamine injected directly into the caudate revealed no significant differences among the three groups, although the cocaine-pretreated groups (regardless of location or injection) showed slightly less stereotypic behavior.

The time course of lidocaine-induced pharmacological kindling is highly similar to that observed with amygdala kindling. Electrophysiological changes also show a certain parallelism. Repeated lidocaine administration produces amygdala spiking that bears a strong resemblance to that achieved with electrical kindling of the amygdala (Post *et al.*, 1975, 1987). In addition, the behavioral seizure intself resembles that observed with amygdala including the involvement of tonic-clonic movements of head, trunk, and forepaws with rearing and falling (Post *et al.*, 1975). However, there are some differences in that lidocaine seizures can recur in repeated episodes where amygdala-kindled seizures usually are confined to a single episode. Studies with deoxyglucose reveal prominent involvement of the amygdala, hippocampus, and perirhinal structures with lidocaine seizures (Post *et al.*, 1984b). While remarkable behavioral effects in terms of aggression are not often observed with amygdala-kindled seizures, we have found that animals having lidocaine seizures can display a syndrome of marked resistance to capture and aggressivity towards handling by human investigators (Figure 23) (Post *et al.*, 1984c).

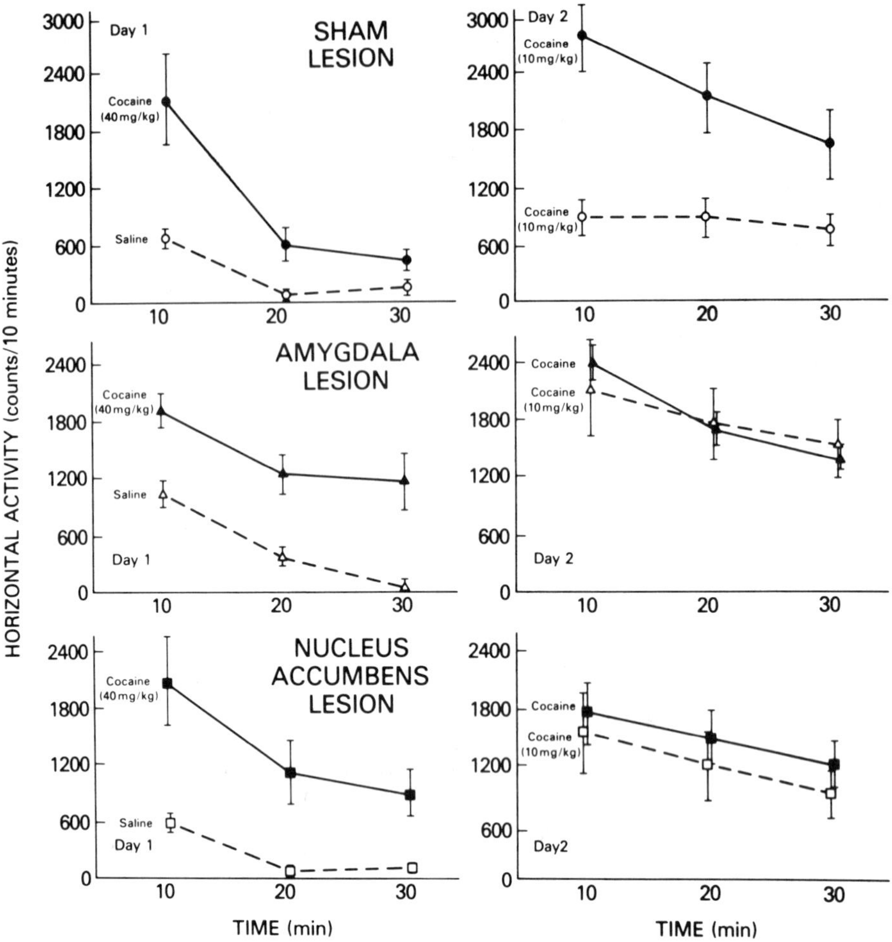

Figure 20. Amygdala and n. accumbens lesions did not significantly attenuate day 1 hypersensitivity in response to cocaine (40 mg/kg i.p.) compared with sham-lesioned animals (top, left). The lack of effect of n. accumbens lesions on cocaine hypersensitivity may be related to either the high dose of cocaine used (40 mg/kg) or the relatively small depletions of dopamine obtained. Nonetheless, following the low-dose cocaine challenge on day 2, both lesions inhibited sensitization compared with sham-lesioned animals (top, right). While the activity of the saline controls shifted upwards more than the cocaine-induced activity was inhibitied; the differential effect of cocaine was not observed as it was in hippocampal-lesioned animals where the lesion also elevated activity in both groups.

As illustrated in Figure 21, administration of cocaine (65 mg/kg i.p.) is associated with a higher incidence of seizures than the equally potent local anesthetic lidocaine; seizures occur in approximately 60% of the cocaine-treated animals on day 1. The remainder of the animals have seizures on days 2 and 3, consistent with a pharmacological kindling effect. However, there is a higher incidence of cocaine-induced lethality, as all of the animals died after the first or second episode of cocaine

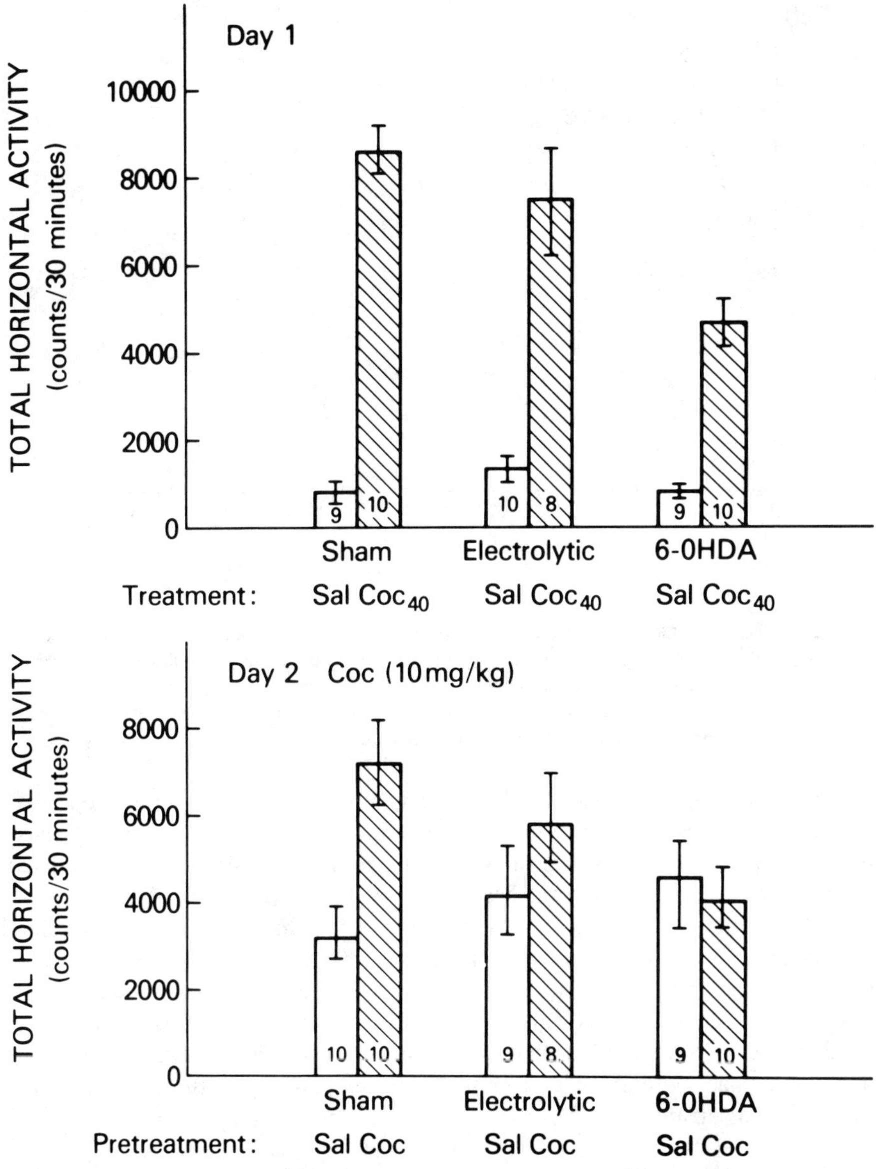

Figure 21. In a replication and extension of the data in Figure 20, either electrolytic or selective dopamine lesions of the amygdala (achieved with 6-hydroxydopamine with desmethylimipramine pretreatment) were observed to block cocaine sensitization as observed by the lack of effect on day two in either group (middle and right, bottom) compared with sham-lesioned animals (bottom, left).

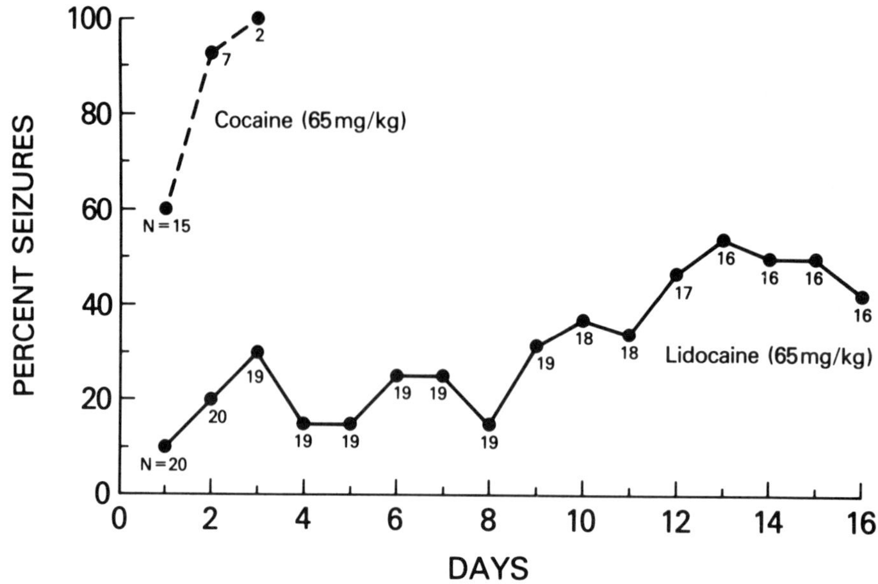

Figure 22. The percentage of rats experiencing lidocaine- or cocaine-induced seizures is shown over days. Rats received a single i.p. injection (65 mg/kg) each day. The study began with 15 or 20 subjects per group, and the number remaining is shown below each day's percentage. Despite the equal local anesthetic potency of these drugs, cocaine is much more convulsant than lidocaine on day one, animals kindle faster to it, and are more likely to die during a seizure.

seizures in this study, in contrast to the general ability of animals to survive repeated lidocaine-induced seizures. This lethality observed with the cocaine seizures appears to be associated with a respiratory failure as opposed to a cardiac arrest, as animals continue to have sustained rapid heart beats. When lower doses of cocaine are utilized, a profile of pharmacological kindling more like that observed with lidocaine (65 mg/kg i.p.) can be observed. For example, when animals are administered cocaine (50 mg/kg i.p.), fewer animals have seizures on day 1 and, when 40 mg/kg i.p.is utilized, none of the animals have seizures on day 1, although 80% have seizures by day 15 (Figure 24). In these instances, cocaine-induced seizures remain highly lethal and more animals succumb to cocaine-induced seizures than to lidocaine-induced seizures.

The Pharmacology of Local-Anesthetic-Induced Kindling

Animals do not develop lidocaine-induced kindled seizures when they are pretreated with diazepam (Figure 25). This parallels the blockade of the development of amygdala-kindled seizures achieved by repeated electrical stimulation with pharmacological blockade with diazepam (Table 4). In electrical kindling of the amygdala, diazepam is capable of blocking both the developmental phase of the seizures as well as blocking the completed or fully developed kindled seizure (Schmutz *et al.*, 1988). In contrast, the anticonvulsant carbamazepine is insufficient to block

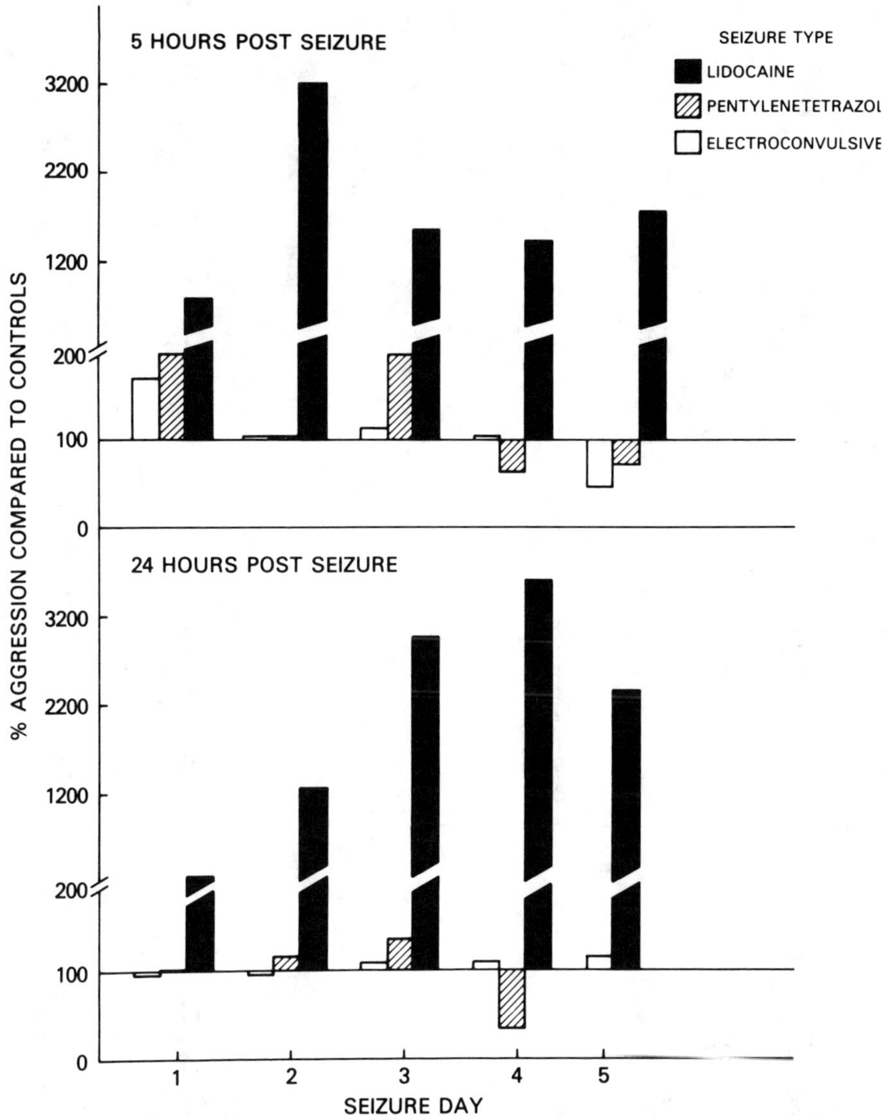

Figure 23. Albert and Richmond scale ratings of aggression were markedly increased compared with controls following repeated once-daily seizures induced by lidocaine (65 mg/kg, i.p.), but not by pentylenetetrazol (80 mg/kg, i.p.), or by electroconvulsive shock (120–140 V for 1 s administered by ear clip electrodes vs. sham controls). These increases were observed both 5 and 24 h following lidocaine, but not the other seizure types.

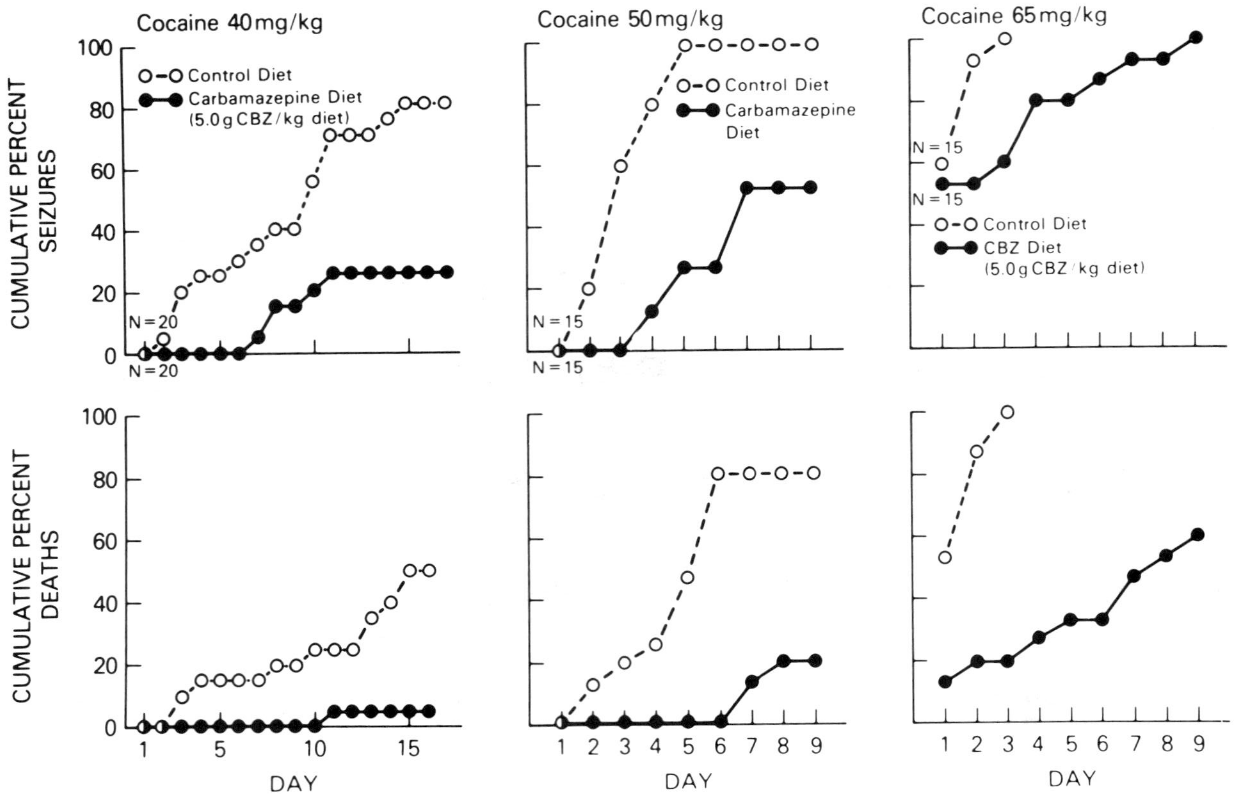

Figure 24. Rats were placed on a control or carbamazepine-containing diet for 4 d prior to the studies and were kept on these diets throughout the period of cocaine injections. Cocaine was administered in doses of 40, 50, or 65 mg/kg. Carbamazepine was effective in inhibiting the cocaine-induced seizures and lethality.

the development of amygdala-kindled seizures electrically (Weiss and Post, 1987), although it is the most potent anticonvulsant for blocking completed amygdala-kindled seizures (Albright and Burnham, 1980).

In complete contrast to its effects on amygdala-kindled seizures, carbamazepine is able to block the development of local-anesthetic-induced seizures but is without effect on completed (Post *et al.*, 1984b) or fully developed (Weiss *et al.*, 1987, 1989b) lidocaine-kindled seizures or acute high-dose cocaine seizures. Thus, there appears to be a double dissociation between the effects of carbamazepine on seizure development and on completed seizures achieved by electrical stimulation of the amygdala as compared to those achieved by the local anesthetics lidocaine or cocaine (Table 4).

Chronic administration of carbamazepine in the diet of animals is able to block the development of cocaine-induced seizures (Figure 24). However, remarkably, repeated administration of carbamazepine (15 mg/kg i.p.) prior to each local anesthetic injection is unable to block the development of pharmacological kindling. This same dose and time frame of injection (15 min prior to seizure induction) is very effective in inhibiting fully developed amygdala-kindling seizures. Moreover, we have been unable to find any dose of carbamazepine, from 10–50 mg/kg i.p., which is sufficient to block a fully developed lidocaine seizure or a high-dose, acute cocaine seizure (Figure 26). Thus, it would appear that chronic carbamazepine administered orally, achieving steady concentrations in the low therapeutic range (but not intermittent carbamazepine injected i.p. achieving equal or higher concentrations of drug), can block the development of local-anesthetic-induced pharmacological kindling (Figure 27), but not the late or completed phases of the response.

These data, taken with those observed in amygdala kindling, illustrate the proposition that different phases of kindling are associated with different pharmacological responsivities (Post *et al.*, 1986a; Schmutz *et al.*, 1988; Mace and Burnham, 1987). They also highlight the notion that pharmacological kindling with local anesthetics has a biochemistry different from electrical kindling of the amygdala (Table 4). In addition, these data also emphasize the difference in pharmacological responsivity between cocaine-induced pharmacological kindling and cocaine-induced behavioral sensitization. As illustrated in Figure 15, carbamazepine is not able to block the development of cocaine-induced behavioral sensitization. (Carbamazepine is similarly ineffective when given in the diet — data not shown). In contrast, it is highly effective in blocking the pharmacological kindling effect, at least when administered chronically in the diet (Figure 24).

Calogero *et al.*, (1988) and Rivier and Vale (1987) have recently reported that cocaine releases the neuropeptide corticotropin-releasing factor (CRH). Local-anesthetic-induced release of CRH has been blocked by the administration of carbamazepine (Calogero et al., 1987). Intracerebroventricular administration of CRH itself can induce the late onset of limbic-like seizures (Figure 28) (Ehlers *et al.*, 1983; Weiss *et al.*, 1986). Thus, it remains a possibility that carbamazepine was interfering with pharmacological kindling produced by local anesthetics through its effects on blockade of CRH. We attempted to test this hypothesis by reversing the

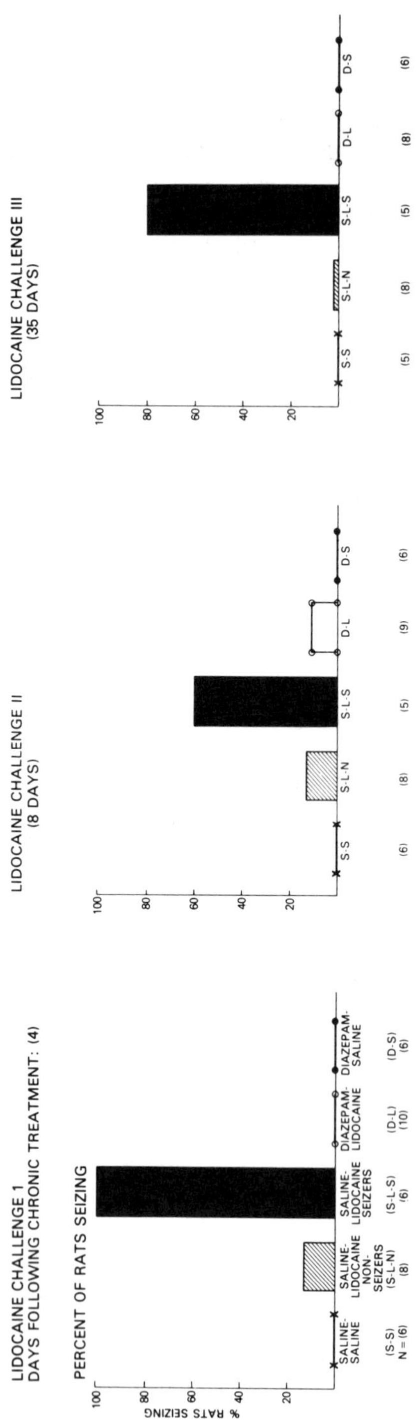
LIDOCAINE CHALLENGE 1
DAYS FOLLOWING CHRONIC TREATMENT: (4)
PERCENT OF RATS SEIZING
% RATS SEIZING
100
80
60
40
20
SALINE-SALINE
SALINE-LIDOCAINE NON-SEIZERS
SALINE-LIDOCAINE SEIZERS
DIAZEPAM-LIDOCAINE
DIAZEPAM-SALINE
(S-S) N = (6)
(S-L-N) (8)
(S-L-S) (6)
(D-L) (10)
(D-S) (6)
LIDOCAINE CHALLENGE II
(8 DAYS)
S-S
S-L-N
S-L-S
D-L
D-S
(6)
(8)
(5)
(9)
(6)
LIDOCAINE CHALLENGE III
(35 DAYS)
S-S
S-L-N
S-L-S
D-L
D-S
(5)
(8)
(5)
(8)
(6)

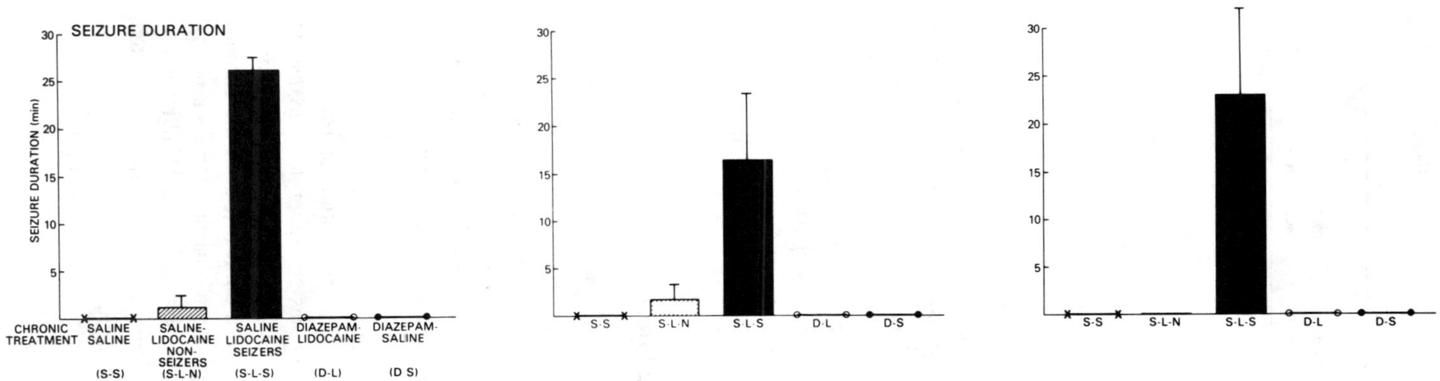

Figure 25. This figure illustrates the effect of an acute lidocaine challenge 4, 8, and 35 d following or four weeks (five times per week) with once-daily injections of lidocaine (65 mg/kg, i.p.) or saline with or without diazepam (1 mg/kg). None of the rats that received diazepam co-treatment developed lidocaine seizures, while 6 of the 14 rats that received saline and lidocaine developed seizures during the chronic treatment phase of the experiment. The upper portion of the figure shows the percentage of rats in each group that seized and the lower portion of the figure shows the seizure duration. Following acute administration of lidocaine alone, seizures were seen only in the lidocaine-pretreated animals that did not receive diazepam co-treatment. This effect persisted when rats were rechallenged with lidocaine alone 8 and 35 d later (Pitem *et al.*, unpublished data). These data indicate that lidocaine-induced seizure sensitization required repeated physiological activation by lidocaine and does not occur with lidocaine plus diazepam (i.e., when the proconvulsant effects are blocked). These findings and the persistent seizure sensitization 35 d after chronic lidocaine administration was terminated support the interpretation that lidocaine is producing a bona fide pharmacological kindling effect.

Table 3
Effects of Drugs and Lesions on Development and Expression of Cocaine-Induced Behavioral Sensitization

	Blocks day 1 activity	Blocks development of sensitization	Blocks expression of sensitization
Neuroleptics	+ +	+ +	0
Diazepam	+ +	+ +	+ +
Clonidine	0	+ +	+ +
Carbamazepine	0	0	0
Lithium	0	+ +	?
Nucleus accumbens lesions	0	(+ +)	?
Amygdala lesions	0	(+ +)	?
Hippocampus lesions	—	0	0

effects of carbamazepine's blockade of local anesthetic kindling by co-administering CRH i.c.v., either 10 or 100 μg (Weiss *et al.*, 1991, in press). While CRH administration did appear to reverse the effects of carbamazepine on cocaine-induced kindled seizures, CRH itself also potentiated cocaine-induced seizures. These data leave ambiguous whether carbamazepine's effects were mediated by inhibition of CRH release and other studies are required. However, it is of considerable interest that an endogenous peptide potentially released during stress may enhance convulsive responsivity to cocaine (and its subsequent lethality) in pharmacological, if not physiological, doses.

Implications of Carbamazepine's Antikindling Effects: Possible Blockade of Kindling and Panic

Clearly, the mechanism by which carbamazepine blocks the development of local-anesthetic-induced seizures requires further exploration. However, from the clinical perspective, it would appear that chronic treatment with carbamazepine is required to block local-anesthetic-induced pharmacological kindling of seizures. However, it should be emphasized that, to the extent that the data in the experimental animals are relevant to observations in man, even very high acute doses of carbamazepine are insufficient to block acute high-dose cocaine seizures and may actually increase the associated lethality (Figure 26). Therefore, there would appear little reason to think that carbamazepine could acutely play a role in prevention or reversal of acute, cocaine-related toxicity. Rather, benzodiazepines and other anticonvulsants would appear more appropriate for acute seizures. Nonetheless, chronic oral administration of carbamazepine would appear to deserve a clinical trial in the chronic cocaine user as a potential inhibitor of the development of pharmacological kindling and some of its associated phenomena. Our data (Weiss *et al.*, 1987, 1989b [Figure 24]) suggest it would block the development of cocaine seizures and their associated lethality. Such a trial is also indicated, given the preliminary data that carbamazepine may change threshold for seizures in man and block cocaine craving (Halikas *et al.*, 1988), as postulated on the basis of our preclinical data.

Table 4
Pharmacological Responsivity as a Function of Stage of Development of Kindling

	Development	Completed	Spontaneous
Seizure Stage / Stimulation			

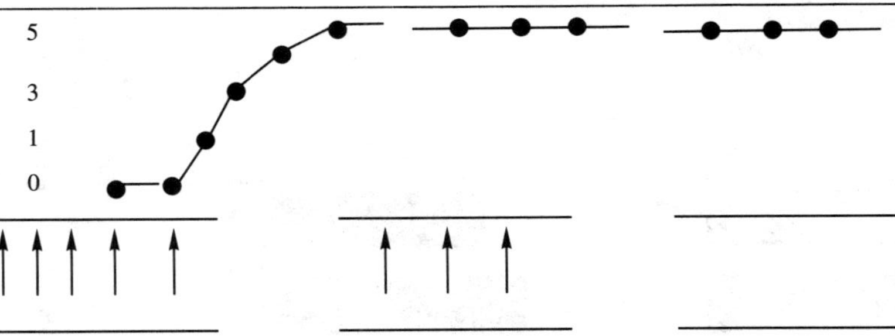

	Development	Completed	Spontaneous
Amygdala Kindling			
Diazepam	+ +	+ +	0
Phenytoin	0	±	+ +
Carbamazepine	0	+ +	?
Clonidine	+ +	0	
Atropine	(+ +)	0	
ECS	+ +	+ +	
Valproate	+ +	+ +	
Opiates	0	0	
MK 801[a]	+ +	±	
Local Anesthetic Kindling			
Diazepam	+ +	+ +	
Carbamazepine (chronic)	+ +	0	
Carbamazepine (intermittent)	0		

[a] McNamara, 1988; see Racine, 1978 and Post *et al.*, 1984 and 1986 for additional references.

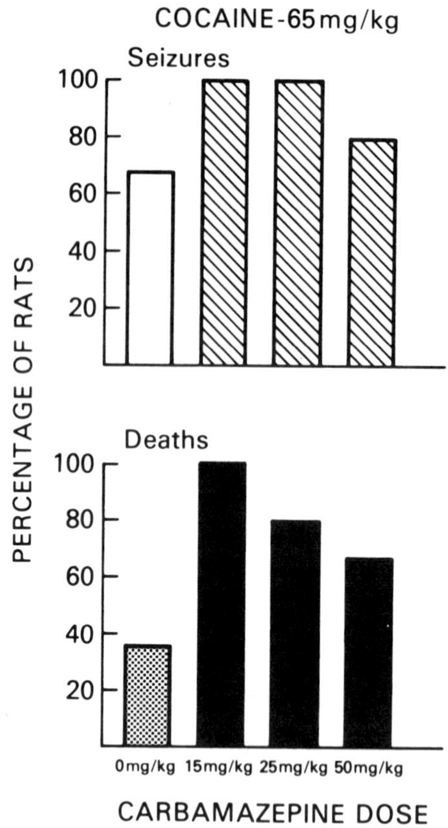

Figure 26. Carbamazepine was administered i.p. 15 min before cocaine, in doses of 15, 25, or 50 mg/kg in an attempt to block acute high-dose (65 mg/kg) cocaine seizures and lethality. No anticonvulsant effect was seen at any of these doses. Cocaine-induced lethality (bottom) either was not affected or was exacerbated by these acute doses.

Among the panoply of behavioral and physiological effects of cocaine that has been reported following chronic use of this agent, there appears to be a high incidence of panic attacks. Washton and Gold (1984) reported a 50% incidence of panic attacks in the first 500 callers to the Cocaine Hotline although other investigators have noted a lower incidence (17%), even among heavy users (Nunes and Quitkin, 1989). Elsewhere, based on the observations of Uhde and associates, we have suggested that cocaine-induced panic attacks may develop in a time course that parallels that of pharmacological kindling; that is, repeated cocaine-induced panic attacks may begin to emerge and increase in regularity following long periods of repeated, intermittent drug use (Post *et al.*, 1987, 1986b). With sufficient repetition of these panic attacks, a phase (perhaps akin to the spontaneity in electrical kindling) occurs when panic attacks occur spontaneously, i.e., without administration of cocaine (Figure 29). Preliminary data in our case (Post *et al.*, 1987) and

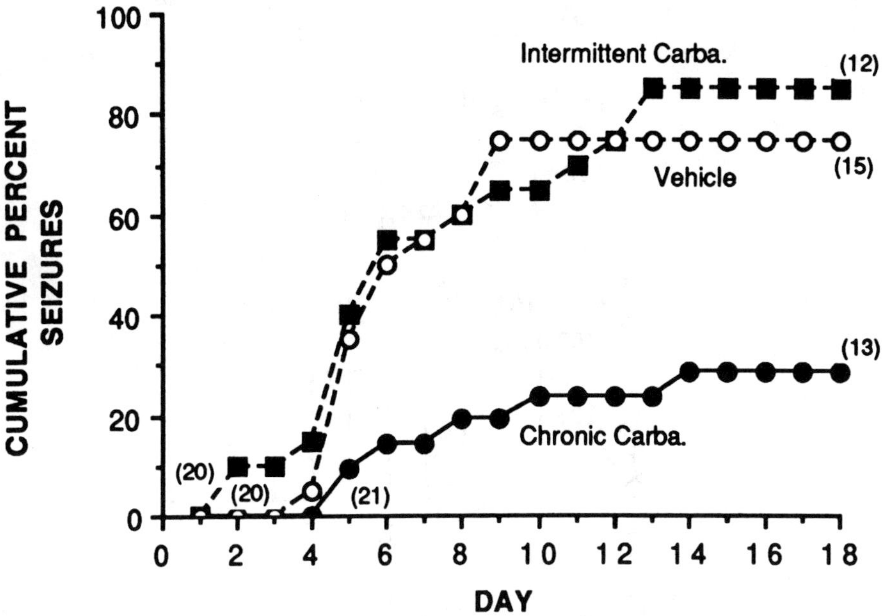

Figure 27. Pretreatment with chronic carbamazepine (carba.) in diet (not repeated injection) is required to block lidocaine-kindled seizures. Rats were treated with chronic carbamazepine (diet), i.p. injected carbamazepine (15 mg/kg), or vehicle for 4 d prior to lidocaine injections and were continued on this regimen throughout the study. The repeated i.p. carbamazepine had no effect on the development of lidocaine-kindled seizures, whereas chronic carbamazepine was highly effective.

from Louie *et al.* (1989) suggest that the anticonvulsants carbamazepine and clonazepam are effective in cocaine-induced panic attacks and that these patients may not be as responsive to the traditional antipanic treatments with tricyclic antidepressants.

It is of further interest that carbamazepine does not block the stimulant-induced euphoria achieved by methylphenidate in man (Meyendorff *et al.*, 1985). This is consistent with its inability to block acute stimulant-induced hyperactivity (Post *et al.*, 1984a,c; Koella *et al.*, 1976) and stimulant-induced behavioral sensitization (Figure 15), except for an effect on maximum stereotypy ratings. To the extent that cocaine-induced panic attacks are related to a pharmacological kindling-like effect as postulated, we would predict that carbamazepine would be able to block the development of cocaine-induced panic as well as cocaine-induced seizures, but not cocaine-induced euphoria.

Clearly, these direct clinical predictions based on extrapolation from preclinical studies in rodents require careful and systematic clinical exploration and testing. Nonetheless, it is apparent that there is an epidemic of cocaine use and cocaine-related deaths, many of which appear related to the occurrence of seizures immediately preceding. Thus, cocaine is being administered by human users in doses that are acutely convulsant or epileptogenic. Seventeen percent of the first 500

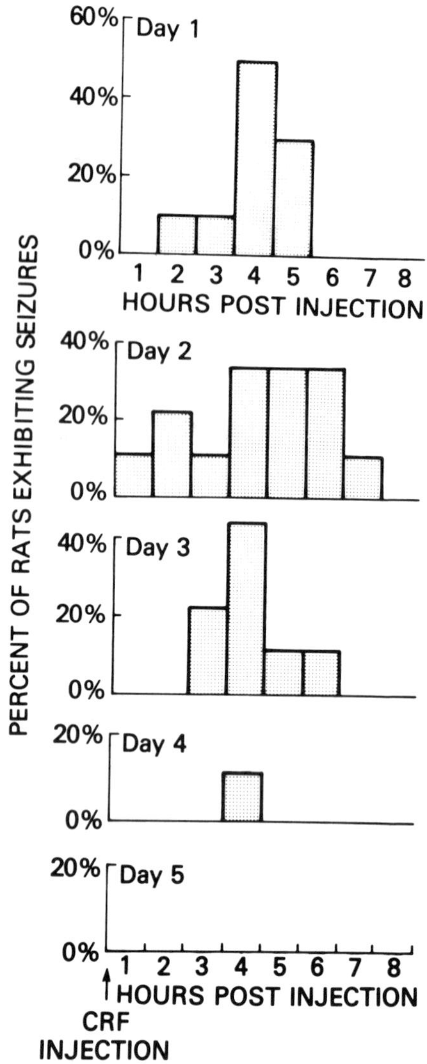

Figure 28. The temporal distribution of seizures on each day of CRF injection is illustrated. On the first day, 60% of the CRF-injected rats showed seizures, and most of these occurred following a delay of 4 h. Although seizure latencies were shorter for some animals on the second day, a 4–6 h delay was still seen in many animals. By the 4th d, only one animal experienced a seizure, and, on the 5th d, no rats had seizures, apparently representing tolerance development to the seizure-inducing effects of daily i.c.v. CRF.

callers to the Cocaine Hotline reported seizures (Washton and Gold, 1984). Obviously, it would be very difficult to ascertain whether, in any of these instances, a pharmacological kindling effect as opposed to an acute overdose would account for the one time seizure induction. Nonetheless, the patterns of cocaine use, including its intermittency, appears ideally suited to produce a pharmacological kin-

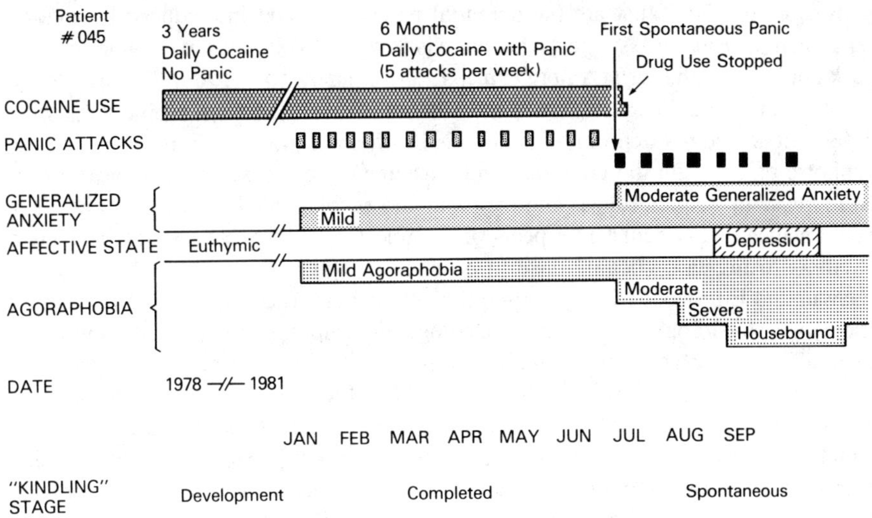

Figure 29. Development of panic disorder following cocaine use. The course of development of panic disorder for an individual patient who used cocaine is illustrated. Following 3 years of cocaine use, essentially daily by the intranasal route, this patient began to develop panic attacks immediately after cocaine use. Panic attacks occurred approximately three times per week for the next 6 months. Following these many repetitions of cocaine-induced panic attacks, he developed the first spontaneous panic attack, at which time the patient discontinued his cocaine use and developed more severe psychiatric symptoms, i.e., generalized anxiety and agoraphobia.

dling-like effect. Since the phenomenon of pharmacological kindling (i.e., the emergence of seizures to repetition of a previously subconvulsant dose) has now been reported in a variety of species, including rodents (Post and Kopanda, 1976), cats (Post *et al.*, unpublished observation), dogs (Tatum and Seevers, 1929), and nonhuman primates (Post *et al.*, 1976, Post, 1977; but see Matsuzaki *et al.* [1976] for contrary data), it is not inconceivable that a similar phenonemon could also occur in man.

Given the clear-cut demonstration of pharmacological kindling to a seizure end point across a variety of species, it would appear that human cocaine users should be warned of such a possibility, particularly given the likelihood of escalation to high doses. It would also appear important to warn individuals who survive a cocaine-induced seizure that they are at grave risk for reexperiencing another such episode, should they ever re-administer cocaine.

FURTHER CLINICAL IMPLICATIONS OF COCAINE-INDUCED SENSITIZATION AND KINDLING

Given the elucidation of the phenomena of cocaine-induced behavioral sensitization and pharmacological kindling in animal models, we are in a position to ask whether similar phenomena may occur in human users and bear on the toxicological profile of cocaine. What are the potential parallels of cocaine-induced behavioral sensitization? Evidence suggests that there is acute tachyphylaxis to cocaine-induced peak euphoria or the rush (Ambre *et al.*, 1988). Sherer *et al.* (1988), Sherer (1988), and Kumor and colleagues (1988) have demonstrated that even maintaining high doses of cocaine with chronic infusions is insufficient to maintain the rush or peak euphoric effects. On the contrary, with sustained cocaine administration, there is an increase in suspiciousness and paranoia and accompanied dysphoria rather than maintenance of the euphoric aspects of cocaine. Therefore, the cocaine user often attempts to increase cocaine dose in order to recapture the peak euphoric effects (Gawin and Ellinwood, 1988). While this attempt is largely unsuccessful, because it appears physiologically and pharmacologically impossible even with high doses maintained under controlled conditions (see above), it may propel the user into high enough doses of cocaine to achieve the dual liability of cocaine as a stimulant and a local anesthetic. High doses of psychomotor stimulants, such as amphetamine, which are not potent local anesthetics in themselves, appear capable of inducing marked behavioral aberrations including manic-like behavior and the development of paranoid psychoses (Post, 1975; Ellinwood and Kilbey, 1977; Bell, 1975; Angrist *et al.*, 1975; Griffith *et al.*, 1970, 1972; Connell, 1958). There is increasing evidence that cocaine can likewise produce similar psychotic states (Post, 1975; Manschreck *et al.*, 1988; Jeri, 1978; Gawin and Ellinwood, 1988; Ellinwood and Kilbey, 1977). In addition, these high doses also may make cocaine's local anesthetic effects pharmacologically relevant. With repeated high-dose use of cocaine, as high as 5–10 g in a night's binge, doses similar to those given to animals are achieved, with the possibility of pharmacological kindling and its associated lethality. In addition, to the extent that cocaine-induced panic attacks follow a pattern of kindling-like progression, the repeated user may be at continually greater risk for the emergence of this subconvulsant phenomenon. Baxter *et al.* (1988) have recently described a case of cocaine-induced panic with amnesia in which PET scans with deoxyglucose revealed a "hot spot" in the temporal lobes. In addition, Merriam *et al.* (1988) have also described an instance of complex partial seizure status induced by cocaine. Stevens *et al.* (1969) have likewise described an epileptic patient with depth electrodes implanted who inhaled cocaine (100 mg) intranasally and spiking in the amygdala was observed in association with marked increases in activation and euphoria. In this instance, spikes in the absence of impairment of consciousness would be equivalent to simple partial seizures. Thus, it appears that subictal or ictal activation of limbic substrates, similar to that observed in animals, can occur in man.

In this fashion, considering cocaine's local-anesthetic properties and the potential for kindling-like phenomena might help explain the wide array of behavioral and physiological toxicities that are increasingly being recognized as a consequence of repeated high-dose cocaine administration (Table 5). Not only are there phenomena typical of psychomotor stimulants such as amphetamine, including depression, dysphoria, and paranoid psychosis, but also a variety of relatively novel phenomena. These include psychosensory illusions and hallucinations (Siegel, 1984; Louie *et al.*, 1989), marked cognitive impairment, the induction of panic attacks, as well

Table 5
Cocaine Effects in Man:
Schematic Relationship to Dose and Duration

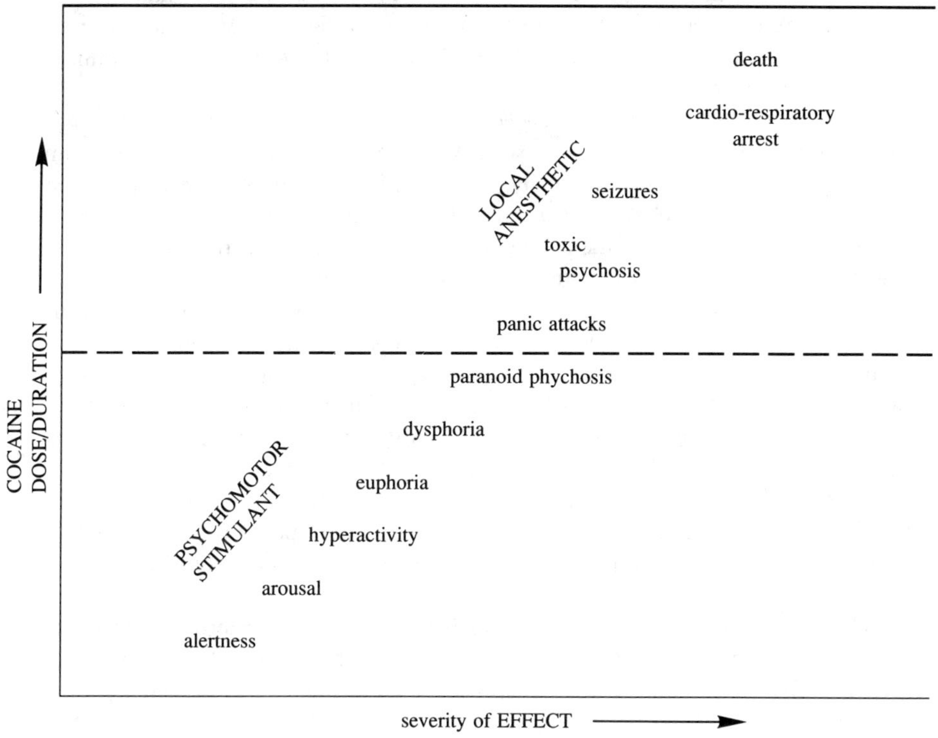

Table 6
Hidden Liabilities of Cocaine

The transition from a benign ''high'' to craving, dysphoria, paranoia, panic, seizures, and lethality

1. Tolerance to euphoria → dose escalation
2. Sensitization to activating, dysphoric effects (stimulant): even same dose → progressively worse effects
3. Seizures may develop following repeated doses (local anesthetic): same safe dose one day, may be lethal the next
4. Panic attacks may develop after repeated use
5. Intense craving: animals will self-administer cocaine until they die
6. Cocaine in young, healthy people is also associated with heart attacks, arrhythmias, stroke
7. Worse craving and side effects as function of route: oral < nasal < i.v. < inhalation (crack)
8. Craving and addiction poorly treatable by current medical and psychiatric methods
9. Costs: money, friends, family; physical and psychological health
10. Unpredictable: who will be able to avoid addiction

ONLY SOLUTION: DON'T PLAY THE GAME, DON'T START

as increases in aggression and violence. In regard to aggression, it is noteworthy that repeated lidocaine administration, when it is sufficient to produce seizures in rodents, is associated with moderate to marked resistance to capture and, in some instances, remarkably vicious attacks (Post *et al.*, 1984b). In very rare instances animals (housed alone) have sprung from their cage at investigators attempting to capture them.

The extent to which cocaine-induced aggression is driven by psychomotor stimulant-induced phenomena and its associated paranoia, or the local-anesthetic component (Post *et al.*, 1984) of cocaine's actions, remains to be further elucidated. Nevertheless, the preclinical data suggest that both pharmacological actions of cocaine could contribute to the phenomena making it a particularly dangerous drug for inducing aggressive proclivities, perhaps in individuals already so predisposed. The local anesthetic component of cocaine could more specifically account for the reports of cognitive dysfunction and difficulty concentrating that are increasingly being reported by cocaine users. It is of interest in relationship to cognitive changes and panic attacks that Baxter *et al.* (1988) reported that cocaine use was associated with an area of increased glucose utilization on PET scan in a human with memory and panic difficulties. This would be further understandable from the perspective that the local anesthetic seizure (at least in the rat) appears to be highly selective for limbic structures, especially amygdala and hippocampus (Post *et al.*, 1984b), which have been closely linked to alterations in learning and memory (Mishkin and Appenzeller, 1987).

When procaine, a relatively pure local anesthetic, is administered in subconvulsant doses to man (0.46–2.3 mg/kg, i.v.) it is associated with (1) cognitive and sensory distortions, illusions, if not outright hallucinations in essentially every sensory modality, (2) a wide range of affective response from euphoria to severe anxiety, (3) activation of fast EEG activity (26–44 Hz) selectively over temporal lobe areas (which correlated with degree of dysphoria), and (4) increases in ACTH, cortisol, and prolactin, but not growth hormone (Kellner *et al.*, 1987). Doom anxiety has likewise been reported with the pure local anesthetic lidocaine (Saravay *et al.*, 1986). Thus, a variety of clinical syndromes may be attributable to cocaine's local anesthetic rather than its psychomotor stimulant properties.

Consideration of the conditioned components of cocaine-induced behavioral sensitization may also yield new perspectives on clinical phenomena. There is increasing clinical recognition of the potent conditioned components of cocaine's actions clinically. Gawin and Kleber (1987), Washton (1988), and Gawin and Ellinwood (1988) have described anecdotal reports of the induction of cocaine-induced craving, euphoria, or withdrawal syndromes upon the presentation of cues associated with habitual cocaine use. O'Brien, Childress, and colleagues have systematically explored this phenomenon and demonstrated high amounts of cocaine craving when former addicts who have been abstinent for at least 1 month are exposed to pictures of cocaine-related paraphernalia (O'Brien *et al.*, 1988). This group has also documented that it is extremely difficult to extinguish such cocaine-induced conditioned cues (Childress *et al.*, 1987). Repeated exposure to pictures of cocaine-related paraphernalia and other potential conditioned stimuli have been employed in attempts to decondition these phenomena. In these instances, it appears that the cocaine-conditioned craving is extremely resistant to extinction. While subjective craving can be attenuated with many repetitions, Childress and O'Brien have observed that conditioned physiological effects, such as hypothermic reactions, can persist after many months of abstinence and systematic deconditioning.

Thus, many of the conditioned effects observed in the behavioral sensitization paradigm may be appropriate models for related phenomena in man. Clearly, the parallels need to be adequately drawn and not overgeneralized. Given the clinical evidence that cocaine-induced euphoria and rush may show acute tolerance or tachyphylaxis, it becomes essential to attempt to understand what are the precise clinical equivalents in man that may occur analogous to cocaine-induced behavioral sensitization in animals. Given the reports that stimulant-induced punding (Ellinwood *et al.*, 1977; Gawin and Ellinwood, 1988), perhaps the equivalent of animal stereotypy, can become an increasing problem with repeated use, it is possible that punding, as well as other aspects of stimulant-induced behavioral disruption, including dysphoria and paranoid psychosis, may be modeled by the behavioral sensitization effect. One chronic high-dose user told us his paranoia was context dependent (D.L., personal communication). When he was using cocaine with friends in a novel environment, the effects would not be as severe as when he was at home, where he habitually used cocaine. No amount of self-reassurance before he administered cocaine would overcome his certainty that police or FBI were hiding behind the trees in his backyard. This delusion did not occur in other contexts.

Furthermore, even the toxic effects of cocaine can be altered by contextual variables. Dworkin *et al.* (1988) demonstrated that rats self-administering cocaine develop less weight loss and a lower mortality rate from cocaine than their "yoked" controls which passively received identical amounts and schedules of administration of cocaine. These data further suggest caution in directly extrapolating from studies involving experimentor-administered drug to ones involving self-administration.

Whatever effects in humans emerge as the precise equivalents to behavioral sensitization, the demonstration of sensitization in animals documents the possibility that a variety of behavioral toxicities may show reverse tolerance and progressively emerge with repeated administration of the same dose rather than becoming attenuated. This effect has been demonstrated for cocaine-induced hyperthermia in the dog (Tatum and Seevers, 1929) and raises the possibility that hyperthermic reactions and their associated potentials for lethality could also demonstrate sensitization-like phenomena. When we administered cocaine repeatedly in high doses to rhesus monkeys, they also demonstrated an increased proclivity for stereotypies in the first weeks of administration, but then showed greater degrees of behavioral inhibition associated with cataplexy and bizarre hallucinatory-like behavior (Post *et al.*, 1976, 1987). In this context, the long-lasting aspects of behavioral sensitization are noteworthy. Reactivation of paranoid psychosis has been reported to occur many years after amphetamine addicts are drug-free (Sato *et al.*, 1983; Shuster, personal communication, 1987). In addition, reactivation of psychosis is reported to occur with very small doses of stimulant re-administration, perhaps also consistent with the conditioned component of stimulant-induced behavioral sensitization.

From the foregoing analysis and preclinical models, we would suggest that public health campaigns attempt to warn potential cocaine users of the several hidden liabilities of the drug; that is, in addition to its initial seductiveness as a drug capable of producing mood elevation and euphoria of seemingly benign proportions and its later ability to induce, for some, an irresistable craving, it also possesses the liability for producing increasingly severe behavioral and physiological effects upon repetition. The preclinical models demonstrate that repeated administration of the same dose is capable of producing increasing behavioral toxicities (behavioral sensitization) as well as convulsive toxicities (pharmacological kindling).

Clinical exprience suggests that the cocaine user is generally not able to maintain the same dose over time and it is likely to engage in dose escalation (Gawin and Ellinwood, 1988). This phenomenon, in conjunction with whatever behavioral sensitization and pharmacological kindling effects may also be engaged, would appear to raise the likelihood of progressive behavioral and physiological toxicities. In particular, the addict cannot be reassured that ability to tolerate a given dose safely on one occasion would necessarily result in a similar safe experience on the next occasion. On the contrary, there is reason to expect that repeated administration of a given dose may lead to augmented behavioral toxicities, including dysphoria, suspiciousness, paranoia, hostility, cognitive impairment, and, possibly, panic attacks.

Given the extreme lethality of cocaine-related seizures in a variety of animal species, the cocaine user should again be forewarned of the potential lethal consequences of a local-anesthetic-related seizure. It is quite clear that many famous athletes and other prominent personalities, including Len Bias, the University of Maryland basketball star who celebrated being drafted by the Boston Celtics, experienced a cocaine-related seizure immediately prior to their demise. While it is unclear what dose was required to produce such an effect, it is equally clear that there is wide individual variability in the ability to tolerate a given dose of cocaine, as well as the virtual impossibility of knowing the relative purity of a given dose of cocaine bought on the street. Even if such dosage issues were not a factor, the kindling data suggests that the concept of an overdose be revised. It is possible, based on the models presented, that a previously tolerated dose on one occasion may become lethal on the next.

While some may argue about the validity of extrapolating preclinical data to man, it would appear that information of such possible critical importance to cocaine users (current and potential) be made available so that informed decisions can be made. It is widely acknowledged that, with a few exceptions, the scientific pharmacological community did not appropriately help to convey the dangers of cocaine to the public over the past 2 decades (Gawin and Ellinwood, 1988). Given this information gap, it would now appear all the more warranted to convey an appropriate respect for this pharmacological agent with its dual properties and hazards as both a psychomotor stimulant and a local anesthetic. On the street, it eventually was recognized with amphetamine that "speed kills". As cocaine-related emergency room visits and deaths mount, "coke kills" would appear even more appropriate. While it might be argued that the sensitization and kindling effects demonstrated in rodents, cats, dogs, and nonhuman primates might not necessarily be directly extrapolatable to man, the burden of proof would certainly seem to rest on those who would choose to ignore such a possibility.

REFERENCES

P.S. Albright and W.M. Burnham. Development of a new pharmacological seizure model: effects of anticonvulsants on cortical- and amygdala-kindled seizures in the rat. *Epilepsia* **21**: 681 (1980).

J.J. Ambre, S.M. Belknap, J. Nelson, T.I. Ruo, S.G. Shin, and A.J. Atkinson, Jr. Acute tolerance to cocaine in humans. *Clin. Pharmacol. Ther.* **44**: 1 (1988).

B. Angrist, G. Sathanathan, S. Wilk, and S. Gershon. Amphetamine psychosis: behavioral and biochemical aspects, in *Catecholamines and Schizophrenia.* S.W. Matthysse and S.S. Kety, eds., Pergamon Press, Oxford, p. 13 (1975).

S.M. Antelman and L.A. Chiodo. Amphetamine as a stressor, in *Stimulants: Neurochemical, Behavioral, and Clinical Perspectives.* I. Cresse, ed., Raven Press, New York, p. 269 (1983).

L.R. Baxter, Jr., J.M. Schwartz, M.E. Phelps, J.C. Mazziotta, J. Barrio, R.A. Rawson, J. Engel, B.H. Guze, C. Selin, and R. Sumida. Localization of neurochemical effects of cocaine and other stimulants in the human brain. *J. Clin. Psychiatry* **49**: 23 (1988).

D.S. Bell. The experimental reproduction of amphetamine psychosis. *Arch. Gen. Psychiatry* **29**: 35 (1975).

R.J. Beninger and B.L. Hahn. Pimozide blocks establishment but not expression of amphetamine-produced environment-specific conditioning. *Science* **220**: 1304 (1983).

R.J. Beninger and R.S. Herz. Pimozide blocks establishment but not expression of cocaine-produced environment-specific conditioning. *Life Sci.* **38**: 1425 (1986).

R. Byck, D. Paly, C. van Dyke, F.R. Jeri, and P. Jatlow. Cocaine smoking, in *Abstracts of the 12th CINP Congress.* Abstr. # 101, Pergamon Press, Elmsford, NY, p. 97 (1980).

A.E. Calogero, M.A. Kling, R. Bernardini, W.T. Galluci, R.M. Post, G.P. Chrousos, and P.W. Gold. Cocaine stimulates rat hypothalamic corticotropin releasing hormone secretion *in vitro. Clin. Res.* **36**: 361 (1988).

A.E. Calogero, M.A. Kling, W.T. Galluci, C. Saoutis, R.M. Post, G.P. Ghrousos, and P.W. Gold. Local anaesthetics procaine and lidocaine stimulate corticotropin releasing hormone secretion *in vitro:* clinical implications. Abstracts, Society for Neuroscience, New Orleans, Nov. 16–20, Abstr. # 319.2, p. 1163 (1987).

A.R. Childress, A.T. McLellan, R.N. Ehrman, and C.P. O'Brien. Extinction of conditioned responses in abstinent cocaine or opioid users, in *NIDA Monograph #76, Problems of Drug Dependence.* U.S. Government Printing Office, Washington, DC, p. 189 (1987).

P.H. Connell. *Amphetamine Psychosis* (Maudsley Monograph No. 5). Oxford University Press, New York (1958).

S.M. Dworkin, C. Volkmer, and S.I. Dworkin. Toxic consequences of cocaine are augmented by noncontingent drug administration. Abstracts, Society for Neuroscience, Toronto, Nov. 13–18, Abstr. # 388.3, p. 961 (1988).

C.L. Ehlers, S.J. Henriksen, M. Wang, J. Rivier, W. Vale, and F.E. Bloom. Corticotropin releasing factor produces increases in brain excitability and convulsive seizures in rats. *Brain Res.* **278**: 332 (1983).

E.H. Ellinwood and M.M. Kilbey. Chronic stimulant intoxication models of psychosis, in *Animal Models in Psychiatry and Neurology,* I. Hanin and E. Usdin, eds., Pergamon Press, Elmsford, NY, p. 61 (1977).

E.H. Ellinwood and M.M. Kilbey. Fundamental mechanisms underlying altered behavior following chronic administration of psychomotor stimulants. *Biol. Psychiatry* **15**: 49 (1980).

E.H. Ellinwood, M.M. Kilbey, S. Castellani, and C. Khoury. Amygdala hyperspindling and seizures induced by cocaine, in *Advances in Behavioral Biology, Vol. 21, Cocaine and Other Stimulants.* E.H. Ellinwood and M.M. Kilbey, eds., Plenum Press, New York, p. 303 (1977).

K. Gale. Catecholamine-independent behavioral and neurochemical effects of cocaine in rats, in *NIDA Research Monograph Series 54, Mechanisms of Tolerance and Dependence,* C. Wm. Sharp, ed., U.S. Government Printing Office, Washington, DC, p. 323 (1984).

F.H. Gawin and E.H. Ellinwood, Jr. Cocaine and other stimulants. Actions, abuse, and treatment. *N. Engl. J. Med.* **318**: 1173 (1988).

F.H. Gawin and H. Kleber. Issues in cocaine abuse treatment research, in *Cocaine: Clinical and Biobehavioral Aspects.* S. Fisher, A. Raskin, and E.H. Uhlenhuth, eds., Oxford University Press, New York, p. 174 (1987).

G.R. Gay, C.W. Sheppard, D.S. Inaba, and J.A. Newmeyer. Cocaine in perspective: "Gift from the Sun God" to "The Rich Man's Drug". *Drug Forum* **2**: 409 (1973).

P. Gloor. Inputs and outputs of the amygdala: what the amygdala is trying to tell the rest of the brain, in *Limbic Mechanisms.* K.E. Livingstone and O. Hornykiewicz, eds., Plenum Press, New York, p. 189 (1978).

G.V. Goddard, D.C. McIntyre, and C.K. Leech. A permanent change in brain function resulting from daily electrical stimulation. *Exp. Neurol.* **25**: 295 (1969).

L.S. Goodman and A. Gilman. *The Pharmacological Basis of Therapeutics.* 3rd ed., Macmillan, New York (1965).

J.D. Griffith, J.H. Cavanaugh, J. Held, and J.A. Oates. Experimental psychosis induced by the administration of d-amphetamine. In *Amphetamines and Related Compounds, Proceedings of the Mario Negri Institute for Pharmacological Research, Milan, Italy.* Raven Press, New York, p. 897 (1970).

J.D. Griffith, J. Cavanaugh, J. Held, and J.A. Oates. Dextroamphetamine. Evaluation of psychomimetic properties in man. *Arch. Gen. Psychiatry* **26**: 97 (1972).

J.A. Halikas. Does carbamazepine cure cocaine craving? Abstracts of the 27th ACNP Annual Meeting. San Juan, Dec. 11–16, p. 146 (1988).

F.R. Jeri, C.C. Sanchez, T. del Pozo, M. Fernandez, and C. Carbajal. Further experience with the syndromes produced by coca paste smoking. *Bull. Narc.* **30**: 1 (1978).

C.H. Kellner, R.M. Post, F. Putnam, R. Cowdry, D. Gardner, M.A. Kling, M.D. Minichiello, J.R. Trettau, and R. Coppola. Intravenous procaine as a probe of limbic system activity in psychiatric patients and normal controls. *Biol. Psychiatry* **22**: 1107 (1987).

W.P. Koella, P. Levin, and V. Baltzer. The pharmacology of carbamazepine and some other antiepileptic drugs. In *Epileptic Seizures — Behavior — Pain.* W. Birkmayer, ed., Hans Huber, p. 32 Bern (1976).

K. Kumor, M. Sherer, L. Thompson, E. Cone, J., Mahaffey, and J.H. Jaffe. Lack of cardiovascular tolerance during intravenous cocaine infusions in human volunteers. *Life Sci.* **42**: 2063 (1988).

L. Lewin. *Phantastica: Narcotic and Stimulating Drugs.* E.P. Dutton and Co., New York (1931).

A.K. Louie, R.A. Lannon, and T.A. Ketter. Treatment of cocaine-induced panic disorder. *Am. J. Psychiatry* **146**: 40 (1989).

J.A. Mace and W.M. Burnham. The effect of repeated seizures on anticonvulsant drug response in the kindling model. *Electroencephalogr. Clin. Neurophysiol.* **67**: 171 (1987).

T.C. Manschreck, J.A. Laughery, C.C. Weisstein, D. Allen, B. Humblestone, M. Naville, H. Podlewsi, and N. Mitra. Characteristics of freebase cocaine psychosis. *Yale J. Biol. Med.* **61**: 115 (1988).

M. Matsuzaki, P.J. Spingler, and A.L. Misra. Cocaine: tolerance to its convulsant and cardiorespiratory stimulating effects in the monkey. *Life Sci.* **19**: 193 (1976).

J.O. McNamara. Biochemical and pharmacologic studies of NMDA responses in the kindling model of epilepsy. Presented at the CIBA-Geigy Symposium: Recent Developments in Kindling. Toronto, Nov. 13, 1988.

A.E. Merriam, A. Medalia, and B. Levine. Partial complex status epilepticus associated with cocaine abuse. *Biol. Psychiatry* **23**: 515 (1988).

E. Meyendorff, B. Lerer, N.C. Moore, J. Bow, and S. Gershon: Methylphenidate infusion in euthymic bioplars: effect of carbamazepine pretreatment. *Psychiatry Res.* **16**: 303 (1985).

M. Mishkin and T. Appenzeller. The anatomy of memory. *Sci. Am.* **256**: 80 (1987).

E.A. Murray and M. Mishkin. Amygdalectomy impairs crossmodal association in monkeys. *Science* **228**: 604 (1985).

E. Nunes and F. Quitkin. Efficacy of imipramine in cocaine abuse. Abstract, Proceedings of the Psychiatric Research Society, Park City, Utah, March 8–11, 1989.

C.P. O'Brien, A.R. Childress, I.O. Arndt, A.T. McLellan, G.E. Woody, and I. Manny. Pharmacological and behavioral treatments of cocaine dependence: controlled studies. *J. Clin. Psychiatry* **49**: 17 (1988).

W.D. Owens. Signs and symptoms presented by those addicted to cocaine. *JAMA* **58**: 329 (1912).

J.P.J. Pinel. Kindling-induced experimental epilepsy in rats: cortical stimulation. *Exp. Neurol.* **71**: 559 (1981).

J.P.J. Pinel and L.I. Rovner. Electrode placement and kindling-induced experimental epilepsy. *Exp. Neurol* **58**: 335 (1978a).

J.P.J. Pinel and L.I. Rovner. Experimental epileptogenesis: kindling-induced epilepsy in rats. *Exp. Neurol.* **58**: 190 (1978b).

R.M. Post. Cocaine psychoses: a continuum model. *Am. J. Psychiatry* **132**: 225 (1975).

R.M. Post. Clinical aspects of cocaine: assessment of acute and chronic effects in animals and man, in *Cocaine: Chemical, Biological, Clinical, Social, and Treatment Aspects.* S.J. Mule, ed., CRC Press, Boca Raton, FL, p. 203 (1976).

R.M. Post. Progressive changes in behavior and seizures following chronic cocaine administration: relationship of kindling and psychosis, in *Advances in Behavioral Biology, Vol. 21, Cocaine and Other Stimulants.* E.H. Ellinwood and M.M. Kilbey, eds., Plenum Press, New York, p. 353 (1977).

R.M. Post. Lidocaine kindled limbic seizures: behavioral implications, in *Kindling 2,* J. Wada, ed., Raven Press, New York, p. 149 (1981).

R.M. Post, J.C. Ballenger, T.W. Uhde, and W.E. Bunney, Jr. Efficacy of carbamazepine in manic-depressive illness: implications for underlying mechanisms, in *Neurobiology of Mood Disorders.* R.M. Post and J.C. Ballenger, eds., Williams & Wilkins, Baltimore, p. 777 (1984a).

R.M. Post and N.R. Contel. Cocaine-induced behavioral sensitization: a model for recurrent manic illness, in *Biological Psychiatry, 1981.* C. Perris, G. Struwe, and B. Jansson, eds., Elsevier, Amsterdam, p. 746 (1981).

R.M. Post and N.R. Contel. Human and animal studies of cocaine: implications for development of behavioral pathology, in *Stimulants: Neurochemical, Behavioral, and Clinical Perspective.* I. Creese, ed., Raven Press, New York, p. 169 (1983).

R.M. Post, C. Kennedy, M. Shinohara, K. Squillace, M. Miyaoka, S. Suda, D.H. Ingvar, and L. Sokoloff. Metabolic and behavioral consequences of lidocaine-kindled seizures. *Brain Res.* **324**: 295 (1984b).

R.M. Post and R.T. Kopanda. Cocaine, kindling, and psychosis. *Am. J. Psychiatry* **133**: 627 (1976).

R.M. Post, R.T. Kopanda, and K.E. Black. Progressive effects of cocaine on behavior and central amine metabolism in rhesus monkeys: relationship to kindling and psychosis. *Biol. Psychiatry* **11**: 403 (1976).

R.M. Post, R.T. Kopanda, and A. Lee. Progressive behavioral changes during chronic lidocaine administration: relationship to kindling. *Life Sci.* **17**: 943 (1975).

R.M. Post, A. Lockfeld, K. Squillace, and N.R. Contel. Drug-environment interaction: context dependency of cocaine-induced behavioral sensitization. *Life Sci.* **28**: 755 (1981).

R.M. Post, D.R. Rubinow, and J.C. Ballenger. Conditioning and sensitization in the longitudinal course of affective illness. *Br. J. Psychiatry* **149**: 191 (1986a).

R.M. Post, T.W. Uhde, R.T. Joffe, and L. Bierer. Psychiatric manifestations and implications of seizure disorders, in *Medical Mimics of Psychiatric Disorders.* I. Extein and M. Gold, eds., APA Press, Washington, DC, p. 35 (1986b).

R.M. Post and S.R.B. Weiss. Stimulant-induced behavioral sensitization: a model for neuroleptic nonresponsiveness, in *Animal Models of Psychiatric Disorders.* P. Simon, P. Soubrie, and D. Widlocher, eds., S. Karger, New York, p. 52 (1988).

R.M. Post, S.R.B. Weiss, and A. Pert. Differential effects of carbamazepine and lithium on sensitization and kindling. *Prog. Neuropsychopharmacol. Biol. Psychiatry* **8**: 425 (1984c).

R.M. Post, S.R.B. Weiss, A. Pert, and T.W. Uhde. Chronic cocaine administration: sensitization and kindling effects, in *Cocaine: Clinical and Biobehavioral Aspects*. A. Raskin and S. Fisher, eds., Oxford University Press, New York, p. 109 (1987).

R. Racine. Kindling: the first decade. *Neurosurgery* **3**: 234 (1978).

C. Rivier and W. Vale. Cocaine stimulates adrenocorticotropin (ACTH) secretion through a corticotropin-releasing factor (CRF)-mediated mechanism. *Brain Res.* **422**: 403 (1987).

S.M. Saravay, J. Marie, M.D. Steinberg, and C.J. Rabiner. "Doom anxiety" and delirium in lidocaine toxicity. *Am. J. Psychiatry* **144**: 159 (1986).

M. Sato, C.C. Chen, K. Akiyama, and S. Otsuki. Acute exacerbation of paranoid psychotic state after long-term abstinence in patients with previous methamphetamine psychosis. *Biol. Psychiatry* **18**: 429 (1983).

J. Scheel-Kruger, C. Braestrup, M. Nielson, K. Golembrowska, and F. Mogilnicka. Cocaine: discussion on the role of dopamine in the biochemical mechanism of action, in *Advances in Behavioral Biology, Vol. 21, Cocaine and Other Stimulants*. E.H. Ellinwood and M.M. Kilbey, eds., Plenum Press, New York, p. 373 (1977).

M. Schmutz, K. Klebs, and V. Baltzer. Inhibition or enhancement of kindling evolution by antiepileptics. *J. Neural Transm.* **72**: 245 (1988).

M.A. Sherer. Intravenous cocaine — psychiatry effects, biological mechanisms. *Biol. Psychiatry.* **24**: 865 (1988).

M.A. Sherer, K.M. Kumor, E.J. Cone, and J.H. Jaffe. Suspiciousness induced by four-hour intravenous infusions of cocaine. Preliminary findings. *Arch. Gen. Psychiatry* **45**: 673 (1988).

R.K. Siegel. Changing patterns of cocaine use: longitudinal observations, consequences, and treatment, in *Cocaine: Pharmacology, Effects and Treatment of Abuse. NIDA Research Monograph No. 50*, J. Grabowski, ed., U.S. Government Printing Office, Washington, DC, p. 92 (1984).

J.R. Stevens, V.H. Mark, F. Ervin, P. Pacheco, and K. Suematsu. Deep temporal stimulation in man. *Arch. Neurol.* **21**: 157 (1969).

S. Tadokoro and H. Kuribara. Reverse tolerance to the ambulation-increasing effect of methamphetamine in mice as an animal model of amphetamine psychosis. *Psychopharmacol. Bull.* **22**: 757 (1986).

A.L. Tatum and M.H. Seevers. Experimental cocaine addiction. *J. Pharmacol. Exp. Ther.* **36**: 401 (1929).

J.M. Van Rossum. Mode of action of psychomotor stimulants. *Int. Rev. Neurobiol.* **12**: 307 (1970).

J.A. Wada and M. Sato. Generalized convulsive seizures induced by daily electrical stimulation of the amygdala in cats. *Neurology* **24**: 565 (1974).

J.A. Wada, M. Sato, and M.E. Corcoran. Persistent seizure susceptibility and recurrent spontaneous seizures in kindled cats. *Epilepsia* **15**: 465 (1974).

A.M. Washton. Preventing relapse to cocaine. *J. Clin. Psychiatry* **49**: 34 (1988).

A.M. Washton and M.S. Gold. Chronic cocaine abuse: evidence for adverse effects on health and functioning. *Psychiatry. Ann.* **14**: 733 (1984).

S.R. Weiss and R.M. Post. Carbamazepine and carbamazepine-10,11-epoxide inhibit amygdala-kindled seizures in the rat but do not block their development. *Clin. Neuropharmacol.* **10**: 272 (1987).

S.R.B. Weiss, M. Costello, R. Woodward, D.J. Nutt, and R.M. Post. Chronic carbamazepine inhibits the development of cocaine-kindled seizures. *Abstracts, Society of Neuroscience 17th Annual Meeting*. New Orleans, Nov. 16–21, 1987, Abstr. # 262.20 p. 950 (1987).

S.R.B. Weiss, J. Nierenberg, R. Lewis, and R. M. Post. Corticotropin-releasing hormone: Potentiation of cocaine kindled seizures and lethality. *Epilepsia* (1991), in press.

S.R.B. Weiss, R.M. Post, P.W. Gold, G. Chrousos, T.L. Sullivan, D. Walker, and A. Pert. CRF-induced seizures and behavior: interaction with amygdala kindling. *Brain Res.* **372**: 345 (1986).

S.R.B. Weiss, R.M. Post, A. Pert, R. Woodward, and D. Murman. Context-dependent cocaine sensitization: Differential effect of haloperidol on development versus expression. *Pharmacol. Biochem. & Behav.* **34**: 655 (1989a).

S.R.B. Weiss, R.M. Post, F. Szele, R. Woodward, and J. Nierenberg. Chronic carbamazepine inhibits the development of local-anesthetic seizures kindled by cocaine and lidocaine. *Brain Res.* **497**: 72 (1989b).

R.A. Wise. Neural mechanisms of the reinforcing action of cocaine, in *Cocaine: Pharmacology, Effects, and Treatment of Abuse*. NIDA Monograph Series 50, J. Grabowski, ed., U.S. Government Printing Office, Washington, DC, p. 15 (1984).

R.J. Wyatt, R.C. Alexander, M.S. Egan, and D.G. Kirsh. Schizophrenia: just the facts. What do we know, how well do we know it? *Schiz. Res.* **1**: 3 (1988).

8

Neuropharmacology of Cocaine: Effects on Dopamine and Serotonin Systems

MATTHEW P. GALLOWAY
Cellular and Clinical Neurobiology Program
Department of Psychiatry
Wayne State University School of Medicine
Neurochemical Pharmacology Research Unit
Lafayette Clinic
Detroit, Michigan

INTRODUCTION

Behavioral studies that measure either drug self-administration or intracranial self-stimulation have established that enhanced dopaminergic transmission is a pivotal component of the neuronal reward pathway. As such, the ability of cocaine to enhance monoaminergic neurotransmission via inhibition of neuronal re-uptake is thought to be one mechanism that contributes to both its psychotropic and reinforcing actions. Since monoamine neurons are subject to several monoamine-sensitive regulatory mechanisms, elevated levels of synaptic dopamine (DA) or serotonin (5-HT) may be expected to activate feedback pathways. Thus, in addition to a transient increase in overall DA or 5-HT transmission, cocaine may actually diminish certain regulated activities in either DA or 5-HT neurons as a compensatory mechanism. In this chapter, I review the involvement of DA and 5-HT systems in cocaine's effects on behavior and the recent evidence demonstrating that cocaine elevates extracellular DA *in vivo*. Additionally, the biochemical response of both DA- and 5-HT-containing neurons to acute exposure to cocaine or amphetamine is evaluated. Finally, the effects of repeated cocaine on parameters relevant to the control of presynaptic DA release are reviewed.

INTERACTION OF COCAINE WITH DOPAMINERGIC SYSTEMS

Behavioral Reinforcement

Questions of cocaine's effects on DA systems have benefited greatly from behaviorial studies that have identified both the neuroanatomy and transmitter systems that constitute circuits critical to drug self-administration and reinforcement. In terms of the pharmacology of reinforcement, numerous studies have shown that nonselective DA antagonists interrupt self-administration of cocaine in rats and nonhuman primates (Wise, 1984). Although the disruption of cocaine self-administration by nonselective DA antagonists appears to be established, the effects of selective D_1 and D_2 antagonists is still emerging. Koob *et al.* (1987) demonstrated that low doses of the D_1 antagonist SCH 23390 increased responding for cocaine self-administration in the rat, suggesting that DA-D_1 receptor blockade decreases the reward component of self-administration. Also, the rewarding effects of electrical brain stimulation in the rat (Nakajima and McKenzie, 1986) and the discriminative stimulus effect of cocaine in monkeys (Kleven *et al.*, 1988a) and rats (Barrett and Appel, 1989) can be antagonized by SCH 23390. Woolverton (1986) and Roberts and Vickers (1984), however, concluded that D_2, but not D_1, receptor systems were involved with cocaine self-administration in rhesus monkeys and rats, respectively. Furthermore, it was concluded that blockade of cocaine self-administration by either D_1 or D_2 antagonists was not selective since responding to food was also blocked in the same subjects (Woolverton and Virus, 1989). Interestingly, both eating (Hernandez and Hoebel, 1988) and cocaine (see references below) increase levels of DA in the nucleus accumbens, suggesting a common mechanism of action for eating and cocaine administration on this particular DA system. Although seemingly contrary, these studies with selective DA antagonists suggest that cocaine self-administration is dependent on simultaneous D_1 and D_2 receptor stimulation, and raise the possibility that cocaine self-administration is an enabled behavior; that is, a D_2-mediated behavior that requires simultaneous D_1 tone for full expression (White, 1987; Carlson *et al.*, 1987; White *et al.*, 1988). Moreover, if stimulant reward pathways are mediated by a balance of D_1 and D_2 receptor activation, repeated use of psychomotor stimulants may be associated with alterations in the balanced responsiveness of specific DA receptor subtypes.

The anatomical locus of the dopaminergic components of reward and reinforcement in the rat has been localized to the mesolimbic DA projection areas. For example, destruction of either dopaminergic terminals in the nucleus accumbens (Roberts *et al.*, 1980) or cell bodies in the ventral tegmental area (VTA) (Roberts and Koob, 1982) with the DA specific toxin 6-hydroxy-DA disrupts cocaine self-administration in the rat. The site of action was further defined by demonstrating that when rats were allowed to self-administer cocaine directly into discrete brain regions, the medial prefrontal cortex (but not the accumbens or VTA) supported initiation of cocaine self-administration in a sulpiride reversible fashion (Goeders and Smith, 1983; Goeders *et al.*, 1986). Further, DA systems were specifically linked to the behavior since selective lesions of the medial prefrontal DA system with 6-hydroxy-DA decreased intracranial self-administration of cocaine to control levels (Goeders and Smith, 1986). However, recent work by Martin-Iverson *et al.* (1986) suggests that 6-OH-DA lesions of the medial prefrontal cortex do not modify the maintenance of peripheral (i.v.) self-administration of cocaine. Similarly, cocaine injected into the nucleus accumbens, but not prefrontal cortex or striatum, substituted for the peripheral training dose of cocaine in rats previously trained to discriminate cocaine (Wood and Emmett-Oglesby, 1989).

Sensitization

Just as studies of cocaine self-administration have clearly identified the importance of DA systems in acute drug reinforcement, other studies have established that repeated exposure to cocaine (and other psychomotor stimulants such as amphetamine) produces an exaggerated motor response to a challenge injection after discontinuation of the chronic administration (reviewed by Post and Contel, 1983; Post *et al.*, 1990). The enhancement of a drug response, known as behavioral sensitization, has been extensively studied in animals treated chronically with amphetamine (reviewed in Robinson and Becker, 1986), yet the underlying neurochemical mechanism of the phenomena remains elusive. Nonetheless, a common effect of drugs that produce sensitization is an acute and transient enhancement of mesolimbic dopaminergic activity. Thus, amphetamine causes DA release, cocaine inhibits DA uptake, and intra-VTA morphine increases DA activity (Gysling and Wang, 1983; Kalivas, 1985; Vezina and Stewart, 1989; Kalivas and Duffy, 1988).

Development of cross-sensitization between systemic cocaine and amphetamine, which indicates the similarity of neuronal effects exerted by these drugs, can be demonstrated in rats repeatedly treated with amphetamine (Shuster *et al.*, 1977; Kalivas and Weber, 1988). Cross-sensitization between systemic cocaine and intra-A10 enkephalin after chronic treatment with either was demonstrated by DuMars *et al.* (1988). Interestingly, mild footshock stress, which is known to selectively activate medial prefrontal DA neurons (Deutch *et al.*, 1986; Roth, 1984), is also related to psychomotor stimulant sensitization. Sensitization to the locomotor effects of amphetamine can be produced by external stress such as footshock and handling (Antelman and Chiodo, 1983; Robinson *et al.*, 1987; Wilcox *et al.*, 1986). Moreover, repeated footshock or cocaine sensitizes the DA system to a footshock challenge (Kalivas and Duffy, 1989). In a related study, Rivet *et al.* (1989) found that proper corticosteroid tone (acting through type II glucocorticoid receptors) is necessary for the development of amphetamine-induced sensitization in rats.

Recently, several groups determined the effects of selective DA antagonists on the development of sensitization. Pretreatment with the D_1 antagonist SCH 23390 prevents the behavioral sensitization induced by repeated amphetamine (Iorio *et al.*, 1983; Christensen *et al.*, 1984). Vezina and Stewart (1989) differentiated the sensitization elicited by either amphetamine or intra-VTA morphine and concluded that the D_1 antagonist SCH 23390 (but not the D_2 blocker sulpiride) blocked development of amphetamine-induced sensitization. Morphine-induced sensitization was insensitive to SCH 23390 pretreatment; however, pretreatment with pimozide (but not Ro 22-2586) prevented sensitization to morphine (Verzina and Stewart, 1989). Additionally, putative blockade of DA autoreceptors with low-dose sulpiride had no effect on opiate- or amphetamine-induced sensitization. This latter observation argues against potential changes in DA autoreceptor responsivity contributing to the development of sensitization, although it remains to be determined if low-dose sulpiride actually produces a selective block of nerve terminal DA autoreceptors under such conditions (see below for discussion of chronic cocaine effects). Ujike *et al.* (1989) reported that both D_1 and D_2 antagonists block the acute locomotor effects of methamphetamine and prevent the behavioral sensitization produced by chronic methamphetamine.

The studies cited above make it clear that D_1 DA receptor stimulation plays a significant role in the expression of both behavioral sensitization and self-administration. As such, the D_1-D_2 enabling mechanisms described by White (1987; White *et al.*, 1988) and Walters (Carlson *et al.*, 1987) have significant implications for understanding the basis of either behavior, and it becomes a primary objective to determine (1) if selective DA agonists can be used to elicit sensitization and (2) if either sensitization or self-administration represents an ''enabled'' phenomenon.

Evidence that Cocaine Increases Synaptic DA

Uptake studies *in vitro* demonstrated that cocaine inhibits the uptake of radiolabeled DA into slices or synaptosomes (Ross and Renyi, 1966; Heikkila *et al.*, 1975; Kennedy and Hanbauer, 1983; Dubcovich and Zahniser, 1985). Whether inhibition of neuronal re-uptake had any consequence on synaptic levels of DA *in vivo* was difficult to ascertain since acute cocaine treatment produced little effect on post-mortem tissue levels of DA or its metabolites DOPAC and HVA (Taylor and Ho, 1977; Bowers and Hoffman, 1986; Kalivas *et al.*, 1988). Thus, even though DA metabolites generally reflect DA activity (Roth *et al.*, 1976), attempts to demonstrate the cocaine-induced increase in DA function by alterations in metabolite levels were equivocal. In constrast, the norepinephrine metabolite 3-methoxy-4-hydroxyphenylglycol (MHPG) is increased 2.5-fold and acetylcholine turnover increased 3-fold in the hippocampus of rats treated with cocaine (90 μmol/kg, 30 min) (Robinson and Hambrecht, 1988).

Not until the advent of brain microdialysis was the cocaine-induced increase in extracellular DA established unequivocally *in vivo* (Church *et al.*, 1987). Hurd and Ungerstedt (1989) demonstrated that i.v. cocaine rapidly and reversibly increased striatal DA levels in a Ca^{++}-dependent fashion. They also concluded that the action of cocaine was elicited at the level of the nerve terminal since intranigral cocaine did not increase striatal DA. Bradberry and Roth (1989) reported that systemic cocaine increased extracellular DA, both at the axon terminal (nucleus accumbens) and the cell body region (VTA), although terminal release was of a greater magnitude than dendritic release.

Using microdialysis, several studies have attempted to resolve the issue of potential differences in the efficacy of cocaine with respect to different DA projection fields. Di Chiara and Imperato (1988) found that, in freely moving rats, a variety of drugs of abuse (e.g., ethanol, opiates, nicotine, amphetamine, and cocaine) preferentially increased DA levels in the nucleus accumbens, when compared to their effect on striatal DA. As noted above, Bradberry and Roth (1989) demonstrated that cocaine had a more pronounced effect on accumbens DA when compared to increases in the VTA. Although most agree that mesolimbic DA regulates reinforcement and reward, the relative contributions of specific DA tracts (such as the mesoaccumbens and mesocortical projections) is a matter of debate (see discussions above; Goeders and Smith, 1983, 1986; Martin-Iverson *et al.*, 1986). Moghaddam and Bunney (1989) addressed the issue by monitoring extracellular DA simultaneously in the medial prefrontal cortex and in the nucleus accumbens. They found that systemic cocaine (1 mg/kg, i.v.) increased accumbens DA without affecting prefrontal DA levels. When the cocaine dose was doubled, prefrontal DA increased threefold, whereas accumbens DA increased by a factor of seven. As pointed out by the authors (Moghaddam and Bunney, 1989), the differential effect of cocaine on prefrontal and accumbens DA neurons may be related to the presence of chloral hydrate and the known sensitivity of medial prefrontal DA neurons to anesthetic and mild stress (Deutch *et al.*, 1986; Knorr *et al.*, 1989). In other words, alterations in the rate of mesocortical DA cell firing would have a greater influence on the cocaine effect, which hinges on impulse-dependent DA release, than for amphetamine, which has a direct DA-releasing effect. In contrast to the differential effects of cocaine, Moghaddam and Bunney (1989) reported that amphetamine (1 mg/kg, i.v.) increased DA levels equally (about threefold) in both mesocortical and mesoaccumbens DA fields. The different dialysis profiles of cocaine and amphetamine is consistent with the distinct mechanism of action of each compound.

Using kinetic data obtained from DA dialysis measurements, Nicolaysen and Justice (1988a) constructed a computer model to predict the dynamics of DA release from DA terminals. Since the increased extracellular DA levels observed with microdialysis can arise from either re-uptake inhibition or enhanced release (both vesicular and nonvesicular), they used the model to resolve the question of cocaine's effect as an uptake inhibitor or releaser. After it was determined that a linear relationship exists between extracellular cocaine levels and DA levels (Nicolaysen *et al.*, 1988b; Hurd *et al.*, 1988), previously reported data from the literature were compared to theoretical constructs to differentiate release and re-uptake inhibition. Based on metabolite levels after cocaine treatment, Nicolaysen and Justice (1988a) concluded that cocaine does not promote DA release; rather, cocaine increases extracellular DA solely through its uptake inhibitor properties.

Further evidence differentiating the mechanism of cocaine from amphetamine was obtained using striatal slices *in vitro* under basal conditions; that is, in the absence of impulse dependent DA release. Data presented in Figure 1 demonstrate that amphetamine, but not cocaine, increases extracellular DA under basal (i.e., nonstimulated) conditions. Moreover, the DA-releasing action of amphetamine is completely blocked by cocaine. These observations support the notion that amphetamine is a substrate for the DA transporter, whereas cocaine is an antagonist at the transporter. Thus, under normal conditions *in vivo*, both compounds increase synaptic levels of DA; however, they do so by distinct mechanisms.

Compensatory Response to Cocaine: DA Synthesis *In Situ*

Because DA and 5-HT neurons are sensitive to autoregulatory feedback pathways (for reviews, see Wolf and Roth, 1987a; Aghajanian *et al.*, 1987), cocaine-induced

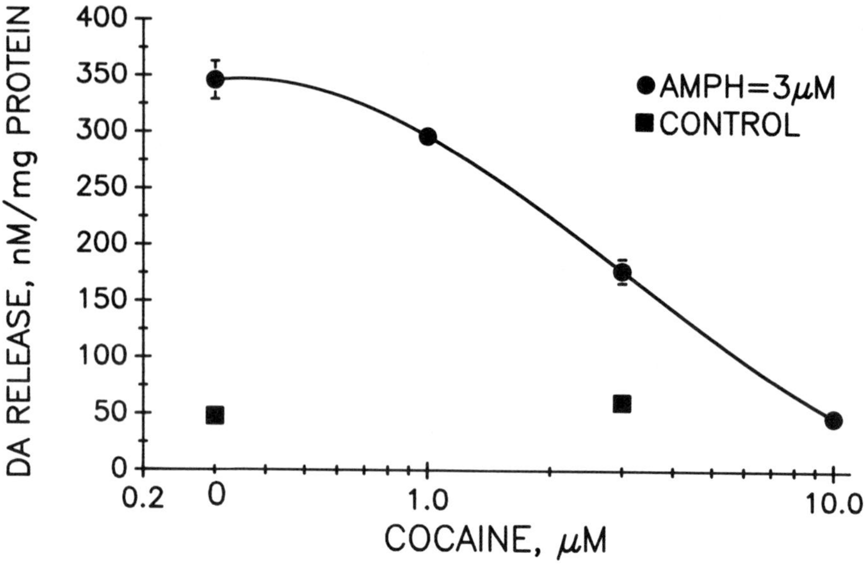

Figure 1. In the absence of impulse dependent release, cocaine does not increase extracellular DA, in contrast to the profound effect of amphetamine under similar conditions. Cocaine effectively blocked the DA releasing effect of amphetamine, supporting the role of cocaine as an uptake inhibitor rather than a DA releaser. Striatal slices were incubated with various doses of cocaine (0–10 μ*M*) or amphetamine (3 μ*M*) for 10 min after which an aliquot of the supernatant was removed for HPLC-EC analysis. Each point is the average ± SEM of four to five incubations.

increases in extracellular transmitter levels might be expected to activate feedback pathways as a means of compensating for enhanced chemical transmission. To test this hypothesis, the most effective experimental design is a measure of DA or 5-HT function (e.g., release, firing, or synthesis) utilizing a preparation that preserves the appropriate feedback mechanisms. In the case of a re-uptake inhibitor such as cocaine, it is difficult to observe a dynamic inhibition of release of DA or 5-HT *in vivo* (e.g., utilizing microdialysis) when increased synaptic DA is an experimental prerequisite. Also, as noted above, post-mortem DA metabolites do not accurately reflect the putative action of cocaine on synaptic DA.

However, the concept that cocaine produces an inhibitory feedback response has been tested extensively using single-unit studies of identified A9 DA (Pitts and Marwah, 1986, 1987), A10 DA (White *et al.*, 1987; Einhorn *et al.*, 1988; White *et al.*, 1991), and 5-HT neurons (Pitts and Marwah, 1986, 1987; Lakoski and Cunningham, 1988; Cunningham and Lakoski, 1988; Lakoski, 1991). Except for identified mesoprefrontal DA neurons (White *et al.*, 1987), the studies cited above have demonstrated that cocaine administration suppresses the unit activity of DA and 5-HT neurons. Pitts and Marwah (1987) also reported that a population of A9 DA neurons increased unit activity in response to cocaine.

In addition to the modulation of DA release and DA unit activity, extracellular DA can regulate its biosynthesis either directly via autoreceptor occupation or

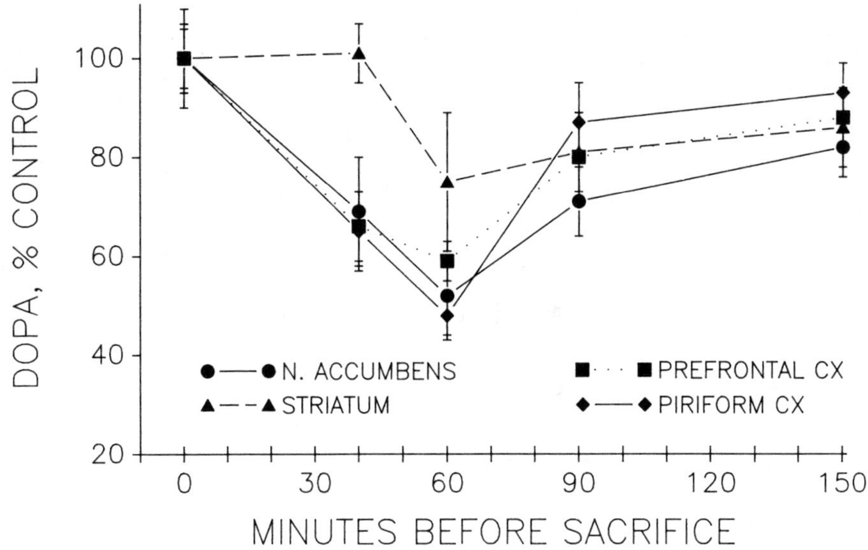

Figure 2. Time course of the cocaine-induced inhibition of DA synthesis *in vivo* illustrates that the effect occurred rapidly and was reversible. Rats were sacrificed various times after administration of cocaine (59 μmol/kg, i.p.), each point is the mean ± SEM for four to five subjects. Significant differences were evident at 40 and 60 min in all tissues, except striatum, 60 min only (ANOVA, least significant difference test, $p < 0.05$). (Adapted from M.P. Galloway. *Synapse* **6**: 63 (1990).)

indirectly by stimulation of long loop striato-nigral feedback pathways (Wolf and Roth, 1987a). The influence of various parameters on DA synthesis *in situ* can be assessed directly by measuring DOPA and 5-HTP accumulation *in vivo* after administration of the decarboxylase inhibitor *m*-hydroxybenzylamine (reviewed by Broadhurst and Briley, 1988). Synthesis can also be measured *in situ* in striatal brain slices *in vitro* which have a more defined and controlled milieu, yet retain the relevant presynaptic regulatory mechanisms affecting synthesis (Wolf *et al.*, 1986; Salah *et al.*, 1989). We recently characterized the normal autoreceptor control of DA synthesis solely as a D_2-mediated phenomenon with no influence from D_1 receptor stimulation (Salah *et al.*, 1989; Wachtel *et al.*, 1989; Brooderson *et al.*, 1990).

Systemic administration of cocaine inhibited DA synthesis *in vivo* in a time-dependent fashion (Figure 2). Acute cocaine suppressed synthesis in the striatum, nucleus accumbens, medial prefrontal cortex, piriform cortex, and hippocampus (Galloway, 1990). The inhibition occurred rapidly (the earliest period tested, 40 min before sacrifice), with a maximum observed at 60 min, and was reversible after 120–150 min. Synthesis was attenuated starting at a dose of 10 μmol/kg (3.4 mg/kg i.p.) and, in the striatum and accumbens, the maximal effect was evident at 10 μmol/kg whereas 60 μmol/kg of cocaine exerted the maximal effect on prefrontal DA synthesis. The greatest inhibition was evident in the mesocortical system, followed by the mesoaccumbens and nigrostriatal systems. It is notable

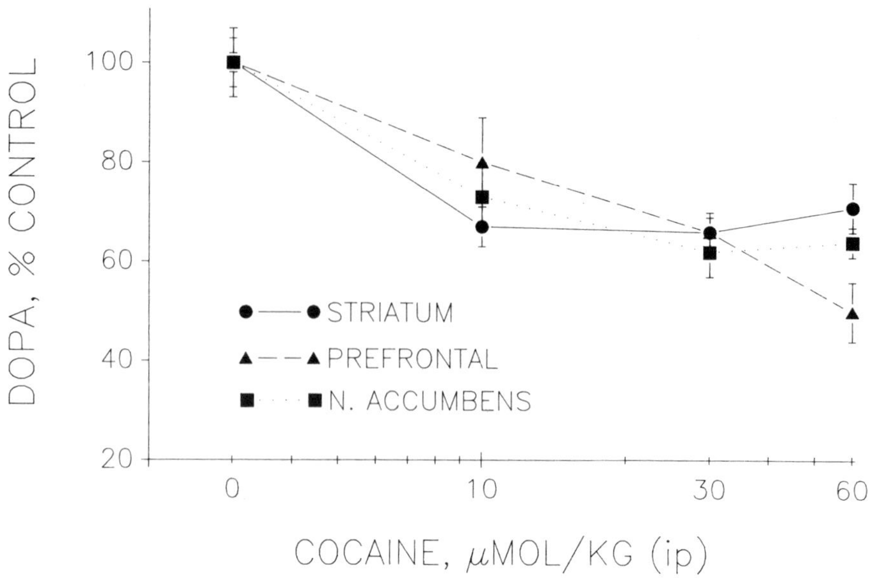

Figure 3. Dose-response relationship between DA synthesis and cocaine in selected DA projection fields illustrates a similar response to increasing cocaine, except in the prefrontal cortex at the highest cocaine dose. Cocaine was administered 60 min before sacrifice; each point is the mean ± SEM for four subjects. Inhibition was significantly different from control at 10 μmol/kg for accumbens and striatum, 30 μmol/kg for prefrontal. (Adapted from M.P. Galloway. Synapse **6**: 63 (1990).)

that cocaine exerted a similar effect on DA synthesis (irrespective of the projection area) since these DA neurons represent several variables, for example, DA content, DA turnover rate, firing rate (White and Wang, 1984; Chiodo *et al.*, 1984), autoreceptor responsiveness (Bannon *et al.*, 1981; Galloway *et al.*, 1986), and long-loop feedback pathways (Wang, 1981). The acute compensatory reponse of DA synthesis (in the medial prefrontal cortex) to increased levels of synaptic DA is in contrast to the absence of an effect of cocaine on unit activity in mesoprefrontal DA neurons (White *et al.*, 1987). Thus, the diminution of prefrontal DA synthesis may represent a major compensatory mechanism since impulse flow in mesoprefrontal DA neurons appears not to be altered by acute cocaine (White *et al.*, 1987). In mesoaccumbens DA neurons, a similar compensatory response is apparent at both the level of unit activity (Einhorn *et al.*, 1988) and DA synthesis (Figure 3).

Further experiments to characterize the acute inhibitory effect of cocaine on synthesis revealed that inhibition of DA, but not 5-HT, synthesis was blocked by pretreatment with the D_2 antagonist sulpiride (Figure 4) and eticlopride (Galloway, 1990), suggesting that inhibition of DA synthesis *in vivo* was mediated by stimulation of D_2 receptors. Sulpiride pretreatment did not block the effect of cocaine on synthesis in noradrenergic projection fields such as the hippocampus or temporal cortex. That cocaine inhibited DA synthesis in depolarized striatal slices *in vitro* suggests that cocaine increases extracellular DA (above the depolarized level) which subsequently stimulates D_2 DA receptors, with an ensuing inhibition of DA synthesis (a comparable conclusion was reached by Wolf *et al.*, 1986).

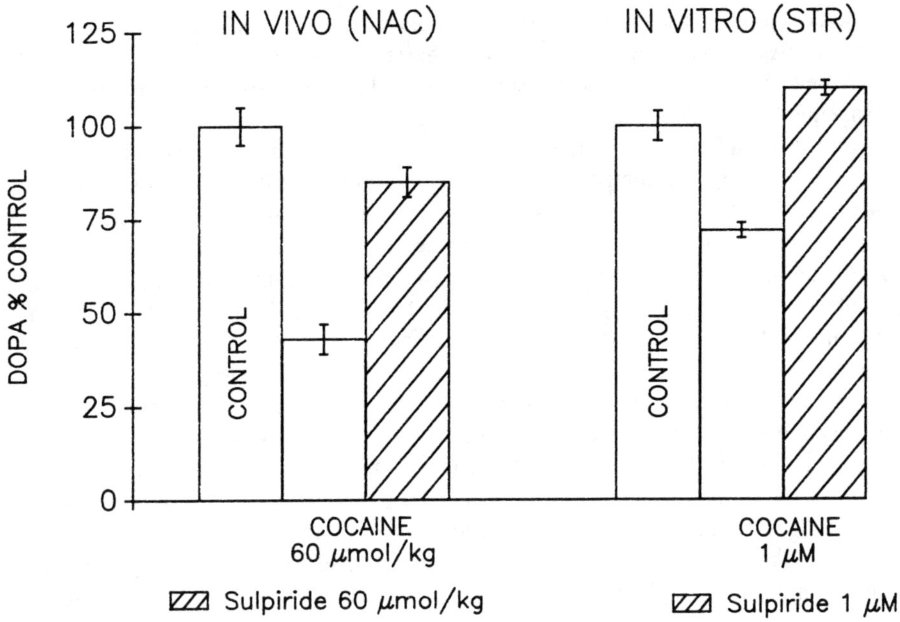

Figure 4. Left: Cocaine-induced inhibition of nucleus accumbens DA synthesis *in vivo* was prevented by prior treatment with the D_2 antagonist sulpiride (60 μmol/kg, 15 min prior to cocaine). Each bar is the mean ± SEM of five subjects. Right: In striatal slices *in vitro*, cocaine-induced inhibition of DA synthesis is blocked by the D_2 antagonist sulpiride (1 μ*M*; K^+ = 30 m*M* for all groups). These data illustrate that stimulation of D_2 receptors is necessary for cocaine to inhibit DA synthesis *in situ*, and that inhibition can occur in the absence of long-loop feedback pathways.

Since long-loop feedback pathways do not regulate synthesis in the slice preparation, the *in vitro* diminution in DA synthesis under these conditions may be mediated by stimulation of nerve terminal DA autoreceptors. It also follows that the effect of cocaine on DA synthesis *in vivo* may be mediated by endogenous DA acting at synthesis-modulating DA autoreceptors. The possibility that cocaine stimulates DA autoreceptors directly was tested by measuring the ability of cocaine to reverse the γ-butyrolactone (GBL)-induced increase in striatal DA synthesis *in vivo* (Walters and Roth, 1976). Cocaine (60 μmol/kg) neither reversed the GBL-induced increase in DA synthesis nor did cocaine interfere with the ability of direct DA agonists (e.g., quinpirole or apomorphine) to reverse the GBL effect (Galloway, 1990). Thus, cocaine is neither a direct DA agonist nor a DA antagonist and, to suppress DA synthesis, cocaine requires impulse-dependent DA release. Additional evidence for the specificity of the cocaine effect on synthesis was the observation that procaine (30 μmol/kg) did not inhibit DA or 5-HT synthesis, and depletion of monamines with reserpine decreased the effectiveness of cocaine in the medial prefrontal cortex, but not in the striatum (Galloway, 1990).

Response to Amphetamine: DA Synthesis *In Situ*

Administration of amphetamine, in contrast to cocaine, produces a differential response on DA synthesis *in vivo*, depending on which area of the brain is examined.

Time-course experiments revealed that amphetamine (10 μmol/kg, s.c.) affected striatal DA synthesis in a temporally biphasic fashion: after a transient twofold increase (with a zenith at 40 and 60 min), striatal DA synthesis eventually decreased to a value less than control. The effect was selective for the striatum because, under identical conditions, DA synthesis *in vivo* is unchanged in the accumbens and prefrontal cortex. The contribution of feedback pathways in this differential response to amphetamine can be demonstrated by prior antagonism of D_2 receptors. For example, in rats pretreated with the D_2 antagonist eticlopride, subsequent amphetamine administration increased DA synthesis in the medial prefrontal cortex, accumbens, and striatum (Tyler *et al.*, 1989; Tyler and Galloway, 1990). The amphetamine dose response reveals a mirror image relationship between the striatum and prefrontal cortex: low doses (1–10 μmol/kg, 45 min before sacrifice) either enhance (striatum) or inhibit (prefrontal) synthesis, whereas higher doses are relatively ineffective in both areas. Interestingly, a high dose of amphetamine (30 μmol/kg), which does not alter basal striatal DA synthesis, can reverse the GBL-induced increase in striatal or accumbens DA synthesis (Galloway *et al.*, unpublished) supporting the hypothesis that amphetamine releases DA independent of impulse flow, as a substrate for the DA transporter. These seemingly paradoxical observations with amphetamine (i.e., the ability to increase basal synthesis coupled with the ability to reverse GBL) can be reconciled by the supposition that amphetamine releases an intraneuronal, regulatory pool of DA. Thus, the time course of amphetamine on DA synthesis would be an increase (relief of end-product inhibition) followed by a subsequent attenuation, possibly as a result of autoreceptor stimulation. Accordingly, the apparent absence of an effect of high-dose amphetamine on basal striatal synthesis may reflect a balance between the intraneuronal activation of synthesis (elimination of feedback inhibition) and the attenuation of synthesis by autoreceptor occupation.

Consequently, although both amphetamine and cocaine are able to increase extracellular levels of DA *in vivo*, opposite effects on DA biosynthesis are observed. As such, a significant difference between the effects of cocaine and amphetamine on presynaptic DA function lies in the subsequent compensatory mechanism; that is, after an initial increase in functional DA transmission, cocaine will decrease DA synthesis (and perhaps transmitter replenishment), whereas after amphetamine, presynaptic DA synthesis will be enhanced. The differences in acute compensatory mechanism may have a substantial impact on functional transmitter availability. With both drugs, activation of inhibitory long-loop feedback pathways will presumably be evident but, depending on the particular DA neurons, this regulatory mechanism may have varying degrees of influence. The consequences of this hypothesis, especially as they relate to long-term drug abuse, remain to be determined.

Regulation of DA Systems: Implications for Cocaine

Since the mesoaccumbens and mesoprefrontal DA systems mediate an important element in reward and reinforcement, it is noteworthy that these tracts differ in their regulatory characteristics.

Control of either nigrostriatal (A9) or mesoaccumbens (A10) unit activity via stimulation of impulse-modulating DA autoreceptor is essentially equivalent in terms of agonist potency (White and Wang, 1984a; Chiodo, 1988), which contrasts the relative insensitivity of identified mesocortical DA neurons to DA agonists (Chiodo *et al.*, 1984; White and Wang, 1984b). Consistent with this difference, Einhorn *et al.* (1988) and White *et al.* (1987) found that systemic cocaine inhibited A10 DA neuronal firing but was without effect on the activity of mesocortical DA cells. Whereas this difference was seemingly predictable based on autoreceptor differences, Einhorn *et al.* (1988) found that a long-loop feedback pathway was primarily responsible for the A10 response to cocaine, which contrasts the inhibition afforded by amphetamine (Wang, 1981; White *et al.*, 1987). Even though the suppressant effect of cocaine on A10 DA neurons is only partly mediated entirely by somatodendritic DA autoreceptors, mesoaccumbens DA neurons do diminish their firing, albeit partially, to compensate for increased extracellular DA, whereas mesocortical DA neurons do not decrease their firing under similar conditions.

Autoregulation of DA release appears to be conserved in all DA neuronal systems, including the medial prefrontal cortex (Altar *et al.*, 1987; Wolf and Roth, 1987b). Thus, elevated DA levels subsequent to cocaine will eventually attenuate DA release in all DA projections. As noted above though, an experimental approach to examine the effect of increased synaptic DA on the actual release process remains to be developed, although the time resolution provided by *in vivo* cyclic voltammetry may provide insight. In this regard, DA projections that are not associated with cocaine-induced increases in extracellular DA [e.g., the medial prefrontal cortex (Moghaddam and Bunney, 1989)] may be immune to this important feedback mechanism which is potentially involved with the termination of cocaine's acute effect.

The other major neuronal function regulated by extracellular DA is the activity of tyrosine hydroxylase (Wolf and Roth, 1987a). Nerve terminal synthesis modulating DA autorecptors on mesoaccumbens DA terminals appear to be more sensitive to DA agonists than their nigrostriatal counterparts (Roth *et al.*, 1979; Demarest *et al.*, 1983), although DA synthesis in both of these structures is tightly coupled to DA autoreceptors. In contrast, mesocortical DA neurons are incapable of directly regulating DA synthesis (Bannon *et al.*, 1981a; Galloway, *et al.*, 1986). Nonetheless, prefrontal DA synthesis is sensitive to intracellular DA levels and, thus, any alteration of DA levels in prefrontal DA neurons may eventually manifest as an altered rate of synthesis (Galloway *et al.*, 1986). As reviewed above, inhibition of prefrontal DA synthesis *in vivo* is a prominent effect of systemically adminstered cocaine (Galloway, 1990). Since the prefrontal DA projection is characterized by a relatively small DA pool and a high turnover rate (e.g., $t_{1/2}$ = 15 min, Galloway *et al.*, 1986; Bannon *et al.*, 1981b), any disruption of DA synthesis would be expected to have a profound effect on transmitter availability. This mode of regulation is remarkable because (1) the mesoprefrontal DA projection is a significant element of the reward pathway (Goeders and Smith, 1983, 1986) and (2) mesoprefrontal DA neurons do not exhibit a diminution of their firing rate in response to cocaine (White *et al.*, 1987). In other words, acute suppression of DA synthesis (and perhaps DA release), in prefrontal DA neurons may represent the major compensatory mechanism in this group of neurons.

INTERACTION OF COCAINE WITH SEROTONIN

Effects of Cocaine on 5-HT Function

The ability of cocaine to modulate 5-HT function *in vivo* has been convincingly demonstrated by Lakoski (1991), who reported that 5-HT unit activity in the dorsal raphe is potently and completely inhibited by cocaine administered systemically (cumulative dose to 50% inhibition = 2 μmol/kg, i.v.) (Cunningham and Lakoski, 1988). As pointed out by Cunningham and Lakoski (1988), the depression of locus coeruleus noradrenergic unit activity by systemic cocaine (Pitts and Marwah, 1986, 1987) may contribute to the potent inhibition of 5-HT units since the noradrenergic system is known to drive 5-HT raphe neurons (Barban and Aghajanian, 1980; Reinhard *et al.*, 1983). Nevertheless, iontophoretically applied cocaine, but not procaine, decreased raphe unit activity by 40%, supporting a direct effect on raphe cells (reviewed in Lakoski and Cunningham, 1988). The rate suppressant effect of systemic cocaine on 5-HT raphe cells was attenuated by prior treatment with reserpine (Pitts and Marwah, 1986, 1987) and systemic administration of the related, local anesthetic procaine was without effect on firing. In the raphe slice *in vitro,* intracellular recordings revealed that cocaine hyperpolarized the membrane by 13 mV in a spiperone reversible fashion leading Pan and Williams (1989) to suggest that the cocaine effect was mediated by endogenous 5-HT. Thus, electrophysiological effects of cocaine are consistent with the reported inhibition of 5-HT uptake by cocaine (Ross and Renyi, 1969; Friedman *et al.*, 1975).

In agreement with the 5-HT re-uptake properties of cocaine, acute administration of cocaine inhibits 5-HT synthesis (after decarboxylase inhibition) in a dose-related, reversible fashion (Figure 5). Procaine did not inhibit 5-HT synthesis, nor was the effect of cocaine on 5-HT synthesis diminished by pretreatment with the DA antagonist sulpiride (Galloway, 1990). The particular locus of cocaine's action on 5-HT synthesis (e.g., terminal 5-HT autoreceptors, impulse flow, long-loop feedback pathways) remains to be determined. The suppressant effect of cocaine on 5-HT synthesis may follow as a consequence of cocaine's ability to decrease 5-HT unit firing (Lakoski and Cunningham, 1988). Thus, similar to DA systems, 5-HT neurons exhibit a compensatory response to cocaine administration. Currently the effect of cocaine on extracellular 5-HT levels, determined with brain microdialysis, remains to be determined.

The Role of 5-HT in Reinforcement: Amphetamine and Cocaine

It is known that balanced 5-HT innervation plays a fundamental role in the treatment and perhaps etiology of affective disorders (reviewed by Meltzer and Lowy, 1987). Thus, the widespread 5-HT innervation of the central nervous system (CNS) has the potential to influence diverse processes such as motivation, pleasure, and perhaps reward and reinforcement. As psychomotor stimulants, both cocaine and amphetamine share the ability to increase DA neurotransmission (albeit through distinct physiological mechanisms), whereas their effect on synaptic levels of 5-HT is noticeably distinct (*vide infra*). Currently, the role of 5-HT systems in the rewarding properties of cocaine remains to be confirmed (*vide infra*). However, 5-

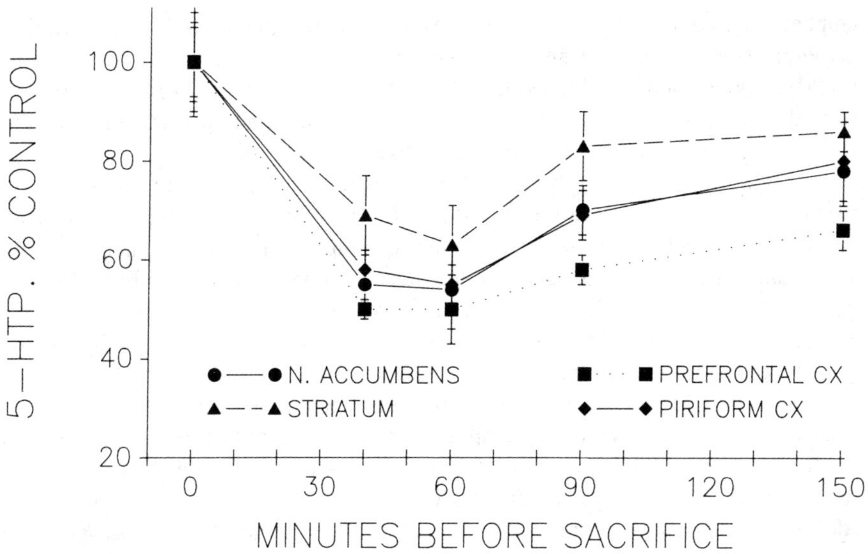

Figure 5. Time course of the cocaine-induced inhibition of 5-HT synthesis *in vivo* illustrates that the effect occurred rapidly and was partially reversible at the longest time measured. Rats were sacrificed at various times after administration of cocaine (59 μmol/kg, i.p.), each point is the mean ± SEM for four to five subjects. Significant differences (vs. control) were evident at all times except at 90 and 150 min in the striatum and at 150 min in the accumbens (ANOVA, least significant difference test, $p < 0.05$). (Adapted from M.P. Galloway. *Synapse* **6**: 63 (1990).)

HT has a definite influence on the self-administration of amphetamine and evidence exists to suggest that 5-HT-containing neurons can modulate amphetamine self-administration in rats. Thus, attenuation of 5-HT function (e.g., destruction of 5-HT neurons with 5,7-DHT) increases, whereas augmentation of 5-HT tone (e.g., administration of fluoxetine or tryptophan) decreases, amphetamine self-administration (Lyness *et al.*, 1979, 1980; Smith *et al.*, 1986, Yu *et al.*, 1986; Porrino *et al.*, 1989). Recently, ethanol self-administration in rats was reduced by prior treatment with fluoxetine (Haraguchi *et al.*, 1990).

Since cocaine and amphetamine share several behavioral properties such as cross-discrimination, behavioral sensitization, and the potential for addiction, it might be assumed that the inverse relationship between 5-HT activity and amphetamine self-administration would extend to the reinforcing properties of cocaine. It might also be speculated that the disruption of cocaine self-administration by the D_1 antagonist SCH 23390 (*vide supra*) may be related to the 5-HT blocking properties of SCH 23390 (Bischoff *et al.*, 1986). These considerations notwithstanding, recent reports are apparently conflicting (Carroll *et al.*, 1990; Porrino *et al.*, 1989) and suggest the opposite: the behavioral effects of amphetamine and cocaine can actually be distinguished based on their response to 5-HT re-uptake inhibitors. In rats trained to self-administer intravenous cocaine (1 mg/kg, FR10) or amphetamine, Porrino *et al.* (1989) found that increasing 5-HT tone by pretreatment with the 5-HT re-uptake fluoxetine (2.50–10 mg/kg, i.p.) decreased the self-administration rate for amphetamine, but not for cocaine. However, after changing the self-administration paradigm (FR4) and lowering the cocaine dose (0.1 or 0.2 mg/kg), Carroll *et al.* (1990) demonstrated that fluoxetine (2.5–10 mg/kg, i.v.) reversibly reduced cocaine self-administration by 50%. Interestingly, when the cocaine dose was increased to 0.4 mg/kg, fluoxetine was no longer effective (Carrol *et al.*, 1990).

An additional difference between 5-HT and DA with respect to reinforcement was recently demonstrated by Ritz and Kuhar (1989) and Kuhar *et al.* (1990), who found that the self-administration potency of amphetamine-type compounds, but not cocaine-type drugs, was uniquely and inversely related to their ability to displace paroxetine from 5-HT transporter sites. These same authors also found that the reinforcing properties of cocaine were strongly correlated with their ability to bind the DA transporter (Ritz *et al.*, 1987).

Although the studies cited above imply that increasing 5-HT tone shows a selectivity for the suppression of amphetamine self-administration, an examination of the neurochemical mechanism of action for cocaine and amphetamine suggests a further interpretation. Microdialysis experiments have shown that, in terms of doses and duration of action, systemic amphetamine preferentially releases DA in the anterior lateral striatum, when compared to its effects on extracellular 5-HT (Kuczenski and Segal, 1989). In this study, dopamine levels were elevated starting at a dose of 0.5 mg/kg (s.c.) amphetamine, whereas extracellular levels of 5-HT did not increase until the amphetamine dose reached 2 mg/kg (s.c.); the duration of the 5-HT increases was transient, lasting 20–30 min and the magnitude of the 5-HT increase was severalfold less than the DA response (Kuczinski and Segal, 1989). Thus, when administered systemically (to mimic the conditions associated with reinforcement), amphetamine can be differentiated from cocaine based on the response of extracellular 5-HT. As expected, infusion of amphetamine directly into the nucleus accumbens (thereby bypassing certain feedback mechanisms) resulted in a profound increase in both DA and 5-HT (Hernandez *et al.*, 1987). However, based on the ability of fluoxetine to reduce systemic amphetamine self-administration, one would predict that the high dose of amphetamine used by Hernandez *et al.* (1987) (i.e., capable of stimulating 5-HT release) would tend to attenuate the rate of amphetamine self-administration.

Cocaine, by virtue of its ability to inhibit both DA and 5-HT re-uptake simultaneously enhances (activates) both dopaminergic and serotonergic transmission. For example, in the nucleus accumbens, the cocaine-induced inhibition of particular type I neurons appears to be the result of the combined effects of increased DA and 5-HT input (White *et al.*, 1987; Einhorn *et al.*, 1988). Furthermore, since 5-HT re-uptake is blocked by cocaine, it is not unreasonable to predict that, under particular conditions (e.g., dose), the presence of an additional 5-HT re-uptake inhibitor (e.g., fluoxetine) would be without effect on cocaine self-administration, a prediction confirmed by Porrino *et al.* (1989). It remains to be resolved whether increasing 5-HT transmission (e.g., with direct, selective 5-HT agonists) beyond the enhancement associated with a 5-HT re-uptake inhibitor will modify cocaine self-administration, as reported for amphetamine self-administration. The detailed characterization *in vivo* of cocaine's effect on extracellular 5-HT levels in the accumbens, and the comparison to the effects of amphetamine, also remain to be determined. It is conceivable that the use of a selective 5-HT re-uptake inhibitor, such as fluoxetine (Prozac) which is currently used as an antidepressant, may be useful in the clinical management of cocaine craving in patients during cocaine abstinence (Martin *et al.*, 1989).

An additional fundamental difference between the acute actions of cocaine and amphetamine is their effect on DA and 5-HT synthesis. Accordingly, amphetamine will exert differential effects both within the heterogenous dopaminergic system and between neuronal systems such as those utilizing DA, 5-HT, and NE. These basic differences on both the initial increase of 5-HT or DA coupled with the compensatory response of each system may underlie differences in the abuse liability of different drugs.

Serotonin Dopamine Interactions

Using the observation that increased 5-HT tone after fluoxetine diminishes amphetamine and cocaine self-administration (references cited above), one can speculate that dual innervation (by DA and 5-HT) of a particular accumbens pathway is additive to satisfy a reward circuit in the accumbens. If both DA and 5-HT innervated the circuit equally, decreased input from either pathway would be balanced by increased activity through the other input. Alternatively, if the primary input to the reinforcement circuit were dopaminergic, facilitation of DA function (release, synthesis, or unit activity) by either axo-axonic or axo-dendritic 5-HT inputs would also be consistent with the available behavioral evidence. Under the latter scenario with DA serving as the primary reward pathway, heightened 5-HT tone would physiologically increase DA transmission and thereby decrease the requirement to drive the DA system with the self-administered drug.

In support of this putative interaction between 5-HT and DA systems, we have recently observed that administration of 5-HT-1 agonists increase DA function in the nucleus accumbens (Figure 6). Administration of trifluoromethylphenylpiperazine (TFMPP) increases DA synthesis in the accumbens, but not prefrontal cortex, when measured *in vivo* after inhibition of DOPA decarboxylase with NSD-1015 (Johansen and Galloway, 1989). Under the same conditions, TFMPP decreased 5-HT synthesis, as would be expected of a 5-HT autoreceptor agonist. However, the elevation of DA synthesis was not a consequence of attenuated 5-HT activity in that the 5-HT-1A agonist 8-OHDPAT (1 to 10 μmol/kg, s.c.) did not stimulate DA synthesis in mesolimbic or mesocortical DA fields.

Further evidence that 5-HT may facilitate DA release was obtained with brain microdialysis *in vivo* (Benloucif and Galloway, 1989). In these experiments (Figure 7), 5-HT agonists (TFMPP, RU 24969, 8-OHDPAT, and 5-HT) increased DA release in a transient and dose-related fashion. As the agonists were administered directly through the dialysis probe, it is feasible that 5-HT facilitates DA release via an interaction at the level of the nerve terminal. Thus, evidence from both DA synthesis and DA release studies *in vivo* support the notion that increased 5-HT tone facilitates DA activity, at least at the level of the DA function in the accumbens and anterior striatum. These observations support the possibility that elevated 5-

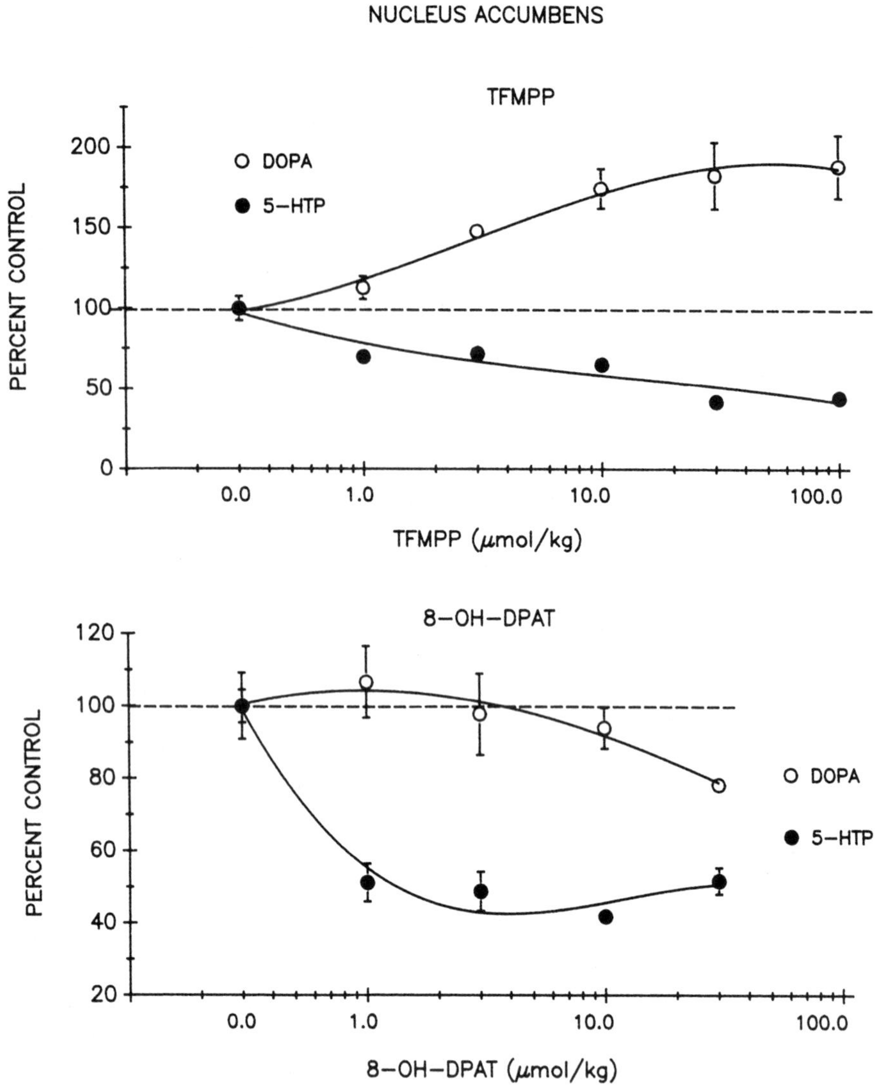

Figure 6. Top: Administration of the 5-HT agonist TFMPP (50 min before sacrifice) produced a dose-dependent activation in nucleus accumbens. DA synthesis *in vivo* (significantly different starting at 3 μmol/kg). As expected for a 5-HT agonist, TFMPP decreased 5-HT synthesis under similar conditions. Bottom: In contrast to TFMPP, the 5-HT-1A agonist 8-OHDPAT did not increase DA synthesis in the accumbens. In the same subjects, 8-OHDPAT profoundly inhibited 5-HT synthesis. (Adapted from P.A. Johansen and M.P. Galloway. *Soc. Neurosci. Abstr.* **15**: 1313 (1989).)

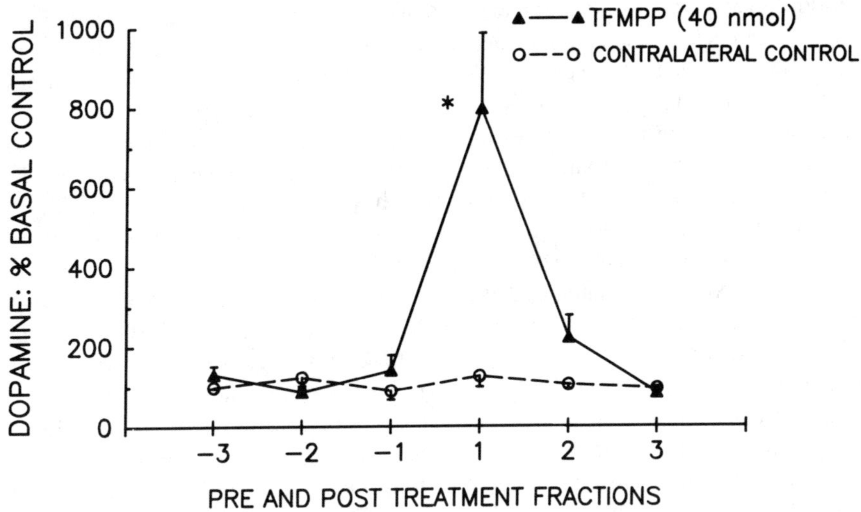

Figure 7. Microdialysis of the anterior striatum revealed that administration of TFMPP directly into the anterior striatum increases extracellular DA ($p < 0.05$, post-drug vs. pre-drug, three way ANOVA, repeated measures). These findings suggest that, *in vivo*, 5-HT agonists facilitate DA release. Each point is the mean ± SEM for 5 (▲) or 2 (○) subjects. (Adapted from S. Benloucif and M.P. Galloway. *Soc. Neurosci Abstr.* **15**: 1313 (1989).)

HT, subsequent to inhibition of re-uptake by cocaine, may enhance dopaminergic transmission at selected synapses and thereby decrease the requirement to drive DA transmission.

CHRONIC COCAINE: FUNCTIONAL CHANGES RELATED TO SENSITIZATION

The major emphasis of studies with repeated cocaine or amphetamine treatment has focused on the mechanisms associated with the expression of behavioral sensitization. Since behavioral enhancement is most likely a consequence of augmented dopaminergic activity in the striatum (stereotypy) or accumbens (locomotor activity), inquiries have centered on the pre- and post-synaptic components of dopamine transmission. Several experimental approaches and diverse hypothesis concerning amphetamine-induced sensitization were reviewed recently by Robinson and Becker (1986).

Metabolites, Receptors, and Uptake

In contrast to the toxicity associated with repeated use of phenethylamines, chronic administration of cocaine to experimental animals does not appear to be neurotoxic to D or 5-HT neurons. Klevin *et al.* (1988b) were unable to detect any major alteration in DA, 5-HT, or metabolites in four representative brain areas after repeated treatment with heroic doses of cocaine. Doses of 20 to 100 mg/kg/d (both i.p. and continuous i.v. administration), followed by either a 3- or 60-d withdrawal period, failed to change monoamine levels. Recently, we reported that plasma HVA levels from cocaine-addicted patients enrolled in a cocaine abstinence program did not differ significantly from control HVA levels (Martin *et al.*, 1989). These findings in rodents and humans do not support the DA depletion hypothesis of chronic cocaine, a theory promulgated by clinical reports suggesting that bromocryptine, a DA agonist, alleviates cocaine craving in cocaine-addicted subjects (Dackis *et al.*, 1987). Nonetheless, the DA depletion hypothesis can accommodate a functional decrease in DA neurotransmission and it remains to be determined if cocaine induces a lesion in a selective DA tract in the human CNS.

Cocaine toxicity, measured as decreased tyrosine hydroxylase immunoreactivity, was reported to occur in rats treated with 20 mg/kg/d for 10 d (Trulson *et al.*, 1987). However, Ryan *et al.* (1988) were unable to detect any axonal degeneration (silver staining) or changes in tyrosine hydroxylase immunoreactivity in tissues obtained from rats treated with substantial doses of cocaine. Trulson and Ulissey (1987) also reported that basal DA synthesis was decreased 30% after repeated cocaine treatment. However, in rats ($n = 6$) repeatedly exposed to cocaine (10 mg/kg, i.p., bid, $\times$ 14 d), we found that basal DA synthesis *in vivo* was slightly increased (10 to 20%) in 6 representative DA projections (Galloway *et al.*, unpublished observations).

Provocative regional differences have been reported in brain tissues obtained from animals treated repeatedly with stimulants. Goeders and Kuhar (1987) reported that treatment with cocaine for 2 weeks (10 mg/kg, i.p.) produced locomotor sensitization, decreased the number of striatal D_2 binding sites by 34%, and increased D_2 receptor sites in the accumbens by 29%. Similary, *in vivo* treatment with amphetamine decreases striatal, but not accumbens, stimulated adenylate cyclase activity and D_1 receptor sites (Roberts-Lewis *et al.*, 1986; Barnett *et al.*, 1987). Missale *et al.* (1985) found substantial regional differences in the efficacy of cocaine to inhibit DA uptake in slices *in vitro*. They reported that 10 μM cocaine inhibited striatal Na^+-dependent DA uptake by 50%, while having little effect on DA uptake in accumbens slices. In slices obtained from rats treated repeatedly with cocaine, the V_{max} for striatal DA uptake was increased by 9% whereas the V_{max} in accumbens tissue was increased 147% (Missale *et al.*, 1985), suggesting that functional alterations in mesoaccumbens DA neurons were substantial. Taken together, the similar effect of repeated cocaine on accumbens D_2 (postsynaptic ?) and accumbens DA uptake mechanisms (presynaptic) may be compensatory, related phenomena. However, Boja and Kuhar (1989) recently reported that cocaine binding sites and cocaine inhibition of DA uptake were equivalent when determined in synaptosomes obtained from either the striatum or accumbens. Thus, the intriguing finding that repeated cocaine selectively alters DA uptake kinetics (Missale *et al.*, 1985) remains to be confirmed. It also remains to be determined if chronic cocaine can selectively modify individual aspects of the transporter (e.g., transport kinetics or cocaine binding site). The recent expression of the human dopamine transporter in frog oocytes (Bannon *et al.*, 1990) provides a useful tool to investigate individual molecules of the DA transporter complex for potential changes that may occur after repeated drug exposure.

DA Release and DA Autoreceptor Sensitivity

When compared to studies with repeated amphetamine, relatively few studies exist to describe DA release parameters after continuous exposure to cocaine. In a recent study using brain slices obtained from rats treated 2–3 weeks prior with cocaine (15 mg/kg, × 3 d), Kalivas and Duffy (1988) found that K^+-stimulated (40 m*M*) DA release from DA cell bodies was significantly diminished, but DA release from DA terminals was either unaltered or increased. Similar differences were noted utilizing DA synthesis *in vivo*. Kalivas *et al.* (1988) found that the acute, cocaine-induced decrease in A10 DOPA synthesis *in vivo* was diminished in rats pretreated with cocaine, whereas the cocaine effect on DA synthesis in terminal fields was unchanged. In contrast to the effect of repeated cocaine on A10 DA release, depolarization-induced DA release after amphetamine treatment is invariably enhanced (e.g., Robinson and Becker, 1986; Castaneda *et al.*, 1988). However, the release-enhancing effects of amphetamine pretreatment have been observed in DA terminal fields, in contrast to cocaine's effect on DA release from dendrites, noted above. Kalivas *et al.* (1988) suggested that chronic cocaine resulted in decreased dendritic DA release with an attendant decrease in somatodendritic DA autoreceptor stimulation (Kalivas *et al.*, 1988). In turn, the activity of meso-accumbens DA neurons would be increased and more responsive to a subsequent cocaine challenge. Although this is a plausible hypothesis, the factor(s) controlling DA release from dendrites still remain(s) elusive, especially the "releasability" component thought to be altered by cocaine. Moreover, dendritic DA synthesis is not subject to DA autoreceptor regulation (Nissbrandt *et al.*, 1988). With a unified theme, multiple theories explain the enhanced DA release that is associated with sensitization: dendritic DA release is decreased by an unknown intrinsic factor with a subsequent decreased interaction at inhibitory autoreceptors, or stimulated release from DA terminals is enhanced by an alteration of an unknown factor.

In rats treated with either cocaine or methamphetamine for 14 d followed by a 7-d holiday, microdialysis of striatal DA demonstrated an enhanced response to a challenge dose of systemic cocaine (Akimoto *et al.*, 1989; Kazahaya *et al.*, 1989). This observation provides direct evidence that cocaine-induced sensitization is associated with increased DA in an intact, *in vivo* system. Robinson *et al.* (1988) provided similar, direct evidence for amphetamine-induced sensitization utilizing dialysis measurements of extracellular DA in the nucleus accumbens of freely moving rats. Although these reports are seemingly congruent, under dialysis conditions, cocaine's effect on extracellular DA is contingent on impulse-dependent DA releases, whereas amphetamine's effect is largely impulse and Ca^{++} independent. This mechanistic distinction illustrates that sensitization of DA release probably involves a series of complex modifications at both the nerve terminal and soma. Nonetheless, microdialysis will provide further insight into the possibility that alterations in DA autoreceptor sensitivity contribute to the changes in DA release afforded by chronic drug treatment.

Augmented DA release can be demonstrated in striatal slices obtained from animals treated only once with cocaine. Peris and Zahniser (1989) reported that after a single exposure to cocaine (10 mg/kg, i.p., 1 to 14-d withdrawal), amphetamine-stimulated [^{3}H]DA release from striatal slices was significantly augmented. Subsequently, they (Peris and Zahniser, 1989) demonstrated that the augmentation of [^{3}H]DA release *in vitro* was critically dependent on DA receptor stimulation *in vivo* at the time of cocaine administration. These biochemical changes are correlated with the ability of a single cocaine exposure to induce locomotor sensitization (Lin-Chu *et al.*, 1985). The sensitivity of release-modulating DA autoreceptors was slightly increased when determined in striatal slices obtained from rats treated chronically with cocaine (10 mg/kg/d, 1–2 weeks, 1-d withdrawal) (Dwoskin *et al.*, 1988). We observed a similar increase in autoreceptor sensitivity using the GBL model to examine the sensitivity of synthesis modulating DA autoreceptors *in vivo*. In the nucleus accumbens of rats treated with escalating doses of amphetamine (1–10 mg/kg, s.c., × 30 d), the ability of the D_2 agonist quinpirole (3–300 μ/kg, s.c.) to reverse the GBL-induced increase in DA synthesis was significantly enhanced (Tyler *et al.*, 1989).

In a direct measure of somatodendritic DA autoreceptor sensitivity, Henry *et al.* (1989) found that repeated cocaine treatment decreased the responsiveness of DA autoreceptors on A10 soma. The desensitization was evident when challenged with either systemic apomorphine or iontophoretic DA, and the number of spontaneously active DA neurons was increased by chronic cocaine (Henry *et al.*, 1989; White *et al.*, 1991). In the nucleus accumbens, dopamine-sensitive neurons demonstrated an enhanced D_1-DA receptor responsiveness in the treated subjects. Decreased inhibitory control of mesoaccumbens DA neurons, combined with supersensitive target cells in the accumbens, would therefore produce an overall increase in DA neurotransmission, providing a basis for the increased DA function associated with behavioral (locomotor) sensitization.

Given the potential role of 5-HT systems in the regulation of stimulant self-administration (see discussion above; Lyness *et al.*, 1980; Smith *et al.*, 1986; Yu *et al.*, 1986; Carroll *et al.*, 1990), surprisingly little information exists concerning 5-HT function after repeated cocaine exposure. Comparable studies on 5-HT release parameters (*in vitro* and *in vivo*), 5-HT autoreceptor sensitivity, and the interaction between 5-HT and DA systems are warranted given the significant effects of acute cocaine on pre- and post-synaptic 5-HT receptors (Cunningham and Lakoski, 1988; White *et al.*, 1987). Since cocaine interacts with the 5-HT transporter, it is conceivable that stimulants which interact with 5-HT neurons, such as MDMA ("ecstasy") or methamphetamine ("ice"), would have a more pronounced effect on 5-HT release in subjects repeatedly exposed to cocaine, akin to the exaggerated response to amphetamine.

SUMMARY

As an inhibitor of both DA and 5-HT re-uptake, cocaine produces a transient potentiation of neurotransmission by increasing synaptic transmitter levels. By activation of either postsynaptic feedback pathways or presynaptic autoreceptors, both 5-HT and DA neurons compensate for increased transmitter levels by decreasing either unit activity and/or transmitter synthesis. The mesoprefrontal DA system is exceptional in that the compensatory response occurs only at the level of attenuated synthesis, not unit activity. As the increase in DA transmission is a determinant in the reinforcing (addictive) property of cocaine, elucidating the regulatory response of DA neurons may provide valuable insight for clinical strategies to reduce the acute effects of cocaine. Moreover, understanding the changes that occur during chronic drug administration may explain the enhancement of DA neuronal activity associated with behavioral sensitization and ultimately the addictive properties of cocaine. Thus, basic studies defining the effects of cocaine on DA function will provide a clinical strategy to reduce cocaine abuse. Similar to the development of naltrexone and naloxone for opiate addiction, introduction of a selective DA antagonist or partial agonist may block the reward pathway activated by increased synaptic DA. Currently, the development of a successful strategy to prevent cocaine abuse is impeded by the absence of a pharmacological antagonist and the inability to precipitate a defined withdrawal behavior in animals.

One of the more promising perspectives for the clinical management of cocaine abstinence has emerged from behavioral studies of the influence of 5-HT on self-administration. The demonstration that augmenting 5-HT tone decreases amphetamine and cocaine self-administration in rats warrants the use of selective 5-HT re-uptake inhibitors (e.g., fluoxetine, Prozac, or sertraline) in clinical trials of cocaine withdrawal. The clinical response to 5-HT re-uptake inhibitors may be more favorable during the abstinence period than the active cocaine usage period because of the inhibition of 5-HT re-uptake attendant with the presence of cocaine. Increasing 5-HT neurotransmission apparently substitutes for a portion of the DA-mediated reward pathway, and thus 5-HT re-uptake inhibitors may be of prophylactic benefit to subdue the strong craving that complicates the latter phases of cocaine abstinence (Martin *et al.*, 1989; Gawin and Kleber, 1986). Additionally, the use of an antidepressant during the craving period would be of therapeutic benefit for the anhedonia associated with withdrawal.

ACKNOWLEDGMENTS

The author is gratefully indebted for the unpublished data and insight provided by Drs. Susan Benloucif, Patricia M. Johansen, and Corrine Tyler. I am also grateful for the expert technical assistance of Edward A. Novak, Benjamin N. Mathews, Dr. Julie S. Stahl, Michael J. Keegan, and the insight provided by Drs. David Clark, Joan M. Lakoski, Raymond S. Salah, and Francis J. White. Work was supported by PHS grants R01-DA-04102, R01-MH-41227, the Pharmaceutical Manufacturer's Association, the United Parkinson Foundation, and the State of Michigan, Department of Mental Health.

REFERENCES

G.K. Aghajanian, J.S. Sprouse, and K. Rasmussen. Physiology of the midbrain serotonin system, in: *Psychopharmacology: The Third Generation of Progress*. H.Y. Meltzer, ed., Raven Press, New York, p. 141 (1987).

K. Akimoto, T. Hamamura, and S. Otsuki. Subchronic cocaine treatment enhances cocaine-induced dopamine efflux, studied by *in vivo* intracerebral dialysis. *Brain Res.* **490**: 339 (1989).

C.A. Altar, W.C. Boyar, E. Oei, and P.L. Wood. Dopamine autoreceptors modulate the *in vivo* release of dopamine in the frontal, cingulate, and entorhinal cortices. *J. Pharmacol. Exp. Ther.* **242**: 115 (1987).

S.M. Antelman and L.A. Chiodo. Amphetamine as a stressor. In: *Stimulants: Neurochemical, Behavioral, and Clinical Perspectives*. I. Creese, ed., Raven Press, New York, p. 269 (1983).

M.J. Bannon, R.L. Michaud, R.H. Roth. Mesocortical dopamine neurons: lack of autoreceptors modulating dopamine synthesis. *Mol. Pharmacol.* **19**: 270 (1981a).

M.J. Bannon, E.B. Bunney, and R.H. Roth. Mesocortical dopamine neurons: rapid transmitter turnover compared to other brain catecholamine systems. *Brain Res.* **218**: 376 (1981b).

M.J. Bannon, C.-H. Xue, K. Shibata, L.J. Dragovic, and G. Kapatos. Expression of a human cocaine-sensitive dopamine transporter in xenopus laevis oocytes. *J. Neurochem.* **54**: 706 (1990).

J.M. Barban and G.K. Aghajanian. Suppression of firing activity of 5HT neurons in the dorsal raphe by alpha-adrenoreceptor antagonists. *Neuropharmacology* **19**: 355 (1980).

J.V. Barnett, D.S. Segal, and R. Kuczenski. Repeated amphetamine pretreatment alters the responsiveness of striatal dopamine stimulated adenylate cyclase to amphetamine induced desensitization. *J. Pharmacol. Exp. Ther.* **242**: 40 (1987).

R.L. Barrett and J.B. Appel. Effects of stimulation and blockade of dopamine receptor subtypes on the discriminative stimulus properties of cocaine. *Psychopharmacology* **99**: 13 (1989).

S. Benloucif and M.P. Galloway. Regulation of dopamine release by serotonin 1a and 1b agonists measured by microdialysis. *Soc. Neurosci. Abstr.* **15**: 1313 (1989).

S. Bischoff, M. Heinrich, J.M. Sonntag, and J. Krauss. The D-1 dopamine receptor antagonist SCH 23390 also interacts potently with brain serotonin (5-HT2) receptors *Eur. J. Pharmacol.* **129**: 367 (1986).

J.W. Boja and M.J. Kuhar. ^{3}[H]Cocaine binding and inhibition of ^{3}[H]dopamine uptake is similar in both the rat striatum and nucleus accumbens. *Eur. J. Pharmacol.* **173**: 215 (1989).

M.B. Bowers, Jr. and F.J. Hoffman, Jr. Regional brain homovanillic acid following tetrahydrocannabinol and cocaine. *Brain Res.* **366**: 405 (1986).

C.W. Bradberry and R.H. Roth. Cocaine increases extracellular dopamine in rat nucleus accumbens and ventral tegmental area as shown by *in vivo* microdialysis. *Neurosci. Lett.* **103**: 97 (1989).

A.M. Broadhurst and M. Briley. Catecholamine and serotonin synthesis *ex vivo* as an index of *in vivo* neuronal activity and regulation. *Trends Pharmacol. Sci.* **9**: 349 (1988).

R.B. Brooderson, F.J. White, and M.P. Galloway. Inhibition of dopamine synthesis in striatal slices by the D_1 agonist SKF 38393 is not mediated by D_1 dopamine receptors. *Synapse* (in press) 1990.

J.H. Carlson, D.A. Bergstrom, and J.R. Walters. Stimulation of both D-1 and D-2 dopamine receptors appears necessary for full expression of postsynaptic effects of dopamine agonists: a neurophysiological study. *Brain Res.* **400**: 205 (1987).

M.E. Carrol, S.T. Lac, M. Asencio, and R. Kragh. Fluoxetine reduces intravenous cocaine self-administration in rats. *Pharmacol. Biochem. Behav.* **35**: 237 (1990).

E. Castaneda, J.B. Becker, and T.E. Robinson. The long-term effects of repeated amphetamine treatment *in vivo* on amphetamine, KCl and electrical stimulation evoked striatal dopamine release *in vitro*. *Life Sci.* **42**: 2447 (1988).

L.A. Chiodo. Dopamine containing neurons in the mammalian central nervous system: electrophysiology and pharmacology. *Neurosci. Biobehav. Rev.* **12**: 49 (1988).

L.A. Chiodo, M.J. Bannon, A.A. Grace, R.H. Roth, and B.S. Bunney. Evidence for absence of impulse-regulating somatodendritic and synthesis-modulating nerve terminal autoreceptors on subpopulations of mesocortical dopamine neurons. *Neuroscience* **12**: 1 (1984).

A.V. Christensen, J. Arnt, J. Hyttel, J.-J. Larsen, and O. Svendsen. Pharmacological effects of the specific dopamine D1 antagonist SCH 23390 in comparison with neuroleptics. *Life Sci.* **34**: 1529 (1984).

W.H. Church, J.B. Justice, and L.D. Byrd. Extracellular dopamine in rat striatum following uptake inhibition by cocaine, nomifensine, and benztropine. *Eur. J. Pharmacol.* **139**: 345 (1987).

K.A. Cunningham and J.M. Lakoski. Electrophysiological effects of cocaine and procaine on dorsal raphe serotonin neurons. *Eur. J. Pharmacol.* **148**: 457 (1988).

C.A. Dackis, M.S. Gold, D.R. Sweeney, J.P. Byron, and R. Climko. Single-dose bromocriptine reverses cocaine craving. *Psychiatr. Res.* **20**: 261 (1987).

K.T. Demarest, K.L. Lawson-Wendling, and K.E. Moore. D-Amphetamine and gamma-butyrolactone alteration of dopamine synthesis in the terminals of nigrostriatal and mesolimbic neurons — possible role of various autoreceptor sensitivities. *Biochem. Pharmacol.* **32**: 691 (1983).

A.Y. Deutch, S.Y. Tam, and R.H. Roth. Footshock and conditioned stress increase 3,4-dihydroxyphenylacetic acid (DOPAC) in the ventral tegmental area but not in the substantia nigra. *Brain Res.* **333**: 143 (1986).

G. Di Chiara and A. Imperato. Drugs abused by humans preferentially increase synaptic dopamine concentrations in the mesolimbic system of freely moving rats. *Proc. Natl. Acad. Sci. U.S.A.* **85**: 5274 (1988).

M.L. Dubcovich and N.R. Zahniser. Binding characteristics of the dopamine uptake inhibitor. *Pharmacology* **34**: 1137 (1985).

L.A. DuMars, L.D. Rodger and P.W. Kalivas. Behavioral cross-sensitization between cocaine and enkephalin in the A10 dopamine region. *Behav. Brain Res.* **27**: 87 (1988).

L.P. Dwoskin, J. Peris, R.P. Yasuda, K. Philpott, and N.R. Zahniser. Repeated cocaine administration results in supersensitivity of striatal D-2 dopamine autoreceptors to pergolide. *Life Sci.* **42**: 255 (1988).

L.C. Einhorn, P.A. Johansen, and F.J. White. Electrophysiological effects of cocaine in the mesoaccumbens dopamine system: studies in the ventral tegmental area. *J. Neurosci.* **8**: 100 (1988).

E. Friedman, S. Gershon, and J. Rotrosen. Effects of cocaine treatment on the turnover of 5-hydroxytryptamine in the rat brain. *Br. J. Pharmacol.* **54**: 61 (1975).

M.P. Galloway, M.E. Wolf, and R.H. Roth. Regulation of dopamine synthesis in the medial prefrontal cortex is mediated by release modulating autoreceptors: studies *in vivo*. *J. Pharmacol. Exp. Ther.* **236**: 689 (1986).

M.P. Galloway. Neurochemical interactions of cocaine with dopaminergic systems. *Trend Phamacol. Sci.* **9**: 451 (1988).

M.P. Galloway. Regulation of dopamine and serotonin synthesis by acute administration of cocaine. *Synapse* **6**: 63 (1990).

F.H. Gawin and H.D. Kleber. Abstinence symptomatology and psychiatric diagnosis in cocaine abusers. *Arch. Gen. Psychiatry* **43**: 107 (1986).

N.E. Goeders and J.E. Smith. Reinforcing properties of cocaine in the medial prefrontal cortex: primary action on presynaptic dopaminergic terminals. *Pharmacol. Biochem. Behav.* **25**: 191 (1986).

N.E. Goeders and J.E. Smith. Cortical dopaminergic involvement in cocaine reinforcement. *Science* **221**: 773 (1983).

N.E. Goeders, S.I. Dworkin, and J.E. Smith. Neuropharmacological assessment of cocaine self-administration into the medial prefrontal cortex. *Pharmacol. Biochem. Behav.* **24**: 1429 (1986).

N.E. Goeders and M.J. Kuhar. Chronic cocaine administration induces opposite changes in dopamine receptors in the striatum and nucleus accumbens. *Alcohol Drug Res.* **7**: 207 (1987).

K. Gysling and R.Y. Wang. Morphine induced activation of A10 dopamine neurons in the rat. *Brain Res.* **277**: 119 (1983).

G.R. Hanson, L.A. Matsuda, and J.W. Gibb. Effects of cocaine on methamphetamine-induced neurochemical changes: characterization of cocaine as a monoamine uptake blocker. *J. Pharmacol. Exp. Ther.* **242**: 507 (1987).

M. Haraguchi, H.H. Samson, and G.A. Tolliver. Reduction of oral ethanol self-administration in the rat by the 5-HT uptake blocker fluoxetine. *Pharmacol. Biochem. Behav.* **35**: 259 (1990).

R.E. Heikkila, H. Orlansky, and G. Cohen. Studies on the distinction between uptake inhibition and release of [^{3}H]dopamine in rat brain tissue slices. *Biochem. Pharmacol.* **24**: 847 (1975).

D.H. Henry, M.A. Green, and F.J. White. Electrophysiological effects of cocaine in the mesoaccumbens dopamine system: repeated administration. *J. Pharmacol. Exp. Ther.* **251**: 833 (1989).

L. Hernandez, F. Lee, and B.G. Hoebel. Simultaneous microdialysis and amphetamine infusion in the nucleus accumbens and striatum of freely moving rats: increase in extracellular dopamine and serotonin. *Brain Res. Bull.* **19**: 623 (1987).

L. Hernandez and B.G. Hoebel. Food reward and cocaine increase extracellular dopamine in the nucleus accumbens as measured by microdialysis. *Life Sci.* **42**: 1705 (1988).

Y.L. Hurd and U. Ungerstedt. Cocaine: an *in vivo* microdialysis evaluation of its acute action on dopaminergic transmission in rat striatum. *Synapse* **3**: 48 (1989).

Y.L. Hurd, J. Kehr, and U. Ungerstedt. *In vivo* microdialysis as a technique to monitor drug transport: correlation of extracellular cocaine levels and dopamine overflow in the rat brain. *J. Neurochem.* **51**: 1314 (1988).

Y.L. Hurd, F. Weiss, G.F. Koob, and U. Ungerstedt. Cocaine reinforcement and extracellular dopamine overflow in rat nucleus accumbens: an *in vivo* microdialysis study. *Brain Res.* **498**: 199 (1989).

L.C. Iorio, A. Bennet, F.H. Leitz, V.P. Houser, and C.A. Korduba. SCH 23390, a potential benzapine antipsychotic with a unique interaction on dopaminergic systems. *J. Pharmacol. Exp. Ther.* **226**: 462 (1983).

P.A. Johansen and M.P. Galloway. Regulation of dopamine synthesis *in vivo* by selective 5HT1 agonists. *Soc. Neurosci. Abstr.* **15**: 1313 (1989).

P.W. Kalivas. Sensitization to repeated enkephalin administration into the ventral tegmental area of the rat. II. Involvement of the mesolimbic dopamine system. *J. Pharmacol. Exp. Ther.* **235**: 544 (1985).

P.W. Kalivas and P. Duffy. Effects of daily cocaine and morphine treatment on somatodendritic and terminal field dopamine release. *J. Neurochem.* **50**: 1498 (1988).

P.W. Kalivas, P. Duffy, L.A. Dumars, and C. Skinner. Behavioral and neurochemical effects of acute and daily cocaine administration in rats. *J. Pharmacol. Exp. Ther.* **245**: 485 (1988).

P.W. Kalivas and P. Duffy. Similar effects of daily cocaine and stress on mesocorticolimbic dopamine neurotransmission in the rat. *Biol. Psychiatry* **25**: 913 (1989).

P.W. Kalivas and B. Weber. Amphetamine injection into the ventral mesencephalon sensitizes rats to peripheral amphetamine and cocaine. *J. Pharmacol. Exp. Ther.* **245**: 1095 (1988).

Y. Kazahaya, K. Akimoto, and S. Otsuki. Subchronic methamphetamine treatment enhances methamphetamine- or cocaine-induced dopamine efflux *in vivo*. *Biol. Psychiatry* **25**: 903 (1989).

L.T. Kennedy and I. Hanbauer. Sodium-sensitive cocaine binding to rat striatal membrane: possible relationship to dopamine uptake sites. *J. Neurochem.* **41**: 172 (1983).

M.S. Kleven, E.W. Anthony, L.I. Goldberg, and W.L. Woolverton. Blockade of the discriminative stimulus effects of cocaine in rhesus monkeys with the D_1 dopamine antagonist SCH 23390. *Psychopharmacology* **95**: 427 (1988a).

M.S. Kleven, W.L. Woolverton, and L.S. Seiden. Lack of long-term monoamine depletions following repeated or continuous exposure to cocaine. *Brain Res. Bull.* **21**: 233 (1988b).

A. Knorr, A.Y. Deutch, and R.H. Roth. The anxiogenic β-carboline FG-7142 increases *in vivo* and *in vitro* tyrosine hydroxylation in the prefrontal cortex. *Brain Res.* **495**: 355 (1989).

G.F. Koob, Le H. Thai, and I. Creese. The D_1 dopamine receptor antagonist SHC 23390 increases cocaine self-administration in the rat. *Neurosci. Lett.* **79**: 315 (1987).

R. Kuczenski and D. Segal. Concomitant characterization of behavioral and striatal neurotransmitter response to amphetamine using *in vivo* microdialysis. *J. Neurosci.* **9**: 2051 (1989).

J.M. Lakoski and K.A. Cunningham. Cocaine interaction with central monoaminergic systems: electrophysiological approaches. *Trend Pharmacol. Sci.* **9**: 177 (1988).

J.M. Lakoski. Cocaine modulation of serotonin function: electrophysiological approaches, in: *Cocaine: Pharmacology, Physiology, and Clinical Strategies*. J.M. Lakoski, M.P. Galloway, and F.J. White, eds., CRC Press, Boca Raton, FL (1991).

G. Lin-Chu, T.E. Robinson and J.M. Becker. Sensitization of rotational behavior produced by a single exposure to cocaine. *Pharmacol. Biochem. Behav.* **22**: 901 (1985).

W.H. Lyness, N.M. Friedle, and K.E. Moore. Increased self-administration of D-amphetamine after destruction of 5-hydroxytryptaminergic neurons. *Pharmacol. Biochem. Behav.* **12**: 937 (1980).

W.H. Lyness, N.M. Friedle, and K.E. Moore. Destruction of dopaminergic nerve terminals in nucleus accumbens: effect on D-amphetamine self-administration. *Pharmacol. Biochem. Behav.* **11**: 553 (1979).

S.R. Martin, V.K. Yergani, R.A. Lohdi, and M.P. Galloway. Clinical ratings and plasma HVA during cocaine abstinence. *Biol. Psychiatry* **26**: 356 (1989).

M.T. Martin-Iverson, C. Szostak, and H.C. Fibiger. 6-Hydroxydopamine lesions of the medial prefrontal cortex fail to influence intravenous self-administration of cocaine. *Psychopharmacology* **88**: 310 (1986).

H.Y. Meltzer and M.Y. Lowy. The serotonin hypothesis of depression, in: *Psychopharmacology: The Third Generation of Progress.* H.Y. Meltzer, ed., Raven Press, New York, p. 513 (1987).

C. Missale, L. Castelletti, S. Govoni, P.F. Spano, M. Trabucchi, and I. Hanbauer. Dopamine uptake is differentially regulated in rat striatum and nucleus accumbens. *J. Neurochem.* **45**: 51 (1985).

B. Moghaddam and B.S. Bunney. Differential effect of cocaine on extracellular dopamine levels in rat prefrontal cortex and nucleus accumbens: comparison to amphetamine. *Synapse* **4**: 156 (1989).

S. Nakajima and G.M. McKenzie. Reduction of the rewarding effect of brain stimulation by a blockade of dopamine D_1 receptor with SCH 23390. *Pharmacol. Biochem. Behav.* **24**: 919 (1986).

L.C. Nicolaysen and J.B. Justice. Effects of cocaine on release and uptake of dopamine *in vivo*: differentiation by mathematical modeling. *Pharmacol. Biochem. Behav.* **31**: 327 (1988a).

L.C. Nicolaysen, H.T. Pan, and J.B. Justice. Extracellular cocaine and dopamine concentrations are linearly related in rat striatum. *Brain Res.* **456**: 317 (1988b).

H. Nissbrandt, G. Engberg, H. Wikstrom, T. Magnusson, and A. Carlsson. NSD 1034: an amino acid decarboxylase inhibitor with a stimulatory action on dopamine synthesis not mediated by classical dopamine receptors. *Naunyn Schmiedeberg's Arch. Pharmacol.* **338**: 148 (1988).

Z.Z. Pan and J.T. Williams. Differential actions of cocaine and amphetamine on dorsal raphe neurons *in vitro. J. Pharmacol. Exp. Ther.* **251**: 56 (1989).

J. Peris and N.R. Zahniser. Persistent augmented dopamine release after acute cocaine requires dopamine receptor activation. *Pharmacol. Biochem. Behav.* **32**: 71 (1989).

D.K. Pitts and J. Marwah. Cocaine modulation of central monoaminergic neurotransmission. *Pharmacol. Biochem. Behav.* **26**: 453 (1987)

D.K. Pitts and J. Marwah. Electrophysiological effects of cocaine on central monoaminergic neurons. *Eur. J. Pharmacol.* **131**: 95 (1986).

L.J. Porrino, M.C. Ritz, N.L. Goodman, L.G. Sharpe, M.J. Kuhar, and S.R. Goldberg. Differential effects of the pharmacological manipulation of serotonin systems on cocaine and amphetamine self-administration in rats. *Life Sci.* **45**: 1529 (1989).

R.M. Post. Kindling effects of chronic cocaine administration, in: *Cocaine: Pharmacology, Physiology, and Clinical Strategies.* J.M. Lakoski, M.P. Galloway, and F.J. White, eds., CRC Press, Boca Raton, FL (1991).

R.M. Post and N.R. Contel. Human and animal studies of cocaine: implicaitons for development and behavioral pathology, in: *Stimulants: Neurochemical. Behavioral, and Clinical Perspectives.* I. Creese, ed., Raven Press, New York, p. 169 (1983).

J.F. Reinhard, M.P. Galloway, and R.H. Roth. Noradrenergic modulation of serotonin synthesis and metabolism. II: stimulation by 3-isobutyl-1-methylxanthine. *J. Pharmacol. Exp. Ther.* **226**: 764 (1983).

M.C. Ritz, R.J. Lamb, S.R. Goldberg, and M.J. Kuhar. Cocaine receptors on dopamine transporters are related to self-administration of cocaine. *Science* **237**: 1219 (1987).

M.C. Ritz and M.J. Kuhar. Relationship between self-administration of amphetamine and monoamine receptors in brain: comparison with cocaine. *J. Pharmacol. Exp. Ther.* **248**: 1010 (1989).

J.-M. Rivet, L. Stinus, M. LeMoal, and P. Mormede. Behavioral sensitization to amphetamine is dependent on corticoid receptor activation. *Brain Res.* **498**: 149 (1989).

D.C.S. Roberts and G.F. Koob. Disruption of cocaine self-administration following 6-hydroxydopamine lesions of the ventral tegmental area in rats. *Pharmacol. Biochem. Behav.* **17**: 901 (1982).

D.C.S. Roberts, G.F. Koob, P. Klonoff, and H.C. Fibiger. Extinction and recovery of cocaine self-administration following 6-hydroxydopamine lesions of the nucleus accumbens. *Pharmacol. Biochem. Behav.* **12**: 781 (1980).

J.M. Roberts-Lewis, P.H. Roseboom, L.M. Iwaniec, and M.Y. Gnegy. Differential down-regulation of D_1 stimulated adenylate cyclase activity in rat forebrain after *in vivo* amphetamine treatments. *J. Neurosci.* **6**: 2245 (1986).

D.C.S. Roberts and G. Vickers. Atypical neuroleptics increase self-administration of cocaine: an evaluation of a behavioral screen for anti-psychotic activity. *Psychopharmacology* **82**: 135 (1984).

T.E. Robinson and J.B. Becker. Enduring changes in brain and behavior produced by chronic amphetamine administration: a review and evaluation of animal models of amphetamine psychosis. *Brain Res. Rev.* **11**: 157 (1986).

T.E. Robinson, J. Becker, E.A. Young, H. Akil, and E. Castaneda. The effects of footshock stress on regional brain dopamine metabolism and pituitary β-endorphin release in rats previously sensitized to amphetamine. *Neuropharmacology* **26**: 679 (1987).

T.E. Robinson, P.A. Jurson, J.A. Bennett, and K.M. Bentgen. Persistent sensitization of dopamine neurotransmission in ventral striatum (nucleus accumbens) produced by prior experience with (+)-amphetamine: a microdialysis study in freely moving rats. *Brain Res.* **462**: 211 (1988).

S.E. Robinson and K.L. Hambrecht. The effect of cocaine on hippocampal cholinergic and noradrenergic metabolism. *Brain Res.* **457**: 383 (1988).

S.B. Ross and A.L. Renyi. Uptake of some tritiated sympathomimetic amines by mouse brain cortex *in vitro*. *Acta Pharmacol. Toxicol.* **24**: 297 (1966).

S.B. Ross and A.L. Renyi. Inhibition of the uptake of tritiated 5-hydroxytryptamine in brain tissue. *Eur. J. Pharmacol.* **7**: 270 (1969).

R.H. Roth, L.C. Murrin, and J.R. Walters. Central dopaminergic neurons: effects of alterations in impulse flow on the accumulation of dihydroxyphenylacetic acid. *Eur. J. Pharmacol.* **36**: 163 (1976).

R.H. Roth. Dopamine autoreceptors: pharmacology, function and comparison with post-synaptic dopamine receptors. *Commun. Psychopharmacol.* 3: 429 (1979).

R.H. Roth. CNS dopamine autoreceptors: distribution, pharmacology, and function. *Ann. N.Y. Acad. Sci.* **430**: 27 (1984).

L.J. Ryan, M.E. Martone, J.C. Linder, and P.M. Groves. Cocaine, in contrast to D-amphetamine, does not cause axonal terminal degeneration in neostriatum and agranular frontal cortex of Long-Evans rats. *Life Sci.* **43**: 1403 (1988).

R.S. Salah, D.M. Kuhn, and M.P. Galloway. Dopamine autoreceptors modulate the phosphorylation of tyrosine hydroxylase in rat striatal slices. *J. Neurochem.* **52**: 1517 (1989).

L. Shuster, G. Yu, and A. Bates. Sensitization to cocaine stimulation in mice. *Psychopharmacology* **52**: 185 (1977).

F.L. Smith, D.S.L. Yu, D.G. Smith, A.P. Lecesse, and W.H. Lyness. Dietary tryptophan supplements attenuate amphetamine self-administration in the rat. *Pharmacol. Biochem. Behav.* **25**: 849 (1986).

D. Taylor and B.T. Ho. Neurochemical effects of cocaine following acute and repeated injection. *J. Neurosci. Res.* **3**: 95 (1977).

C.B. Tyler, M.J. Keegan, and M.P. Galloway. Regulation of dopamine synthesis *in vivo* after acute or chronic amphetamine. *Soc. Neurosci. Abstr.* **15**: 429 (1989).

C.B. Tyler and M.P. Galloway. Acute administration of amphetamine: differential regulation of dopamine snythesis in dopamine projection fields. *J. Pharmacol. Exp. Ther.* (submitted) 1990.

M.E. Trulson, J.C. Joe, S. Babb, and J.D. Raese. Chronic cocaine administration depletes tyrosine hydroxylase immunoreactivity in the meso-limbic dopamine system in rat brain: quantitative light microscopic studies. *Brain Res. Bull.* **19**: 39 (1987).

M.E. Trulson and M.J. Ulissey. Chronic cocaine administration decreases dopamine synthesis rate and increases [3H]-spiroperidol binding in rat brain. *Brain Res. Bull.* **19**: 35 (1987).

H. Ujike, T. Onoue, A. Kazufumi, T. Hamamura, and S. Otsuki. Effects of selective D-1 and D-2 dopamine antagonists on development of methamphetamine induced behavioral sensitization. *Psychopharmacology* **98**: 89 (1989).

P. Vezina and J. Stewart. The effect of dopamine receptor blockade effects of amphetamine and morphine. *Brain Res.* **499**: 108 (1989).

S.R. Wachtel, X.-T. Hu, M.P. Galloway, and F.J. White. D_1 Dopamine receptor stimulation enables the postsynaptic, but not autoreceptor, effects of D_2 dopamine agonists in nigrostriatal and mesoaccumbens dopamine systems. *Synapse* **4**: 327 (1989).

J.R. Walters and R.H. Roth. Dopaminergic neurons: an *in vivo* system for measuring drug interactions with presynaptic receptors. *Naunyn-Schmiedeberg's Arch. Pharmacol.* **296**: 5 (1976).

R.Y. Wang. Dopaminergic neurons in the rat ventral tegmental area. III: Effects of D and L amphetamine. *Brain Res. Rev.* **3**: 153 (1981).

F.J. White and R.Y. Wang. A10 dopamine neurons: role of autoreceptors in determining firing rate and sensitivity to dopamine agonists. *Life Sci.* **34**: 1161 (1984).

F.J. White, S.R. Wachtel, P.A. Johansen, and L.C. Einhorn. Electrophysiological studies of the rat mesoaccumbens dopamine system: focus on dopamine receptor subtypes, interactions, and the effects of cocaine, in: *Neurophysiology of Dopaminergic Systems-Current Status and Clinical Perspectives.* L.A. Chiodo and A.S. Freeman, eds., Lakeshore Publishing, Detroit, p. 317 (1987).

F.J. White and R.Y. Wang. A10 dopamine neurons: role of autoreceptors in determining firing rate and sensitivity to dopamine agonists. *Life Sci.* **34**: 1161 (1984).

F.J. White *et al. Cocaine: Pharmacology, Physiology, and Clinical Strategies.* J.M. Lakoski, M.P. Galloway, and F.J. White, eds., CRC Press, Boca Raton, FL (1991).

R.A. Wilcox, T.E. Robinson, and J.B. Becker. Enduring enhancement in amphetamine stimulated striatal dopamine release *in vitro* produced by prior exposure to amphetamine or stress *in vivo. Eur. J. Pharmacol.* **124**: 375 (1986).

R.A. Wise. Neural mechanisms of the reinforcing actions of cocaine. *NIDA Res. Monographs* **50**: 15 (1984).

M.E. Wolf, M.P. Galloway, and R.H. Roth. Regulation of dopamine synthesis in the medial prefrontal cortex: studies in brain slices. *J. Pharmacol. Exp. Ther.* **236**: 699 (1986).

M.E. Wolf and R.H. Roth. Dopamine autoreceptors, in: *Dopamine Receptors.* I. Creese and C.M. Fraser, eds., Alan R. Liss, New York, p. 45 (1987a).

M.E. Wolf and R.H. Roth. Dopamine neurons projecting to the medial prefrontal cortex possess release-modulating autoreceptors. *Neuropharmacolgy* **26**: 1053 (1987b).

D.M. Wood and M.W. Emmett-Oglesby. Mediation in the nucleus accumbens of the discriminative stimulus produced by cocaine. *Pharmacol. Biochem. Behav.* **33**: 453 (1989).

W.L. Woolverton. Effects of a D1 and D2 dopamine antagonist on the self-administration of cocaine and piribedil by rhesus monkeys. *Pharmacol. Biochem. Behav.* **24**: 541 (1986).

W.L. Woolverton and R.M. Virus. The effects of a D_1 and D_2 dopamine antagonist on behavior maintained by cocaine or food. *Pharmacol. Biochem. Behav.* **32**: 691 (1989).

D.S. Yu, F.L. Smith, D.G. Smith, and W.H. Lyness. Fluoxetine-induced attenuation of amphetamine self-administration in rats. *Life Sci.* **39**: 1383 (1986).

N.R. Zahniser and J. Peris. Neurochemical mechanisms of cocaine-induced sensitization, in: *Cocaine: Pharmacology, Physiology, and Clinical Strategies.* J.M. Lakoski, M.P. Galloway, and F.J. White, eds., CRC Press, Boca Raton, FL (1991).

9

A Cocaine Receptor Associated with Dopamine Transport and Drug Self-Administration

MICHAEL J. KUHAR, MARY C. RITZ, DIMITRI GRIGORIADIS, AND ROBERT LEW
Addiction Research Center
National Institute on Drug Abuse
Baltimore, Maryland

JOHN SHARKEY
Department of Clinical Neurosciences
Western General Hospital
Crewe Road
Edinburgh, Scotland

INTRODUCTION

Cocaine, a psychostimulant drug, has a variety of physiological effects. Its powerful, seductive reinforcing properties have made it a widely abused drug and a severe sociological, medical, and personal problem. How does cocaine produce its reinforcing effects? What are the neuroanatomical and biochemical mechanisms underlying its reinforcing properties? These are some of the questions that this chapter will attempt to answer.

A COCAINE BINDING SITE ON THE DOPAMINE TRANSPORTER: A RECEPTOR RELATED TO DRUG SELF-ADMINISTRATION

Brain dopamine-containing neuronal systems, particularly mesolimbic pathways, have been implicated in the mechanism of action of cocaine (Wise, 1984). As a result of these studies, it has often been assumed that the reinforcing properties of cocaine are related to the inhibition of dopamine uptake. However, prior to our experiments, there were no receptor binding data that supported this notion. A key question is, therefore, what is the cocaine binding site or receptor related to its reinforcing properties? By definition, the cocaine receptor would be the molecular site in the brain with which cocaine interacts to produce all of its subsequent reinforcing effects. Indeed, several binding sites for cocaine have been identified in brain, but none of these had been tested for their possible association with the reinforcing properties of the drug (Kennedy and Hanbauer, 1983; Reith *et al.*, 1980; Calligaro and Eldefrawi, 1987).

Drug self-administration by primates and other species is considered to be a useful model of abuse liability. Most drugs that humans abuse are self-administered in these animal studies. Because this experimental paradigm of self-administration is widely believed to be the best model for abuse liability testing, we utilized it to form a basis for identifying relative potencies of abused drugs.

Our strategy for identifying the cocaine receptor related to its reinforcing properties was to attempt to relate the potencies of cocaine-like drugs in drug self-administration studies with the potencies of the same drugs at various cocaine binding sites in brain. Table 1 lists various compounds along with their relative behavioral potencies and their potencies in inhibiting ^{3}H-mazindol binding to the dopamine transporter. The potencies of these same drugs in several other receptor binding assays were studied as well (Ritz *et al.*, 1987). We found that a statistically significant relationship could only be found between dopamine transporter binding and behavioral studies (Table 1 and Figure 1). Cocaine binding at other sites did not correlate with behavior. Thus, these data suggest that the cocaine receptor related to substance abuse is the one associated with dopamine uptake inhibition or the one found on the dopamine transporter.

Are all drugs that inhibit dopamine uptake potent reinforcers? Theoretically, all drugs that inhibit dopamine uptake should be reinforcers according to the model proposed here. However, other properties of drugs may interfere with the reinforcing properties. For example, tricyclic antidepressant drugs such as imipramine are not self-administered but do inhibit dopamine transport. A potential explanation for this apparent contradiction might be related to the fact that imipramine is a weak inhibitor of dopamine transport, about 30-fold weaker than cocaine. If imipramine were administered so that its blood levels were 30-fold higher than efficacious blood levels of cocaine, then imipramine levels would be in the range for human lethality. Thus, imipramine would not be self-administered because high-enough blood levels could not be obtained *in vivo*. Another dopamine uptake inhibitor, benztropine, is similarly not self-administered in animals. Benztropine is a potent anticholinergic drug (greater potency at muscarinic receptors than at dopamine transporters) and, if blood levels of benztropine were adequate to inhibit dopamine uptake, it would be a powerfully acting anticholinergic under these conditions. At these levels, benztropine would produce anticholinergic delirium (Tune *et al.*, 1981). Thus, drugs that are known to be dopamine uptake inhibitors *in vitro* have other properties which would seriously interfere with their reinforcing properties *in vivo*.

Which dopaminergic pathways in brain are related to its reinforcing properties? While there are several well-known dopaminergic pathways in brain, the mesolimbic

Table 1
Potencies of Cocaine and Related Compounds in Self-Administration and Biochemical Studies

Compound	Relative behavioral potency	Dopamine uptake K_i μM	Dopamine uptake Relative potency
1. WIN 35,065-2	0.26	0.26	0.41
2. WIN 35,981	0.58	0.36	0.56
3. (−)-Cocaine	1	0.64	1
4. Dimethocaine	0.67	1.29	2.02
5. (+/−)Norcocaine	2.82	1.21	1.89
6. Procaine	14.1	104	164
7. Chloroprocaine	29	65	102
8. (+)Pseudococaine	33	116	183
9. Mazindol	0.17	0.023	0.036
10. Methylphenidate	1	0.39	0.61
11. (+)-Amphetamine	0.20	3.6	5.63
12. Lidocaine	ND	3298	5153
13. Procainamide	ND	1943	3036
14. DEAE	ND	18197	28433
15. WIN 35,065-3	ND	385	602
16. (+)-Cocaine	ND	136	213

Note: The relative behavioral potencies in self-administration studies were determined by averaging values obtained from studies of cocaine reinforcement. These could not be determined (ND) for the compounds lidocaine, procainamide, DEAE, WIN 35,065-3, and (+)-cocaine due to their low potencies or toxic side effects. Except in two cases, the animals utilized were monkeys. [^{3}H] Mazindol was used to label dopamine (10^{-5} *M* nomifensin blank) transport sites in striatum. Tissues were homogenized, then incubated for 1 h in buffer (50 m*M* tris, 120 m*M* NaCl, 5 m*M* KCl; pH 7.8; 4°C) containing a final [^{3}H]mazindol concentration of 4 n*M*. K_i values were determined from analyses of competition curves. Standard errors for K_i values (mean of three to five assays) were less than 10% of the values. Relative potencies were calculated so that cocaine had a value of one. DEAE, diethylaminoethanol. (From M.C. Ritz, R.J. Lamb, S.R. Goldberg, and M.J. Kuhar. *Science* **237**: 1219 (1987). With permission.)

pathway appears to be related to cocaine's reinforcing properties. Extensive lesion and direct injection studies provide powerful support for this notion (Goeders and Smith, 1986; Goeders *et al.*, 1986; Roberts *et al.*, 1980; Roberts and Koob, 1982). Moreover, the dopaminergic mesolimbic pathway has been shown to be part of an electrical self-stimulation reward system in brain (Fibiger, 1978; Wise, 1978).

The experiments referred to above strongly support the proposal that cocaine receptors related to dopamine transport inhibition are the ones related to drug reinforcement. There are other studies which also support this suggestion. For example, dopaminergic agonists such as bromocryptine reduce craving for cocaine. Also, blood levels of cocaine in humans during a "high" are similar to the K_i values for cocaine at the dopamine related binding site (Fischman *et al.*, 1983).

If cocaine binding to the dopamine transporter, with resulting inhibition of dopamine transport, is the fundamental mechanism of action of the reinforcing properties of cocaine, it is reasonable to assume that potentiation of dopaminergic transmission in mesolimbic pathways is related to reinforcement. In support of this,

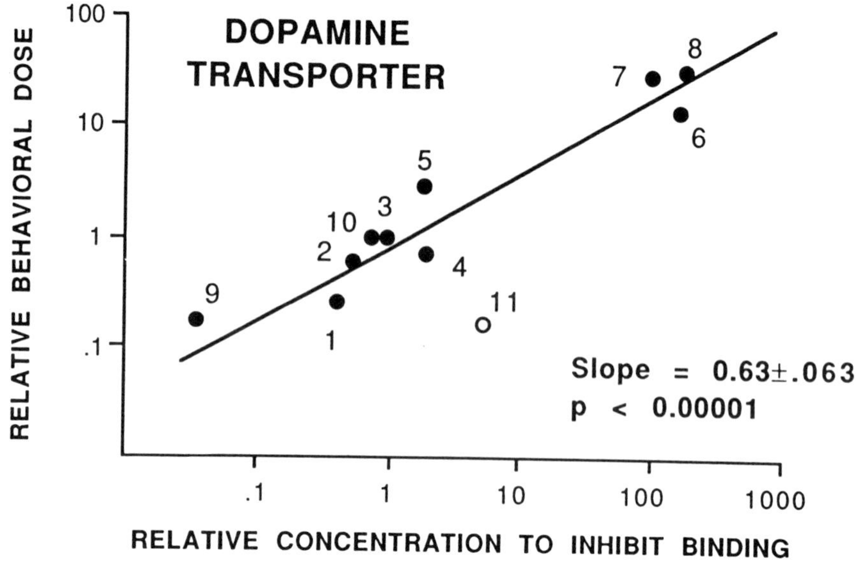

Figure 1. Relation between the relative behavioral doses of cocaine and related compounds in self-administration studies and their relative inhibitory concentrations for [^{3}H]mazindol binding at the dopamine transporter. Linear regression of logarithms of relative behavioral doses on logarithms of relative inhibitory concentrations for compounds 1 through 10 from Table 1 were calculated with BMDP. Value shown is slope ± SD. Compound 11 (open circle) was not included in the regression analysis (see text for discussion). A two-tailed test of significance was applied ($p < 0.00001$). (From M.C. Ritz, R.J. Lamb, S.R. Goldberg, and M.J. Kuhar. *Science* **237**: 1219 (1987). With permission.)

it has been found that dopamine receptor blockers alter cocaine self-administration (DeWit and Wise, 1977; Goeders *et al.*, 1986).

DOES AMPHETAMINE USE THE SAME RECEPTOR AS COCAINE?

Amphetamine is similar to cocaine in that it is also an abused psychostimulant drug. Amphetamine produces physiologic and behavioral effects similar to those of cocaine in both animals and humans (Fischman *et al.*, 1976; Ellinwood and Kilbey, 1977; Bejerot, 1970). Does amphetamine produce its reinforcing effects through the same receptor site as cocaine?

By following a strategy similar to that described for cocaine, that is, testing for a significant relationship between potencies of amphetamine-like compounds in drug self-administration studies and potencies at the dopamine transporter (Ritz and Kuhar, 1989), we were unable to establish such a relationship (Tables 2 and 3). The slope of the line relating biochemical potencies and behavioral potencies for cocaine was 0.72 with a p value of 0.001. The slope of the line relating amphetamine-like drugs was 0.84 but the p value was 0.13. An issue more serious and difficult than the nonsignificant p value for amphetamine is the fact that the amphetamine-like drugs (Table 2) are relatively weak at this binding site. By contrast, the potencies of cocaine and cocaine-like drugs at the receptor are similar to the

Table 2
Potencies of Amphetamine and Related Compounds in Self-Administered and Biochemical Studies

	Relative behavioral potency	Dopamine transporter K_i values (μM)	Serotonin transporter K_i values (μM)
1. (−)-Cocaine	1	0.64	0.14
2. (+)-Amphetamine	0.21	3.6	10.50
3. D,L-Cathinone	0.26	19.36	150.5
4. (−)-Amphetamine	0.64	31.2	50.1
5. Phentermine	1.94	42.6	38.30
6. D-Ephedrine	2.63	148.80	138.2
7. Benzphetamine	1.93	18.20	4.4
8. Phenmetrazine	2.01	14.80	119.5
9. Diethylproprion	5.20	43.0	377.3
10. Phendimetrazine	6.25	94.23	108.1
11. L-Ephedrine	12.34	107.4	125.5
12. Clortermine	29.06	56.52	5.87
13 D,L-MDA	29.63	25.92	4.72
14. Chlorphentermine	38.75	11.3	1.14
15. Methamphetamine	0.31	2.86	40.0
16. *N*-Ethylamphetamine	0.78	37.48	2.62
17. Phenylpropanolamine	ND	102.8	204.6
18. PMA (paramethoxy-amphetamine	ND	176.5	4.96
19. DOM	ND	161.5	16.4
20. Fenfluramine	ND	196.0	1.94

Note: The relative behavioral potencies of amphetamine and related drugs were determined from studies of drug reinforcement as described in the text. These could not be determined (ND) for the compounds phenylpropanolamine, PMA, DOM, and fenfluramine due to their low potencies or toxic properties. Standard error values for K_i values (mean of three to five assays). were less than 10% of the values. (From M.C. Ritz and M.J. Kuhar. *J. Pharmacol. Exp. Ther.* **248**: 1010 (1989). With permission.)

blood levels of these drugs during active self-administration. However, the blood levels for amphetamines are substantially lower during behavioral studies than the K_i values found in the binding inhibition studies. Therefore, binding to the cocaine receptor on the dopamine transporter is apparently not the primary or sole mechanism of action for amphetamine. Presumably, an additional mechanism is required, or the receptor for amphetamine is completely different. It could still be located on the dopaminergic nerve terminal, of course, but would presumably be a different molecular binding site.

There are several studies suggesting that amphetamine and cocaine have different neurochemical mechanisms of action. For example, the locomotor activity of effects of cocaine can be blocked by pretreatment with reserpine, while those of amphetamine cannot (Langer and Arbilla, 1984). Also, some studies suggest that cocaine interacts with vesicular pools of dopamine, while amphetamine does not (McMillen, 1983). Thus, while these two drugs produce similar behavioral effects, their molecular mechanisms of action are probably different.

Table 3
Multiple Regression Analysis: Potency in Self-Administration Studies vs. Receptor Binding Studies

	Transporters	
	DA	5-HT
	Amphetamine	
Slope	0.84	−0.66
p value	0.13	0.03
	Cocaine	
Slope	0.72	−0.08
p value	0.001	0.45

Note: Results of multiple regresson analyses of the relationship between relative drug potencies in studies of drug reinforcement and relative binding potencies at the binding sites indicated. (From M.C. Ritz and M.J. Kuhar. *J. Pharmacol. Exp. Ther.* **248**: 1010 (1989). With permission.)

SEROTONIN UPTAKE INHIBITION MAY OPPOSE DRUG SELF-ADMINISTRATION

In studies with amphetamine, we found that the inhibitory properties of amphetamine-like drugs at the serotonin transporter were inversely related to drug self-administration (Table 3 and Figure 2). Consistent with this finding are studies showing that agents that enhance serotonergic transmission (such as fenfluramine, imipramine, and LSD) are not reinforcing in animals (Griffiths *et al.*, 1979, 1980; Johanson and Schuster, 1980; Woods, 1978). Further, lesions of serotonergic nerve terminals increase self-administration of amphetamine (Lecesse and Lyness, 1984) and administration of serotonin uptake blockers decreases amphetamine-reinforced self-injections (Yu *et al.*, 1986). Thus, it seems interesting that drugs such as cocaine have effects at sites which promote both reinforcement and self-administration (dopamine transporter) and at sites which inhibit drug self-administration (serotonin transporter). Obviously, the net effects at the reinforcing sites are more powerful, in the balance. Alternatively, this simple model may not be adequate to explain the reinforcing and/or aversive properties of these drugs.

COCAINE BINDING AT SIGMA RECEPTORS

After examining cocaine binding at several receptor sites, it was clear that (−)-cocaine inhibited binding of ^{3}H-haloperidol at sigma receptors in brain (Table 4).

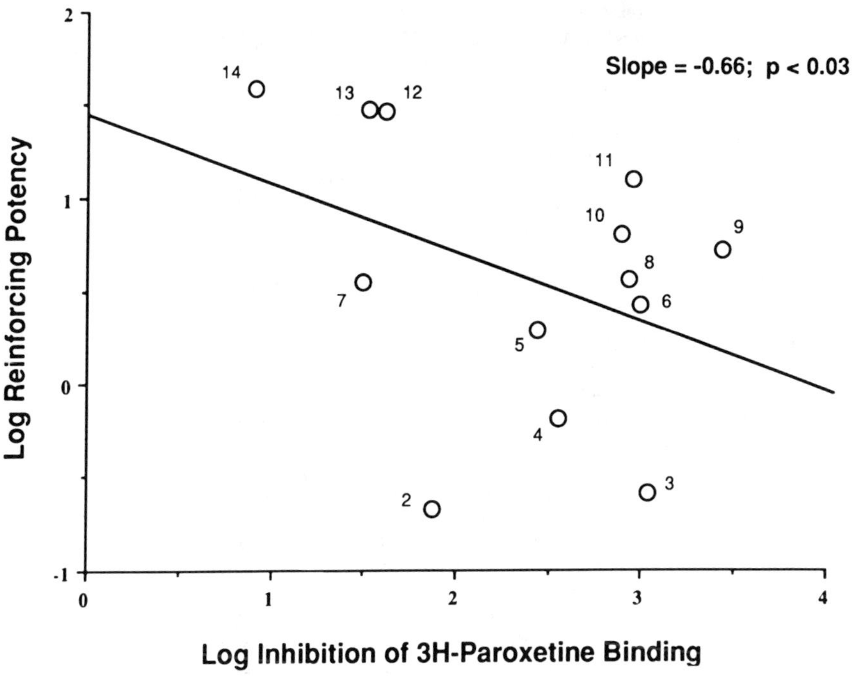

Figure 2. Inverse relationship between the potencies of amphetamine-related compounds for drug reinforcement and their relative binding potencies at the ^{3}H-paroxetine binding sites in rat midbrain. Amphetamine-related drugs are compounds 2 through 14 shown in Table 2. Slope and P values were determined by the multiple regression analysis. (From M.C. Ritz and M.J. Kuhar. *J. Pharmacol. Exp. Ther.* **248**: 1010 (1989). With permission.)

However, the potency of cocaine at the sigma receptor site was tenfold less than its potency at the dopamine transporter. Also, after examining a large number of other drugs, it was found that the reinforcing properties of cocaine and these drugs could not be correlated to binding at the sigma site (Sharkey *et al.*, 1988a). Nevertheless, cocaine binding at the sigma site could be important for some of its effects since sigma sites have been related to psychotomimetic effects of drugs in humans (Martin *et al.*, 1976; Keats and Telford, 1964).

COCAINE IS A WEAK ANTIMUSCARINIC DRUG

Cocaine inhibits receptor binding at the muscarinic cholinergic receptor in brain and heart. However, it tends to be a weak inhibitor of ligand binding at muscarinic receptors (Table 5). Cocaine has the properties of an antagonist rather than an agonist at these sites (Sharkey *et al.*, 1988b). In addition to binding studies, bioassays have shown that cocaine is a cholinergic muscarinic antagonist in the guinea pig atria (Sharkey *et al.*, 1988b). Even though cocaine is a weak inhibitor at these sites, its concentration in the blood of people who have died from an apparent

Table 4
Binding Affinity of Cocaine-Related to σ Sites in Rat Cerebellum

r	K_i value (μM)
(−)-Cocaine	6.7 ± 0.3
(+)-Cocaine	25.9 ± 2.4
(−)-Norcocaine	13.6 ± 1.3
Benzoylecognine	>100
Ecognine	>100
Ecognine methyl-ester	>100
Procaine	3.6 ± 0.1
DEAE	3.2 ± 0.3
Chloroprocaine	1.2 ± 0.3
Dimethocaine	0.8 ± 0.03
Lidocaine	8.7 ± 0.6
Procainamide	26.0
(+)-Amphetamine	12.7 ± 1.5

Note: [^{3}H]Haloperidol competition curves were analyzed using the LIGAND computer programs. For each of the compounds examined, the binding curve could best be described by a kinetic model assuming one binding site. Values are expressed as the mean K_i (μM ± SEM for three to five separate experiments. (From J. Sharkey, K.A. Glen, S. Wolfe, and M.J. Kuhar. *J. Pharmacol.* **149**: 171 (1988). With permission.)

overdose of cocaine is in the range of the inhibitory constant for cocaine at these sites. Thus, it is possible that the antimuscarinic properties of cocaine become relevant at higher, toxic doses of the drug. Additional studies will be required to establish this.

STRUCTURE-ACTIVITY STUDIES: MOLECULAR REQUIREMENTS OF THE COCAINE RECEPTOR

By employing a series of cocaine analogs in binding studies, we were able to identify various features of cocaine and cocaine-like molecules that were required for significant affinity for the cocaine receptor on the dopamine transporter. Presumably, these structural features reflect the nature of the molecular site where cocaine binds. It was found (Ritz *et al.*, 1990) that the following parts of the cocaine molecule are required for appreciable binding affinity and subsequent behavioral activity: the phenyl ring, the methylester in the beta conformation on the tropane ring, and at least some parts of the tropane moiety (data not shown). The structural requirements for cocaine binding to other cocaine binding sites (i.e., other transporters) were somewhat different (Ritz *et al.*, 1990).

Table 5
Potency of Cocaine and Related Compounds at Muscarinic Receptor

	Heart (M_2)	Medulla (M_2)	Hippocampus (M_1)
(+)-Cocaine	2.4 ± 0.2	2.4 ± 0.4	2.2 ± 0.1
(−)-Cocaine	18.8 ± 2.0[a]	24.1 ± 2.2[a]	39.5 ± 2.6
Cocaine metabolites:			
Norcocaine	21.5 ± 1.8	20.9 ± 0.7	17.5 ± 1.5
Benzoylecognine	>100	>100	>100
Ecognine	>100	>100	>100
Cocaine analogs:			
Procaine	5.5 ± 0.5	4.4 ± 1.3	3.8 ± 0.6
Chlorprocaine	1.2 ± 0.1	0.9 ± 0.1	0.6 ± 0.03
Dimethocaine	3.8 ± 0.1	2.9 ± 0.1	3.3 ± 0.3
Lidocaine	19.1 ± 1.6[a]	24.0 ± 1.4[a]	52.9 ± 7.4
WIN 35,981	>100	>100	>100
d-Amphetamine	52.2 ± 7.2	51.2 ± 7.3	36.2 ± 8.1

[a] Significantly different from value in hippocampus, an M_1 receptor subtype: $p = 0.005$.

K_i values for cocaine and related compounds at [^{3}H]QNB binding sites in the rat heart, medulla, and hippocampus are shown. In each case the displacement curve could be best described using a model assuming one binding site. Values are expressed as mean (micromolar) ± S.E.M. for 3 to 5 separate experiments. (From J. Sharkey, M.C. Ritz, J.A. Schenden, R.C. Hanson, and M.J. Kuhar, *J. Pharmacol. Exp. Ther.* **246**: 1048 (1988).)

SOLUBILIZATION OF THE COCAINE RECEPTOR/ DOPAMINE TRANSPORTER

In order to more fully characterize the cocaine receptor, it will be important to purify the molecule and, ultimately, to identify the gene for the cocaine binding molecule and the entire dopamine transporter. Accordingly, we developed a photoaffinity probe which is highly selective for the dopamine transporter (Van der Zee, 1980) and binds with high affinity. This probe, [^{125}I]-Deep {^{125}I-1-[-2-(diphenylmethoxy)ethyl]-2-(4-azido-3-iodophenyl)ethyl]piperazine} covalently labels a molecule in the rat striatum which has the pharmacological properties of the cocaine receptor/dopamine transporter (Grigoriadis *et al.*, 1989). Binding of [^{125}I]-DEEP to the molecule is blocked by several inhibitors of dopamine transport such as mazindol, nomifensine, GBR 12935 (Berger *et al.*, 1985; Janowsky *et al.*, 1986), and (−) cocaine; but binding is not affected by desipramine or citalopram, compounds which have higher affinity for the norepinephrine transporter and serontonin transporter, respectively. Also, the cerebellum, which does not have the dopamine transporter, does not have the binding molecule. This receptor molecule labeled by [^{125}I]-DEEP migrates on SDS-PAGE with a molecular weight of approximately 58,000 Da (Figure 3). The next step will be to purify the receptors and to further characterize them, for example, by obtaining information regarding the primary sequence.

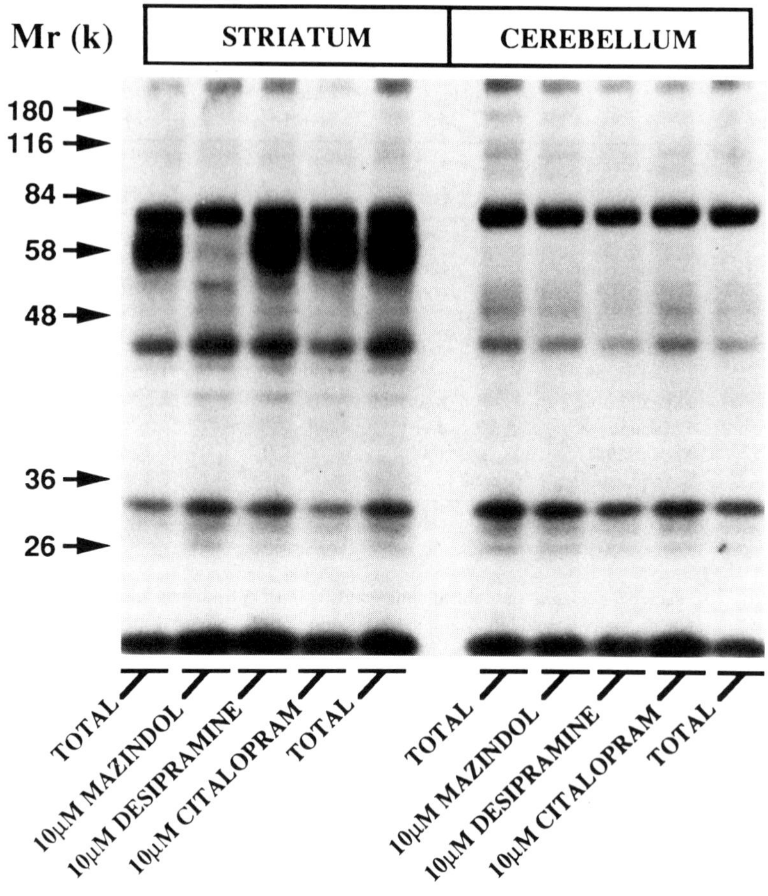

Figure 3. Covalent incorporation of [^{125}I]DEEP in striatal and cerebellar homogenates and autoradiographic visualization of bound [^{125}I]DEEP in SDS gels. Rat striatal and cerebellar homogenates were prepared in parallel, incubated with 1 to 2 n*M* [^{125}I]DEEP in the presence or absence of various compounds. Binding assays were conducted for 60 min at 0 to 4°C and then photolyzed. The autoradiogram shown is overexposed to bring out minor peaks of radioactivity in order to more thoroughly assess specificity; the bulk of radioactivity however is in the 58,000-Da band. Although photoincorporation of [^{125}I]DEEP to many nonspecific proteins was evident in both tissues, there was a conspicuous absence of any labeled proteins in the cerebellum with an apparent molecular weight of 58,000 Da. (From D.E. Grigoriadis, A.A. Wilson, R. Lew, J.S. Sharkey, and M.J. Kuhar. *J. Neurosci.* **9**: 2664 (1989). With permission.)

CONCLUSIONS

While several bindings sites for cocaine in brain have been identified, only one of these sites appears to be related to the reinforcing properties of the drug. This site is the cocaine receptor associated with the dopamine transporter. Amphetamine, a drug which shares many behavioral properties of cocaine but which has been shown to differ from cocaine in some of its neurochemical properties, does not appear to exert its reinforcing effects through the same receptor site. In agreement with earlier experiments suggesting that inhibition of serotonin uptake opposes reward systems in the brain, binding of amphetamine-like compounds at serotonin transporter was inversely related to its drug self-administration properties. Cocaine also bound to sigma receptors and muscarinic cholinergic receptors, but at lower affinities; while its reinforcing properties are not related to these receptors, some actions of cocaine may be mediated through these sites. The molecular properties of the cocaine receptor are being elucidated through detailed structure-activity relationships and through studies which allow the solubilization and purification of the receptor. Thus, receptor binding studies and other biochemical techniques will play a major role in elucidating the mechanism of action of cocaine and related psychostimulant drugs.

REFERENCES

N. Bejerot. A comparison of the effects of cocaine and central stimulants. *Br. J. Addic.* **65**: 35 (1970).

P. Berger, A. Janowsky, F. Vocci, P. Skolnik, M.M. Schweri, and S.M. Paul. [^{3}H]GBR-12395: a specific high affinity ligand for labeling the dopamine transport complex. *Eur. J. Pharmacol.* **107**: 289 (1985).

D.O. Calligaro and M.E. Eldefrawi. Central and peripheral cocaine receptors. *J. Pharmacol. Exp. Ther.* **243**: 61 (1987).

DeWit and R.A. Wise. Blockade of cocaine reinforcement in rats with the dopamine blocker pemozide but not with the nonradrenergic blockers phentolamine or phenoxybenzamine. *Can. J. Psychol.* **31**: 195 (1977).

E.H. Ellinwood and M.M. Kilbey. *Cocaine and Other Stimulants, Advances in Behavioral Biology.* Vol. 21. E.H. Ellinwood and M.M. Kilbey, eds., Plenum Press, New York (1977).

H.C. Fibiger. Drugs and reinforcement mechanisms: a critical review of the catecholamine theory. *Annu. Rev. Pharmacol. Toxicol.* **18**: 37 (1978).

M. Fischman, C.R. Schuster, and Y. Hitano. A comparison of the subjective and cardiovascular effects of cocaine and lidocaine in humans. *Pharmacol. Biochem. Behav.* **18**: 123 (1983).

M.W. Fischman, C.R. Schuster, L. Resnekov, J.F.E. Schick, N.A. Krasnegor, W. Fennell, and D.X. Freedman. Cardiovascular and subjective effects on intravenous cocaine administration in humans. *Arch. Gen. Psychiatry* **33**: 983 (1976).

N.E. Goeders, S.I. Dworkin, and J.E. Smith. Neuropharmacological assessment of cocaine self-administration into the medial prefrontal cortex. *Pharmacol. Biochem. Behav.* **24**: 1429 (1986).

N.E. Goeders and J.E. Smith. Reinforcing properties of cocaine in the medial prefrontal cortex: primary action on presynaptic dopaminergic terminals. *Pharmacol. Biochem. Behav.* **25**: 191 (1986).

R.R. Griffiths, G.E. Bigelow, and J.E. Henningfield. Similarities in animal and human drug taking behavior, in: *Advances in Substance Abuse.* Vol. 1, N.K. Mello, ed., JAI Press, Greenwich, CT, p. 1, (1980).

R.R. Griffiths, J.V. Brady, and L.D. Bradford. *Advances in Behavioral Pharmacology.* Vol. 2 T. Thompson and P.B. eds., Academic Press, New York, p. 163 (1979).

D.E. Grigoriadis, A.A. Wilson, R. Lew, J.S. Sharkey, and M.J. Kuhar. Dopamine transport sites selectively labled by a novel photoaffinity probe [^{125}I]-DEEP. *J. Neurosci.* **9(8)**: 2664 (1989).

A. Janowsky, P. Berger, F. Vocci, R. Labarca, P. Skolnick, and S.M. Paul. Characterization of sodium-dependent [^{3}H]GBR-12935 bindng in brain: a radioligand for selective labeling of the dopamine transport complex. *J. Neurochem.* **46**: 1272 (1986).

C.E. Johanson and C.R. Schuster. Animal models of drug self-administration, in: *Advances in Substance Abuse.* Vol. 1, N.K. Mello, ed., JAI Press, Greenwich, CT, p. 219 (1980).

A.S. Keats and J. Telford. Narcotic antagonists as analgesics: clinical aspects, in: *Molecular Modifications in Drug Design*. Advances in Chemistry Series. R.F. Gould, ed., American Chemical Society, Washington, DC, (1964).

L.T. Kennedy and I. Hanbauer. Sodium-sensitive cocaine binding to rat striatal membrane: possible relationship to dopamine uptake sites. *J. Neurochem.* **41**: 172 (1983).

S.Z. Langer and S. Arbilla. The amphetamine paradox in dopaminergic neurotransmission. *Trends Pharmacol. Sci.* **4**: 387 (1984).

A.P. Lecesse and W.H. Lyness. The effects of putative 5-hydroxytryptamine receptor active agents on D-amphetamine self-administration in controls and rats with 5,7-dihydroxytryptamine median forebrain bundle lesions. *Brain Res.* **303**: 153 (1984).

W.R. Martin, C.G. Eades, J.A. Thompson, and P.E. Gilbert. The effects of morphine- and nalorphine-like drugs in the nondependent- and morphine-dependent chronic spinal dog. *J. Pharmacol. Exp. Ther.* **197**: 517 (1976).

B.A. McMillen. CNS stimulants: two distinct mechanisms of action for amphetamine-like drugs. *Trends Pharmacol. Sci.* **4**: 429 (1983).

M.E.A. Reith, H. Sershen, and A. Lajtha. Saturable ^{3}H-cocaine binding in central nervous system of mouse. *Life Sci.* **27**: 1055 (1980).

M.C. Ritz, E.J. Cone, and M.J. Kuhar. Cocaine inhibition of ligand binding at dopamine, norepinephrine and serotonin transporters: a structure-activity study. *Life Sci.* **46**: 635 (1990).

M.C. Ritz and M.J. Kuhar. Relationship between self-administration of amphetamine and monoamine receptors in brain: comparison with cocaine. *J. Pharmacol. Exp. Ther.* **248**: 1010 (1989).

M.C. Ritz, R.J. Lamb, S.R. Goldberg, and M.J. Kuhar. Cocaine receptors on dopamine transporters are related to self-administration of cocaine. *Science* **237**: 1219 (1987).

D.C.S. Roberts and G.F. Koob. Disruption of cocaine self-administration following 6-hydroxydopamine lesions of the ventral tegmental area in rats. *Pharmacol. Biochem. Behav.* **17**: 901 (1982).

D.C.S. Roberts, G.F. Koob, P. Klonofl, and H.C. Fibigen. Extinction and recovery of cocaine self-administration following 6-hydroxydopamine lesions of the nucleus accumbens. *Pharmacol. Biochem. Behav.* **12**: 781 (1980).

J. Sharkey, K.A. Glen, S. Wolfe, and M.J. Kuhar. Cocaine binding at σ receptors. *J. Pharmacol.* **149**: 171 (1988a).

J. Sharkey, M.C. Ritz, J.A. Schenden, R.C. Hanson, and M.J. Kuhar. Cocaine inhibits muscarinic cholinergic receptors in heart and brain. *J. Pharmacol. Exp. Ther.* **246**: 1048 (1988b).

L.E. Tune, N.F. Damlouji, A. Holland, T.J. Gardner, M.F. Folstein, and J.T. Coyle. Association of postoperative delirium with raised serum levels of anticholinergic drugs. *Lancet* **2**: 651 (1981).

P. Van der Zee, H.S. Koger, J. Gootjes, and W. Hespe. Aryl 1,4-dialk(en)ylpiperazines as selective and very potent inhibitors of dopamine uptake. *Eur. J. Med. Chem.* **15**: 363 (1980).

R.W. Wise, Catecholamine theories of reward: a critical review. *Brain Res.* **152**: 215 (1978).

R.W. Wise. Neural mechanism of the reinforcing action of cocaine, in: *Cocaine: Pharmacology, Effects and Treatment of Abuse*. Vol. 50. J. Grabowski, ed., NIDA Research Monograph Series, p. 15 (1984).

J. Woods. Behavioral pharmacology of drug self-administration, in: *Psychopharmacology: A Generation of Progress*. M.A. Lipton, A. DiMascio, and K.F. Killam, eds., Raven Press, New York, p. 595 (1978).

D.S.L. Yu, F.L. Smith, D.G. Smith, and W.H.J. Lyness. Fluoxetine-induced attenuation of amphetamine self-administration in rats. *Life Sci.* **29**: 1383 (1986).

10

Central and Peripheral Cocaine Binding Sites

MAARTEN E.A. REITH
Center for Neurochemistry
N. S. Kline Institute for Psychiatric Research
Orangeburg, New York

INTRODUCTION

Cocaine has multiple effects such as vasoconstriction (Trendelenburg, 1972), central stimulation (Scheel-Krüger *et al.*, 1977), local anesthesia (Ritchie and Green, 1980), liver toxicity (Thompson *et al.*, 1979), and myocardial changes (Tazelaar *et al.*, 1987). It is therefore to be expected that different populations of cocaine binding sites exist that subserve various actions of the drug. This chapter will focus on those sites that have been measured by high-affinity binding of [^{3}H] cocaine and their relationship with sites measured by related tritiated ligands. The characterization of the [^{3}H]cocaine binding sites and their function by no means rules out other actions of cocaine. For example, [^{3}H]cocaine binding to sodium channels is not detected in assays of [^{3}H]cocaine binding under the conditions tested so far; still, blockade of sodium channels by cocaine is a very important action of the compound. Despite this limitation, the development of receptor binding assays for [^{3}H]cocaine has opened up new experimental approaches. For instance, the intriguing possibility can now be addressed that cocaine binding sites are regulated by exposure to cocaine, and that there are endogenous ligands acting upon these sites (section on Regulation and Endogenous Ligands). In addition, the availability of many other tritiated drugs that block monoamine uptake makes it possible to study the relationship between the recognition sites for these different compounds and the cocaine sites (section on Comparison with Binding Sites for Tritiated Ligands Other than Cocaine). The question can now be addressed whether the binding sites for [^{3}H]cocaine or other uptake blockers are separate from the substrate recognition sites for monoamine uptake (section on Equal to or Allosterically Linked with Monoamine Recognition Sites?). Finally, the study of cocaine binding sites in the periphery may give insight on how cocaine affects tissues other than the brain; for instance, liver binding may be involved in the protection against cocaine toxicity. Cocaine binding to blood platelets may represent a site for cocaine's interference with the removal of serotonin from plasma.

CENTRAL COCAINE BINDING SITES

Association with Monoaminergic Nerve Terminals

The initial characterization of [^{3}H]cocaine binding in mouse cerebral cortex was carried out under experimental conditions commonly used in receptor studies, such as the use of crude membrane fractions and 50 m*M* Tris-HCl at pH 7.4-7.7 (Reith *et al.*, 1980a, 1981; Sershen *et al.*, 1980). Under these conditions at 21°C, [^{3}H]cocaine binds with a K_d between 0.3 and 0.8 μM in a sodium-insensitive manner. The binding is only slightly higher at 21°C than at 2 or 37°C, and is destroyed by trypsin treatment or heating of membranes at 100°C. Both association and dissociation at 21°C are extremely rapid, but binding assays can be terminated by the filtration method, provided it is carried out extremely rapidly (i.e., by a single manifold and not a cell harvester) (Reith *et al.*, 1980a, 1981, 1986a). Cerebrocortical cocaine binding is not inhibited by monoamines, GABA, acetylcholine, or morphine (Reith *et al.*, 1980a). It is, however, very sensitive to blockers of neuronal uptake of serotonin; there is a significant correlation between the potencies of various drugs in inhibiting [^{3}H]cocaine binding and their potencies in inhibiting serotonin uptake (Reith *et al.*, 1983a). This, together with the reduction in cocaine binding by neurotoxins that lesion serotonergic nerve terminals, suggests that part of the [^{3}H]cocaine binding in rodent cerebral cortex is associated with serotonergic nerve terminals. In these experiments, rat cerebral cortex also displayed a low-affinity component under conditions that gave only a high-affinity component in the mouse cerebral cortex (Reith *et al.*, 1983a). Consonant with an association of the high-affinity binding component with nerve terminals is the fact that an appreciable amount of [^{3}H]cocaine binding occurs to the synaptosomal fraction prepared by sucrose gradient centrifugation of the P_2 fraction (Reith *et al.*, 1984b). In this context, it is of interest that high-affinity binding of [^{3}H]cocaine occurs to blood platelets (Reith *et al.*, 1983a), a model system for central serotonergic terminals (Stahl and Meltzer, 1978).

Under the same conditions as used in the above studies with cerebral cortex (Tris-HCl buffer, crude membranes), [^{3}H]cocaine binding in rat striatum has a K_d of 360 n*M* and is strongly stimulated by Na^+ (Kennedy and Hanbauer, 1983.) The Na^+-dependent binding is eliminated by lesioning of the dopaminergic nerve terminals, and the inhibitory potencies of various drugs correlate with their potencies in inhibiting dopamine uptake, suggesting an association of striatal cocaine binding with dopaminergic terminals (Kennedy and Hanbauer, 1983). A dopaminergic relation has also been reported for [^{3}H]cocaine binding in the presence of Na^+ in human putamen and rat striatum (Schoemaker *et al.*, 1985). In the latter study, it was also shown that the human putamen and rat striatum have a low-affinity component of [^{3}H]cocaine binding, measured in the presence of Na^+, with a K_d of 26

μM, the pharamacological relevance of which is unknown. This binding may represent accumulation of cocaine in lipid membranes similar to the "binding" of [^{3}H]cocaine that occurs to artificial liposome membranes (Reith *et al.*, 1984d). The low-affinity binding in the rat cerebral cortex (Reith *et al.*, 1983a) may be of the same nature. Interestingly, there is high affinity [^{3}H]cocaine binding in rodent striatum in the absence of Na^+ (Kennedy and Hanbauer, 1983; Reith *et al.*, 1985a). A separate study addressed the question of whether this binding represents the unstimulated Na^+-dependent site or the Na^+-independent site of the serotonergic type observed in other brain areas (Reith *et al.*, 1985a). The pharmacological profile of the binding and the reduction caused by the serotonin neurotoxin 5,7-dihydroxytryptamine indicated it to be of the Na^+-independent serotonergic type.

After the finding that Tris-HCl buffer is inhibitory to [^{3}H]cocaine binding in mouse cerebral cortex (Reith *et al.*, 1984a), the binding was examined with different buffer systems and, instead of crude membranes, the P_2 fraction was used. The effect of varying Na^+ concentrations was assessed with dilute Tris-HCl buffer (5mM), sodium phosphate buffer, and HEPES buffer. In agreement with the results of Kennedy and Hanbauer (1983) observed with Tris buffer for rats, the striatum was the only region of mouse brain in which [^{3}H]cocaine binding was stimulated by Na^+. The striatal binding was maximally stimulated (fourfold) at 25 mM Na^+ and declined gradually with higher concentrations of Na^+; at 150 mM Na^+, there still was a two- to threefold higher binding than that with no sodium (Reith, 1988). In mouse brain regions other than the striatum, the binding did not require Na^+ and was not stimulated or inhibited by Na^+ concentrations below 100–200 mM. With 25 mM sodium phosphate buffer (48 mM Na^+) the K_d values for [^{3}H]cocaine binding in most brain regions were between 120 and 190 nM (Reith *et al.*, 1986b). An exception was the cerebellum, with a K_d of 396 nM and a strongly inhibitory effect of Na^+ (Reith, 1988). The B_{max} values ranged from 1 to 2 pmol/mg of protein in various brain regions and the density was approximately 7 pmol/mg of protein in the striatum. The potencies of cocaine analogs in inhibiting cerebrocortical and striatal [^{3}H]cocaine binding in sodium phosphate buffer were consonant with a serotonergic and dopaminergic association, respectively, as observed in Tris buffer (Reith *et al.*, 1986b). The dissociation of [^{3}H]cocaine binding in the mouse cerebral cortex had a half-life of 25 s; the dissociation in the striatum was heterogenous and could be resolved into two components with half lives of 2 and 27 s, comprising 88 and 12%, respectively, of the total binding sites (Reith *et al.*, 1986a). The slowly dissociating component in the striatum with the half-life of 27 s, observed also in the cerebral cortex, probably represents the serotonin-related sites; and the rapidly dissociating component with the half-time of 2 s the dopamine-related sites. A previous estimate from saturation analysis was that 75% of the striatal cocaine binding sites were dopamine related (Reith *et al.*, 1986b).

The inhibition of Na^+-independent [^{3}H]cocaine binding by serotonin uptake blockers suggests that cocaine, which is an uptake blocker itself, binds to the carrier involved in serotonin uptake. The lack of stimulation of [^{3}H]cocaine binding by Na^+ is surprising, but not unprecedented, for ligands that bind to uptake sites. For instance, 8-OHDPAT [8-hydroxy-2-(di-*N*-propylamino)tetralin] binds to serotonin uptake sites in the striatum in the absence of sodium (Schoemaker and Langer, 1986). [^{3}H]Cocaine binding is also not stimulated by Na^+ in dopaminergic areas other than the striatum, such as the nucleus accumbens (Missale *et al.*, 1985) and olfactory tubercle (Kennedy and Hanbauer, 1983; Reith *et al.*, 1986b), but the nature of this binding is not known. It has been suggested that absence of Na^+-dependent cocaine binding in the nucleus accumbens explains the lack of inhibitory activity of cocaine on [^{3}H]dopamine uptake into nucleus accumbens slices (Missale *et al.*, 1985). More work is needed to clarify this point, because [^{3}H]dopamine uptake into nucleus accumbens synaptosomes is inhibited potently by cocaine (Berger *al.*, 1988). In addition, [^{3}H]GBR-12935, a ligand that binds Na^+-dependently with high affinity (K_d of 1 n*M*) to dopamine uptake sites in the striatum, has been shown recently to bind also in the nucleus accumbens in a Na^+-dependent manner with a similar affinity (K_d of 1 n*M*; Berger *et al.*, 1988). Most likely, the proper definition of nonspecific uptake is crucial, as shown in a study comparing the effect of cocaine on dopamine uptake into synaptosomes from the striatum, prefrontal cortex, and olfactory tubercle (Hadfield and Nugent, 1983).

Even though cocaine is as potent a blocker of norepinephrine uptake as of dopamine and serotonin uptake (Koe, 1976), no evidence has been found for an association between [^{3}H]cocaine binding and norepinephrine uptake sites (Reith, 1983a). In contrast, [^{3}H]mazindol has been shown to label both the dopamine uptake sites in the striatum and the norepinephrine sites in the cerebral cortex of rats (Javitch *et al.*, 1984). A possible explanation is that [^{3}H]cocaine dissociates too rapidly from the norepinephrine sites at 0°C to be preserved during the wash with cold buffer in the filtration assay.

The K_d value (150 n*M*) for [^{3}H]cocaine binding in the mouse cerebral cortex measured with sodium phosphate buffer (pH 7.7) is lower than that (706 n*M*) with Tris-HCl buffer (pH 7.7) (Reith *et al.*, 1984a). It is possible that Tris-HCl competes for the cocaine binding sites with a K_d in the millimolar range. Consonant with this, Tris-HCl acted like a competitive inhibitor in a saturation analysis, decreasing the K_d without changing the B_{max} (Reith *et al.*, 1984a). In addition, the dissociation rate of [^{3}H]cocaine binding was the same in Tris-HCl buffer as in sodium phosphate buffer, as expected for a competitive inhibitor (Reith *et al.*, 1986a). Tris-HCl also inhibits the Na^+-dependent [^{3}H]cocaine binding in the rat striatum (Calligaro and Eldefrawi, 1988). The inhibition seems to be due to a decrease in B_{max} and is counteracted by Na^+ (see section on Comparison with Binding Sites for Tritiated Ligands other than Cocaine).

A recent report described [^{3}H]cocaine binding to rat brain membranes (P_2 fraction) in 10 m*M* sodium phosphate buffer containing 0.32 *M* sucrose (Calligaro and Eldefrawi, 1987). In all brain regions, including striatum, the binding was heterogeneous, consisting of a high-affinity component with a K_d of 14–17 n*M* and a low-affinity component with a K_d of 68–660 n*M*. Although the high-affinity binding in cerebral cortex and striatum was blocked by (+)-pseudococaine, norcocaine, and cinnamoylcocaine with a similar rank order of potency as the cortical [^{3}H]norepinephrine uptake and striatal [^{3}H]dopamine uptake, respectively, the limited number of drugs tested does not allow a firm conclusion about the nature of these cocaine binding sites. More work is needed to clarify the precise relationship between the rat cerebrocortical [^{3}H]cocaine binding sites reported in this study and those described above for the mouse. Lesion studies would be helpful in this respect.

The high-affinity cocaine site in rat striatum described by Calligaro and Eldefrawi (1987) appears to be the same dopamine-related site as described above as judged from the inhibitory potencies of various drugs correlating with inhibition of dopamine uptake (El-Maghrabi *et al.*, 1988; Calligaro and Eldefrawi, 1988). Other properties of the high-affinity site are similar to those of the striatal sites discussed above, such as the dependency upon temperature and pH, the extremely rapid association and dissociation, and the rank order of inhibitory potencies of (+)-pseudococaine, norcocaine, and ecgonine. The K_d of 17 to 35 n*M* observed in the above studies for cocaine binding in the rat striatum is appreciably lower than that observed in the mouse striatum (122 n*M*) (Reith *et al.*, 1986b). It is conceivable that rats and mice are different in this respect; in addition, the studies were done with different buffers (10 m*M* sodium phosphate containing 0.32 *M* sucrose vs. 25 m*M* sodium phosphate). It is not clear why the density of the rat striatal cocaine sites in the study of Calligaro and Eldefrawi (1988) (2–3 pmol/mg of protein) is lower than that of the cocaine sites observed in other studies with or without Tris-HCl buffer (6–9 pmol/mg of protein) in both rats (Hanbauer *et al.*, 1984; Schoemaker *et al.*, 1985) and mice (Reith *et al.*, 1986b), and also lower than that of binding sites labeled with other dopamine uptake ligands (see below).

Affinity for Cocaine Analogs: Relationship with Behavioral Effects

The effect of cocaine analogs on [^{3}H]cocaine binding was studied with P_2 membranes from mouse brain with sodium phospate buffer (Reith *et al.*, 1986b). Extremely helpful tools were the Winthrop phenyltropane analogs of cocaine synthesized by Clark *et al.* (1973) (see Figure 1). The structure-activity relationships of cocaine analogs in inhibiting Na^+-independent (cortical) and Na^+-dependent (striatal) [^{3}H]cocaine binding were generally similar. Removal of the ester linkage between the tropane and phenyl rings of cocaine (WIN 35,428 and WIN 35,065-2) did not reduce the affinity of these molecules for cocaine binding sites (Reith *et al.*, 1986b). Moving the carbomethoxy group (R_2) on C_2 from an axial (cocaine and WIN 35,065-2) (see Table 1) to an equatorial [(+)-pseudococaine and WIN 35,140] position diminished the potency 20- to 80-fold. Likewise, moving the *O*-benzoyl group (R_4) on C_3 from an equatorial (benzoylpseudotropine) to an axial (benzoyltropine) position reduced the activity of the molecule four- to sevenfold. Therefore, the position of the C_2 constituent seems to be more important than that of the C_3 constituent (Reith *et al.*, 1986b). Removal of the C_2 constituent (benzoylpseudotropine vs. cocaine) appreciably reduced the activity (10- to 20-fold), but had a less profound effect than removal of the methylester from the C_2 position of cocaine (benzoylecgonine), which rendered the molecule inactive, presumably by making it more polar. WIN 35,065-3 is a pure enantiomer of WIN 35,065-2; the two compounds are D- and L-isomers, respectively (Clarke *et al.*, 1973). WIN 35,065-3 was 200- to 800-fold weaker than WIN 35,065-2, indicating sterospecificity.

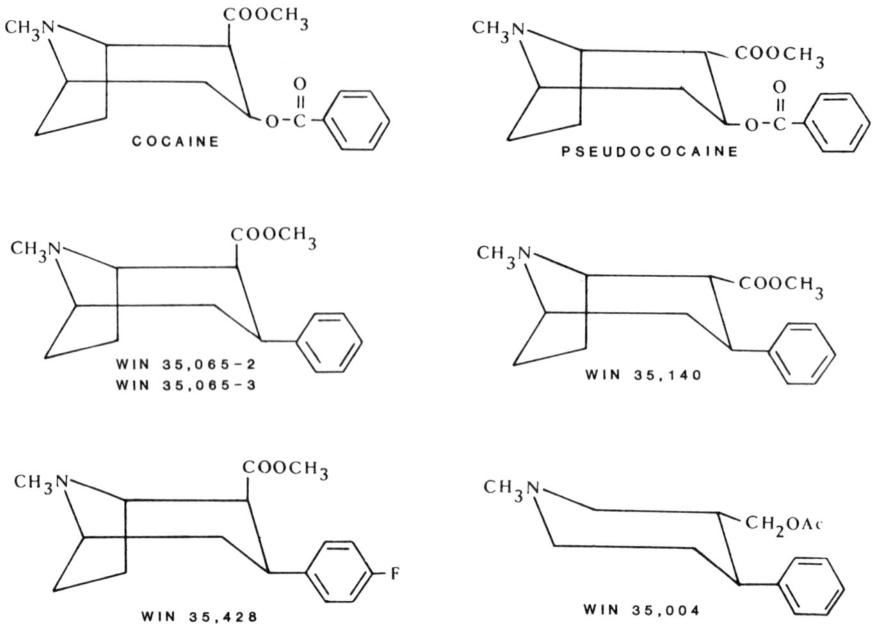

Figure 1. Structures of cocaine, pseudococaine, and phenyltropane analogs (WIN) of cocaine prepared by Clarke *et al.* (1973) at Winthrop.

There were some differences between the potencies of cocaine congeners in the inhibition of Na^+-independent and Na^+-dependent cocaine binding. For inhibition of Na^+-dependent [^{3}H]cocaine binding in the striatum, removal of the ester linkage between the tropane and phenyl rings (WIN 35,428 and WIN 35,065-2) made the molecule more potent (in contrast to the lack of change in affinity for Na^+-independent sites), whereas *N*-demethylation (norcocaine) resulted in some loss (twofold) of activity (in contrast to a 1.5-fold increase in affinity for Na^+-independent sites. In addition, C_3 epimerization of benzoylpseudotropine (yielding benzoyltropine) had a greater effect in reducing the potency toward the Na^+-dependent [^{3}H]cocaine binding sites (sevenfold) than toward the Na^+-independent sites (fourfold). Such differences may indicate that cocaine and cocaine congeners interact at basically different sites on the dopamine and serotonin transporter, or that the environment is different at the cocaine binding sites on these transporters.

When binding and uptake were compared in the same study (Reith *et al.*, 1986b), there was a highly significant correlation between the potencies of cocaine analogs in inhibiting Na^+-independent and Na^+-dependent [^{3}H]cocaine binding and their potencies in inhibiting uptake of [^{3}H]serotonin into cerebrocortical synaptosomes and of [^{3}H]dopamine into striatal synaptosomes, respectively. Table 1 summarizes the changes in potency for uptake inhibition by structural changes as compared with the parent compound cocaine. Some of these structure-activity relationships for cocaine congeners have also been found for inhibition of [^{3}H]mazindol binding to rat striatal dopamine uptake sites, such as the reduction in activity by isomerization

Table 1
Changes in Potency as Compared with Cocaine Itself by Introducing Varying Changes in the Molecular Structure.

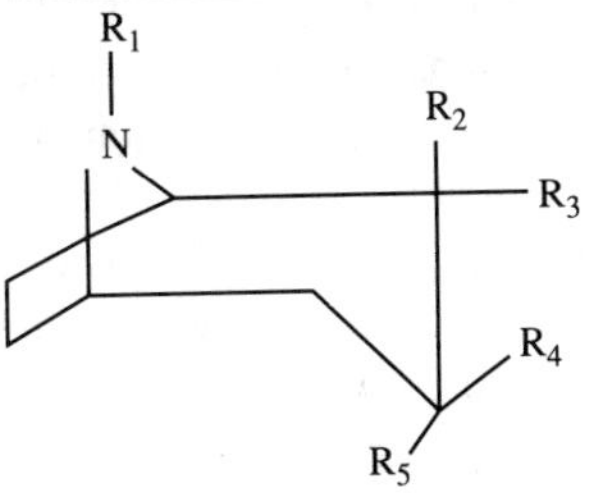

	Change in Potency		
	Sodium channel	5-HT uptake	DA uptake
N-demethylation			
cocaine to norcocaine	4 × ↑	10 × ↑	2 × ↓
Removal of ester linkage			
cocaine to WIN 35,065-2		1 ×	5 × ↑
cocaine to WIN 35,428	4 × ↓	3 × ↑	11 × ↑
R_2 from axial to equatorial			
cocaine to (+)-pseudococaine	4 × ↑	15 × ↓	40 × ↓
WIN 35,065-2 to WIN 35,140		150 × ↓	30 × ↓
WIN 35,428 to WIN 35,140	4 × ↑	400 × ↓	70 × ↓
R_4 from equatorial to axial			
benzoylpseudotropine to benzoyltropine	3 × ↓	2 × ↓	6 × ↓
Removal of R_2			
cocaine to benzoylpseudotropine	8 × ↑	7 × ↓	21 × ↓
Removal of R_4			
cocaine to ecgonine methylester	1000 × ↓		
Extension R_3 in neopseudococaine			
cocaine to (+)-neopseudococaine	100 × ↑	1 ×	20 × ↓

All data were collected from Balb/cBy mice. Sodium channel inhibitory activity was measured by blockade of [^{3}H]batrachotoxinin A20-α-benzoate binding to cerebrocortical purified synaptosomes (Reith *et al.*, 1986c). The potency in blocking serotonin (5-HT) uptake was measured by inhibition of [^{3}H]serotonin uptake into cerebrocortical synaptosome-containing fractions, and the potency in blocking dopamine (DA) uptake by inhibition of [^{3}]dopamine uptake into striatal synaptosome-containing fractions. (From M.E.A. Reith, B.E. Meisler, H. Sershen, and A. Lajtha. *Biochem. Pharmacol.* **35**: 1123 (1986). With permission.)

to D-isomers, epimerization of the C_2 substituents, and hydrolysis of the C_2 substituent to a more polar form (Ritz *et al.*, 1988). In addition, the binding of [^{3}H]cocaine in monkey striatum was inhibited by cocaine analogs with the following rank order of potency: WIN 35,065-2 > (−)-cocaine > WIN 35,065-3 > (+)-cocaine (Madras *et al.*, 1988). These structure-activity relationships were different from those observed for local anesthetic activity, as measured by inhibition of binding of [^{3}H]batrachotoxinin-A20-α-benzoate to voltage-dependent sodium channels (Reith *et al.*, 1986c), or by counteraction of the depolarization induced by veratridine resulting in a change of the distribution of the lipophilic cation [^{3}H]triphenylmethylphosphonium (Zimanyi *et al.*, 1989). For sodium channel activity, the ester linkage between the tropane and phenyl rings of cocaine was not necessary: WIN 35,140 had the same potency as cocaine, whereas WIN 35,428 and WIN 35,065-3 were approximately three- to fourfold less potent than cocaine (see also Table 1). In addition, *N*-demethylation (norcocaine) increased the inhibitory activity fourfold, whereas moving the carbomethoxy group on C_2 from the axial (cocaine and WIN 35,428) to an equatorial [(+)-pseudococaine and WIN 35,140] postion increased the potency by a factor of 4. (+)Neopseudococaine was approximately 100-fold more potent than cocaine, in contrast to the 5- to 35-fold reduction observed for inhibition of cocaine binding in the cerebral cortex and striatum, respectively (Reith *et al.*, 1986b).

The cocaine analogs that were potent in inhibiting [^{3}H]cocaine binding to monoamine uptake sites had been shown to be behaviorally active in earlier studies. For instance, WIN 35,065-2 and WIN 35,428 were three to ten times more potent than cocaine in affecting scheduled-controlled behavior of squirrel monkeys (Spealman *et al.*, 1977). WIN 35,065-2 was also three to ten times more potent than cocaine in maintaining self-administration, whereas its enantiomer, WIN 35,065-3, was inactive in squirrel monkeys (Spealman and Kelleher, 1981). WIN 35,428 and WIN 35,065-2 generalized to cocaine more potently than cocaine itself in a drug discrimination paradigm in pigeons (Järbe, 1981). Dopamine systems have been implicated in these behavioral effects, and striatal dopamine appears to be particularly important in cocaine-induced increases in stereotyped behavior (Beninger, 1983). Thus, the correlation between the potencies of cocaine congeners in inhibiting Na^+-dependent cocaine binding in the striatum and in inducing stereotyped sniffing in mice (Reith *et al.*, 1986b) is in agreement with the notion that the Na^+-dependent cocaine binding sites are involved in cocaine-induced increases in stereotyped behavior.

It was necessary to administer the cocaine congeners introcerebroventricularly in order to induce the stereotyped effects (Reith *et al.*, 1986b). In fact, intraperitoneal (i.p.) administration of most cocaine congeners exhibited a depressing rather than a central stimulatory effect. In such a case, after i.p. administration of a cocaine congener, the animal was less alert, but had the eyes open and did not assume postures normally associated with sleep (Reith *et al.*, 1985b) There was a reasonable agreement between the rank order of potencies of cocaine congeners and other local anesthetics in depressing locomotor behavior and in inhibiting [^{3}H]BTX-B binding to sodium channels (Reith *et al.*, 1986c). Such a comparison has to be judged with caution, since drugs applied *in vivo* must be transported to their presumed sites of action, are susceptible to metabolizing enzymes, and may have multiple effects. Nevertheless, the observed correlation is consonant with the implication of sodium channels in the inhibition of locomotor behavior by cocaine congeners. The present results do not indicate whether central sodium channels are involved. There is some evidence suggesting action at the peripheral level (Reith and Lajtha, 1986). It is conceivable that multiple targets at the peripheral level are involved in the effect of local anesthetics, such as cardiovascular system (decrease in force of contraction and arteriolar dilation), the neuromuscular junction (decrease in conduction), and the lungs (depression of respiration) (Ritchie and Green, 1980). In this context, it is of interest that the notable side effects of lidocaine in humans are sleepiness and dizziness (Ritchie and Green, 1980).

The locomotor depression induced by norcocaine was not antagonized by low doses of α_2-adrenoceptor antagonists or neuroleptics aimed at presynaptic α_2-adrenoceptors or presynaptic dopamine receptors (Reith and Lajtha, 1986). Norcocaine is approximately four times more potent as a local anesthetic (Reith *et al.*, 1986c) and almost as potent as cocaine in inhibiting dopamine uptake (Reith *et al.*, 1986b). The fact that norcocaine, when administered i.p. to mice, is a powerful depressant drug at doses (1–10 mg/kg) substantially lower than the doses of cocaine (20–40 mg/kg i.p.) required for locomotor activation (Reith *et al.*, 1985b) may be related to the high potency of norcocaine in inhibiting serotonin uptake (approximately ten times that of cocaine, Reith *et al.*, 1986b). In this context, it should be recalled that activation of serotonergic systems often inhibits the locomotor-activating property of cocaine (Scheel-Krüger *et al.*, 1977: Pradhan *et al.*, 1978).

A recent study by Ritz *et al.* (1987) showed a significant correlation between the potencies of various cocaine-like drugs in inhibiting [^{3}H]mazindol binding to dopamine uptake sites in the rat striatum and their potencies in maintaining self-administration in squirrel monkeys trained to self-administer cocaine. In contrast, there was no relationship between the binding to dopamine uptake sites and their potencies in binding to a large number of other presynaptic and postsynaptic sites. This is consonant with the implication of dopamine uptake blockade by cocaine-like drugs in their capability of maintaining self-administration.

An interesting possibility to irreversibly inactivate the dopamine uptake sites opened up with the finding that metaphit, an isothiocyanate analog of phencyclidine (Contreras *et al.*, 1986), produced a long-term inactivation of Na^+-dependent [^{3}H]cocaine binding in mouse striatum (Berger *et al.*, 1986), of [^{3}H]mazindol binding in mouse striatum (Zimanyi *et al.*, 1988), and of [^{3}H]methylphenidate binding to rat striatum (Schweri *et al.*, 1987) *in vitro*. In addition, *in vivo* administration of metaphit to mice antagonized cocaine-induced locomotor stimulation measured 24 h after the metaphit treatment (Berger *et al.*, 1986). However, subsequent efforts to detect *in vivo* blockade of dopamine uptake sites by metaphit were unsuccessful (Sershen *et al.*, 1988). The failure of cocaine and other blockers of dopamine uptake to stimulate locomotor activity in metaphit-pretreated animals was attributed to a greater rate of dopamine catabolism reflected in relatively high levels of homovanillic acid (Sershen *et al.*, 1988).

Regulation and Endogenous Ligands

The development of binding assays for drugs that block monoamine uptake enables us to study the relationship between these sites and the substrate recognition sites at which the monoamines bind before they are translocated. If the interesting possibility, that these sites are distinct and are coupled by an allosteric linkage, is true, it could represent a novel regulatory mechanism for monoamine uptake by a site that is acted upon by an endogenous ligand different from the monamine itself. A possible example is the down-regulation of imipramine binding sites in rat hippocampus by chronic imipramine treatment accompanied by an increase in the V_{max} of uptake of serotonin into hippocampal slices (Barbaccia *et al.*, 1983). Evidence for regulation of dopamine-related sites has come from studies in which rats were treated daily with cocaine for 3 weeks and examined 48 h after the last injection: whereas [^{3}H]dopamine uptake into striatal slices was unchanged as compared with control animals (Missale *et al.*, 1985), there was an increase in Na^+-dependent [^{3}H]cocaine binding sites in the striatum (Hanbauer *et al.*, 1984). These results show that uptake of the monoamine and binding of the inhibiting drug do not always change in parallel.

Missale *et al.* (1985) also reported that 20 min after one injection of cocaine, dopamine uptake into striatal slices was decreased (reduction in V_{max}), whereas uptake into nucleus accumbens slices was increased (elevation in $V_{max.}$). In animals that were treated daily with cocaine for 3 weeks and examined 48 h after the last injection, there was no change in the [^{3}H]dopamine uptake into striatal slices as compared with saline-treated rats; in contrast, uptake into nucleus accumbens slices showed an increased V_{max} value (Missale *et al.*, 1985). In comparing the acute and chronic experiments, it is tempting to conclude that tolerance to cocaine's effect occurs in the striatum but not in the nucleus accumbens. However, we do not know what the uptake response is in animals treated chronically with cocaine and examined 20 min after an acute cocaine challenge. It would therefore be better to avoid the term tolerance in describing the chronic effects. In addition, more work is needed to clarify the reported lack of effect of cocaine in inhibiting *in vitro* the dopamine uptake into accumbens slices (Missale *et al.*, 1985), since cocaine is a good uptake inhibitor in accumbens synaptosomes (Berger *et al.*, 1988).

Na^+-dependent [^{3}H]cocaine binding to rat striatal membranes was shown to be a function of age (Missale *et al.*, 1986). The B_{max} was significantly lower in 25-month-old rats than in 3-month-old rats. In contrast, the V_{max} of [^{3}H]dopamine uptake into striatal slices was unchanged, and the K_m of dopamine uptake was higher in senescent rats. This could be interpreted as a separate regulation of cocaine sites and some other aspect of the dopamine uptake system, but other explanations are also possible (see section on Monoamine Recognition Sites). Many steps are involved in the neuronal uptake of dopamine, and it is conceivable that changes in the uptake rate of [^{3}H]dopamine result from alterations somewhere in the uptake system other than at the dopamine recognition site.

Mouse brain extracts were found to contain material that inhibits the Na^+-independent binding of [^{3}H]cocaine (Reith *et al.*, 1980b). Whole brain tissue was homogenized in 1% NaCl and centrifuged at 50,000 G. The supernatant was heated at 100°C and, after removal of the denatured protein by centrifugation, the material was filtered through an Amicon UM-10 membrane ultrafilter. After gel filtration of the ultrafiltrate on a Sephadex G-25 column, the cocaine binding inhibitory material was retained by the column and its elution pattern suggested a low molecular weight (<5000). The material was sensitive to acid hydrolysis and peptidase action, suggesting the presence of peptide(s). The inhibitor had effects similar to that of cocaine on the specific binding of various ligands to opiate and nonopiate receptors in mouse brain membranes (Reith *et al.*, 1980b). The purification was approximately 100-fold compared with whole homogenate protein. A further enrichment could be achieved by gel filtration on a Sephadex G-10 column separating the inhibitory activity from amino acids. Subsequent attempts at purifying the material by thin-layer chromatography, ion-exchange separation, or HPLC were unsuccessful, resulting in loss of cocaine binding inhibitory activity.

Hanbauer *et al.* (1984) reported the partial purification of endogenous material that inhibits Na^+-dependent [^{3}H]cocaine binding to striatal membranes. It was extracted in acetic acid; after gel filtration on a Biogel P-10 column, the proteins in the void volume (mol wt > 20,000) were precipitated with ammonium sulfate and the cocaine binding inhibitory activity was distributed over 3 bands upon isoelectric focusing. The inhibitory activity in the ammonium sulfate precipitate was lost upon treatment with pronase, suggesting its peptidergic nature, but not with trypsin, making it unlikely that the endogenous inhibitor consisted of multiple active subunits. Later, this endogenous material with cocaine binding inhibitory activity was identified as GABA-modulin (Hanbauer, 1988).

Comparison with Binding Sites for Tritiated Ligands other than Cocaine

Ligands other than [^{3}H]cocaine have been reported to bind to serotonin uptake sites. [^{3}H]Imipramine was the first ligand found to bind with high affinity to serotonin uptake sites in various brain regions, including the cerebral cortex, hypothalamus, hippocampus, and striatum (Raisman *et al.*, 1980; Langer *et al.*, 1980; Kinnier *et al.*, 1981). The binding was reduced by serotonin neurotoxins and was inhibited by various drugs with potencies correlating with their serotonin uptake inhibitory potencies; this, together with the high-affinity [^{3}H]imipramine binding to blood platelets, indicated an association with serotonergic nerve terminals (for review, see Langer *et al.*, 1987). Similar characteristics have been shown for other ligands such as [^{3}H]RO 11-2465 (Dumbrille-Ross and Tang, 1983), [^{3}H]norzimelidine (Hall, 1984), and [^{3}H]citalopram (D'Amato *et al.*, 1987), ligands that are no longer commercially available. The use of [^{3}H]imipramine for labeling serotonin uptake sites is confounded by the existence of low-affinity binding of imipramine (Reith *et al.*, 1983b; Hrdina, 1984; Marcusson *et al.*, 1986; Reith, 1986), possibly representing an interaction with lipid constituents of brain membranes (Reith *et al.*, 1984d). A more selective ligand appears to be [^{3}H]paroxetine, which binds with very high affinity to brain membranes (K_d of 0.15 n*M*) (Habert *et al.*, 1985). The differences in the potency (0.1–0.7 μM) reported for serotonin in inhibiting the binding of these various ligands are probably due to the instability of serotonin during the long incubation times. Other differences between the ligands could arise from their binding to nonidentical regions of the receptor protein that overlap on the serotonin transporter. These regions share a common site that mediates the uptake inhibitory action and may or may not be identical to the substrate recognition site for serotonin uptake. Thus, the ligands can bind to subsites on different portions of the membrane and the binding of various ligands can be influenced in different ways by ions such as Na^+, by other drugs, or by temperature. The B_{max} values would be expected to be the same for the different ligands. Indeed, all the above ligands show a similar maximal binding of 0.3–0.6 pmol/mg of protein in the cerebral cortex, slightly less than that observed for [^{3}H]cocaine binding, 1.2 pmol/mg of protein. This could mean that [^{3}H]cocaine labels additional sites with a pharmacological profile similar to the sites of serotonergic nerve terminals that are labeled with [^{3}H]imipramine, [^{3}H]RO 11-2465, [^{3}H]norzimelidine, [^{3}H]citalopram, and [^{3}H]paroxetine. The involvement of a common site involved in the binding of [^{3}H]cocaine and [^{3}H]imipramine is also suggested by the vulnerability of both bindings to treatment of cerebrocortical membranes with *N*-ethylmaleimide, trypsin, and phospholipase A_2, and the resistance of both bindings to exposure to dithiothreitol (Reith *et al.*, 1984b). The relatively greater sensitivity of cocaine binding to heat inactivation could be explained by the involvement of heat-sensitive subsites in the binding of cocaine. It is an unresolved question why the Hill number is considerably less than unity for the inhibition of [^{3}H]imipramine binding by cocaine and nontricyclic antidepressant drugs such as fluoxetine (Reith *et al.*, 1984b, 1984c). Perhaps the subunits involved in cocaine or fluoxetine binding show negative cooperativity at the low temperature (0–4°C) used for the [^{3}H]imipramine binding assay; it is of interest that the Hill number is unity for the inhibition by fluoxetine of the binding of [^{3}H]paroxetine (Habert *et al.*, 1985) and of [^{3}H]citalopram (D'Amato *et al.*, 1987) assayed at 22°C.

Since the finding that [^{3}H]cocaine binds to dopamine uptake sites in the striatum (Kennedy and Hanbauer, 1983), many other tritiated ligands have been reported to bind to striatal membranes, such as [^{3}H]mazindol (Javitch *et al.*, 1984), [^{3}H]methylphenidate (Schweri *et al.*, 1985), [^{3}H]nomifensin (Dubocovich and Zahniser, 1985), [^{3}H]GBR 12935 (Berger *et al.*, 1985), [^{3}H]GBR 12783 (Bonnet *et al.*, 1988), and [^{3}H]BTCP, a novel tritiated PCP analog (Vignon *et al.*, 1988). Their binding characteristics are similar, such as the requirement for sodium, inhibition by various drugs correlating with dopamine uptake inhibition, and sensitivity to lesioning of dopamine terminals by 6-hydroxydopamine: the B_{max} values for these ligands, 6–13 pmol/mg of protein, are close to the density reported for the high-affinity cocaine binding sites in the striatum, 6–9 pmol/mg of protein (Hanbauer *et al.*, 1984; Schoemaker *et al.*, 1985; Reith *et al.*, 1986b). As outlined above for serotonin-related ligands, differences in binding properties can be explained by overlapping regions on the dopamine carrier where the various ligands bind. Even if the different subsites involved in the binding of various ligands are regulated differently, the observed change in striatal [^{3}H]cocaine binding without a change in [^{3}H]GBR 12935 binding as a result of cortical ablation is intriguing (Grilli *et al.*, 1988). An important piece of evidence in support of striatal cocaine binding sites being distinct from GBR 12935 sites would be the demonstration, under identical assay conditions (buffer, temperature), of a difference between the potency of cocaine (and GBR 12935) in inhibiting [^{3}H]cocaine binding and that in inhibiting [^{3}H]GBR 12935 binding. So far, such evidence has not been reported.

A recent study by Bonnet *et al.* (1988) makes it necessary to reconsider the commonly accepted notion that the K_m for the Na^+ dependency of the dopamine carrier is in the neighborhood of 10^{-1} *M*. They found that the binding of [^{3}H]GBR 12783 was half-maximal at a concentration of approximately 5 m*M* in buffers in which Na^+ was the only cation. Ions such as K^+, Ca^{2+}, Mg^{2+}, and $Tris^+$ competitively inhibited [^{3}H]GBR 12783 binding, most likely by competing with Na^+ for the Na^+ binding site. This explains the relatively high concentrations of Na^+ (30–240 m*M*) required for half-maximal stimulation of tritiated ligands to dopamine uptake sites in previous studies with Tris or Krebs-Ringer buffers (Kennedy and Hanbauer, 1983; Javitch *et al.*, 1984; Schweri *et al.*, 1985; Dubocovich and Zahniser, 1985). Saturation analysis of the stimulatory effect of Na^+ showed it to be of a competitive nature for the binding of [^{3}H]mazindol (Javitch *et al.*, 1984), [^{3}H]GBR12935 (Janowsky *et al.*, 1986), and [^{3}H]GBR 12783 (Bonnet *et al.*, 1988). In contrast, the increase in [^{3}H]cocaine binding to striatal membranes by Na^+ was due to an apparent increase in the B_{max} (Kennedy and Hanbauer, 1983). A recent explanation has been given by Calligaro and Eldefrawi (1988) based on a rather complex interaction between Na^+ and $Tris^+$ for [^{3}H]cocaine binding to rat striatal membranes. In their study, addition of $Tris^+$ to a buffer containing Na^+ as the only cation reduced [^{3}H]cocaine binding by lowering the B_{max}; it was suggested that Na^+ can interact with $Tris^+$ by removing the inhibitory action of $Tris^+$. This $Tris^+$ (and Na^+) binding site, however, is not the same as the site discussed above mediating the increase in [^{3}H]GBR 12783 binding, because Calligaro and Eldefrawi (1988) found Na^+ (50 and 100 m*M*) to be inhibitory when added to 10 m*M* sodium phosphate buffer (i.e., 20 m*M* Na^+ and no $Tris^+$); the binding values were somewhat higher than those in Tris-HCl buffer containing 50 and 100 m*M* Na^+. The reduction in [^{3}H]cocaine binding by adding Na^+ to sodium phosphate buffer was due to an increase in K_d rather than a reduction in B_{max} as for $Tris^+$, suggesting that these effects are not mediated by the same site. Thus, we conclude that either [^{3}H]cocaine binding sites in the striatum interact with Na^+ and $Tris^+$ in a highly complex manner not observed for other ligands, or other factors complicate the saturation analysis of [^{3}H]cocaine binding in the striatum. In regard to the latter possibility, a portion of the [^{3}H]cocaine binding in the striatum most likely represents serotonin uptake sites (Na^+ independent) displaying a K_d similar to the dopamine-related sites (Reith *et al.*, 1985a; Reith *et al.*, 1986a, 1986b). Thus, any condition that reduces the affinity of [^{3}H]cocaine for the dopamine sites will increase the contribution of the serotonin sites, complicating the analysis.

Equal to or Allosterically Linked with Monoamine Recognition Sites?

The work by Langer *et al.* (1987) and Barbaccia *et al.* (1983) raised the interesting possibility that there are separate binding sites for drugs that block serotonin uptake and for serotonin itself. A similar situation could exist for the dopamine transporter (Hanbauer *et al.*, 1984; Missale *et al.*, 1985). If indeed the drug binding sites and the monoamine recognition sites for uptake are distinct and are coupled by an allosteric linkage, it is possible to envision (1) regulation of monamine uptake by a site that is acted upon by an endogenous ligand different from the monoamine itself and (2) a putative cocaine antagonist that prevents the action of cocaine but does not interfere with the normal uptake of monoamines (see above section on Regulation and Endogenous Ligands).

Much of the evidence in favor of an allosteric linkage in the case of the serotonin transporter has come from experiments on blood platelets, indicating a decrease in the dissociation rate of [^{3}H]imipramine binding in the presence of serotonin (Wennogle and Meyerson, 1985; Segonzac *et al.*, 1985). However, it is doubtful whether the high concentrations of serotonin used in these studies (100–500 μM) are meaningful. Moreover, a more recent study has shown that serotonin decelerates the dissociation rate of [^{3}H]imipramine binding in the platelet in a Na^+-independent manner, suggesting that it is an unrelated effect (Humphreys *et al.*, 1988). It should be emphasized also that, despite the similarities in serotonin uptake between platelet and brain, there are important differences as well (Abel *et al.*, 1985; O'Reilly and Reith, 1988); in particular, with regard to the sensitivity of the dissociation rate of [^{3}H]imipramine binding to serotonin uptake sites (Kim and Reith, 1986). Similarly, the dissociation rate of [^{3}H]citalopram binding to brain membranes was not affected by concentrations of serotonin up to 10 μM (D'Amato *et al.*, 1987). The difference in the potency of imipramine (4–15 nM; see references above) in inhibiting [^{3}H]imipramine binding to brain membranes and that in inhibiting synaptosomal [^{3}H]serotonin uptake (140–229 nM) (Ross, 1982; Reith *et al.* 1985a) was found to be due to the difference in temperature at which these two assays were carried out (Reith, 1984c).

Although the low potency of serotonin in inhibiting the binding of tritiated uptake blockers has been advanced as a major argument in support of an allosteric linkage, such a "Michaelis-Menten view" of a transport process does not take into account the complexities of binding, translocation, and dissociation steps (Christensen, 1975). In fact, the low potency of serotonin in inhibiting binding of ligands to the uptake carrier is consonant with the uptake model developed for serotonin by Nelson and Rudnick (1979). This model, shown in Figure 2, takes into account a wealth of information obtained with plasma membrane vesicles, a model system derived from bovine platelet membranes (Rudnick, 1977; Rudnick and Nelson, 1978; Nelson and Rudnick, 1979) or from purified synaptosomes from the cerebral cortex of rats (Kanner, 1978; Kanner and Marva, 1982; Keynan and Kanner, 1987) or mice (O'Reilly and Reith, 1988). Plasma membrane vesicles enable us to study the properties of transport in the absence of storage granules and to control the composition of the internal and external medium. The uptake model is outlined in Figure 2 for serotonin because most information has been collected for this carrier, but many aspects probably apply to the dopamine carrier as well. The first step in transport is the binding of Na^+, Cl^-, and serotonin to sites on the transporter exposed to the membrane exterior (see Figure 2). Na^+, Cl^-, and serotonin are then translocated to the interior (with a rate constant k_{in}), where they dissociate from the carrier. The transporter can return to the state in which the binding sites for

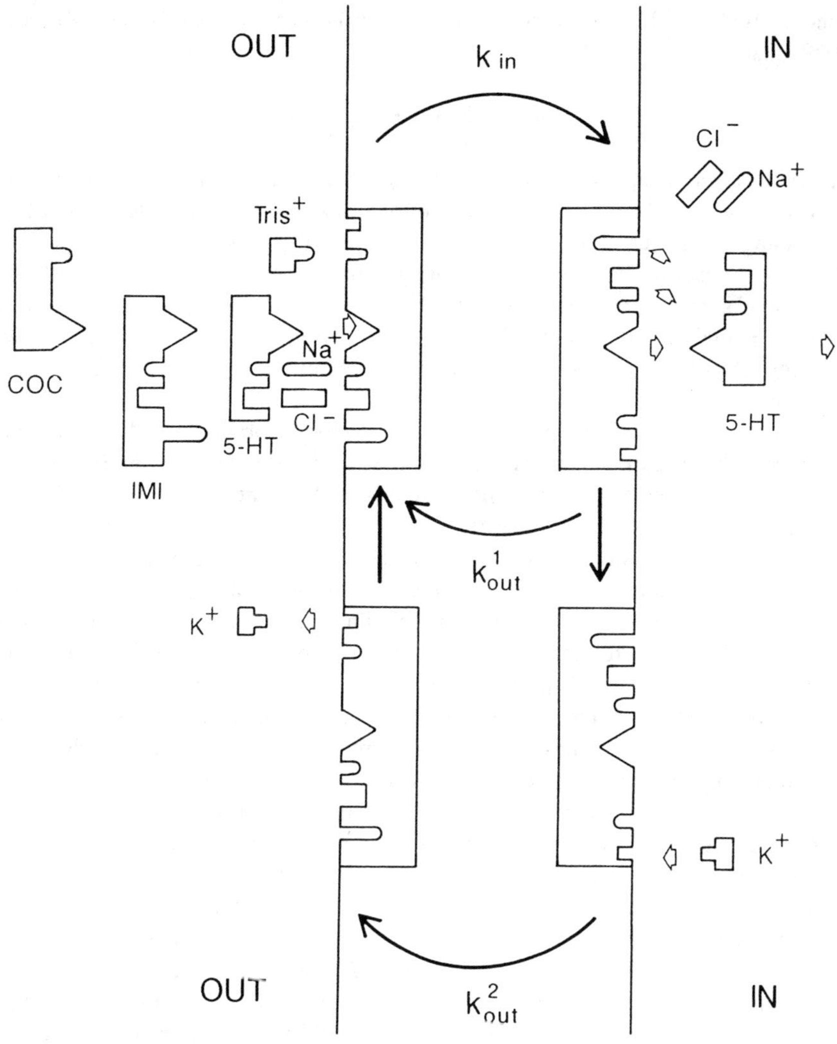

Figure 2. Model for neuronal uptake of serotonin (5-HT). Inhibition by cocaine (COC) and imipramine (IMI) is depicted as occurring at the recognition site for serotonin with the two drugs binding at overlapping regions of the receptor protein. Inward translocation has the rate constant k_{in}. Outward translocation can occur without bound K^+ with the rate constant k^1_{out}, or more rapidly with bound K^+ characterized by the rate constant k^2_{out}. The apparent affinity of uptake $k_m = K_S [1 + (k_{in}/k_{out})]$, in which K_S is the equilibrium dissociation constant for serotonin binding to the substrate recognition site facing the outside.

Na^+, Cl^-, and serotonin are exposed to the outside by two pathways: a slow one, characterized by k^1_{out}, and a relatively faster one, characterized by k^2_{out}, requiring the binding of internal K^+. The actual k_{out} is a composite of the two terms; the higher the internal K^+ concentration, the more k_{out} approaches k^2_{out}. The return of the carrier is the rate-limiting step, that is $k_{out} < K_{in}$. As a special case of the more general model of Levine *et al.* (1965), it can be derived (Nelson and Rudnick, 1979) that

$$K_m = K_s/[1 + (k_{in}/k_{out})]$$

in which K_m is the apparent affinity of the substrate S in the transport process, and K_S is the true equilibrium dissociation constant for serotonin; that is, K_d for serotonin in binding to the substrate recognition site. From $k_{out} < k_{in}$ it follows that $K_m < K_S$. Thus, because there is a rate-limiting step in the transport process after the binding of serotonin to the carrier, the observed K_m for serotonin uptake is smaller than the actual K_S for serotonin binding to the carrier. This would explain the low potency of serotonin in inhibiting imipramine binding without the necessity of invoking two separate sites.

Figure 2 also visualizes a possible explanation for the observation that imipramine binding to the serotonin carrier requires Na^+ (Marcusson *et al.*, 1986; Talvenheimo *et al.*, 1983), whereas cocaine binding does not (Reith *et al.*, 1985a; Reith *et al.*, 1986b) based on the location of the Na^+ binding site. Along the same lines, a $Tris^+$ binding site is depicted (Figure 2) within the binding region of the membrane for cocaine, consonant with the inhibition of cocaine binding by $Tris^+$ (Reith *et al.*, 1984a), but outside the imipramine binding region because there is no information in the literature indicating an inhibitory effect of $Tris^+$ on [^{3}H]imipramine binding. A recent study (Humphreys *et al.*, 1988) with plasma membrane vesicles from blood platelets showed a widely different Na^+ dependency for the effect of various antidepressant drugs in inhibiting serotonin uptake; it was concluded that different drugs bind to different subsites in or near the substrate site and that the availability of these subsites is affected differently by Na^+ binding. The model shown in Figure 2 is an attempt at visualizing these various features. It explains the importance of external Na^+ and Cl^-, and of internal K^+. It shows how ion gradients, in the absence of membrane potential, can drive uptake. The model takes into account the importance of the translocation steps rather than viewing uptake as a special case of Michaelis-Menten kinetics. Furthermore, it explains how a substrate for the serotonin carrier can have a K_m value lower (or higher, depending on the ratio of k_{in}/k_{out}) than its true binding affinity constant, presumably measured as the potency in inhibiting binding of nontransportable radioligands to the serotonin carrier. Finally, the model shows that inhibitors of the transporter that are not translocated themselves block the uptake of serotonin with a K_i value equal to the K_d of their binding to the substrate recognition site. Most likely, cocaine itself belongs to the latter category because no evidence has been found that [^{3}H]cocaine is transported into cerebrocortical slices by an active process (Sershen *et al.*, 1982). In addition, the IC_{50} values of cocaine and its congeners in inhibiting [^{3}H]cocaine binding are comparable to the IC_{50} values in inhibiting [^{3}H]serotonin (cortex) or [^{3}H]dopamine (striatum) uptake into synaptosomes; the slightly higher values obtained in the uptake assays are the result of the inhibitory effect of $Tris^+$, present in the uptake assays but not in the binding assays (Reith *et al.*, 1986b). Also consonant with the notion of identical sites for cerebrocortical cocaine binding and serotonin recognition is the marginal effect of a high concentration (100 μ*M*) of serotonin on the dissociation rate of cerebrocortical [^{3}H]cocaine binding (Reith *et al.*, 1986a). In comparison with the serotonin carrier, less information is available for the dopamine carrier by the lack of studies with model systems; however, it is likely that many aspects are similar to those of the serotonin carrier. For instance, dopamine is more potent in inhibiting the uptake of [^{3}H]dopamine into striatal synaptosomes than in inhibiting the binding of radioligands to the dopamine carrier (Javitch *et al.*, 1984; Dubocovich and Zahniser, 1985; Reith *et al.*, 1986b; Janowsky *et al.*, 1986; Vignon *et al.*, 1988). Norepinephrine and tyramine, which are also substrates for the dopamine carrier in the striatum, are more potent in inhibiting striatal [^{3}H]dopamine uptake than in inhibiting striatal [^{3}H]nominfensin binding (Dubocovich and Zahniser, 1985).

PERIPHERAL COCAINE BINDING SITES

Liver

The initial characterization of [^{3}H]cocaine binding in mouse liver was done with 50 m*M* Tris-HCl buffer (Sershen *et al.*, 1982). With free [^{3}H]cocaine concentrations ranging from 1n*M* to 1.5 μ*M*, no saturation of [^{3}H]cocaine binding was observed; a free concentration range of 10 n*M* to 30 μ*M* revealed a binding site with a K_d of 4–5 μ*M* and a B_{max} of 30 pmol/mg of protein. Compared with cerebrocortical [^{3}H]cocaine binding sites, those in the liver had only moderate stereospecificity, and they discriminated less between the centrally active and centrally inert analogs. Recently, [^{3}H]cocaine binding to rat liver membranes was studied with 10 m*M* sodium phosphate buffer (Calligaro and Eldefrawi, 1987). Saturation analysis showed a high density (66 pmol/mg of protein) of binding sites with a K_d of 1.7 n*M*. Association and dissociation rates were lower than in brain and upon subcellular fractionation, the bulk of the binding was in the microsomal fraction. In addition, the K_d of cocaine binding in liver from mice and humans was much higher (160 and 257 n*M*, respectively) and the B_{max} lower (23 pmol/mg of protein in both cases) than in liver from rats (El-Maghrabi *et al.*, 1988). Since mice are more susceptible to cocaine hepatotoxicity than rats, the suggestion was advanced that the high-affinity cocaine binding sites, relatively more abundant in the rat with higher affinity (lower K_d), play a protective role against cocaine hepatotoxicity by trapping cocaine (El-Maghrabi *et al.*, 1988).

The absence of an effect of phenobarbital on cocaine binding made it unlikely that the cocaine binding sites in liver represent cytochrome P-450 or FAD-containing monooxygenase (El-Maghrabi *et al.*, 1988). The sensitivity to cations, biogenic amines, and their uptake inhibitors was different from that observed in brain. It is possible that Tris-HCl is inhibitory to cocaine binding in the liver as in brain; the site with a K_d of 160 n*M* detetected in the mouse with 10 m*M* sodium phosphate buffer in the study of El-Maghrabi *et al.* (1988) may be the same as the site with a K_d of 4 μ*M* detected in mouse liver with 50 m*M* Tris-HCl in the report of Sershen *et al.* (1982); their B_{max} values, 30 and 22 pmol/mg of protein, are remarkably similar. Chronic treatment of rats with cocaine for 15 d resulted in an appreciable reduction in the density of cocaine binding sites in the liver — from 83 to 25 pmol/mg of protein, the level normally seen in mouse liver (El-Maghrabi, 1988). It is therefore possible that a protective mechanism against cocaine hepatotoxicity is diminished upon repeated exposure to cocaine.

Tissues other than Liver

Calligaro and Eldefrawi (1987) reported high-affinity binding [^{3}H]cocaine to membranes from several peripheral tissues. No binding was found to human red blood cells and human platelets with 2 n*M* [^{3}H]cocaine in the assay (Calligaro and Eldefrawi, 1987). In contrast, with a higher concentration of [^{3}H]cocaine (15 n*M*), high-affinity binding of cocaine to human platelets was detectable (Reith *et al.*, 1983a). This binding was inhibited by cocaine congeners and serotonin uptake blockers with potencies very similar to those for inhibiting [^{3}H]cocaine binding to mouse cerebrocortical membranes. Most likely, the platelet binding is associated with the high-affinity uptake system for serotonin, as in the high-affinity cocaine binding to cerebrocortical membranes. Thus, cocaine can interfere with one of the functions of blood platelets; that is, the removal of serotonin from plasma. The platelet is often used as a model for central serotonin uptake and sometimes mirrors long-term changes occurring in the brain. As an initial approach in human studies, it is therefore of interest to use blood platelet as an indicator for possible changes in the brain. Dackis *et al.* (1985) reported a reduction in V_{max} of serotonin uptake into blood platelets of cocaine abusers measured within 5 d of admission to the clinic. This reduction may be related to depressive symptoms after cocaine withdrawal or may reflect preexisting differences between cocaine addicts and control subjects.

CONCLUDING REMARKS

There are many more sites of action of cocaine than there are binding sites that are labeled by [^{3}H]cocaine with high affinity under the conditions of receptor binding assays. Furthermore, there are probably more binding sites for [^{3}H]cocaine than those demonstrated so far. It is possible that there are discrete areas in the brain where cocaine binds with characteristics diffferent from those of the known sites. For instance, recent reports of cocaine self-administration in the frontal cortex make it of interest to reinvestigate [^{3}H]cocaine binding in this area. The recently reported interaction between Na^+ and $Tris^+$ has important implications for further work on Na^+-dependent cocaine binding sites, and reopens the question of how $Tris^+$ inhibits cocaine binding that does not require Na^+. The Na^+-dependent dopamine-related cocaine binding sites in the striatum probably mediate the activity of cocaine in inducing stereotyped behavior in rodents. Na^+-dependent sites in the nucleus accumbens have not yet been measured by [^{3}H]cocaine binding, but a recent report clearly demonstrated high-affinity binding of [^{3}H]GBR-12935 and potent inhibition by cocaine of dopamine uptake in this region. These Na^+-dependent sites in the nucleus accumbens probably mediate various activities of cocaine involving dopamine, such as locomotor stimulation, drug-discrimination, and self-administration.

The results to date have not ruled out the possibility of an allosteric linkage between separate cocaine binding sites and monoamine recognition sites on the uptake carrier. However, no firm evidence has been obtained in favor of the concept. In fact, various lines of evidence originally advanced to support an allosteric relationship between drug binding sites and monoamine recognition sites have been weakened by newer findings. First, the difference between the potency of serotonin uptake blockers in inhibiting the binding of [^{3}H]imipramine or [^{3}H]paroxetine to the carrier and their potency in inhibiting uptake of [^{3}H]serotonin was found to be due to the difference in temperature at which the binding and uptake assays were carried out. Second, the effect of serotonin in decelerating the dissociation rate of [^{3}H]imipramine binding was found to require excessively high concentrations of serotonin in brain, as compared with those in platelets; in addition, the effect of serotonin on the dissociation rate in platelets was found to be Na^+ independent, arguing against the implication of the recognition sites for serotonin uptake. Third, the low potency of serotonin (or dopamine) in inhibiting binding of [^{3}H]cocaine and other ligands to serotonin (or dopamine) uptake sites compared with its potency in inhibiting [^{3}H]serotonin (or [^{3}H]dopamine) uptake is fully explained by current uptake models that take into account the translocation of the subsrate by the uptake carrier; there is an important difference between *inhibitors* of uptake and *substrates* for uptake. Finally, long-lasting blockade of striatal [^{3}H]cocaine binding sites by metaphit caused a decrease in both [^{3}H]cocaine binding and [^{3}H]dopamine uptake. The present results can be accommodated by a single-site model in which uptake-blocking drugs such as cocaine bind to the same site that recognizes the monoamine substrate for uptake; differences in binding properties (such as ion dependency) between various drugs can be explained by the binding of ligands to overlapping sites; for substrates of uptake, differences in potency in inhibiting ligand binding and monoamine uptake are accounted for by the inward and outward translocation steps involved in uptake. Undoubtedly, the allosterism hypothesis has opened up a great deal of information on monoamine uptake carriers. It has initiated the development of binding assays for those carriers, allowing determinations previously not possible, such as the measurement of uptake sites in frozen tissues. It will probably lead to purification of transporter molecules, and it will be of great interest to see whether the binding sites for radioligands are preserved throughout the purification procedure and what the differences are between the binding sites for the various radioligands.

ACKNOWLEDGMENTS

Much of the work described in this chapter was supported by Grant DA 03025 from the National Institute on Drug Abuse and by a grant for cocaine research from the New York State Department of Mental Hygiene.

NOTE ADDED IN PROOF

Because of the rapid developments that have taken place during the period of time since this review was written, it is necessary to direct attention toward the following recent findings not covered above:

1. WIN 35,428 has become available in tritiated form and is now widely used as a probe for cocaine binding sites. Its affinity is higher and its dissociation rate slower than that of cocaine itself, making it a more practical ligand. (Madras et al., *Mol. Pharmacol.* **36**: 518 (1989); Canfield et al., *Synapse* **6**: 189 (1990)).
2. More structure-activity relationship studies on cocaine binding sites have been carried out, extending the initial observations (Ritz et al., *Life Sci.* **46**: 635 (1990); Kline et al., *J. Med. Chem.* **33**: 2024 (1990); Madras et al., *Pharmacol. Biochem. Behav.* **35**: 949 (1990); Boja et al., *Eur. J. Pharmacol.* **184**: 329 (1990); Carroll et al., *J. Med. Chem.* **34**: 883 (1991); Boja et al., *Mol. Pharmacol.* **39**: 339 (1991)).
3. Cocaine binding to sigma receptors has been reported (Sharkey et al., *Eur. J. Pharmacol.* **149**: 171 (1988)), the importance of sulfhydryl groups for binding sites on the dopamine transporter has been described (Cao et al., *Membrane Biochem.* **8**: 207 (1989); Schweri, *Neuropharmacology* **29**: 901 (1990)), and the relationship between binding sites for uptake blockers and the substrate recognition site has been further explored (Zimanyi et al., *Naunyn-Schmiedeberg's Arch. Pharmacol.* **340**: 626 (1989)).
4. Positron emission tomography studies with [^{11}C]cocaine have been performed in in human and baboon subjects showing the feasibility of *in vito* labeling of the corpus striatum (Fowler et al., *Synapse* **4**: 371 (1989); Logan et al., *J. Cerebr. Blood Flow Metab.* **10**: 740 (1990)).
5. Advances have been made in solubilizing and photoaffinity labeling of the dopamine transporter with [^{3}H]GBR 12935 and radioiodinated photoaffinity probes based on azido-derivatives of GBR (Lew et al., *Synapse* **3**: 372 (1989); Grigoriadas et al., *J. Neurosci.* **9**: 2664 (1989); Sallee et al., *FEBS-Lett.* **256**: 219 (1989); Niznik et al., *Arch. Biochem. Biophys.* **276**: 424 (1990); Berger et al., *Mol. Pharmacol.* **39**: 429 (1991)), of considerable importance for the ultimate isolation of the cocaine receptor associated with dopamine uptake and cloning of the cDNA coding for this protein. In addition, a human cocaine-sensitive dopamine transporter has been expressed in oocytes (Bannon et al., *J. Neurochem.* **54**: 706 (1990)), and a cocaine-sensitive human norepinephrine transporter has been cloned by an expression cloning strategy (Pacholczyk et al., *Nature* **350**: 350 (1991)). As far as the serotonin transporter as a target for cocaine is concerned, citalopram-derived affinity purification has been reported from human blood platelets (Biessen et al., *Biochem.* **29**: 3349 (1990)) and rat cerebral cortex (Graham et al., *Biochem. Soc. Transact.* **19**: 99 (1991)); studics with cerebrocortical plasma membrane vesicles have been extended (Reith et al., *Biochem. Pharmacol.* **38**: 2091 (1989)).

6. The possibility of the existence of full (cocaine) and partial (GBT 12909) agonists for the dopamine uptake site have been suggested based on the observed attenuation of cocaine-induced striatal extracellular dopamine release measured by microdialysis following pretreatment with GBR 12909 (Rothman et al., *FEBS-Lett.* **257**: 341 (1989)); however, more evidence is needed before accepting this novel concept.

REFERENCES

M.S. Abel, D.E. Clody, L.P. Wennogle, and L.R. Meyerson. Effects of chronic desmethylipramine or electroconvulsive shock on selected brain and platelet neurotransmitter recognition sites. *Biochem. Pharmacol.* **34**: 679 (1985).

M.L. Barbaccia, O. Gandolfi, D.M. Chuang, and E. Costa. Modulation of neuronal serotonin uptake by a putative endogenous ligand of imipramine recognition sites. *Proc. Natl. Acad. Sci. U.S.A.* **80**: 5134 (1983).

R.J. Beninger. The role of dopamine in locomotor activity and learning. *Brain Res. Rev.* **6**: 173 (1983).

P. Berger, A. Janowsky, F. Vocci, P. Skolnick, M.M. Schweri, and S.M. Paul. [^{3}H]GBR-12935: a specific high affinity ligand for labeling the dopamine transport complex. *Eur. J. Pharmacol.* **107**: 289 (1985).

P. Berger, A.E. Jacobson, K.C. Rice, R.A. Lesser, and M.E.A. Reith. Metaphit, a receptor acylator, inactivates cocaine binding sites in striatum and antagonizes cocaine-induced locomotor stimulation in rodents. *Neuropharmacology* **25**: 931 (1986).

P.S. Berger, J.D. Elsworth, J. Arroyo, and R.H. Roth. GBR binding in the nucleus accumbens. *Soc. Neurosci.* Abstr. **14**: 765 (1988).

J.J. Bonnet, P. Protais, A. Chagraoui, and J. Costentin. High affinity [^{3}H]GBR 12783 binding to a specific site associated with the neuronal dopamine uptake complex in the central nervous system. *Eur. J. Pharmacol.* **126**: 211 (1988).

D.O. Calligaro and M.E. Eldefrawi. Central and peripheral cocaine receptors. *J. Pharmacol. Ex. Ther.* **243**: 61 (1987).

D.O. Calligaro and M.E. Eldefrawi. High affinity stereospecific binding of [^{3}H]cocaine in striatum and its relationship to the dopamine transporter. *Membr. Biochem.* **7**: 87 (1988).

H.N. Christensen. *Biological Transport.* 2nd ed. W.A. Benjamin, Adv. Book Program, London, p. 113 (1975).

R.L. Clarke, S.J. Daum, A.J. Gambino, M.D. Aceto, J. Pearl, M. Levitt, W.R. Cumiskey, and E.F. Bogado. Compounds affecting the central nervous system. 4.3 β-phenyltropane-2-carboxylic esters and analogs. *J. Med. Chem.* **16**: 1260 (1973).

P.C. Contreras, S. Johnson, R. Freedman, B. Hoffer, K. Olsen, M.F. Rafferty, R.A. Lessor, K.C. Rice, A.E. Jacobson, and T.L. O'Donohue. Metaphit, an acylating ligand for phencyclidine receptors: characterization of *in vivo* actions in the rat. *J. Pharmacol. Exp. Ther.* **238**: 1101 (1986).

C.A. Dackis, M.A. Pasternack Dackis, D. Martin, A.L.C. Pottash, and M.S. Gold. Platelet serotonin transporter in cocaine patients. In: *Problems of Drug Dependence 1984, Research Monograph Series 55.* L.S. Harris, ed., U.S. Government Printing Office, Washington, DC, p. 164 (1985).

R.J. D'Amato, B.L. Largent, A.M. Snowman, and S.H. Snyder. Selective labeling of serotonin uptake sites in rat brain by [^{3}H]citalopram contrasted to labeling of multiple sites by [^{3}H]imipramine. *J. Pharmacol. Exp. Ther.* **242**: 364 (1987).

M.L. Dubocovich and N.R. Zahniser. Binding characteristics of dopamine uptake inhibitor [^{3}H]nomifensine to striatal membranes. *Biochem. Pharmacol.* **34**: 1137 (1985).

A. Dumbrille-Ross and S.W. Tang. Binding of [^{3}H]Ro 11-2465. Possible identification of subclass of [^{3}H]imipramine binding sites. *Mol. Pharmacol.* **23**: 607 (1983).

E.A. El-Maghrabi, D.O. Calligaro, and M.E. Eldefrawi. High affinity binding of [^{3}H]cocaine to rat liver microsomes. *Life Sci.* **42**: 1675 (1988).

M. Grilli, E. Sanna, and I. Hanbauer. Role of corticostriatal nerve projections in the regulation of binding sites for dopamine uptake blockers. *Soc. Neurosci. Abstr.* **14**: 929 (1988).

E. Habert, D. Graham, L. Tahraoui, Y. Claustre, and S.Z. Langer. Characterization of [^{3}H]paroxetine binding to rat cortical membranes. *Eur. J. Pharmacol.* **118**: 107 (1985).

M.G. Hadfield and E.A. Nugent. Cocaine: comparative effect of dopamine uptake in extrapyramidal and limbic systems. *Biochem. Pharmacol.* **32**: 744 (1983).

H. Hall. Characterization of the binding of [^{3}H]norzimeldine, a 5-HT uptake inhibitor, to rat brain homogenates. *Acta Pharmacol. Toxicol.* **55**: 33 (1984).

I. Hanbauer, L.T. Kennedy, M.C. Missale, and E.C. Bruckwick. Cocaine binding sites located in striatal membranes are regulatory sites for dopaminergic synapses, in: *New Vistas in Depression.* Vol. 40, S.Z. Lanager, R. Takahashi, T. Segawa, and M. Briley, eds., Pergamon Press, Oxford, p. 41 (1984).

I. Hanbauer. Modulation of cocaine receptors. In: *Mechanisms of Cocaine Abuse and Toxicity.* National Institute on Drug Abuse Research Monograph **88**. (1988).

P.D. Hrdina. Differentiation of two components of specific [^{3}H]imipramine binding in rat brain. *Eur. J. Pharmacol.* **102**: 481 (1984).

C.J. Humphreys, J. Levin, and G. Rudnick. Antidepressant binding to the porcine and human platelet serotonin transporters. *Mol. Pharmacol.* **33**: 657 (1988).

A. Janowsky, P. Berger, F. Vocci, R. Labarca, P. Skolnick, and S.M. Paul. Characterization of sodium-dependent [^{3}H]GBR-12935 binding in brain: a radioligand for selective labeling of the dopamine transport complex. *J. Neurochem.* **46**: 1272 (1986).

T.U.C. Järbe. Cocaine cue in pigeons: time course studies and generalization to structurally related compounds (norcocaine, WIN 35,428 and 35,065-2) and (+)-amphetamine. *Br. J. Pharmacol.* **73**: 843 (1981).

J.A. Javitch, R.O. Blaustein, and S.H. Snyder. [^{3}H]Mazindol binding associated with neuronal dopamine and norepinephrine uptake sites. *Mol. Pharmacol.* **26**: 35 (1984).

B.I. Kanner. Active transport of γ-aminobutryic acid by membrane vesicles isolated from rat brain. *Biochemistry* **17**: 1207 (1978).

B.I. Kanner and E. Marva. Effux of L-glutamate by synaptic plasma membrane vesicles isolated from rat brain. *Biochemistry* **21**: 3143 (1982).

L.T. Kennedy and I. Hanbauer. Sodium-sensitive cocaine binding to rat striatal membrane: possible relationship to dopamine uptake sites. *J. Neurochem.* **41**: 172 (1983).

S. Keynan and B.I. Kanner. γ-Aminobutyric acid transport in reconstituted preparations from rat brain: coupled sodium and chloride fluxes. *Biochemistry* **27**:12 (1987).

S.S. Kim and M.E.A. Reith. Effect of serotonin on the dissociation of high-affinity binding of [^{3}H]imipramine in mouse cerebral cortex. *Neurosci. Lett.* **67**: 123 (1986).

W.J. Kinnier, D.M. Chuang, G. Gwynn, and E. Costa. Characteristics and regulation of high affinity [^{3}H]imipramine binding to rat hippocampal membranes. *Neuropharmacology* **20**: 411 (1981).

B.K. Koe. Molecular geometry of inhibitors of the uptake of cathecholamines and serotonin in synaptosomal preparations of rat brain. *J. Pharmacol. Exp. Ther.* **199**: 649 (1976).

S.Z. Langer, A.M. Galzin, M.F. Poirier, H. Loo, D. Sechter, and E. Zarifian. Association of [^{3}H]-imipramine and [^{3}H]-paroxetine binding with the 5HT transporter in platelets: relevance to studies in depression. *J. Recept. Res.* **7**: 499 (1987).

S.Z. Langer, C. Moret, R. Raisman, M.L. Dubocovich, and M. Briley. High affinity [^{3}H]imipramine binding in rat hypothalamus: association with uptake of serotonin but not of norepinephrine. *Science* **210**: 1133 (1980).

M. Levine, D.L. Oxender, and W.D. Stein. The substrate-facilitated transport of the glucose carrier across the human erythrocyte membrane. *Biochim. Biophys. Acta* **109**: 151 (1965).

B.K. Madras, J. Bergman, D. Canfield, and R.D. Spealman. [^{3}H](−)-Cocaine binding to membranes of monkey striatum. *FASEB J.* **2**: A1137 (1988).

J.O. Marcusson, I.T. Backsrom, and S.B. Ross. Single-site model of the neuronal 5-hydroxytryptamine uptake and imipramine-binding site. *Mol. Pharmacol.* **30**: 121 (1986).

C. Missale, L. Castelletti, S. Govoni, P.F. Spano, M. Trabucchi, and I. Hanbauer. Dopamine uptake is differentially regulated in rat striatum and nucleus accumbens. *J. Neurochem.* **45**: 51 (1985).

M. Missale, S. Govoni, G. Pasinetti, C. Assini, P.F. Spano, F. Battaini, and M. Trabucchi. Age-dependent changes in the mechanisms regulating dopamine uptake in the central nervous system. *J. Gerontol.* **41**: 136 (1986).

P.J. Nelson and G. Rudnick. Coupling between platelet 5-hydroxtryptamine and potassium transport. *J. Biol. Chem.* **254**: 10084 (1979).

P.J. Nelson and G. Rudnick. The role of chloride ion in platelet serotonin transport. *J. Biol. Chem.* **257**: 6151 (1982).

C.A. O'Reilly and M.E.A. Reith. Uptake of [^{3}H]serotonin into plasma membrane vesicles from mouse cerebral cortex. *J. Biol. Chem.* **263**: 6115 (1988).

S.N. Pradhan, A.K. Battacharyya, and S. Pradhan. Serotoninergic manipulation of the behavioral effects of cocaine in rats. *Commun. Psychopharmacol.* **2**: 481 (1978).

R. Raisman, M.S. Briley, and S.Z. Langer. Specific tricyclic antidepressant binding sites in a rat brain characterised by high-affinity ^{3}H-imipramine binding. *Eur. J. Pharmacol.* **61**: 373 (1980).

M.E.A. Reith, H. Sershen, and A. Lajtha. Saturable [^{3}H]cocaine binding in central nervous system of mouse. *Life Sci.* **27**: 1055 (1980a).

M.E.A. Reith, H. Sershen, and A. Lajtha. Endogenous peptide(s) inhibiting [^{3}H]cocaine binding in mouse brain. *Neurochem. Res.* **5**: 1291 (1980b).

M.E.A. Reith, H. Sershen, and A. Lajtha. Binding of [^{3}H]cocaine in mouse brain: kinetics saturability. *J. Recept. Res.* **2**: 233 (1981).

M.E.A. Reith, H. Sershen, D.L. Allen, and A. Lajtha. A portion of [^{3}H]cocaine binding in brain is associated with serotonergic neurons. *Mol. Phamacol.* **23**: 600 (1983a)

M.E.A. Reith, H. Sershen, and A. Lajtha. High- and low-affinity binding of [^{3}H]imipramine in mouse cerebral cortex. *J. Neurochem.* **40**: 389 (1983b)

M.E.A. Reith, B.E. Meisler, H. Sershen, and A. Lajtha. [^{3}H]cocaine binding in brain is inhibited by Tris (hydroxymethyl)aminomethane. *J. Neurosci. Meth.* **12**: 151 (1984a).

M.E.A. Reith, D.L. Allen, H. Sershen, and A. Lajtha. Similarities and differences between high affinity binding sites for cocaine and imipramine in mouse cerebral cortex. *J. Neurochem.* **43**: 249 (1984b).

M.E.A. Reith, H. Sershin, and A. Lajtha. Thermodynamics of the interactions of tricyclic drugs with binding sites for [^{3}H]imipramine in mouse cerebral cortex. *Biochem. Pharmacol.* **33**: 4101 (1984c)

M.E.A. Reith, H. Sershen, and A. Lajtha. Binding of imipramine and cocaine to a model lipid membrane. *Neurochem. Res.* **9**: 965 (1984d).

M.E.A. Reith, B.E. Meisler, H. Sershen, and A. Lajtha. Sodium-independent binding of [^{3}H]cocaine in mouse striatum is serotonin related. *Brain Res.* **342**: 145 (1985a)

M.E.A. Reith, B.E. Meisler, and A. Latha. Locomotor effects of cocaine, cocaine congeners, and local anesthetics in mice. *Pharmacol. Biochem. Behav.* **23**: 831 (1985b).

M.E.A. Reith. Binding of [^{3}H]imipramine to mouse cerebrocortical membranes and to glass fiber filters. *J. Neurochem.* **46**: 760 (1986).

M.E.A. Reith and A. Lajtha. Locomotor depression in mice by norcocaine does not involve central α_2-adrenergic or presynaptic dopamine receptors. *Pharmacol. Biochem. Behav.* **24**: 305 (1986).

M.E.A. Reith, H. Sershen, and A. Lajtha. Binding sites for [^{3}H]cocaine in mouse striatum and cerebral cortex have different dissociation kinetics. *J. Neurochem.* **46**: 309 (1986a).

M.E.A. Reith, B.E. Meisler, H. Sershen, and A. Lajtha. Structural requirements for cocaine congeners to interact with dopamine and serotonin uptake sites in mouse brain and to induce stereotyped behavior. *Biochem. Pharmacol.* **35**: 1123 (1986b).

M.E.A. Reith, S.S. Kim, and A. Lajtha. Structural requirements for cocaine congeners to interact with [^{3}H]bratachotoxinin A20-α-benzoate binding sites on sodium channels in mouse brain synaptosomes. *J. Biol. Chem.* **261**: 7300 (1986c).

M.E.A. Reith. Cocaine receptors on monoamine transporters and sodium channels. In: *Mechanism of Cocaine Abuse and Toxicity*. National Institute on Drug Abuse Monograph **88**.(1988).

J.M. Ritchie and N.M. Green. Local anesthetics, in *The Pharmacological Basis of Therapeutics*. 6th ed., A. Goodman Gilman, L.S. Goodman, and A. Gilman, eds., MacMillan, New York, p. 300 (1980).

M.C. Ritz, R.J. Lamb, S.R. Goldberg, and M.J. Kuhar. Cocaine receptors on dopamine transporters are related to self-administration of cocaine. *Science* **237**: 1219 (1987).

M.C. Ritz, E.J. Cone, J. Sharkey, and M.J. Kuhar. Structure activity relationship of cocaine and related compounds in binding to dopamine transporters. *Soc. Neurosci. Abstr.* **14**: 963 (1988).

S.B. Ross. The characteristics of serotonin uptake systems, in: *Biology of Serotonergic Transmission.* N.N. Osborne, ed., John Wiley & Sons, New York, p. 159 (1982).

G. Rudnick. Active transport of 5-hydroxytryptamine by plasma membranes vesicles isolated from human blood platelets. *J. Biol. Chem.* **252**: 2170 (1977).

G. Rudnick and P.J. Nelson. Platelet 5-hydroxytryptamine transport, an electroneutral mechanism coupled to potassium. *Biochemistry* **22**: 4739 (1978).

J. Scheel-Krüger, C. Braestrup, M. Nielson, K. Golembiowska, and E. Mogilnicka. Discussion on the role of dopamine in the biochemical mechanism of action. In: *Cocaine and Other Stimulants.* E.H. Ellinwood and M.M. Kilbey, eds., Plenum Press, New York, p. 373 (1977).

H. Schoemaker, C. Pimoule, S. Arbilla, B. Scatton, F. Javoy-Agid, and S.Z. Langer. Sodium dependent [^{3}H]cocaine binding associated with dopamine uptake sites in the rat striatum and human putamen decrease after dopaminergic denervation and in Parkinson Disease. *Naunyn-Schmiedeberg's Arch. Pharmacol.* **329**: 227 (1985).

H. Schoemaker and S.Z. Langer. [^{3}H]-8-OH-DPAT labels the serotonin transporter in the rat striatum. *Eur. J. Pharmacol.* **124**: 371 (1986).

M.M. Schweri, P. Skolnick, M.F. Rafferty, K.C. Rice. A.J. Janowsky, and S.M. Paul. [^{3}H]Threo-(+)-methylphenidate binding to 3-4-dihydroxyphenylethyl-amine uptake sites in corpus striatum: correlation with the stimulant properties of ritalinic acid esters. *J. Neurochem.* **45**: 1062 (1985).

M.M. Schweri, A.E. Jacobson, R.A. Lessor, and K.C. Rice. Metaphit irreversibly inhibits [^{3}H]threo-(+)-methylphenidate binding to rat striatal tissue. *J. Neurochem.* **48**: 102 (1987).

A. Segonzac, H. Schoemaker, T. Tateishi, and S.Z. Langer. 5-Methoxytryptoline, a competitive endocoid acting at [^{3}H]imipramine recognition sites in human platelets. *J. Neurochem.* **45**: 249 (1985).

H. Sershen, M.E.A. Reith, and A. Lajtha. The pharmacological relevance of the cocaine binding site in mouse brain. *Neuropharmacology* **19**: 1145 (1980).

H. Sershen, M.E.A. Reith, and A. Lajtha. Comparison of the properties of central and peripheral binding sites for cocaine. *Neuropharmacology* **21**: 469 (1982).

H. Sershen, P. Berger, A.E. Jacobson, K.C. Rice, and M.E.A. Reith. Metaphit prevents locomotor activation induced by various phychostimulants and interferes with the dopaminergic system in mice. *Neuropharmacology* **27**: 23 (1988).

R.D. Spealman, S.R. Goldberg, R.T. Kelleher, D.M. Goldberg, and P.J. Charlton. Some effects of cocaine and two cocaine analogs on schedule-controlled behavior of squirrel monkeys. *J. Pharmacol. Exp. Ther.* **202**: 500 (1977).

R.D. Spealman and R.T. Kelleher. Self-administration of cocaine derivatives by squirrel monkeys. *J. Pharmacol. Exp. Ther.* **216**: 532 (1981).

S.M. Stahl and H.Y. Meltzer. A kinetic and pharmacologic analysis of 5-hydroxytryptamine transport by human platelets and platelet storage granules: comparison with central serotonergic neurons. *J. Pharmacol. Exp. Ther.* **205**: 118 (1978).

J. Talvenheimo, H. Fises, P.J. Nelson and G. Rudnick. The serotonin transporter-imipramine "receptor". *J. Biol. Chem.* **258**: 6115 (1983).

H.D. Tazelaar, S.B. Karch, B.G. Stephens, and M.E. Billingham. Cocaine and the heart. *Hum. Pathol.* **18**: 195 (1987).

M.L. Thompson, L. Shuster, and K. Shaw. Cocaine-induced hepatic necrosis in mice — the role of cocaine metabolism. *Biochem. Pharmacol.* **28**: 2389 (1979).

U. Trendelenburg. Classification of sympathomimetic amines, in: *Handbook of Experimental Pharmacology, Catecholamines.* Vol. 33. H. Blaschko and E. Muscholl, eds., Springer-Verlag, Berlin, p. 336 (1972).

J. Vignon, V. Pinet, J.M. Kamenka, and R. Chicheportiche. [^{3}H]N-[1-(2-benzo(b)thiophenyl)cyclohexyl] piperidine ([^{3}H]BTCP): a new phencyclidine analog selective for the dopamine uptake complex. *Eur. J. Pharmacol.* **148**: 427 (1988).

L.P. Wennogle and L.R. Meyerson. Serotonin uptake inhibitors differentially modulate high affinity imipramine dissociation in human platelet membranes. *Life Sci.* **36**: 1541 (1985).

I. Zimanyi, E. Wang, A.Lajtha, and M.E.A. Reith. The effect of cocaine and cocaine congeners on veratridine-induced depolarization in mouse cerebrocortical synaptoneurosomes. *J. Neurosci. Res.* **22**: 201 (1989).

I. Zimanyi, A.E. Jacobson, K.C. Rice, A. Lajtha, and M.E.A. Reith. Long-term inhibition of the dopamine uptake complex by metaphit. *Soc. Neurosci. Abstr.* **14**: 681 (1988).

11

Neurochemical Mechanisms of Cocaine-Induced Sensitization

NANCY R. ZAHNISER
Department of Pharmacology
University of Colorado Health Sciences Center
Denver, Colorado

JOANNA PERIS
Department of Pharmacodynamics
University of Florida Health Center
Gainesville, Florida

INTRODUCTION

It has long been recognized that, with repeated administration, certain of the behavioral effects produced by cocaine become progressively greater (Tatum and Seevers, 1929; Downs and Eddy, 1932). Enhanced responding to repeated presentations of a drug is referred to as "sensitization" or "reverse tolerance". Cocaine-induced sensitization has been observed in a number of species, including mice, rats, cats, dogs, nonhuman primates, and man (see Stripling and Ellinwood, 1977; Post and Contel, 1983).

Drug-induced sensitization is certainly not unique to cocaine. This phenomenon has been best characterized for the central nervous system (CNS) stimulant amphetamine (Segal and Mandell, 1974). In addition to amphetamine, sensitization is most often associated with other CNS stimulants such as cocaine and methylphenidate and with opioid drugs (see Algeri *et al.*, 1980; Kuczenski, 1983; Segal and Schuckit, 1983; Kolta *et al.*, 1985; Robinson and Becker, 1986; Kalivas *et al.*, 1988). Cross-sensitization has also been studied in relationship to a wide variety of drugs, including those used in the treatment of depression, schizophrenia, and anxiety (see Antelman, 1988). Intermittent, repeated exposure to stressful situations can also produce behavioral sensitization (Antelman and Chiodo, 1983). These observations involving drug administration and stress may be related to one another. Antelman (1988) has proposed that the ability of drugs to produce sensitization is due to their foreign, and therefore stressful, properties. In support of this idea, cross-sensitization between stress and certain of these drugs has been demonstrated (see Robinson, 1988).

The purpose of this chapter is to discuss cocaine-induced sensitization and, in particular, to focus on possible neurochemical bases for this phenomenon. It should be noted that, although sensitization is the focus here, tolerance can also be induced in response to repeated administration of cocaine. Whether sensitization or tolerance is observed depends in part on the response measured and the administration regimen used. The influence of these variables will first be discussed in detail; several neurochemical theories for sensitization will then be introduced. Next, the evidence for and against these theories will be presented. To conclude, a brief discussion of the clinical implications of cocaine-induced sensitization will be given.

COCAINE-INDUCED BEHAVORIAL SENSITIZATON

Description of Behaviors

The most-often studied behaviors which show sensitization are unconitioned, notably motor hyperactivity and stereotyped responses (Figure 1). It should be emphasized that only some behaviors induced by cocaine administration exhibit sensitization and that the behaviors elicited depend on the dose of cocaine administered. Motor hyperactivity, or running, is the predominant behavior manifested at low doses of cocaine. As the dose of cocaine is increased, stereotyped behaviors become more apparent. These behaviors, because of their focused nature, necessarily preclude motor hyperactivity. Stereotypic behaviors included focused sniffing, rearing, repetitive head bobbing, and oral movements (chewing, licking, and biting). Cocaine-induced sensitization has been reported when horizontal motor activity (Shuster *et al.*, 1977), duration of sniffing and head bobbing, or the number of rearing events are measured (Barr *et al.*, 1983). Similar findings have been reported with amphetamine administration, and the complexity of behavioral sensitization with respect to the particular behavior measured is even more apparent from this literature. For example, in contrast to the increased horizontal locomotor activity, sniffing and head and limb movements observed with repeated amphetamine administration, oral behaviors such as licking and biting show tolerance rather than sensitization (Eichler *et al.*, 1980; Rebec and Segal, 1980). Furthermore, not only the types of behaviors elicited but also the patterns of these behaviors may change as sensitization develops (Kuczenski and Segal, 1988). Finally, high doses of cocaine can induce kindling and convulsions (see Post and Weiss, 1988).

Other behavioral measures have also been used to study cocaine-induced changes in behavior. Peripheral injections of cocaine to either intact rats (Glick and Hinds, 1984) or rats with unilateral 6-hydroxydopamine lesions of the nigrostriatal pathway (Lin-Chu *et al.*, 1985) causes them to rotate. This rotational behavior shows cocaine-induced sensitization. Cocaine administration increases acoustic startle responding in mice; however, tolerance, rather than sensitization, to this effect develops after chronic administration of cocaine (Kokkinidis, 1986).

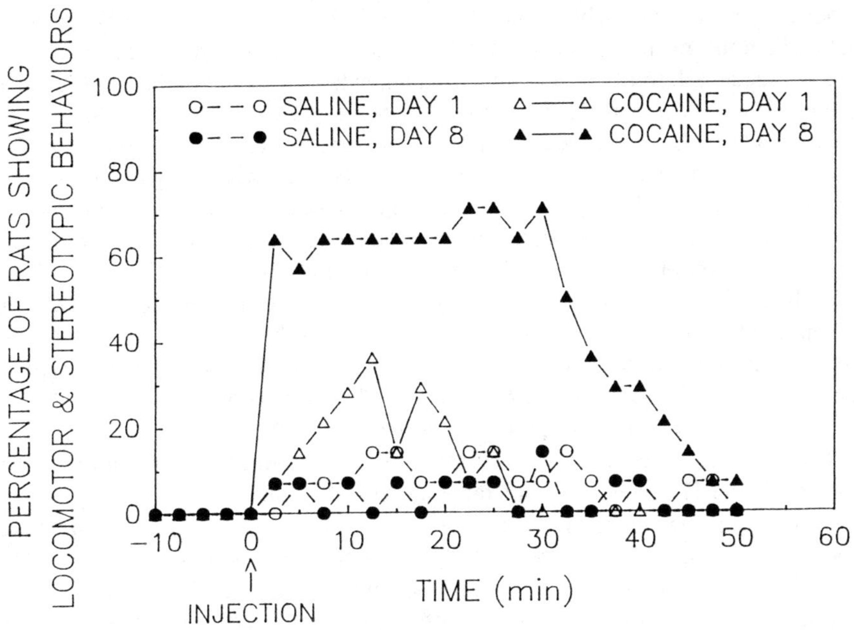

Figure 1. Time course of increased locomotion and stereotypy following repeated injections of cocaine. Male Sprague-Dawley rats received an i.p. injection of cocaine (10 mg/kg) once daily for 8 d. The behavior was rated on days 1 and 8 by an observer who was not aware of the drug treatment and was scored for 10-s periods every 2.5 min beginning at 10 min prior to the injection and continuing until 50 min postinjection. A single score was given based on an expansion of the scale published by Ernst (1967). Behavioral sensitization is evident in the rats that received repeated cocaine injections. Furthermore, the intensity of locomotor and stereotyped behaviors was significantly greater in the 8-d cocaine rats compared to the 1-d cocaine rats (data not shown). Mean values for N = 14 are shown.

Table 1
Characteristics of Behavioral Sensitization to Cocaine

Increased magnitude of behavioral response to the same dose
Can be produced by only a single pretreatment dose
Persistent effect
Intermittent, as opposed to continuous, administration more likely to produce
Dose-related effect
Large individual variation
Influenced by genetics, gender, and age
Influenced by classical conditioning to environmental cues
Cross-sensitization occurs between cocaine and amphetamine

Characteristics of Behavioral Sensitization

Certain properties characteristic of cocaine-induced behavioral sensitization (Table 1) are also associated with sensitization to other psychomotor stimulants (see Post and Weiss, 1988). Although repeated administration of cocaine is often used to produce behavioral sensitization, even a single dose of cocaine is sufficient to produce sensitization (Glick and Hinds, 1984; Lin-Chu *et al.*, 1985). Another hallmark of this phenomenon is that it is long lasting. The behavioral responsiveness to a subsequent dose of cocaine remains augmented for weeks to months (Post and Contel, 1983; Glick and Hinds, 1984; Lin-Chu *et al.*, 1985). As with other drugs, the time interval between cocaine injections also appears to be an important variable (Post, 1980; Antelman, 1988). With longer intervals between drug exposure, sensitization is more likely to occur. Thus, intermittent administration of cocaine produces behavioral sensitization, whereas behavioral tolerance results from continuous infusion of cocaine (Reith *et al.*, 1987). Because cocaine has a relatively short half-life, even once- or twice-daily injections of cocaine are considered to be intermittent. Cocaine-induced sensitization has been demonstrated to be dose dependent in mice injected intraperitoneally (i.p.) with doses of cocaine between 5 and 30 mg/kg (Shuster *et al.*, 1977). In rats, doses of cocaine from 10 to 40 mg/kg i.p. are generally used to produce sensitization.

Cocaine-induced behavioral sensitization is also influenced by a number of other factors (Table 1). There is considerable individual variation in the development of behavioral sensitization to the psychomotor stimulant (Glick *et al.*, 1986; Kuczenski and Segal, 1988; Robinson, 1988). Factors identified as contributing to individual variation in behavioral sensitization include genetics, age, sex differences, gonadal hormones, and environment (Robinson, 1988). Shuster and colleagues (1977) showed the influence of genetic determinants when they found that recombinant-inbred lines of mice derived from C57B1/6By and BALB/cBy strains showed different degrees of sensitization in response to cocaine administration. Furthermore, it is interesting to note that individual variability in the *initial response* to cocaine and in the *development of sensitization* to cocaine do not appear to be related. For example, C57B1/6J mice show good running activity to an initial dose of cocaine, but only a slight degree of sensitization with repeated administration of cocaine: C57B1/6By mice show good running activity and appreciable sensitization; BALB/cBy mice show moderate running activity but no significant sensitization; and A/J mice show very little running but considerable sensitization (Shuster *et al.*, 1977). The age of the animal appears to influence the initial response more than development of sensitization to cocaine. Older rats show greater cocaine-induced behaviors when injected for the first time (Kilbey and Ellinwood, 1977; Post and Contel, 1983); however, sensitization develops similarly in all age groups (Kilbey and Ellinwood, 1977). Gender of the animals is also an important variable in the development of cocaine-induced behavioral sensitization. Female rats show greater initial sensitivity to cocaine and also enhanced development of behavioral sensitization (Post and Contel, 1983; Glick and Hinds, 1984). Although hormonal factors have been investigated in detail for amphetamine-induced sensitization (see Robinson, 1988), to our knowledge, these have not been investigated for cocaine-induced sensitization.

There are several lines of evidence showing that learning and classical conditioning can be involved in cocaine-induced behavioral sensitization. Vasopressin-deficient rats exhibit impaired behavioral sensitization to cocaine and this has been attributed to the role of vasopressin in learning and memory (Post *et al.*, 1982).

Electrolytic lesions of the amygdala, but not of hippocampus or cerebellum, block cocaine-induced sensitization (Post *et al.*, 1987b), and the amygdala is suggested to be involved in the components of sensitization requiring conditioning and memory. That classical, or Pavlovian, conditioning contributes to the sensitization produced by cocaine is evidenced by the fact that sensitization is more pronounced when cocaine administration is coupled with cues associated with previous injections of cocaine (Hinson and Poulos, 1981; Barr *et al.*, 1983; see Stewart and Vezina, 1988). These cues include environmental context. Post and colleagues (1981) have reported that repeated administration of cocaine induces sensitization only if rats receive their injections of cocaine in the same environment in which their behavior is rated. It has also been suggested that higher doses or more repetitions with lower doses are necessary in order for the context-independent component of sensitization to become apparent (Post and Weiss, 1988). However, we have observed cocaine-induced sensitization in rats given a moderate dose of cocaine once in the test cage, then injected once daily for 6 d in their home cages, and finally injected once again in the test cage on day 8 (Figure 1; Zahniser *et al.*, 1988c). Clearly, classical conditioning to environmental cues does influence cocaine-induced behavioral sensitization, but cannot totally account for this phenomenon.

Cross-sensitization among the pyschomotor stimulants, between the psychomotor stimulants and opioid drugs and between the psychomotor stimulants and stressors has again been studied to the greatest extent for amphetamine (see Robinson, 1988; Stewart and Vezina, 1988). It has been shown that local injections of amphetamine into the ventral tegmental area (VTA), the region containing the cell bodies of the mesolimbic dopaminergic tract, produce behavioral sensitization to subsequent systemic administration of cocaine (Kalivas and Weber, 1988). Although it remains to be tested, these results suggest that cross-sensitization may occur with cocaine and many of the agents which show cross-sensitization with amphetamine. Further, it has been suggested that stimulant drugs produce their long-lasting effects by their action as stressors (Antelman, 1988). Stimulants and environmental stressors— such as tail-pinch, footshock, or food deprivation — exhibit cross-sensitization (see Antelman, 1988; Robinson, 1988). Such results suggest that daily handling and injections can contribute to the manifestation of sensitization. Even saline injections are stressful, and this fact emphasizes the importance of including the proper controls in this type of experiment.

There appear to be many similarities in the behavioral sensitization produced by cocaine, amphetamine, and apomorphine even though each of these drugs has a different mechanism of activating dopaminergic transmission. Cocaine prolongs the action of dopamine (DA) in the synaptic cleft by blocking DA uptake. Amphetamine both releases DA and blocks DA uptake, whereas apomorphine is a direct acting DA receptor agonist. Nevertheless, perhaps because of their different mechanisms of action, the enhanced behavioral responsiveness produced by one of these drugs is not always affected in the same way by subsequent or concurrent administration of another of these agents. For example, cocaine-sensitized rats show reduced, rather than enhanced, stereotyped behavior to subsequent systemic apomorphine injection (Epstein and Altshuler, 1979; Gale, 1984). Also, concurrent administration of apomorphine given repeatedly with either amphetamine or cocaine inhibits only the amphetamine-induced behavioral sensitization (Riffee *et al.*, 1988). That apomorphine antagonizes the action of amphetamine is also somewhat surprising. Such findings emphasize that cocaine, amphetamine, and apomorphine may produce similar behavioral sensitization phenomena that can be distinguished when probed with additional drugs.

Pathways and Neurotransmitters Involved

Most of the experimental evidence to date has implicated brain dopaminergic systems as central in cocaine-induced behavioral sensitization. First, the behaviors induced by cocaine have been specifically linked to these systems. Locomotor activity, sniffing, and rearing, all of which show sensitization, can be induced by administration of selective D_2, but not D_1, DA receptor agonists (see Clark and White, 1987). Increased locomotor activity and stereotyped behaviors primarily involve activation of the mesolimbic and nigrostriatal dopaminergic systems, respectively (Costall and Naylor, 1977). Consistent with this idea, selective lesions of DA-containing fibers projecting to the nucleus accumbens substantially decrease cocaine-induced locomotor activity (Kelly and Iversen, 1976). However, despite the apparent dichotomy of behaviors mediated by the two dopaminergic pathways, there is also evidence that the mesolimbic and nigrostriatal dopaminergic pathways interact in producing locomotor and stereotyped behaviors (see Kokkinidis *et al.*, 1989). This may explain why 6-hydroxydopamine-induced lesions of the substantia nigra block the increase in locomotor activity induced by systemic cocaine administration (Creese and Iversen, 1975) and why similar lesions of the nucleus accumbens block cocaine-induced sensitization of both locomotor activity and stereotyped behaviors (Post *et al.*, 1987b).

Studies using DA receptor antagonists have provided further evidence for the involvement of dopaminergic systems in the *development* of cocaine-induced sensitization; however, they have also suggested that the *expression* of this phenomenon is not entirely mediated through dopaminergic pathways. It should be noted that the DA receptor antagonists used have not been selective for either of the two DA receptor subtypes. An early study showed that acute pretreatment with either haloperidol (0.1–0.25 mg/kg, i.p.,) or pimozide (0.25–1.0 mg/kg, i.p.) completely blocks stereotyped behavior in cocaine-sensitized rats (Borison *et al.*, 1979). In contrast, a later study showed that pimozide pretreatment (0.4 mg/kg, i.p.) has no effect on a cocaine challenge in cocaine-sensitized rats; however, pretreatment with pimozide prior to the first treatment with cocaine does block cocaine-induced sensitization (Beninger and Herz, 1986). In the latter study, the sensitization meaasured was an environment-specific conditioning produced by cocaine, and it was concluded that pimozide administration blocks the development but not the expression of this type of cocaine sensitization. Similar findings were reported using haloperidol to antagonize DA receptors. Pretreatment with haloperidol (0.2 or 0.5 mg/kg, i.p.) before the initial dose of cocaine blocks the establishment of cocaine-induced sensitization of locomotor activity; however, when the same dose of haloperidol is administered prior to the challenge dose of cocaine, the haloperidol pretreatment does not affect the expression of sensitization (Post *et al.*, 1987a). Interestingly, this group also reported that DA receptor blockade may not even be necessary to prevent the development of sensitization because doses of diazepam that depress cocaine-induced behaviors to a similar extent as haloperidol also prevent the development of cocaine-induced sensitization (Post *et al.*, 1987a).

Kuczenski and Segal (1988) have proposed that the complex behavioral pattern, which includes components of both sensitization and tolerance, produced in response to repeated administration of amphetamine cannot be explained by a single neurochemical change or even a single neurotransmitter system. This is also likely the case with cocaine-induced sensitization. There is no doubt that cocaine stimulates dopaminergic transmission. For example, when cocaine is administered peripherally, extracellular concentrations of DA rapidly increase in the striatum; this is also true if cocaine is locally applied in the striatum (Hurd and Ungerstedt, 1989). Increased synaptic concentrations of DA result from the inhibition of the neuronal DA uptake carrier by cocaine. Cocaine inhibits the uptake of ^{3}H-DA into striatal synaptosomes with an approximate affinity of 1 μM. However, cocaine inhibits ^{3}H-norepinephrine and ^{3}H-serotonin (^{3}H-5-HT) uptake with the same, or even a slightly higher, affinity (Hyttel, 1982; Ritz *et al.*, 1987) and potentially can produce effects by prolonging the synaptic lifetimes of these neurotransmitters as well. Therefore, actions mediated by all three monoamine neurotransmitters may be important.

Inhibition of the neuronal monoamine uptake carriers is thought to be the primary mechanism by which lower doses of cocaine act. At higher concentrations such as 10 μM and above, cocaine is also a local anesthetic (Yasuda *et al.*, 1984). Given repeatedly, high doses of local anesthetics (including cocaine) lower the threshold for seizures. This effect, known as kindling, is also a type of sensitization (see Post and Weiss, 1988; also chapter herein). Thus, there appears to be a type of dose-response relationship between the concentration of cocaine to which the neurons are exposed and the mechanism by which cocaine acts and the particular type of sensitization produced; that is, increased locomotor activity and stereotyped behavior or increased seizure susceptibility.

NEUROCHEMICAL THEORIES OF SENSITIZATION

Changes in Bioavailability of Cocaine

The most trivial explanation for cocaine-induced behavioral sensitization is that repeated administration alters the metabolic disposition of subsequent doses of cocaine and a higher effective concentration is achieved. This possibility has been ruled out for several reasons. First, the short half-life of cocaine and the persistent nature of sensitization are not compatible. Reasonable doses of cocaine are completely cleared from brain in a matter of hours (Nayak *et al.*, 1975; Benuck *et al.*, 1987), whereas sensitization has a much longer duration and is manifested days to weeks after cocaine treatment. That cross-sensitization can occur between cocaine and other structurally different drugs also makes this explanation unlikely. As Post and Contel (1983) have pointed out, another compelling argument against the possibility that a change in drug disposition explains sensitization is that, depending on the endpoint measured, both sensitization and tolerance can occur in the same animal following repeated cocaine administration.

Sites for Potential Changes in DA Neurochemistry

There are undoubtedly several underlying biochemical mechanisms for sensitization. This discussion will focus on possible changes in central dopaminergic systems because these are the systems that have been implicated most often and studied in most detail. However, as discussed above, it is likely that cocaine-induced sensitization may involve other monoamine neurotransmitters and similar arguments could apply to changes in these other pathways.

Because cocaine inhibits DA uptake into nerve terminals and because of the behaviors induced by cocaine administration, the search for neurochemical correlates of cocaine-induced sensitization has focused on the striatum and nucleus accumbens, regions containing the terminals of the nigrostriatal and mesolimbic pathways, respectively. In contrast to the terminal regions, direct actions of cocaine in the cell body regions have not been consistently observed. Local application of cocaine into the substantia nigra does not increase DA content in the extracellular fluid (Hurd and Ungerstedt, 1989). However, small effects of cocaine are seen on DA cells during iontophoresis of cocaine into VTA (Einhorn *et al.*, 1988). Interestingly, sensitization to amphetamine occurs only if amphatamine is locally injected in the cell body regions of these two dopaminergic tracts, but not if it is locally injected into the terminal regions (Dougherty and Ellinwood, 1981; Kalivas and Weber, 1988; Stewart and Vezina, 1988). Similar findings have been reported for morphine and opioid peptides (Joyce and Iversen, 1979; Vezina and Stewart, 1984; Kalivas *et al.*, 1985, Vezina *et al.*, 1987). This type of experiment has not been carried out with cocaine. However, based on the literature for amphetamine and opioid drugs, it appears important to consider the possibility that neurochemical changes can be induced by cocaine in both terminal and cell body regions of the dopaminergic tracts.

Cocaine-induced changes in dopaminergic neurochemistry could involve changes in the dopaminergic neurons themselves (presynaptic changes) or in receptor-effector systems beyond the dopaminergic neurons (postsynaptic changes; Figure 2). To be a basis for sensitization, the change would be expected to augment dopaminergic function. Such neurochemical changes at the level of the dopaminergic neuron could include an increase in the pools of releasable DA, augmentation of DA release, diminished DA uptake, and/or decreased DA metabolism. At the postsynaptic level, enhanced DA receptor responsiveness could be responsible. Based on the behavioral literature, it is expected that persistent neurochemical changes may be present following even a single dose of cocaine. Following repeated dosing, these changes would be expected to be enhanced. It is debatable whether baseline parameters or only parameters measured in the presence of cocaine should be changed. Because of the behaviors involved, it is most pertinent to focus on the nigrostriatal and mesolimbic dopaminergic systems. Where appropriate, data from

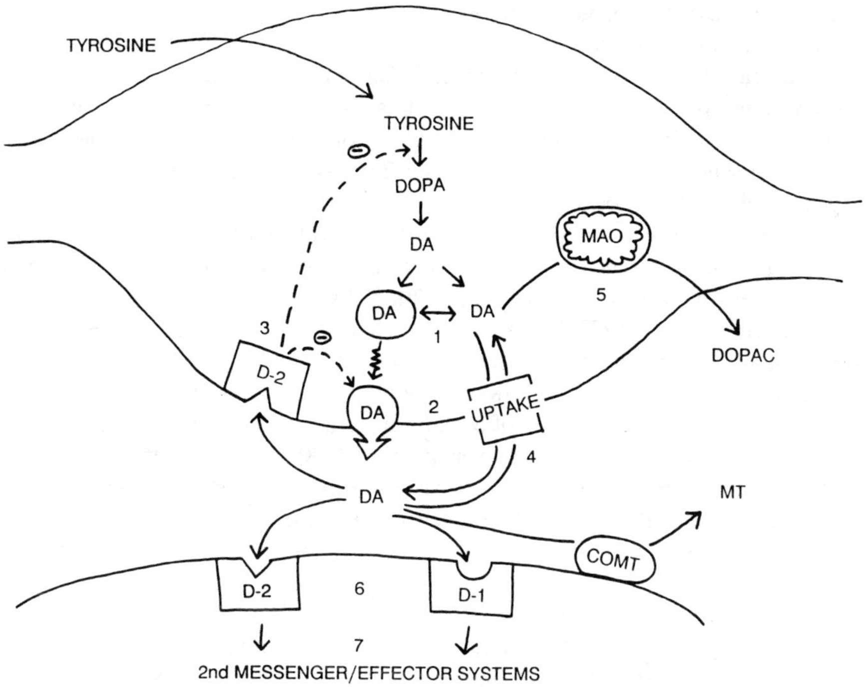

Figure 2. Possible neurochemical mechanisms at dopaminergic synapses underlying cocaine-induced behavioral sensitization. (1) Increased releasable pools of DA. (2) Enhanced release of DA either via vesicular release or a reversal of the uptake pump. (3) Desensitization or down-regulation of inhibitory D_2 DA autoreceptors either at cell bodies (impulse-regulating) or at terminals (synthesis- or release-modulating; shown as the same receptor for the sake of simplicity). (4) Reduced high affinity uptake. (5) Decreased DA metabolism. (6) Supersensitivity or up-regulation of postsynaptic DA receptors. (7) Enhanced DA receptor coupling to and/or responsiveness of second messengers and effector systems. DOPA, dihydroxyphenylalanine; MAO, monoamine oxidase; DOPAC, dihydroxyphenylacetic acid; COMT catechol-*O*-methyl transferase; MT, 3-methoxytyramine; D_1, D_1 DA receptor; D_2, D_2 DA receptor.

other dopaminergic systems, notably the mesocortical and tuberoinfundibular systems, will be mentioned.

The hypotheses most often proposed to explain stimulant-induced sensitization include subsensitivity of dopaminergic autoreceptors, supersensitivity of postsynaptic dopaminergic receptors, and increased release of DA (Baudry *et al.*, 1977; Martres *et al.*, 1977; Muller and Seeman, 1979; Antelman and Chiodo, 1981; Post and Contel, 1983; Robinson and Becker, 1986). Robinson and Becker (1986) reviewed the evidence for each of these hypotheses in the case of amphetamine-induced sensitization and have concluded that the evidence to date supports the hypothesis of increased release of DA as the neural basis of behavioral sensitization. Until recently, DA autoreceptor subsensitivity and postsynaptic DA receptor supersensitivity have been considered to be the most probable explanation for cocaine-induced sensitization (see Post and Contel, 1983).

Presynaptic Dopaminergic Changes

Changes in DA Content

In general, there is little indication that tissue levels of DA are changed following cocaine administration. Even if they were, this might or might not indicate a change in releasable pools of DA. In striatum, levels of DA and its metabolites remain unchanged when measured between 3 and 24 h after either a single injection or repeated injections of cocaine (Taylor and Ho, 1977; Zahniser *et al.*, 1986; Hanson *et al.*, 1987; Kleven *et al.*, 1988). Following 9 d of repeated cocaine injections, an increase has been reported in DA content in both terminal and cell body regions of the nigrostriatal tract when measured 20, but not 60 min after the last cocaine injection (Roy *et al.*, 1978). However, DA content is not increased either at 20 or 60 min after the 18th or 30th consecutive day of injection. The most complete study of this question conclusively showed no long-lasting depletions of DA even following prolonged administration of cocaine and longer periods after cocaine administration has been terminated (Kleven *et al.*, 1988). These results indicate that cocaine does not produce neuronal destruction as has been seen with amphetamine (see Robinson and Becker, 1986); however, they also show no indication of neurochemical changes which might explain sensitization.

Changes in DA Synthesis

The activity of tyrosine hydroxylase (TH), the rate-limiting enzyme in the synthesis of DA, has been measured in cell-free extracts following cocaine administration. Acute injection of cocaine increases striatal TH activity up to 40 min after cocaine injection (Pradhan, 1983); but no effect is observed by 3 h after injection (Hanson *et al.*, 1987). Striatal TH activity is also increased 1 d after repeated cocaine adminstration (Taylor and Ho, 1977). Although TH appears to be transiently activated following cocaine administration, it is difficult to assess the relevance of these measurements given the numerous regulatory parameters impacting on TH activity *in vivo*. Others, measuring accumulations of dihydroxyphenylalanine (DOPA) *in vivo* after inhibition of decarboxylase activity with DSD1015, have shown that acute cocaine administration decreases tyrosine hydroxylation (see chapter by Galloway herein). This decrease is transient, dose-dependent, and blocked by sulpiride or eticlopride. Thus, *in situ* the activity of this rate-limiting enzyme is diminished shortly after a single dose of cocaine and seems an unlikely candidate to explain sensitization. Nonetheless, following repeated cocaine administration, a modest increase in the rate of DA synthesis is observed (Keegan *et al.*, 1987, 1988). This appears to be due to subsensitivity of synthesis-modulating autoreceptors and is discussed below.

Changes in DA Release

Our laboratory has looked for changes in DA release that might be related to sensitization. Amphetamine-induced tritium release from rat striatal slices preloaded with ^{3}H-DA is augmented by 30 to 50% within 1 d after a single injection of cocaine (10 mg/kg, i.p.), and this effect persists for at least 2 weeks after the single injection (Figure 3; Peris and Zahniser, 1987). Pretreatment with another catecholamine

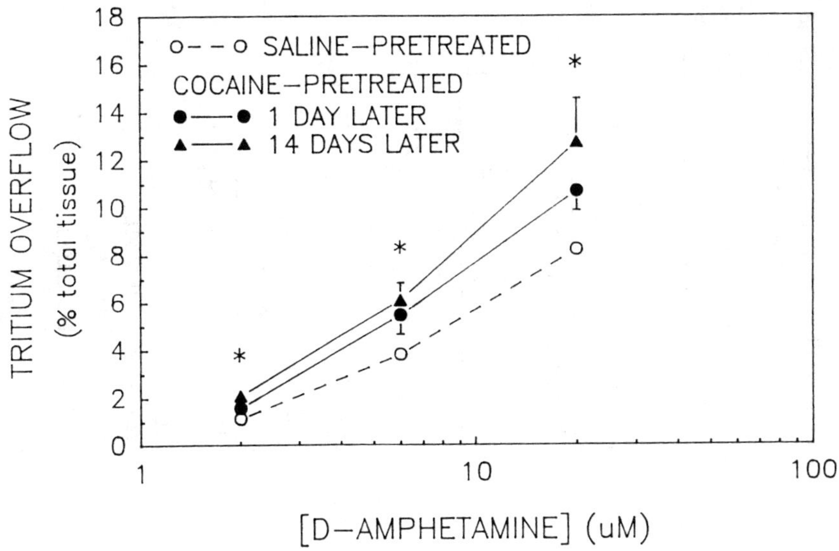

Figure 3. Persistent increase in amphetamine-stimulated ^{3}H-DA release from rat striatal slices following a single dose of cocaine. Rats were pretreated with saline or cocaine (10 mg/kg, i.p.), and release was measured either 1 or 14 d later. Striatal slices were preloaded with ^{3}H-DA. Tritium overflow, or release evoked above basal,was stimulated by the addition of 2, 6, or 20 μM D-amphetamine to the superfusion buffer for 2.5 min. Mean values ± SEM are shown for N = 4 to 6. * $p < 0.05$ for both cocaine-pretreatment groups compared to the saline-pretreatment groups (analysis of variance followed by Newman-Keuls).

uptake blocker, mazindol, or another direct-acting DA receptor agonist, apomorphine, mimics the effect of cocaine, whereas the local anesthetic lidocaine exerts no effect (Figure 4). Pretreatment 15 min before the cocaine injection with antagonists selective for either D_1 or D_2 DA receptors — SCH 23390 or sulpiride, respectively—as well as the nonselective antagonist fluphenazine, abolishes the augmentation of release (Figure 4). Therefore, activation of both D_1 and D_2 DA receptors appears to be necessary for cocaine-induced augmentation of amphetamine-stimulated ^{3}H-DA release (Peris and Zahniser, 1989). This increased release is consistent with increased dopaminergic transmission in the striatum and has several characteristics in common with behavioral sensitization, including its ability to be induced by a single dose of cocaine, its persistent nature, and its lack of development when cocaine and DA receptor blocking agents are administered together.

Cocaine-induced sensitization of stereotyped behavior is more readily apparent after 8 d as compared with 1 d of cocaine treatment; and it persists for at least 7 d after the 8-d repeated treatment is terminated (Figure 5). If the cocaine-induced augmentation of amphetamine-stimulated ^{3}H-DA release were the basis for sensitization, release should be augmented to a greater extent following repeated cocaine administration. In contrast to this prediction, augmentation of amphetamine-stimulated ^{3}H-DA release is not observed 1 and 3 d after cessation of an 8-d, once-

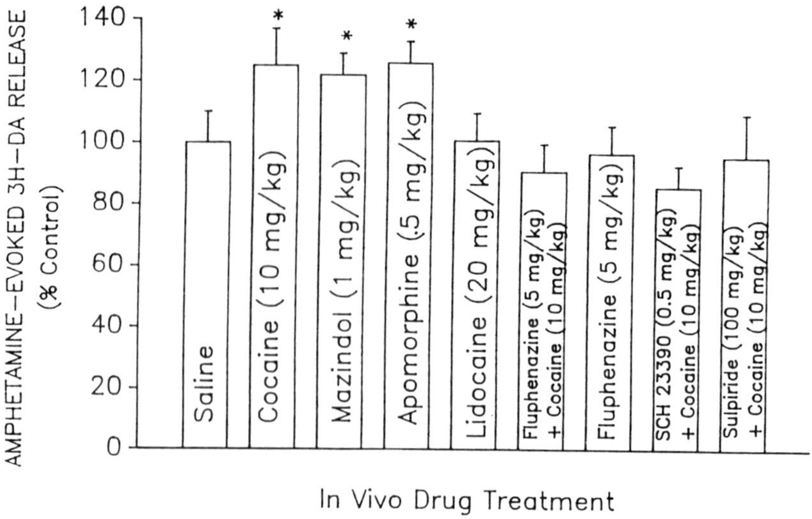

Figure 4. Dopamine receptor stimulation is required for persistent augmentation of amphetamine-stimulated release of ^{3}H-DA from striatal slices. Rats were injected with a single dose of the drugs indicated and sacrificed 1 week later. Tritium overflow was stimulated by addition of 6 μM D-amphetamine to the superfusion buffer for 2.5 min. Mean values ± SEM are shown for N = 5 to 7. * $p < 0.05$ for the cocaine-, mazindol-, and apomorphine-pretreatment groups compared to the saline-pretreatment group (analysis of variance followed by Newman-Keuls).

daily injection regimen with cocaine (10 mg/kg, i.p.). However, following 7 d of withdrawal, a significant 20–30% increase in amphetamine-stimulated ^{3}H-DA release is again observed (Figure 5; Zahniser *et al.*, 1988a). Others have reported that 3 d after repeated cocaine administration is terminated, amphetamine-stimulated ^{3}H-DA release is increased twofold over control values (Johnson and Snell, 1987). Thus, it appears that release is persistently augmented following repeated administration of cocaine but that, at short times after repeated treatment, either tolerance develops to this effect or this effect is obscured by other changes. More importantly, however, these observations produce an apparent dissociation between the augmentation of amphetamine-stimulated ^{3}H-DA release and behavioral sensitization (Figure 5). Behavioral sensitization is also dose related (Shuster *et al.*, 1977); however, we find no apparent relationship between doses of cocaine administered (5-40 mg/kg i.p.) and the degree of augmentation of amphetamine-stimulated ^{3}H-DA release (unpublished observations). Thus, these observations weaken the apparent correlation between the augmented release and behavioral sensitization.

Although, in some cases, tritium and endogenous release measurements can be used interchangeably, this is not always true, particularly when release is stimulated by amphetamine (Herdon *et al.*, 1985). Amphetamine and electrical stimulation release DA from different pools; amphetamine primarily releases DA from the cytoplasmic pool, whereas electrical stimulation promotes vesicular release (Chiueh and Moore, 1975; Parker and Cubeddu, 1986a, 1986b). We observe that relatively more tritium is recovered in the superfusate following exposure to 20 μM am-

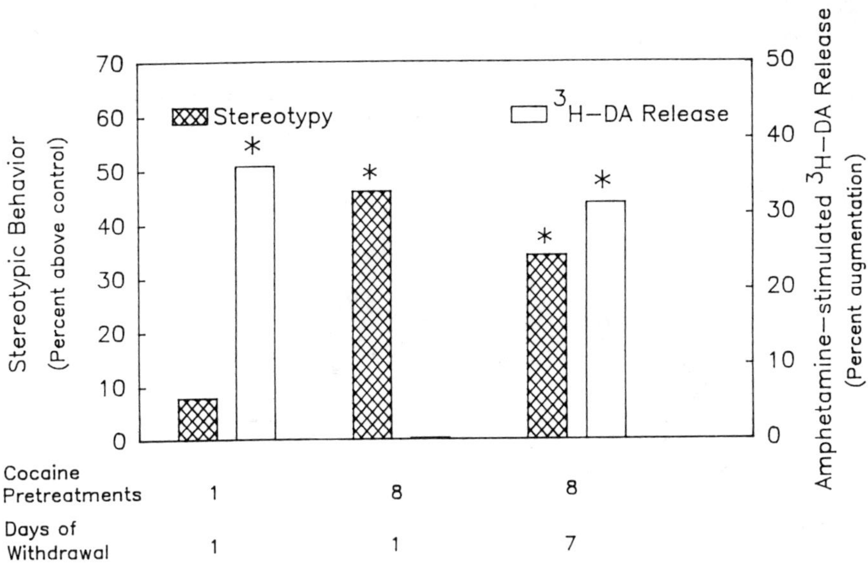

Figure 5. Lack of correlation between cocaine-induced stereotypy and augmentation of amphetamine-stimulated release of ^{3}H-DA. Cocaine (10 mg/kg, i.p.) was administered once daily for the pretreatment times indicated. Stereotypy and ^{3}H-DA release were measured either 1 or 7 d after the last pretreatment. Cocaine-induced stereotypy was rated as in Figure 1. ^{3}H-DA release was stimulated from striatal slices by addition of 20 μM D-amphetamine in the superfusion buffer for 2.5 min. Mean values for N = 4 to 6 are shown. * $p < 0.05$ for behavior compared to cocaine treatment on day 1 (Friedman's test followed by multiple Wilcoxon tests) and for release compared to the respective saline-treated controls (analysis of variance followed by Newman-Keuls).

phetamine for 2.5 min than following 60 pulses of electrical stimulation in the presence of the uptake blocker nomifensine (Figure 6). The tritium signal is essentially all due to ^{3}H-DA following amphetamine stimulation. With electrical stimulation, the signal is due to both ^{3}H-DA and its major metabolite ^{3}H-dihydroxyphenylacetic acid (^{3}H-DOPAC). The ^{3}H-DOPAC is produced because, in slices, this concentration of nomifensine (10 μM) does not completely block reuptake of the released ^{3}H-DA; thus, some ^{3}H-DA is taken back up into the nerve terminals and metabolized. When release is measured in striatal slices that have not been prelabeled (Gerhardt *et al.*, 1989), these two stimulation parameters produce approximately equal amounts of endogenous DA release (Figure 6). Taking into account the contribution from the DOPAC that is produced, the endogenous release produced by the electrical stimulation is greater than that produced by the amphetamine stimulation (Figure 6). Therefore, at least with these stimulation parameters, our results are consistent with the idea that the tracer amount of ^{3}H-DA used to prelabel slices is preferentially released by amphetamine.

These observations lead us to ask whether cocaine administration results in the same changes in amphetamine-stimulated release of endogenous DA, as compared with ^{3}H-DA. As already discussed, amphetamine-stimulated ^{3}H-DA release is persistently augmented following cocaine administration. In contrast, there is no change

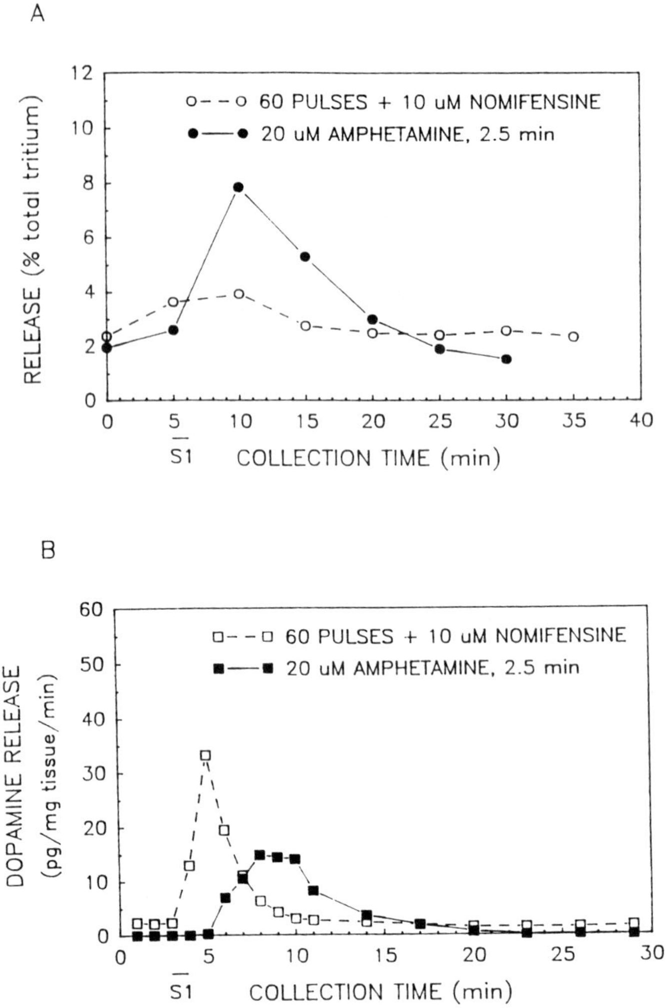

Figure 6. Comparison between amphetamine- and electrical-stimulated release of tritium from ^{3}H-DA preloaded striatal slices (A) and of endogenous DA (B) and DOPAC (C) from naive striatal slices. To stimulate release, 20 μM amphetamine was added to the superfusion buffer for 2.5 min. The electrical stimulation was 60 pulses (1 Hz) applied in the presence of 10 μM nomifensine to block uptake. The stimulation was applied at S1, and the experimental parameters were the same for the tritium and endogenous release measurements with the exception that 5- and 1-min fractions were collected, respectively. Mean values for N = 4 are shown.

in endogenous DA release evoked by amphetamine in striatal slices from the rats pretreated 1 d earlier with cocaine (10 mg/kg, i.p.; Figure 7). Neither treatment group shows enhanced DOPAC release; thus, this cannot explain the different observations with tritium and endogenous release. These results are consistent with the ideas that, in response to cocaine pretreatment, the specific activity of the

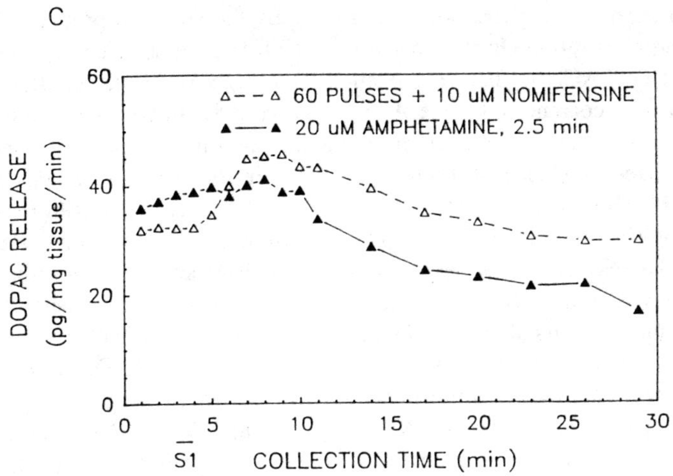

Figure 6C.

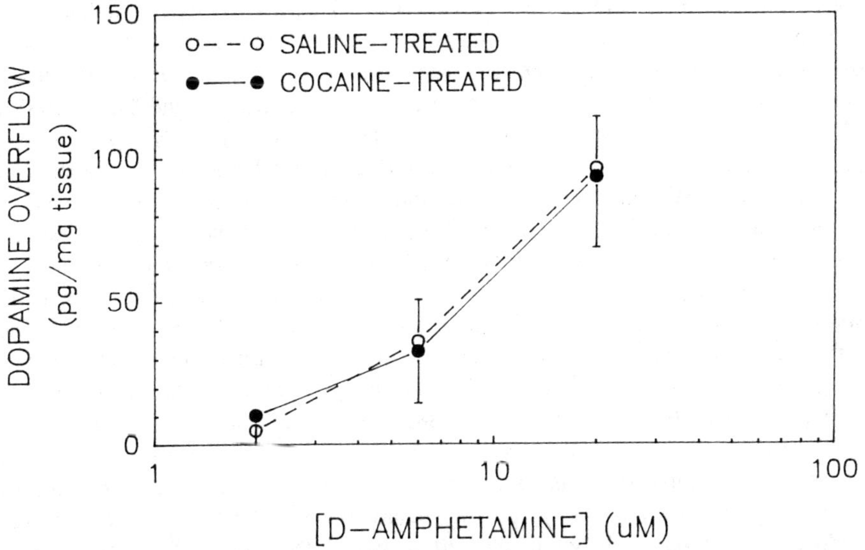

Figure 7. Amphetamine-stimulated release of endogenous DA from striatal slices is not changed 1 d after a single injection of cocaine (10 mg/kg, i.p.). DA overflow was stimulated by addition of 2, 6, or 20 μM D-amphetamine to the superfusion buffer for 2.5 min. Mean values ± for N = 4 are shown.

releasable pool of ^{3}H-DA is increased or that ^{3}H-DA is being released from non-dopaminergic neurons. Johnson and collaborators have also suggested that cocaine administration induces a shift in the releasable and/or metabolically active pools of DA (Johnson and Snell, 1987; Yi and Johnson, 1988).

Electrical stimulation delivered in the presence of cocaine, rather than amphetamine stimulation, might simulate more closely the situation produced by *in vivo* cocaine administration. In the presence of 10 μM cocaine, electrically stimulated release of DOPAC, but not DA, from striatal slices is decreased after a 1-d pretreatment with cocaine (compare A and B, Figure 8). In the presence of the same concentration of the more potent uptake blocker nomifensine, both DA and DOPAC release evoked by electrical stimulation are diminished (Figure 8A and B). Similar experiments showed decreased electrically stimulated release of tritium induced by cocaine pretreatment (Figure 8C). The persistence of these changes must still be investigated. Nonetheless, these changes are consistent with diminished dopaminergic activity and the development of tolerance rather than sensitization. Interestingly, these results show cocaine-induced changes in vesicular release and implicate cocaine-induced decreases in releasable pools of DA. Decreased release of DA following chronic exposure to cocaine has also been suggested by findings in another dopaminergic pathway. Repeated cocaine administration increases basal prolactin release, which is tonically inhibited by DA released from the tuberoinfundibular neurons (Pilotte and Sharpe, 1988). Pyschomotor stimulant-induced alterations in releasable pools of striatal DA are also implicated following repeated administration of amphetamine; however, in this case, increases, rather than decreases, in releasable pools were suggested (Casteñeda *et al.*, 1988). Another group has observed an increase in K^+-stimulated release of endogenous DA from slices of striatum and nucleus accumbens, but this increase is observed only with certain concentrations of K^+ and after certain doses of cocaine (Kalivas and Duffy, 1988). The questions of if and how DA release is changed in behaviorally sensitized rats has become even more intriguing with a recent report of a long-lasting enhancement of DA efflux in striatum, measured with intracerebral microdialysis in response to cocaine administration 1 week after 2 weeks of daily cocaine administration (20 mg/kg, i.p.; Akimoto *et al.*, 1989). Curiously, this augmented release of DA persisted for longer than 3 h following the cocaine administration. In another *in situ* study employing a different treatment paradigm, the results obtained were quite different. Self-administration of cocaine in daily sessions for 10 d leads to tolerance in the ability of cocaine to increase DA and its metabolites in the microdialysate in nucleus accumbens (Weiss *et al.*, 1988). Further experimentation is obviously needed to resolve these different observations. Nonetheless, it appears that, in striatal slices, augmentation of amphetamine-stimulated ^{3}H-DA release following cocaine administration does not always parallel behavioral sensitization and that evoked endogenous DA release is not augmented, as would be expected if it were the basis for sensitization.

Dendrites also release DA in the vicinity of the cell bodies, and Kalivas' group has investigated the effect of cocaine administration on this somatodendritic release. When rats are injected with cocaine (15 or 45 mg/kg, i.p.) for 3 d and then withdrawn for 2 to 3 weeks, a consistent decrease in both amphetamine- and K^+-stimulated release of endogenous DA is observed in ventromedial mesencephalon, the region containing the cell bodies of the nigrostriatal, mesolimbic and mesocortical DA-containing pathways (Kalivas and Duffy, 1988). The significance of these results is discussed in the next section; however, these results stress the importance of

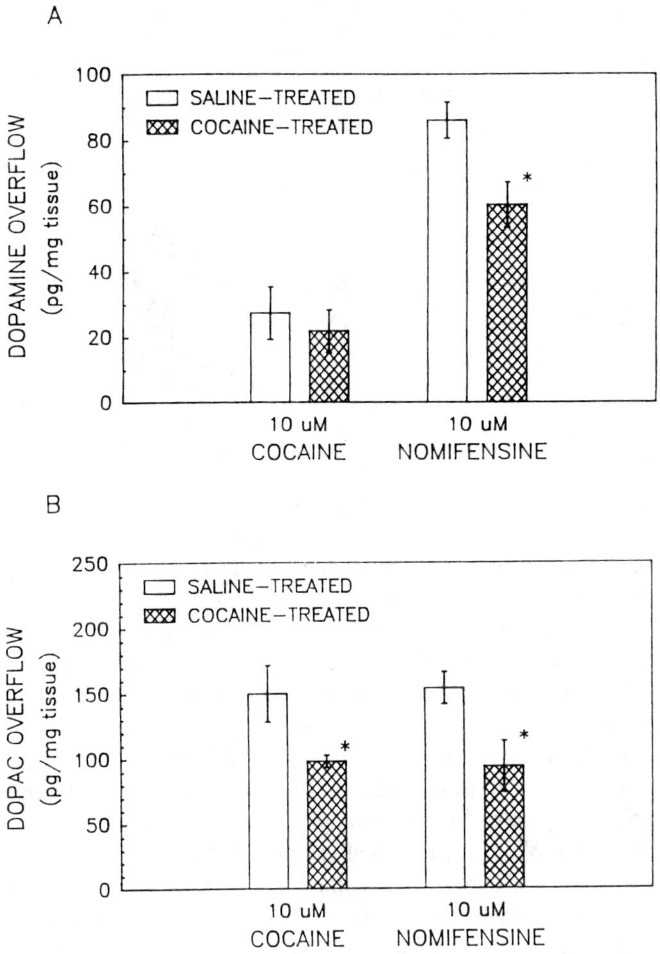

Figure 8. In general, electrically stimulated release of DA and DOPAC from striatal slices measured in the presence of an uptake blocker is reduced following a 1-d cocaine pretreatment. Rat were pretreated with saline or 10 mg/kg cocaine. Overflow of DA (A), DOPAC (B), or tritium (C) was stimulated by 60 pulses (1 Hz) of electrical stimulation in the presence of either 10 μM cocaine or nomifensine. Mean values $\pm$ SEM for N = 4 are shown. * $p < 0.05$ compared to respective control (two-tailed *t*-test).

measuring changes in evoked DA release in the cell body regions as well as the terminals.

Changes in DA Autoreceptors in the Terminals and Cell Bodies

D_2 DA receptors localized on dopaminergic neurons are termed "autoreceptors". Three types of DA autoreceptors have been proposed: impulse-regulating autoreceptors associated with cell bodies and dendrites, synthesis-regulating autoreceptors associated with nerve terminals, and release-modulating autoreceptors also associated with nerve terminals (Jackisch *et al.*, 1980: Bannon and Roth, 1983). DA

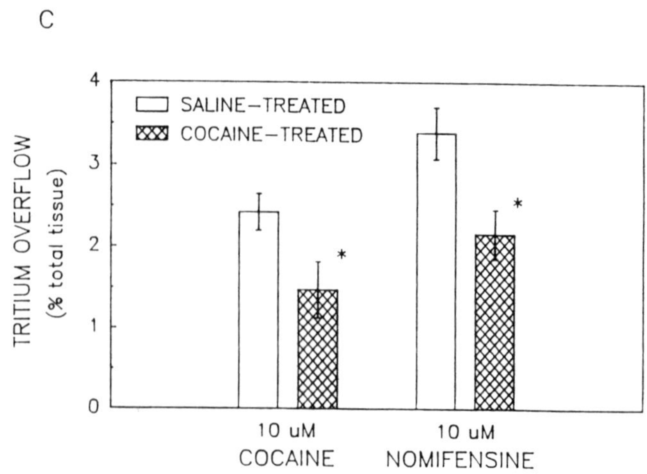

Figure 8C.

autoreceptor subsensitivity was one of the first mechanisms proposed to explain sensitization produced by repeated administration of agents such as apomorphine or amphetamine (Baudry *et al.*, 1977; Martres *et al.*, 1977; Muller and Seeman, 1979; Antelman and Chiodo, 1981). In response to cocaine administration, the data for the impulse-regulating and perhaps the synthesis-modulating DA autoreceptors support this hypothesis.

DA released from the dendrites can interact with the impulse-regulating autoreceptors. Electrophysiological studies in the VTA indicate that these DA autoreceptors are down-regulated by repeated cocaine administration (Henry *et al.*, 1989; see also chapter by White herein). Kalivas and Duffy (1988) have also suggested that their findings of a persistent cocaine-induced decrease in evoked release of ventromedial mesencephalic DA could result in less autoreceptor inhibition at the cell bodies and, thereby, in an increase in the firing of the nigrostriatal tract and release of DA from the terminals. Impulse-regulating autoreceptor subsensitivity is consistent with sensitization and again stresses the importance of cocaine-induced changes that occur at the level of the cell bodies.

Quantitative receptor autoradiography can be used to measure the binding properties of these somatodendritic autoreceptors in substantia nigra pars compacta and in VTA. In one experiment, repeated cocaine treatment for 8 d followed by 1-d withdrawal increased D_2 DA receptors in the substantia nigra pars compacta (Zahniser *et al.*, 1988b); however, in a subsequent experiment, this increase was not consistently observed. These results emphasize that functional changes in receptors may not be accompanied by changes in receptor binding properties.

There is evidence that synthesis-modulating autoreceptors may also become subsensitive with repeated administration of cocaine. Acute cocaine administration leads to decreased DA synthesis in the striatum, nucleus accumbens, and prefrontal cortex via stimulation of inhibitory synthesis-modulating D_2 DA autoreceptors (Keegan *et al.*, 1987). In contrast, repeated administration of cocaine produces no consistent change in basal levels of DA or its metabolites; but a modest increase in the rate of DA synthesis is observed (Keegan *et al.*, 1987). This increased rate of synthesis appears to be due to subsensitivity of the synthesis-modulating autoreceptors (Keegan *et al.*, 1988). Although others' data supporting the persistent nature of this effect are equivocal (see Post-Synaptic Dopaminergic Changes), it is possible that this change could contribute to sensitization.

The activity of D_2 release-modulating autoreceptors can be measured by determining the ability of a D_2 DA receptor agonist such as pergolide to inhibit the tritium overflow evoked by electrical field stimulation from tissue slices preloaded with ^{3}H-DA (Dwoskin *et al.*, 1988). In the striatum, these receptors must be studied using functional assays because they represent such a small proportion of the D_2 DA receptors and are not detected with normal quantitative autoradiographic analysis (Filloux *et al.*, 1988). The activity of these striatal autoreceptors is unchanged when measured 24 h after one injection of cocaine; however, 24 h after 8 or 14 d of cocaine treatment, these autoreceptors show a small but consistent increase in sensitivity (Dwoskin *et al.*, 1988). This supersensitivity would inhibit stimulation-evoked release of DA to a greater extent and may be a compensatory response to limit release. In any case, this change in autoreceptor sensitivity is consistent with the development of tolerance rather than sensitization.

Changes in DA Uptake

Because cocaine blocks high-affinity DA uptake, it was anticipated that this site might be specifically regulated by repeated cocaine administration. When high-affinity uptake of ^{3}H-DA into striatal synaptosomes is measured 20 min after *in vivo* cocaine administration, there is, as expected, a dose-related decrease in DA uptake (Missale *et al.*, 1985). This transient decrease returns to control levels by 60 min. In contrast, 20 min after the last dose of repeated cocaine administration, no changes in uptake are observed; therefore, repeated administration appears to result in tolerance to the ability of cocaine to block uptake (Missale *et al.*, 1985). However, others observe an increased V_{max} for striatal ^{3}H-DA uptake in cocaine-sensitized rats 30 min after the last dose of cocaine (Yi and Johnson, 1988). We find no persistent differences in ^{3}H-DA uptake into either striatal slices or synaptosomes when the uptake is measured 1 d after either a single injection or 8 to 14 d of repeated injections of cocaine (Zahniser *et al.*, 1986, 1988b). ^{3}H-Nomifensine can be used to label specific recognition binding sites for the DA uptake pump (Dubocovich and Zahniser, 1985). Saturation analysis of specific ^{3}H-nomifensine binding to striatal membranes from these same rats also reveals no differences when compared with controls (Zahniser *et al.*, 1986).

Differences in uptake parameters for DA in the striatum and nucleus accumbens have been described. In synaptosomes prepared from nucleus accumbens, the V_{max} for ^{3}H-DA uptake is only one-fourth that found in the striatal synaptosomes (Marshall, 1988); cocaine is also a less potent inhibitor of DA uptake in limbic areas (Hadfield and Nugent, 1983; Missale *et al.*, 1985). Interestingly, the maximal velocity of ^{3}H-DA uptake into the synaptosomes from nucleus accumbens is increased 20 min after either single or repeated administration of cocaine (Missale *et al.*, 1985). Despite these apparent differences between the transport of DA into striatum and nucleus accumbens measured *in vitro*, the conclusion drawn from experiments using *in vivo* microdialysis in the nucleus accumbens is similar to those found in the striatum using *in vitro* preparations; for example, tolerance develops to the effect of cocaine to increase synaptic levels of DA. In the first session of cocaine self-administration, the concentration of DA and metabolites in the *in vivo* microdialysate from the nucleus accumbens increases to 250% of controls values; however, in the tenth session, no increase is observed (Weiss *et al.*, 1988). Taken together, these results suggest that transient changes in the DA uptake pump occur in response to cocaine administration and that these changes, when they persist, are more consistent with the development of tolerance rather than sensitization.

Changes in DA Metabolism

Whereas a number of psychotropic agents are potent inhibitors of monoamine oxidase (MAO), the effect of cocaine on MAO activity is less pronounced (Gol'dina and Gankina, 1988). Acute administration of cocaine (20 mg/kg, i.p.) produces a modest, biphasic effect on the activity of MAO-A, the enzyme primarily responsible for metabolized DA in the striatum; the activity is reduced at 10 min but subsequently increased at 20 min (Pradhan, 1983). The level of DOPAC, one of the major metabolites of DA, decreases in the striatum 60 min after an acute injection of cocaine (Kalivas *et al.*, 1988). Levels of the other major metabolite, homovanillic acid (HVA), are not altered in striatum 30 min after a single cocaine injection (Taylor and Ho, 1977), but are increased 3 h after injection (Hanson *et al.*, 1987). Following repeated injections of cocaine, no consensus exists as to long-lasting changes in DA metabolism. A persistent decrease (Karoum *et al.*, 1987), a slight increase (Taylor and Ho, 1977), or no change (Yeh, 1987; Kleven *et al.*, 1988) in DA metabolism in the striatum have all been observed; thus, the significance of these observations is unclear.

Postsynaptic Dopaminergic Changes

Changes in Postsynaptic D_1 and D_2 DA Receptors

Initially, it seemed likely that an increase in postsynaptic DA receptors was the explanation for cocaine-induced sensitization. Until recently, studies have most often focused on the D_2 DA receptor subtype. An increased density of D_2 DA receptors in striatum in the absence of any changes in receptor affinity is observed 60 mins after a single injection of cocaine; curiously, however, the same change is observed when striatal slices are incubated *in vitro* for 60 min with μM cocaine (Memo *et al.*, 1981). In contrast, when striatal D_2 DA receptors are measured 1 d after a single injection of cocaine, no differences are seen in either receptor density or affinity (Dwoskin *et al.*, 1988). Thus, it appears that no long-lasting changes are induced in striatal postsynaptic D_2 DA receptors that could explain the persistent behavioral changes seen after a single exposure to cocaine.

Despite several studies, no consensus has been reached as to the regulation of D_2 DA receptors by repeated cocaine administration; this situation is also true for several other compounds that stimulate DA receptors (see Creese and Sibley, 1981).

Initial studies reported dose-related increases in striatal D_2 DA receptor number (Borison *et al.*, 1979; Taylor *et al.*, 1979). More recently, Goeders and Kuhar (1987) found that repeated cocaine administration for 15 d resulted in a decreased number of D_2 DA receptors in striatum measured 20 m after the last dose. We find no differences in the affinity or density of these receptors in the striatum 1 d after 8 or 14 d of cocaine administration (Dwoskin *et al.*, 1988; Zahniser *et al.*, 1988b, 1988c). Further evidence against postsynaptic DA receptor supersensitivity in the striatum of cocaine sensitized rats is provided by functional studies in which apomorphine did not increase stereotyped behaviors in these animals (Epstein and Altshuler, 1979; Gale, 1984). In nucleus accumbens, both Goeders and Kuhar (1987) and our group (Zahniser *et al.*, 1988b) find an increased number of D_2 DA receptors, but this appears to be a transient increase because we find that the receptor number returns to normal by 7 d after ceasing cocaine administration. The results of functional studies also indicate that an up-regulation of DA receptors occurs in nucleus accumbens after 2 weeks of cocaine administration; however, preliminary evidence points to a change in D_1, rather than D_2, DA receptors (Henry *et al.*, 1989).

One explanation for these apparently discrepant results from radioligand binding studies in the striatum may be linked to how specific binding to D_2 DA receptors was defined in each study. In the studies of Memo *et al.* (1981) and Taylor *et al.* (1979), specific binding was defined as the difference in the binding of the radioligand ^{3}H-spiperone (^{3}H-spiroperidol) in the absence and presence of 1 μM (+)-butaclamol. It is now recognized that ^{3}H-spiperone binds with high affinity to both D_2, DA and S_2 serotonin (5-HT_2) receptors in brain, and that 1 μM (+)-butaclamol does not distinguish between these two receptors (Zahniser and Dubocovich, 1983; Hamblin *et al.*, 1984). Thus, the increased density of receptors may be attributed to D_2 DA and/or 5-HT_2 receptors. In contrast, S-sulpiride binds selectively to D_2 DA receptors (Zahniser and Dubocovich, 1983). Interestingly, the two studies in which either ^{3}H-S-sulpiride was used directly (Goeders and Kuhar, 1987), or S-sulpiride was used to define the specific binding of ^{3}H-spiperone (Zahniser *et al.*, 1988b), no increase in the density of D_2 DA receptors was observed in the striatum.

We also investigated whether or not D_1 DA receptors are regulated by repeated cocaine administration. In the striatum and nucleus accumbens, the D_1 DA receptors are thought to be localized postsynaptically, whereas in the substantia nigra reticulata the D_1 receptors appear to be localized on the striatonigral terminals (Altar and Hauser, 1987). No changes in D_1 DA receptors were found in any of these areas 1 d after eight daily injections of cocaine (10 mg/kg; Zahniser *et al.*, 1988b). Taken together, there is no convincing evidence that cocaine administration induces long-lasting increases in postsynaptic DA receptors. Once again, however, these studies do not rule out functional changes in receptors.

Changes in Second Messengers and Effector Systems Coupled to DA Receptors

DA acting at postsynaptic receptors in the striatum exhibits dual regulation of adenylate cyclase activity. Activation of D_1 DA receptors stimulates cyclic AMP production while activation of D_2 DA receptors inhibits cyclic AMP formation (Stoof and Kebabian, 1981; Cooper *et al.*, 1986). Both of these actions are mediated via GTP-binding proteins. The potency of DA to stimulate striatal cyclic AMP formation is decreased 60 min after one cocaine injection (Memo *et al.*, 1981). Long-lasting effects of acute exposure to cocaine on striatal adenylate cyclase activity have not been reported. It is interesting to note that acute administration of amphetamine promotes a rapid, transient desensitization of striatal DA-stimulated adenylate cyclase activity (Barnett and Kuczenski, 1986). The potency of DA to stimulate adenylate cyclase activity in the striatum also decreases after repeated amphetamine exposure, and this desensitization occurs concomitantly with behavioral sensitization (Barnett *et al.*, 1987). More recently, other second messenger-effector systems have also been linked to DA receptors; that is, activation of phospholipase C activity for D_1 DA receptors (Felder *et al.*, 1989) and an increased potassium conductance for D_2 DA receptors (Lacey *et al.*, 1987). As yet, no one has measured the effect of repeated cocaine administration on the regulation of adenylate cyclase activity or any of these other potential effector systems by either D_1 or D_2 DA receptors in the brain. The one report following acute exposure, however, is consistent with diminished D_2 DA receptor-mediated stimulation.

Changes in Nondopaminergic Systems

That not all aspects of cocaine-induced sensitization are blocked by DA receptor antagonists emphasizes the importance of considering other neurotransmitter systems that may be involved in this phenomenon. Some possibilities are briefly mentioned here.

Noradrenergic Systems

It is well known that acute exposure to cocaine also blocks the neuronal uptake of norepinephrine, prolongs the lifetime of that neurotransmitter in the synaptic cleft, and potentiates noradrenergic neurotransmission (for example, Yasuda *et al.*, 1984). With repeated cocaine treatment, there is evidence for increased metabolism of norepinephrine in the hypothalamus/thalamus (Taylor and Ho, 1977). Although Roy and colleagues (1978) concluded that the behavioral changes induced by cocaine could be correlated with changes in the levels of norepinephrine in the midbrain and brain stem, there has been no additional evidence for a direct link between changes in noradrenergic systems with cocaine-induced behavioral sensitization.

Serotonergic Systems

Cocaine also potently blocks the neuronal uptake of 5-HT and potentiates its actions (Lakoski and Cunningham, 1988). Inhibition of 5-HT synthesis and blockade of 5-HT receptors potentiate the increase in locomotor activity produced by cocaine (Scheel-Kruger *et al.*, 1977), whereas administration of the synthetic precursor of 5-HT attenuates this action (Pradhan *et al.*, 1978). Therefore, diminution of serotonergic tone could mediate or at least potentiate the behavioral effects of cocaine. Following acute administration of cocaine, transient decreases have been noted in the activity of the rate-limiting enzyme in 5-HT synthesis (i.e., tryptophan hydroxylase), in the concentrations of 5-HT, and in its major metabolite 5-hydroxyindoleacetic acid (Pradhan *et al.*, 1978; Pradhan, 1983; Hanson *et al.*, 1987). Similar decreases in the synthesis of 5-HT and the steady-state concentrations of 5-HT and its metabolites have been observed in response to repeated cocaine administration although these changes may be short-lived (Taylor and Ho, 1977; Roy *et al.*, 1978; Hanson *et al.*, 1987). The possibility that these inhibitory actions on serotonergic function may contribute to cocaine-induced behavioral sensitization should be considered.

Opioid Systems

Upon repeated administration, sensitization occurs to the locomotor-activating effects of opioid receptor agonists such as morphine, and the opioid peptides D-Ala-Met-enkephalinamide (DALA) and Tyr-D-Ala-Gly-NMe-Phe-Gly-ol (DAGO; Shuster *et al.*, 1975; Joyce and Iverson, 1979; Vezina and Stewart, 1984; Kalivas, 1985a; Vezina *et al.*, 1987). This effect is thought to be mediated by μ opioid receptors in the VTA. As noted above, cross-sensitization occurs between cocaine and opioids so that it is likely that common pathways are involved. However, rather than opioids mediating the behavioral sensitization induced by cocaine, it has been postulated that opioids may produce some of the same changes in responsiveness as cocaine by enhancing dopaminergic transmission. Thus, it has been demonstrated that DA receptor antagonists block the behavioral effects of some of these μ opioid receptor agonists and that some of these agents increase DA metabolism in the terminal fields of the mesolimbic pathway (see Kalivas *et al.*, 1988; Stewart and Vezina, 1988).

Other Neuropeptide Systems

Kalivas and colleagues have reviewed the evidence that several other neuropeptides, when injected into the VTA, produce behavioral sensitization and neurochemical consequences similar to that produced by the μ opioid peptides (Kalivas, 1985b; Kalivas *et al.*, 1988). These peptides include neurotensin, substance P (stable analogue), and neuromedin N, but not cholecystokinin 8 or corticotropin-releasing factor. Since neurons containing these compounds have cell bodies or fibers localized in the same regions containing the dopaminergic cell bodies, their role in producing behavioral sensitization and perhaps contributing to that produced by cocaine is currently an active area of investigation.

GABAergic Systems

Repeated administration of a high dose of cocaine (160 mg/kg s.c.) results in an enhanced turnover rate of GABA and decreased ^{3}H-GABA binding in the striatum; these changes persist for at least 10 d after the cocaine administration and are specific to the striatum (Gale, 1984). When the DA receptor antagonist droperidol is concurrently administered with cocaine, neither the cocaine-induced increase in GABAergic activity nor behavioral sensitization is blocked (Gale, 1984). These results lead to the suggestion that cocaine may induce sensitization via a catecholamine receptor-independent mechanism. However, subsequent results discussed below do not support a role for GABA mediation of cocaine-induced sensitization.

The projection area of the GABAergic nigrotegmental pathway has been shown to be critical for expression of certain cocaine-induced stereotyped behaviors (Bachus and Gale, 1986). Additionally, intranigral injection of the $GABA_A$ receptor agonist muscimol evokes stereotyped behaviors. Three days after cessation of a treatment regimen with cocaine that produces behavioral sensitization, the intranigral injection of muscimol produces less stereotypy than in control rats (Bachus and Gale, 1988). These data rule out the possibility that an increase in GABA-receptive outflow from the nigra is responsible for cocaine-induced sensitization and implicate changes in the pathway before the nigra (e.g., striatum) or in other dopaminergic pathways.

Correlation of Neurochemical Changes with Behavior

Of the changes in dopaminergic neurochemistry induced by cocaine administration, only a few appear to be consistent with the development of behavioral sensitization. Increased release of DA is still an intriguing possibility, but needs more rigorous confirmation. The desensitization of the impulse-regulating D_2 DA autoreceptors appears to be a consistent finding; however, augmented DA release in the terminal fields would have to be coupled with this observation and this has not been consistently observed. The possibility that synthesis-modulating autoreceptors become desensitized should be further characterized. The inconsistencies in the striatal postsynaptic DA receptor up-regulation should be resolved; the time course of functional, as well as binding, changes in DA receptors in the nucleus accumbens should be investigated. The correlation between any of these changes and behavioral sensitization has not been extensively tested. The initial apparent correlation between behavioral sensitization and cocaine-induced augmentation of amphetamine-stimulated ^{3}H-DA release was not confirmed with further experimentation. Even if a particular change does not correlate with behavioral sensitization, the next, more difficult step is to determine whether or not it is causal in producing sensitization (Kuczenski and Leith, 1981).

RELEVANCE TO HUMANS AND CLINICAL IMPLICATIONS

In humans, as in other animals, both sensitization and tolerance are observed in response to cocaine depending on the end point measured. When cocaine is abused chronically, stereotyped compulsive behavior may develop (Post, 1975). According to Post, intermittent high doses of cocaine are more likely to cause sensitization, and humans are more likely to administer cocaine in high, binge-type runs (Gawin and Ellinwood, 1988). Not only increasing the dose, but also increasing the duration of cocaine administration, is associated with increasingly severe affective and cognitive alterations: euphoria can lead to dysphoria, which in turn can lead to paranoid psychosis (Post, 1975). Thus, with repeated administration, the effects produced can become more severe. However, it seems that sensitization must eventually lead to tolerance, or there would be no stimulus for repeated administration and drug taking would decrease. On the other hand, acute tolerance to the cardiovascular and subjective effects of cocaine quickly develop (Fischman *et al.*, 1985; Ambre *et al.*, 1988). That the pleasurable, euphorigenic subjective effects eventually decline even with repeated administration of cocaine has important implications for cocaine abusers and confirms anecdotal reports that this is indeed the case. Physiological tolerance may also explain why heavy users can self-administer up to 10 g/d without severe toxic or lethal reactions. However, Washton and Gold (1986) propose that patterns of more frequent and escalating use of cocaine may not be due to development of true tolerance but, instead, to the fact that cocaine must be readministered every 20 to 30 min in order to maintain the euphoric effects and to avoid the rebound dysphoria.

The possibility that sensitization will occur in persons using infrequent, relatively small doses of cocaine underscores the potential seriousness of this problem. With the next dose of drug administered, the possibility always exists that it may produce a more severe response than the previous doses. With amphetamine, there are reports of psychotic episodes being produced by an isolated dose of amphetamine taken years after previous doses (see Snyder, 1973). The contribution of cocaine-induced behavioral sensitization to the response of individuals taking cocaine has yet to be fully determined.

Clinical studies related to cocaine craving suggest two possible long-lasting changes in dopaminergic systems induced by cocaine administration: (1) supersensitive autoreceptors and decreased DA release/turnover and (2) reversible depletion of intraneuronal stores of DA (Dackis and Gold, 1985; Gawin and Ellinwood, 1988). These suggestions have led to the treatment of cocaine addicts with bromocriptine or desipramine. Prolonged hyperprolactinemia in male chronic cocaine abusers following abstinence from cocaine usage indicates a lack of tonic inhibition by infundibular dopaminergic neurons and is consistent with diminished release of DA (Dackis and Gold, 1985; Dax *et al.*, 1988). It is speculated that euphoria induced by cocaine results from acute stimulation of dopaminergic neurotransmission and that craving for cocaine results from depletion of synaptic DA (Dackis and Gold, 1985). A better understanding of how cocaine acts in other mammals may contribute to our understanding of the actions of cocaine in humans.

SUMMARY AND FUTURE DIRECTIONS

At present, there seems to be no simple neurochemical explanation for sensitization involving central dopaminergic systems (Table 2). DA release, impulse-regulating DA autoreceptors, synthesis-modulating autoreceptors, and certain postsynaptic DA receptors show some changes that are consistent with increased dopaminergic transmission and behavioral sensitization. However, none of these changes have been proven to be the basis for sensitization. It is also interesting to note that several of the changes such as those involving some types of DA release, release-modulating autoreceptors, and DA uptake are more consistent with the development of tolerance, rather than sensitization. The fact that at this time there appears to be no simple neurochemical correlate in the dopaminergic system to explain cocaine-

Table 2
Summary of Persistent Cocaine-Induced Changes in Brain Dopaminergic Neurochemistry

Dopamine content — no change
Dopamine synthesis — influenced via synthesis-modulating autoreceptors
Dopamine release
^{3}H-Dopamine — persistent increase in amphetamine-stimulated release
Endogenous Dopamine — persistent increase observed under certain conditions
Dopamine autoreceptors
Impulse-regulating — desensitization
Synthesis-modulating — preliminary data show desensitization
Release-modulating — supersensitivity
Dopamine uptake — no change or an increase
Dopamine metabolism — no consistent change
Postsynaptic D_1 dopamine receptors — no change
Postsynaptic D_2 dopamine receptors — changes appear transient
Effector systems coupled to dopamine receptors — not studied

induced behavioral sensitization may be due to several factors. It may be because the behaviors produced are complex and the neurochemical changes relate to only one particular aspect of the behavior that is manifested. It may also be due to the use of *in vitro* systems which do not maintain or reflect changes that occur *in vivo*. For example, rather than using *in vitro* slices to assess changes in DA release, release could be measured using either *in vivo* dialysis (Robinson *et al.*, 1988; Akimoto *et al.*, 1989; Hurd and Ungerstedt, 1989) or *in vivo* electrochemistry (Gerhardt *et al.*, 1988; Gratton *et al.*, 1988), ideally in freely moving animals. Another possibility is that neurotransmitters other than DA are also involved. The question of the underlying mechanisms that produced behavioral sensitization in response to cocaine administration remains a challenge to be answered in the future.

ACKNOWLEDGMENTS

Our work has been supported by U.S. Public Health Service Grant DA 04216.

REFERENCES

K. Akimoto, T. Hamamura, and S. Otsuki. Subchronic treatment enhances cocaine-induced dopamine efflux, studied by *in vivo* intracerebral dialysis. *Brain Res.* **490**: 339 (1989).

S. Algeri, N. Brunello, and G. Vantini. Different adaptive responses by rat striatal dopamine synthetic and receptor mechanisms after repeated treatment with D-amphetamine, methylphenidate and nomifensine. *Pharmacol. Res. Commun.* **12**: 675 (1980).

C.A. Altar and K. Hauser. Topography of substantia nigra innervation by D_1 receptor-containing striatal neurons. *Brain Res.* 410: 1 (1987).

J.J. Ambre, S.M. Belknap, J. Nelson, T.I. Ruo, S.-G. Shin, and A.J. Atkinson, Jr. Acute tolerance to cocaine in humans. *Clin. Pharmacol. Ther.* **44**: 1 (1988).

S.M. Antelman. Time-dependent sensitization as the cornerstone for a new approach to pharmacotherapy: drugs as foreign/stressful stimuli. *Drug Dev. Res.* **14**: 1 (1988).

S.M. Antelman and L.A. Chiodo. Dopamine autoreceptor subsensitivity: a mechanism common to the treatment of depression and the induction of amphetamine psychosis? *Biol. Psychiatry* **16**: 717 (1981).

S.M. Antelman and L.A. Chiodo. Amphetamine as a stressor. In *Stimulants: Neurochemical, Behavioral, and Clinical Perspectives.* I. Creese, ed., Raven Press, New York, p. 269 (1983).

S.E. Bachus and K. Gale. Muscimol microinfused into the nigrotegmental target area blocks selected components of behavior elicited by amphetamine or cocaine. *Naunyn-Schmiedeberg's Arch. Pharmacol.* **333**: 143 (1986).

S.E. Bachus and K. Gale. Differential effects of chronic treatment with chlorpromazine versus cocaine on behavioral responsiveness to nigral GABA receptor stimulation. *Psychopharmacology* **95**: 56 (1988).

M.J. Bannon and R.H. Roth. Pharmacology of mesocortical dopamine neurons. *Pharmacol. Rev.* **35**: 53 (1983).

J.V. Barnett and R. Kuczenski. Desensitization of rat striatal dopamine-stimulated adenylate cyclase after acute amphetamine administration. *J. Pharmacol. Exp. Ther.* **237**: 820 (1986).

J.V. Barnett, D.S. Segal, and R. Kuczenski. Repeated amphetamine pretreatment alters the responsiveness of striatal dopamine-stimulated adenylate cyclase to amphetamine-induced desensitization. *J. Pharmacol. Exp. Ther.* **242**: 40 (1987).

G.A. Barr, N.S. Sharpless, S. Cooper, S.R. Schiff, W. Paredes, and W.H. Bridger. Classical conditioning, decay and extinction of cocaine-induced hyperactivity and stereotypy. *Life Sci.* **33**: 1341 (1983).

M Baudry, J. Costentin, H. Marcais, M.P. Martres, P. Protais, and J.C. Schwartz. Decreased responsiveness to low doses of apomorphine after dopamine agonists and the possible involvement of hyposensitivity of dopamine "autoreceptors". *Neurosci. Lett.* **4**: 203 (1977).

R.J. Beninger and R.S. Herz. Pimozide blocks establishment but not expression of cocaine-produced environment-specific conditioning. *Life Sci.* **38**: 1425 (1986).

M. Benuck, A. Lajtha, and M.E.A. Reith. Pharmacokinetics of systemically administered cocaine and locomotor stimulation in mice. *J. Pharmacol. Exp. Ther.* **243**: 144 (1987).

R.L. Borison, A. Hitri, H.L. Klawans, and B.I. Diamond. A new model for schizophrenia: behavioral and receptor binding studies. In: *Catecholamines: Basic and Clinical Frontiers.* E. Usdin, I.J. Kopin and J. Barchas, eds., Permagon Press, Elmsford, NY, p. 719 (1979).

E. Castañeda, J.B. Becker, and T.E. Robinson. The long-term effects of repeated amphetamine treatment *in vivo* on amphetamine, KCl and electrical stimulation evoked striatal dopamine release *in vitro*. *Life Sci.* **42**: 2447 (1988).

C.C. Chiueh and K.E. Moore. D-Amphetamine-induced release of "newly synthesized" and "stored" dopamine from the caudate nucleus *in vivo*. *J. Pharmacol. Exp. Ther.* **192**: 642 (1975).

D. Clark and F.J. White. Review: D_1 dopamine receptor — the search for a function: a critical evaluation of the D_1/D_2 dopamine receptor classification and its functional implications. *Synapse* **1**: 347 (1987).

D.M.F. Cooper, C.M. Bier-Laning, M.K. Halford, M.K. Ahlijanian, and N.R. Zahniser. Dopamine, acting through D-2 receptors, inhibits rat striatal adenylate cyclase by a GTP-dependent process. *Mol. Pharmacol.* **29**: 113 (1986).

B. Costall and R.J. Naylor. Mesolimbic and extrapyramidal sites for the mediation of sterotyped behavior patterns and hyperactivity by amphetamine and apomorphine in the rat. *Adv. Behav. Biol.* **21**: 47 (1977).

I. Creese and S.D. Iversen. The pharmacological and anatomical substrates of the amphetamine response in the rat. *Brain Res.* **83**: 419 (1975).

I. Creese and D.R. Sibley. Receptor adaptations to centrally-acting drugs. *Ann. Rev. Pharmacol. Toxicol.* **21**: 357 (1981).

C.A. Dackis and M.S. Gold. New concepts in cocaine additions: the dopamine depletion hypothesis. *Neurosci. Biobehav. Rev.* **9**: 469 (1985).

E.M. Dax, W.W. Weddington, N.S. Pilotte, and J.J. Jaffe. Release of prolactin and cortisol following cocaine cessation in men. *Soc. Neurosci. Abstr.* **14**: 1051 (1988).

G.G. Dougherty, Jr. and E.H. Ellinwood. Chronic D-amphetamine in nucleus accumbens: lack of tolerance or reverse tolerance of locomotor activity. *Life Sci.* **28**: 2295 (1981).

A.W. Downs and N.B. Eddy. The effect of repeated doses of cocaine on the rat. *J. Pharmacol. Exp. Ther.* **46**: 199 (1932).

M.L. Dubocovich and N.R. Zahniser. Binding of the dopamine uptake inhibitor ^{3}H-nomifensine to striatal membranes. *Biochem. Pharmacol.* **34**: 1137 (1985).

L.P. Dwoskin, J. Peris, R.P. Yasuda, K. Philpott, and N.R. Zahniser. Repeated cocaine administration results in supersensitivity of striatal D-2 dopamine autoreceptors to pergolide. *Life Sci.* **42**: 255 (1988).

A.J. Eichler, S.M. Antelman, and C.A. Black. Amphetamine stereotypy is not a homogeneous phenomenon: sniffing and licking show distinct profiles of sensitization and tolerance. *Psychopharmacology* **68**: 287 (1980)

L.C. Einhorn, P.A. Johansen, and F.J. White. Electrophysiological effects of cocaine in the mesoaccumbens dopamine system: studies in the ventral tegmental area. *J. Neurosci.* **8**: 100 (1988).

P.N. Epstein and H.L. Altshuler. Altered response to apomorphine and haloperidol after nine days of cocaine injection. *Pharmacol. Biochem. Behav.* **10**: 189 (1979).

A.M. Ernst. Mode of action of apomorphine and dexamphetamine on gnawing compulsion in rats. *Psychopharmacology* **10**: 316 (1967).

C.C. Felder, P.A. Jose, and J. Axelrod. The dopamine-1 agonist, SKF 82526 stimulates phospholipase-C activity independent of adenylate cyclase. *J. Pharmacol. Exp. Ther.* **248**: 171 (1989).

F. Filloux, T.M. Dawson, and J.K. Wamsley. Localization of nigrostriatal dopamine receptor subtypes and adenylate cyclase. *Brain Res. Bull.* **20**: 447 (1988).

M.W. Fischman, C.R. Schuster, J. Javaid, Y. Hatano, and J. Davis. Acute tolerance development to the cardiovascular and subjective effects of cocaine. *J. Pharmacol. Exp. Ther.* **235**: 677 (1985).

K. Gale. Catecholamine-independent behavioral and neurochemical effects of cocaine in rats. *NIDA Research Monograph 54: Mechanisms of Tolerance and Dependence.* C.W. Sharp, ed., U.S. Government Printing Office, Washington, DC, p. 323 (1984).

F.H. Gawin and E.H. Ellinwood, Jr. Cocaine and other stimulants: actions, abuse and treatment. *New Engl. J. Med.* **318**: 1173 (1988).

G.A. Gerhardt, L.P. Dwoskin, and N.R. Zahniser. Outflow and overflow of picogram levels of endogenous dopamine and DOPAC from rat striatal slices: improved methodology for studies of stimulus-evoked release and metabolism. *J. Neurosci. Meth.* **26**: 217 (1989).

G.A. Gerhardt, A. Gratton, and G.M. Rose. *In vivo* electrochemical studies of the effects of cocaine on dopamine nerve terminals in rat neostriatum. *Physiologia Bohemoslovaca* **37**: 249 (1988).

S.D. Glick and P.A. Hinds. Sex differences in sensitization to cocaine-induced rotation. *Eur. J. Pharmacol.* **99**: 119 (1984).

S.D. Glick, R.M. Shapiro, K.L. Drew, P.A. Hinds, and J.N. Carlson. Differences in spontaneous and amphetamine-induced rotational behavior, and in sensitization to amphetamine, among Sprague-Dawley derived rats from different sources. *Physiol. Behav.* **38**: 67 (1986).

N.E. Goeders and M.J. Kuhar. Chronic cocaine administration induces opposite changes in dopamine receptors in the striatum and nucleus accumbens. *Alcohol Drug Res.* **7**: 207 (1987).

O.A. Gol'dina and E.M. Gankina. The effect of psychotropic substances on the deamination of monoamines in the rat brain. *Farmakol. Toksikol. (U.S.S.R.)* **51**: 22 (1988).

A. Gratton, B.J. Hoffer, and G.A. Gerhardt. Effects of electrical stimulation of brain reward sites on release of dopamine in rat: an *in vivo* electrochemical study. *Brain Res. Bull.* **21**: 319 (1988).

M.G. Hadfield and E.A. Nugent. Cocaine: comparative effect on dopamine uptake in extrapyramidal and limbic systems. *Biochem. Pharmacol.* **32**: 744 (1983).

M.W. Hamblin, S.E. Leff, and I. Creese. Interactions of agonists with D-2 dopamine receptors: evidence for a single receptor population existing in multiple agonist affinity-states in rat striatal membranes. *Biochem. Pharmacol.* **33**: 877 (1984).

G.R. Hanson, L.A. Matsuda, and J.W. Gibb. Effects of cocaine on methamphetamine-induced neurochemical changes: characterization of cocaine as a monoamine uptake blocker. *J. Pharmacol. Exp. Ther.* **242**: 507 (1987).

D.J. Henry, M.A. Greene, and F.J. White. Electrophysiological effects of cocaine in the mesoaccumbens dopamine system: repeated administration. *J. Pharmacol. Exp. Ther.* in press (1989).

H. Herdon, J. Strupish, and S.R. Nahorski. Differences between the release of radio-labelled and endogenous dopamine from superfused rat brain slices: effects of depolarizing stimuli, amphetamine and synthesis inhibition. *Brain Res.* **348**: 309 (1985).

R.E. Hinson and C.X. Poulos. Sensitization to the behavioral effects of cocaine: modification by Pavlovian conditioning. *Pharmacol. Biochem. Behav.* **15**: 559 (1981).

Y.L. Hurd and U. Ungerstedt. Cocaine: an *in vivo* microdialysis evaluation of its acute action on dopamine transmission in rat striatum. *Synapse* **3**: 48 (1989).

J. Hyttel. Citalopram — pharmacological profile of a specific serotonin uptake inhibitor with antidepressant activity. *Prog. Neuro-Psychopharmacol. Biol. Psychiatry* **6**: 277 (1982).

R. Jackisch, A. Zumstein, G. Hertting, and K. Starke. Interneurones are probably not involved in the presynaptic dopaminergic control of dopamine release in rabbit caudate nucleus. *Naunyn-Schmiedeberg's Arch. Pharmacol.* **314**: 129 (1980).

K.M. Johnson and L.D. Snell. Sensitization to the behavioral effects of cocaine is associated with altered dopamine metabolism and release in rat brain. *Soc. Neurosci. Abstr.* **13**: 1718 (1987).

E.M. Joyce and S.D. Iversen. The effect of morphine applied locally to mesencephalic dopamine cell bodies on spontaneous motor activity in the rat. *Neurosci. Lett.* **14**: 207 (1979).

P.W. Kalivas. Sensitization to repeated enkephalin administration into the ventral tegmental area of the rat. II. Involvement of the mesolimbic dopamine system. *J. Pharmacol. Exp. Ther.* **235**: 544 (1985a).

P.W. Kalivas. Interactions between neuropeptides and dopamine neurons in the ventromedial mesencephalon. *Neurosci. Biobehav. Rev.* **9**: 573 (1985b).

P.W. Kalivas and P. Duffy. Effects of daily cocaine and morphine treatment on somatodendritic and terminal field dopamine release. *J. Neurochem.* **50**: 1498 (1988).

P.W. Kalivas, P. Duffy, R. Abhold, and R.P. Dilts. Sensitization of mesolimbic dopamine neurons by neuropeptides and stress, in: *Sensitization in the Nervous System.* P.W. Kalivas and C.D. Barnes, eds., Telford Press, Caldwell, NJ, p. 119 (1988).

P.W. Kalivas, S. Taylor, and J.S. Miller. Sensitization to repeated enkephalin administration into the ventral tegmental area of the rat. I. Behavioral characterization. *J. Pharmacol. Exp. Ther.* **235**: 537 (1985).

P.W. Kalivas and B. Weber. Amphetamine injection into the ventral mesencephalon sensitizes rats to peripheral amphetamine and cocaine. *J. Pharmacol. Exp. Ther.* **245**: 1095 (1988).

F. Karoum, R. Fawcett, and R.J. Wyatt. The effect of chronic cocaine treatment on catecholamine disposition in different brain regions in rat: a specific target effect on frontal cortex dopamine. *Soc. Neurosci. Abstr.* **13**: 1460 (1987).

M.J. Keegan, J. Stahl, and M.P. Galloway. Action of cocaine on dopamine (DA) and serotonin (5-HT) containing neurons. *Soc. Neurosci. Abstr.* **13**: 481 (1987).

M.J. Keegan, J.B. Stahl, and M.P. Galloway. Effects of repeated cocaine or amphetamine on dopamine synthesis modulating autoreceptors (ARs). *Soc. Neurosci. Abstr.* **14**: 739 (1988).

P.H. Kelly and S.D. Iversen. Selective 6OHDA-induced destruction of mesolimbic dopamine neurons: abolition of psychostimulant-induced locomotor activity in rats. *Eur. J. Pharmacol.* **40**: 45 (1976).

M.M. Kilbey and E.H. Ellinwood. The effect of age-related factors on behavior induced by cocaine. *Life Sci.* **20**: 1847 (1977).

M.S. Kleven, W.L. Woolverton, and L.S. Seiden. Lack of long-term monoamine depletions following repeated or continuous exposure to cocaine. *Brain Res. Bull.* **21**: 233 (1988).

L. Kokkinidis. Sensitization to amphetamine and tolerance to cocaine and phencyclidine stimulation in mice. *Pharmacol. Biochem. Behav.* **25**: 1175 (1986).

L. Kokkinidis, R.D. Kirkby, B.D. McCarter, and T.B. Borowski. Alterations in amphetamine-induced locomotor activity and stereotypy after electrical stimulation of the nucleus accumbens and neostriatum. *Life Sci.* **44**: 633 (1989).

M.G. Kolta, P. Shreve, and N.J. Uretsky. Effect of methylphenidate pretreatment on the behavioral and biochemical responses to amphetamine. *Eur. J. Pharmacol.* **117**: 279 (1985).

R. Kuczenski. Biochemical actions of amphetamine and other stimulants, in: *Stimulants: Neurochemical, Behavioral, and Clinical Perspectives.* I. Creese, ed., Raven Press, New York, p. 31 (1983).

R. Kuczenski and N.J. Leith. Chronic amphetamine: is dopamine a link in or a mediator of the development of tolerance and reverse tolerance? *Pharmacol. Biochem. Behav.*, **15**: 405 (1981).

R. Kuczenski and D.S. Segal. Psychomotor stimulant-induced sensitization: behavioral and neurochemical correlates. *Sensitization in the Nervous System.* P.W. Kalivas and C.D. Barnes, eds., Telford Press, Caldwell, NJ, p. 175 (1988).

M.G. Lacey, N.B. Mercuri, and R.A. North. Dopamine acts on D_2 receptors to increase potassium conductance in neurones of the rat substantia nigra zona compacta. *J. Physiol. (London)* **392**: 397 (1987).

J.M. Lakoski and K.A. Cunningham. The interaction of cocaine with central serotonergic neuronal systems: cellular electrophysiologic approaches. *NIDA Research Monograph 88: Mechanisms of Cocaine Abuse and Toxicity.* D. Clouet, K. Asghar, and R. Brown, eds., U.S. Government Printing Office, Washington, DC p. 78 (1988).

G. Lin-Chu, T.E. Robinson, and J.B. Becker. Sensitization of rotational behavior produced by a single exposure to cocaine. *Pharmacol. Biochem. Behav.* **22**: 901 (1985).

J.F. Marshall. Limbic and non-limbic rat striatal regions differ markedly in their high-affinity dopamine transport. *Soc. Neurosci. Abstr.* **14**: 157 (1988).

M.P. Martres, J. Costentin, M. Baudry, H. Marcais, P. Protais, and J.C. Schwartz. Long-term changes in the sensitivity of pre-and postsynaptic dopamine receptors in mouse striatum evidenced by behavioural and biochemical studies. *Brain Res.* **136**: 319 (1977).

M. Memo, S. Pradhan and I. Hanbauer. Cocaine-induced supersensitivity of striatal dopamine receptors: role of endogenous calmodulin. *Neuropharmacology* **20**: 1145 (1981)

C. Missale, L. Castelletti, S. Govoni, P.F. Spano, M. Trabucchi, and I. Hanbauer. Dopamine uptake is differentially regulated in rat striatum and nucleus accumbens. *J. Neurochem.* **45**: 51 (1985).

P. Muller and P. Seeman. Presynaptic subsensitivity as a possible basis for sensitization by long-term dopamine mimetics. *Eur. J. Pharmacol.* **55**: 149 (1979).

P.J. Munson and D. Rodbard. LIGAND: a versatile computerized approach for characterization of ligand-binding systems. *Anal. Biochem.* **107**: 220 (1980).

P.K. Nayak, A.L. Misra, and S.J. Mule. Physiological disposition and biotransformation of ^{3}H-cocaine in acute and chronically-treated rats. *J. Pharmacol. Exp. Ther.* **196**: 556 (1975).

E.M. Parker and L.X. Cubeddu. Effects of D-amphetamine and dopamine synthesis inhibitors on dopamine and acetylcholine neurotransmission in the striatum. I. Release in the absence of vesicular transmitter stores. *J. Pharmacol. Exp. Ther.* **237**: 179 (1986a).

E.M. Parker and L.X. Cubeddu. Effects of D-amphetamine and dopamine synthesis inhibitors on dopamine and acetylcholine neurotransmission in the striatum. I. Release in the presence of vesicular transmitter stores. *J. Pharmacol. Exp. Ther.* **237**: 193 (1986b).

J. Peris and N.R. Zahniser. One injection of cocaine produces a long-lasting increase in [^{3}H]-dopamine release. *Pharmacol. Biochem. Behav.* **27**: 533 (1987).

J. Peris and N.R. Zahniser. Persistent augmented dopamine release after acute cocaine requires dopamine receptor activation. *Pharmacol. Biochem. Behav.* **32**: 71 (1989).

N.S. Pilotte and L.G. Sharpe. Chronic cocaine (C) alters the release of prolactin (PRL) in male rats. *Soc. Neurosci. Abstr.* **14** : 73 (1988).

R.M. Post. Cocaine psychoses: a continuum model. *Am. J. Psychiatry* **132**: 225 (1975).

R.M. Post. Intermittent versus continuous stimulation: effect of time interval on the development of sensitization or tolerance. *Life Sci.* **26**: 1275 (1980).

R.M. Post and N.R. Contel. Human and animal studies of cocaine: implications for development of behavioral pathology, in: *Stimulants: Neurochemical, Behavioral, and Clinical Perspectives.* I. Creese, ed., Raven Press, New York, p. 169 (1983).

R.M. Post, N.R. Contel, and P. Gold. Impaired behavioral sensitization to cocaine in vasopressin deficient rats. *Life Sci.* **31**: 2745 (1982).

R.M. Post, A. Lockfield, K.M. Squillace, and N.R. Contel. Drug-environment interaction: context dependency of cocaine-induced behavioral sensitization. *Life Sci.* **28** : 755 (1981).

R.M. Post and S.R.B. Weiss. Sensitization and kindling: implications for the evolution of psychiatric symptomatology, in: *Sensitization in the Nervous System.* P.W. Kalivas and C.D. Barnes, eds., Telford Press, Caldwell, NJ, p. 257 (1988).

R.M. Post, S.R.B. Weiss, and A. Pert. The role of context and conditioning in behavioral sensitization to cocaine. *Psychopharmacol. Bull.* **23**: 425 (1987a).

R.M. Post, S.R.B. Weiss, J. Smith, Jr., T.L. Sullivan, and A. Pert. Amygdala and nucleus accumbens lesions impair cocaine sensitization. *Soc. Neurosci. Abstr.* **13**: 661 (1987b).

S. Pradhan. Effect of cocaine on rat brain enzymes. *Arch. Int. Pharmacodyn.* **266**: 221 (1983).

S.N. Pradhan, A.K. Bhattacharyya, and S. Pradhan. Serotonergic manipulation of the behavioral effects of cocaine in rats. *Commun. Psychopharmacol.* **2**: 481 (1978).

G.V. Rebec and D.S. Segal. Apparent tolerance to some aspects of amphetamine stereotypy with long-term treatment. *Pharmacol. Biochem. Behav.* **13**: 793 (1980).

M.E.A. Reith, M. Benuck, and A. Lajtha. Cocaine disposition in the brain after continuous or intermittent treatment and locomotor stimulation in mice. *J. Pharmacol. Exp. Ther.* **243**: 281 (1987).

W.H. Riffee, E. Wanek, and R.E. Wilcox. Apomorphine fails to inhibit cocaine-induced behavioral hypersensitivity. *Pharmacol. Biochem. Behav.* **29**: 239 (1988).

M.D. Ritz, R.J. Lamb, S.R. Goldberg, and M.J. Kuhar. Cocaine receptors on dopamine transporters are related to self-administration of cocaine. *Science* **237**: 1219 (1987).

T.E. Robinson. Stimulant drugs and stress: factors influencing individual differences in the susceptibility to sensitization, in: *Sensitization in the Nervous System.* P.W. Kalivas and C.D. Barnes, eds., Telford Press, Caldwell, NJ, p. 145 (1988).

T.E. Robinson and J.B. Becker. Enduring changes in brain and behavior produced by chronic amphetamine administration: a review and evaluation of animal models of amphetamine psychosis. *Brain Res. Rev.* **11**: 157 (1986).

T.E. Robinson, P.A. Jurson, J.A. Bennett, and K.M. Bentgen. Persistent sensitization of dopamine neurotransmission in ventral striatum (nucleus accumbens) produced by prior experience with (+)-amphetamine: a microdialysis study in freely moving rats. *Brain Res.* **462**: 211 (1988).

S.N. Roy, A.K. Bhattacharyya, S. Pradhan, and S.N. Pradhan. Behavioral and neurochemical effects of repeated administration of cocaine in rats. *Neuropharmacology* **17**: 559 (1978).

J. Scheel-Kruger, C. Braestrup, M. Nielson, K. Golembiowska, and E. Mogilnicka. Cocaine: discussion on the role of dopamine in the biochemical mechanism of action. In: *Cocaine and Other Stimulants.* E.H. Ellinwood, Jr. and M.M. Kilbey, eds., Plenum Press, New York, p. 373 (1977).

D.S. Segal and A.J. Mandell. Long-term administration of amphetamine: progressive augmentation of motor activity and stereotypy. *Pharmacol. Biochem. Behav.* **2**: 249 (1974).

D.S. Segal and M.A. Schuckit. Animal models of stimulant-induced psychosis. In: *Stimulants: Neurochemical, Behavioral, and Clinical Perspectives.* I. Creese, ed., Raven Press, New York, p. 131 (1983).

L. Shuster, G.W. Webster, and G. Yu. Perinatal narcotic addiction in mice: sensitization to morphine stimulation. *Addict. Dis.* **2**: 277 (1975).

L. Shuster, G. Yu, and A. Bates. Sensitization to cocaine stimulation in mice. *Psychopharmacology* **52**: 185 (1977).

S.H. Snyder. Amphetamine psychosis: a 'model' schizophrenia mediated by catecholamines. *Am. J. Psychiatry* **130**: 61 (1973).

J. Stewart and P. Venzina. Conditioning and behavioral sensitization, in: *Sensitization in the Nervous System.* P.W. Kalivas and C.D. Barnes, eds., Telford Press, Caldwell, NJ, p. 207 (1988).

J.C. Stoof and J.W. Kebabian. Opposing roles for D-1 and D-2 dopamine receptors in efflux of cyclic AMP from rat neostriatum. *Nature* **294**: 366 (1981).

J.S. Stripling and E.H. Ellinwood, Jr. Sensitization to cocaine following chronic administration in the rat, in: *Cocaine and Other Stimulants.* E.H. Ellinwood, Jr. and M.M. Kilbey, eds., Plenum Press, New York, p. 327 (1977).

A.L. Tatum and M.H. Seevers. Experimental cocaine addiction. *J. Pharmacol. Exp. Ther.* **36**: 401 (1929).

D. Taylor and B.T. Ho. Neurochemical effects of cocaine following acute and repeated injection. *J. Neurosci. Res.* **3**: 95 (1977).

D. Taylor, B.T. Ho, and J.D. Fagan. Increased dopamine receptor binding in rat by repeated cocaine injection. *Commun. Psychopharmacol.* **3**: 137 (1979).

P. Venzina, P.W. Kalivas, and J. Stewart. Sensitization occurs to the locomotor effects of morphine and the specific μ agonist, DAGO, administered repeatedly into the ventral tegmental area but not to the nucleus accumbens. *Brain Res.* **417**: 51 (1987).

P. Venzina and J. Stewart. Conditioning and place-preference sensitization increases in activity induced by morphine in the VTA. *Pharmacol. Biochem. Behav.* **20**: 925 (1984).

A.M. Washton and M.S. Gold. Chronic cocaine abuse: evidence for adverse effects on health and functioning. *Psychiatry Annals* **14**: 733 (1986).

F. Weiss, Y.L. Hurd, G.F. Koob, N.-E. Anden, and U. Ungerstedt. *In vivo* microdialysis study of DA function in the rat nucleus accumbens during acute and repeated cocaine self-administration. *Soc. Neurosci. Abstr.* **14**: 658 (1988).

R.P. Yasuda, N.R. Zahniser, and T.V. Dunwiddie. Electrophysiological effects of cocaine in the rat hippocampus *in vitro*. *Neurosci. Lett.* **45**: 199 (1984).

S.Y. Yeh. Effects of repeated cocaine administration on brain monoamines in rats. *Fed. Proc* **46**: 404 (1987).

S.-J. Yi and K.M. Johnson. Effects of chronic cocaine treatment on the uptake and release of striatal dopamine. *Soc. Neurosci. Abstr.* **14**: 962 (1988).

N.R. Zahniser and M.L. Dubocovich. Comparison of dopamine receptor sites labeled by [^{3}H]-S-sulpiride and [^{3}H]-spiperone in striatum. *J. Pharmacol. Exp. Ther.* **227**: 592 (1983).

N.R. Zahniser, J. Peris, P. Curella, and L.P. Dwoskin. Delayed increase in amphetamine-induced release of dopamine from striatum following repeated cocaine administration. *Soc. Neurosci. Abstr.* **14**: 659 (1988a).

N.R. Zahniser, J. Peris, P. Curella, L.P. Dwoskin, L. O'Keefe, and S.J. Boyson. Repeated cocaine administration results in supersensitive nigrostriatal D-2 dopamine autoreceptors, in: *Pharmacology and Functional Regulation of Dopaminergic Neurons*. P.M. Beart, G. Woodruff, and D.M. Jackson, eds., Macmillan, London, p. 168 (1988b).

N.R. Zahniser, J. Peris, L.P. Dwoskin, P. Curella, R.P. Yasuda, L. O'Keefe, and S.J. Boyson. Sensitization to cocaine in the nigrostriatal dopamine system. *NIDA Research Monograph 88: Mechanisms of Cocaine Abuse and Toxicity*. D. Clouet, K. Asghar, and R. Brown, eds., U.S. Government Printing Office, Washington, DC, p. 55 (1988c).

N.R. Zahniser, R.P. Yasuda, L.P. Dwoskin, K. Philpott, and T.V. Dunwiddie. Repeated cocaine administration: comparison between central dopaminergic and noradrenergic systems. *Soc. Neurosci. Abstr.* **12**: 142 (1986).

12

Electrophysiological Effects of Cocaine in the Mesoaccumbens Dopamine System

FRANCIS J. WHITE, DOUGLAS J. HENRY, XIU-TI HU, MICHAEL JEZIORSKI AND JANICE M. ACKERMAN

Department of Psychiatry
Cellular and Clinical Neurobiology Program
Wayne State University School of Medicine
Neuropsychopharmacology Laboratory
Lafayette Clinic
Detroit, Michigan

INTRODUCTION

The initial finding that rats increase the frequency of behaviors that are immediately followed by (rewarded with) electrical stimulation of circumscribed brain sites led to the concept of ''pleasure centers'' within the brain which mediate the reinforcing effects of stimuli (Olds and Milner, 1954). Importantly, Olds (1956) hypothesized that such ''pleasure centers'' are activated by both natural (food, water, etc.) and artificial (electrical stimulation, drugs) reinforcers. In the decades since Olds' initial work on brain reward, it has become apparent that stimulation of many brain sites may support operant responding. Accordingly, current formulations of brain reward theory emphasize a multisynaptic reward circuit model (e.g., see Wise, 1980; Wise and Bozarth, 1984). This model has been invoked to explain not only the results of self-stimulation experiments, but also the rewarding nature of cocaine and other drugs of abuse. Consequently, investigations of the physiological effects of cocaine relevant to its rewarding properties have closely followed research into the properties of other positively reinforcing stimuli, most notably electrical self-stimulation.

Early studies identified a possible role for catecholamine in brain reward circuitry, based primarily upon the close correspondence between self-stimulation sites and the anatomical distribution of catecholamine neurons and terminals (Crow, 1972;

German and Bowden, 1974). With continued experimentation, it became apparent that such sites correlated more precisely with the known distribution of dopamine (DA) cells than with norepinephrine (NE) systems, leading to hypotheses emphasizing the role of DA in reward processes (Fibiger, 1978; Wise, 1987; Wise and Rompre, 1989).

As with self-stimulation, there is now considerable evidence that the rewarding properties of many drugs of abuse, most notably the psychomotor stimulants cocaine and amphetamine and the opioids, also involve central DA-containing neuronal systems (see Chapter 5; Wise, 1987). The first support for this notion was provided by the finding that DA receptor antagonists, but not adrenoceptor antagonists, attenuate the rewarding effects of psychomotor stimulants (de Wit and Wise, 1977; Ettenberg *et al.*, 1982; Pettit *et al.*, 1984) and heroin (Bozarth and Wise, 1981; Spyraki *et al.*, 1983; however, also see Ettenberg *et al.*, 1982). In addition, direct-acting DA agonists such as apomorphine and bromocriptine also possess rewarding properties (Baxter *et al.*, 1974; Woolverton *et al.*, 1984; Yokel and Wise, 1978). Finally, selective lesions of DA neurons with the catecholamine neurotoxin 6-hydroxydopamine (6-OHDA) disrupt cocaine, amphetamine, and heroin self-administration (Lyness *et al.*, 1979; Phillips *et al.*, 1983; Roberts *et al.*, 1980; Roberts and Koob, 1982; Pettit *et al.*, 1984; Spyraki *et al.*, 1983), whereas lesions of the ascending noradrenergic fiber projections do not (Roberts *et al.*, 1977).

More specifically, it is now generally agreed that the mesoaccumbens DA system, projecting from A10 DA neurons within the ventral tegmental area (VTA) to the nucleus accumbens (NAc), is the DA system primarily involved in psychomotor stimulant and opiate reward processes. First, self-administration of cocaine is antagonized by infusion of DA antagonists into the NAc, but not the caudate-putamen (Phillips *et al.*, 1983), the primary recipient of DA afferents from the substantia nigra. Second, 6-OHDA lesions of DA terminals within the NAc attenuate the rewarding effects of D-amphetamine and cocaine (Lyness *et al.*, 1979; Roberts *et al.*, 1977, 1980; Pettit *et al.*, 1984); similar effects on opiate reward have been reported by some investigators (Phillips *et al.*, 1983; Spyraki *et al.*, 1983), but not by others (Pettit *et al.*, 1984). Third, kainic acid lesions of intrinsic neurons in the NAc will block the rewarding effects of cocaine, apomorphine, and heroin (Zito *et al.*, 1985). Fourth, rats will work for direct injections of amphetamine into the NAc (Hoebel *et al.*, 1983), as well as for heroin or opioid peptides into the VTA (Bozarth and Wise, 1981).

A role has also been suggested for the mesocortical DA system, in particular the VTA-medial prefrontal cortex (mPFC) projection, in the rewarding actions of cocaine. Direct infusions of cocaine into the mPFC will reinforce self-administration behavior in rats, while cocaine infusions into the NAc or VTA do not support self-administration (Goeders and Smith, 1983). The self-administration of cocaine into the mPFC is blocked by co-infusion of the selective D_2 DA antagonist sulpiride (Goeders *et al.*, 1986). In addition, 6-OHDA lesions of DA terminals in the mPFC reduce direct self-administration of cocaine, but not DA, into the cortex (Goeders and Smith, 1986), although DA terminal loss in the mPFC does not block systemic self-administration of cocaine (Martin-Iverson *et al.*, 1986).

The apparently contrasting evidence regarding the site at which cocaine exerts its positively reinforcing effects has led some researchers to propose differing roles for the NAc and the mPFC in the rewarding properties of cocaine. For example, the inability of rats to self-administer cocaine into the NAc suggests that the mPFC may be involved in initiation of the reinforcing effect of cocaine, whereas the NAc may be indirectly activated within the reward pathway and play an integral role in maintaining self-administration behavior (Goeders *et al.*, 1986).

Given the apparent involvement of the mesoaccumbens and mesocortical DA systems in reward processes, we have recently begun to study the effects of cocaine and other drugs of abuse on single cells in the VTA and in its terminal fields within the NAc and mPFC. This chapter will review our recent findings regarding the effects of acute and repeated cocaine administration on single neurons within the rat mesoaccumbens DA system.

EFFECTS OF ACUTE COCAINE ADMINISTRATION ON MESOACCUMBENS DA NEUROTRANSMISSION

A10 DA Neurons in the VTA

Effects of Intravenous Cocaine on Mesoaccumbens DA Neurons

Our first experiments sought to determine the effects of i.v. administration of cocaine on antidromically identified mesoaccumbens A10 DA neurons. In these and all other studies, chloral hydrate (400 mg/kg, i.p.) anesthetized rats were used as subjects. Rats were fixed in a stereotaxic apparatus and the scalp was incised and retracted such that burr holes could be drilled in the skull overlying the VTA and NAc, according to the atlas of Paxinos and Watson (1982). Stimulating electrodes were used to activate neurons antidromically and to demonstrate collision of antidromic impulses with spontaneous action potentials, thereby identifying individuals VTA A10 DA cells which project to the NAc. Recording electrodes were single-barrel glass micropipettes broken back to a tip diameter of 1 μm and filled with a 2 *M* NaCl solution to obtain an impedance of 2–4 MΩ. Single A10 DA neurons were identified using established electrophysiological criteria (Bunney *et al.*, 1973; Grace and Bunney, 1983; Wang, 1981a). Cocaine and other drugs were administered via a catheterized tail vein on a dose regimen in which each dose doubled the immediately preceding dose, at 60 to 90-s intervals. Details of these procedures have been previously published (Einhorn *et al.*, 1988; Henry *et al.*, 1989).

It is now well established that DA and exogenous DA agonists can inhibit the firing of midbrain DA neurons (Aghajanian and Bunney, 1977; Bunney *et al.*, 1973) by stimulating receptors for DA located on the DA cell itself, i.e., autoreceptors (Carlsson, 1975). These impulse-regulating DA autoreceptors are thought to exist along the somatodendritic extent of the neuron and to regulate membrane excitability by controlling a K^+ conductance (Lacey *et al.*, 1987), perhaps via an inhibitory guanine nucleotide binding protein such as G_i (Innis and Aghajanian, 1987). Somatodendritic DA autoreceptors exhibit the pharmacological characteristics of the

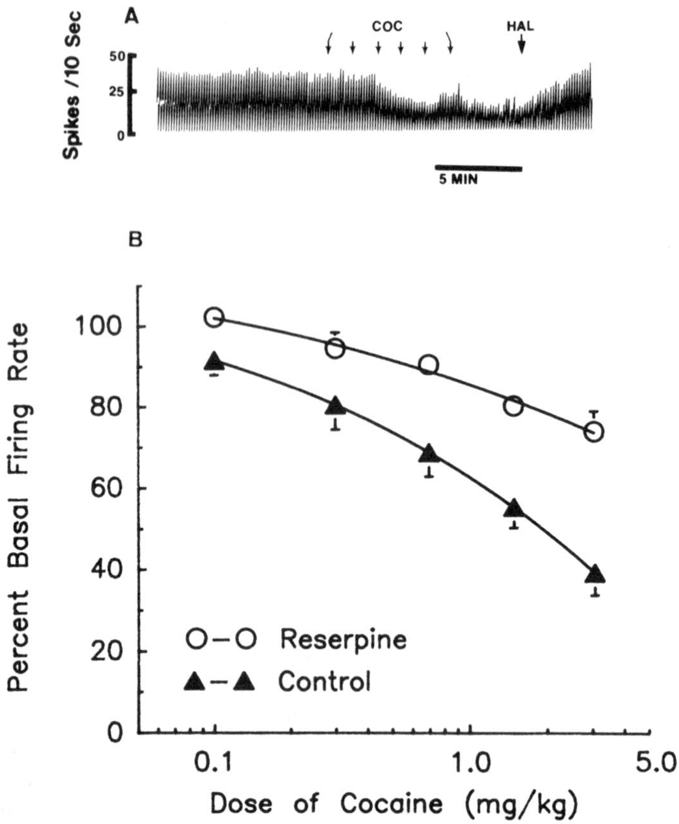

Figure 1. Inhibitory effects of i.v. cocaine on A10 DA neurons are attenuated by reserpine pretreatment. A: The typical partial inhibitory effect of cocaine (COC, 0.1, 0.2, 0.4, m/kg, etc., at arrows) on an antidromically identified mesoaccumbens A10 DA neuron and the reversal of that effect by the DA antagonist haloperidol (HAL, 0.2 mg/kg). B: Intravenous injection of cocaine produced a dose-dependent decrease in the firing of antidromically identified mesoaccumbens DA neurons in the rat ventral tegmental area (VTA). Note that the inhibition was partial, typically 60 to 70% below basal firing rate. In rats pretreated with reserpine (n = 9), the partial inhibitory effects of cocaine were significantly attenuated. Data points represent means; bars represent SEM. The data points for the reserpine-treated group are all significantly different from the control groups except at the first dose of cocaine ($p < 0.01$). (Adapted from L.C. Einhorn, P.A. Johansen, and F.J. White. *J. Neurosci.* **8**: 100 (1988). With permission.)

D_2 DA receptor (Lacey *et al.*, 1987; White and Wang, 1984b; see Clark and White, 1987 for review of DA receptor subtypes).

In our initial experiments, we found that, like DA agonists, systemic administration of cocaine caused a dose-dependent inhibition of the electrical activity of antidromically identified mesoaccumbens A10 DA neurons (Einhorn *et al.*, 1988). However, unlike most other DA agonists, the inhibition produced by maximal (sublethal) doses of cocaine was partial in that cell firing rates were seldom suppressed below 30–40% of the basal firing rate (Figure 1). Of the 12 mesoaccumbens A10 DA cells tested, only 2 were completely inhibited by cocaine. The DA antagonist haloperidol always reversed cocaine-induced suppression (Figure 1). Single bolus injections of cocaine (either 0.5 or 1.0 mg/kg, i.v.) also caused a partial inhibition of most A10 DA neurons (one cell was excited by cocaine) which was followed by a gradual recovery during the 10- to 23-min post-injection period (Einhorn *et al.*, 1988). Thus, the inhibitory effects of cocaine on mesoaccumbens DA neurons are most pronounced immediately following its administration and recover gradually over the succeeding 10–20 min. This time course is consistent with the known subjective effects of cocaine in humans which are maximal immediately following ingestion and which are relatively short lasting, typically on the order of 15–30 min (Fischman *et al.*, 1983).

The second aim of our study was to perform a pharmacological analysis of the mechanisms by which cocaine inhibited the activity of A10 DA cells. Because it is generally recognized that cocaine inhibits the uptake of not only DA, but also serotonin (or 5-hydroxytryptamine, 5-HT) and NE (Ross and Renyi, 1967a,b; Snyder and Coyle, 1969), and exerts local anesthetic actions (Carney, 1955), a series of experiments was conducted to determine the mechanism(s) responsible for the effects of i.v. cocaine on A10 DA neurons. Neither the selective 5-HT uptake inhibitor fluoxetine (Wong *et al.*, 1974) nor the selective noradrenergic uptake inhibitor desmethylimipramine altered the activity of A10 DA neurons (Einhorn *et al.*, 1988). In contrast, other agents which are known to inhibit DA uptake, including the cocaine metabolite norcocaine, the catecholamine uptake inhibitor nomifensine, and the selective DA uptake inhibitor GBR 12909 (Van der Zee *et al.*, 1980), suppressed the activity of these neurons with a potency and efficacy similar to that of cocaine. Procaine, which is a local anesthetic agent lacking significant psychomotor stimulant properties, caused only a slight nonsignificant increase in the firing of A10 DA cells (Einhorn *et al.*, 1988). Taken together, these findings suggest that the electrophysiological effects of cocaine on A10 DA neurons are not related to its local anesthetic actions nor to its ability to inhibit the uptake of NE or 5-HT, but probably reflect its ability to inhibit DA re-uptake via the DA transporter (Ritz *et al.*, 1987).

Additional evidence for this action of cocaine on A10 DA neurons was obtained by reserpine pretreatment studies. Since reserpine is a monoamine-depleting agent which reduces vesicular stores of DA (Carlsson *et al.*, 1957), it should attenuate effects of cocaine if they are due to potentiation of autoreceptor stimulation by endogenous DA. Indeed, reserpine pretreatment significantly attenuated the effects of i.v. cocaine (Figure 1B). That a slight inhibition could still be observed in reserpinized rats might be attributed to release of newly synthesized DA and/or incomplete depletion of vesicular stores. With respect to dendritic release of DA (Chéramy *et al.*, 1981; Korf *et al.*, 1976), it is interesting to note that DA may be stored in vesicles only within proximal dendrites (Groves and Linder, 1983), given that there is little evidence for storage vesicles within distal DA dendrites (Wassef *et al.*, 1981). Therefore, it seems likely that both vesicular and nonvesicular processes may contribute to the release of dendritic DA (Chéramy *et al.*, 1981).

Effects of Iontophoretic Cocaine on A10 DA Neurons

To test whether cocaine inhibited A10 DA neurons by blocking DA transport

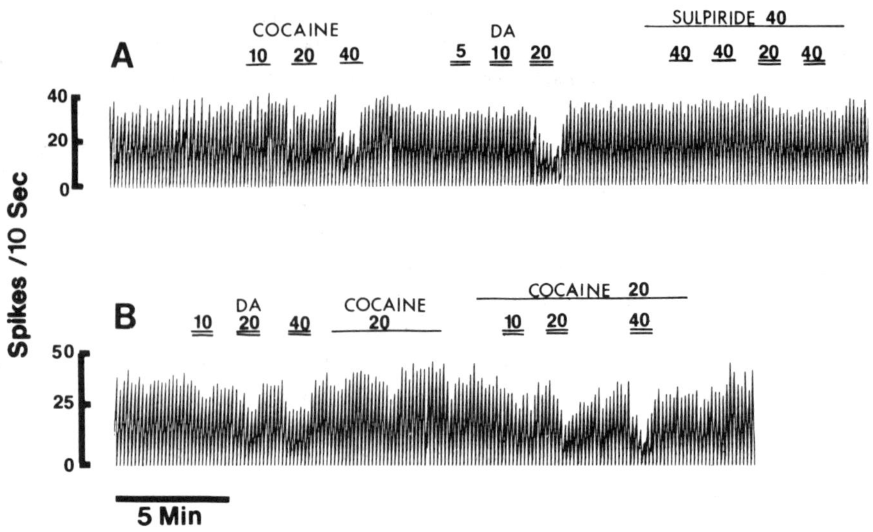

Figure 2. Cumulative rate histograms illustrating the effects of iontophoretic cocaine and DA on A10 DA neurons within the rat VTA. A: The inhibitory effects of cocaine and DA on this A10 DA neuron were almost completely blocked by administration of the D_2 DA receptor antagonist, L-sulpiride. B: Cocaine (20 nA) only slightly inhibited the firing of this neuron, but increased both the amount and the duration of inhibition caused by DA. Lines and numbers represent the duration of iontophoretic current and the amount of current in nanoamperes. (From L.C. Einhorn, P.A. Johansen, and F.J. White. *J. Neurosci.* **8**: 100 (1988). With permission.)

along the somatodendritic extent of the cells, thereby potentiating the stimulation of autoreceptors by extracellular DA, cocaine was administered directly onto A10 DA neurons via microiontophoresis. In these experiments, five-barrel micropipettes were used to record from DA neurons while cocaine (and other drugs) were locally ejected onto the neuron (see Einhorn *et al.*, 1988 for experimental details). Iontophoretic cocaine maximally produced only a 20% inhibition of firing of A10 DA neurons at the currents tested (5–40 nA). Apparent local anesthetic effects of cocaine (diminished amplitude and flattened waveform of individual spikes) were often observed at currents above 40 nA. However, the inhibitory effects of cocaine observed at currents of 20–40 nA appeared to be mediated by DA autoreceptors because they were completely blocked by concurrent iontophoretic administration of the D_2 DA receptor antagonist (−)sulpiride (Figure 2). Simultaneous iontophoretic administration of cocaine and DA produced a significant enhancement of the ability of DA to inhibit A10 DA cells, as indicated by increases in both the efficacy and duration of DA-induced inhibition (Figures 2 and 3).

To determine whether cocaine could potentiate the effects of endogenously released DA (as opposed to iontophoretically applied DA) on DA autoreceptors, we used electrical stimulation of the NAc because it has been demonstrated that such antidromic invasion of DA neurons can enhance dendritic release of DA (Korf *et al.*, 1976; Wang, 1981b). The peristimulus time histogram in Figure 4 illustrates the inhibition of an antidromically identified mesoaccumbens DA neuron following electrical stimulation of the NAc (2.0 mA for 0.5 ms). During iontophoretic ap-

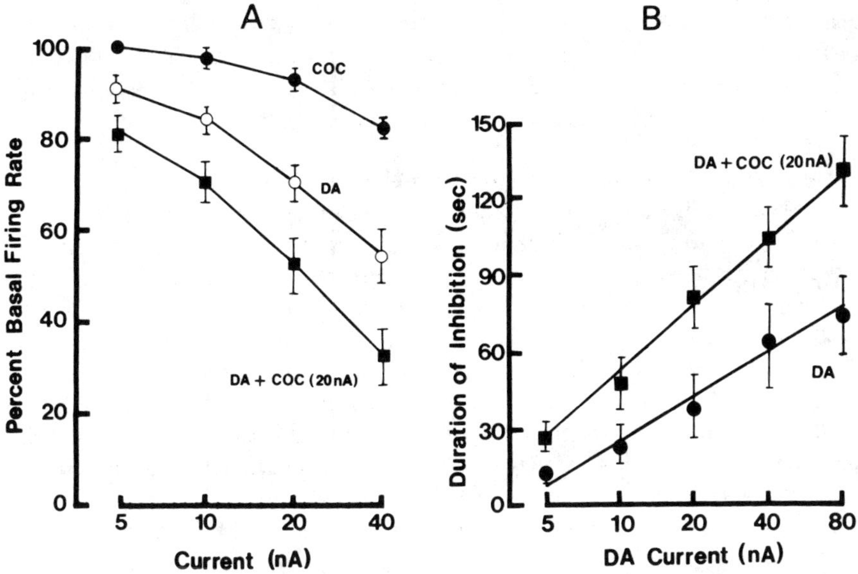

Figure 3. The effects of iontophoretically administered cocaine (0.01 *M*) on the activity of A10 dopamine (DA) neurons within the rat VTA. A: Current-response curves for the inhibitory effects of cocaine (COC), DA (0.01 *M*) and DA + COC (20 nA). Filled circles indicate the weak inhibitory effects of increasing iontophoretic currents applied through the COC barrel (n = 12 cells in 7 rats). Open circles indicate the inhibitory effects of DA alone (n = 10 cells in 5 rats), whereas filled squares indicate the effects of co-administration of DA and a 20-nA current of COC on those same 10 cells. B: COC also significantly increased the duration of the inhibitory effect of DA on those same 10 cells shown in A. The 80-nA current includes only four of the ten cells. For these determinations, the duration of inhibition was defined as the time required for the cell to recover to ≥85% of its basal rate for three consecutive 10-s epochs. Data points represent means; bars indicate SEM. (From L.C. Einhorn, P.A. Johansen, and F.J. White. *J. Neurosci.* **8**: 100 (1988). With permission.)

plication of cocaine at a 20-nA current, the duration of inhibition produced by NAc stimulation was significantly prolonged. Averaged data obtained from eight mesoaccumbens DA cells are shown in Figure 4C. Note that cocaine (20 nA) caused a significant prolongation of this stimulation-induced inhibition (Figure 4C), suggesting that iontophoretic application of cocaine inhibits A10 neuronal activity by blocking the DA transporter along the somatodendritic extent of these neurons (Beart and McDonald, 1980; Björklund and Lindvall, 1975). This blockade would increase extracellular levels of DA in the VTA, as recently shown by microdialysis techniques (Bradberry and Roth, 1989), and prolong the inhibitory effects of DA at somatodendritic impulse-regulating autoreceptors. Taken together, the results obtained with iontophoretic administration suggest that cocaine partially inhibits A10 DA neuronal activity by increasing the level of somatodendritic DA autoreceptor stimulation by endogenous DA.

Effects of Cocaine on NAc Neurons

Background

In view of the extensive literature implicating the NAc in the rewarding effects

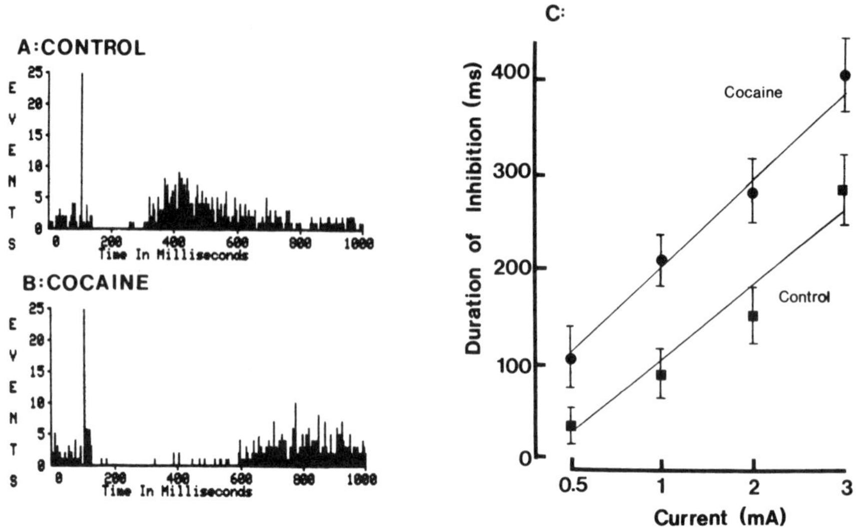

Figure 4. The ability of iontophoretically administered cocaine (0.01 *M*, 20 nA) to enhance the inhibition of firing of antidromically identified mesoaccumbens A10 DA neurons caused by electrical stimulation of the nucleus accumbens (NAc). A: Peristimulus histogram (PSTH) illustrating the inhibition of firing of a mesoaccumbens DA neuron during NAc stimulation (2.0 mA for 0.5 ms). B: PSTH obtained from the same cell shown in A during iontophoretic administration of cocaine (20 nA). Each PSTH was obtained from 25 sweeps initiated 100 ms before the onset of the electrical stimulation (bin width = 4 ms). Shock artifacts are evident at 100 ms. C: Averaged data obtained from six cells tested in the manner shown in A and B. The difference between the control and cocaine conditions was significant at all current intensities ($p < 0.01$). (From L.C. Einhorn, P.A. Johansen, and F.J. White. *J. Neurosci.* **8**: 100 (1988). With permission.)

of cocaine and many other drugs of abuse, we have also studied the effects of cocaine on single unit activity within the rat NAc. The NAc is a heterogeneous structure with respect to its cytoarchitecture and its afferent and efferent connections (see Domesick, 1981 for review). In addition, the NAc possesses a variety of neuronal subtypes as determined by extracellular recording techniques. We have recently identified two subtypes of NAc neurons, based upon their extracellular waveform (White *et al.*, 1987), and have labeled them as type I and II NAc neurons in accordance with previous studies which distinguished subtypes of neurons within the rat caudate-putamen (Skirboll and Bunney, 1979; Nisenbaum *et al.*, 1988). The majority of NAc neurons are type I cells, which exhibit a negative/positive waveform and are seldom spontaneously active in the cloral hydrate anesthetized preparation; the remaining type II neurons exhibit positive/negative waveforms and are more often spontaneously active. In order to study both of these neuronal populations during iontophoretic studies, we have used the excitatory amino acid glutamate to "drive" quiescent neurons. Glutamate-driven units are then studied by determining the ability of cocaine and other agents to influence the excitatory effects of the amino acid.

Intravenous Administration of Cocaine

In our first set of experiments, we determined the effects of cumulative i.v. doses

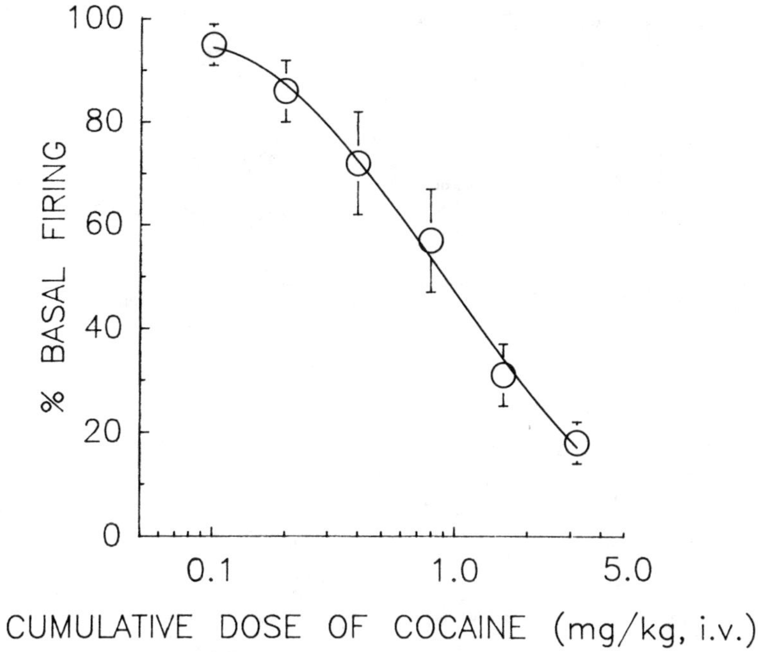

Figure 5. Inhibition of spontaneously active nucleus accumbens (NAc) neurons by i.v. administration of cocaine. Cocaine was administered as described in Figure 1 caption.

of cocaine on the firing of spontaneously active NAc cells. Cocaine caused a dose-dependent suppression of firing of NAc neurons at sublethal doses (Figure 5). Although the potency of cocaine at inhibiting NAc neurons was similar to that for A10 DA cells, cocaine was considerably more efficacious on NAc neurons as compared to A10 DA cells (compare Figures 1 and 5). The inhibitory effects of cocaine on NAc neuronal activity were readily reversed by both the nonselective DA antagonist haloperidol (n = 4) and the selective D_1 DA antagonist SCH 23390 (n = 3). Interestingly, the D_1 DA receptor antagonist was typically more potent. This finding fits well with the report of Koob *et al.* (1987) indicating that SCH 23390 was a more effective antagonist of cocaine self-administration than the D_2 antagonist spiperone. Moreover, this finding is also in congruence with the established enabling role of D_1 receptors for D_2 receptor-mediated behavioral and electrophysiological effects (see Clark and White, 1987 for review).

Iontophoretic Administration of Cocaine

Iontophoretic administration of cocaine was observed to inhibit the firing of both type I and II neurons within the rat NAc, whether they were spontaneously active or quiescent. Interestingly, cocaine was somewhat more potent at inhibiting type I as compared to type II neurons (Figure 6A). This differential effect on type I and II NAc neurons contrasts to that obtained with iontophoretic DA, which was equipotent at inhibiting the two neuronal subtypes (Figure 6B). Cocaine was also ob-

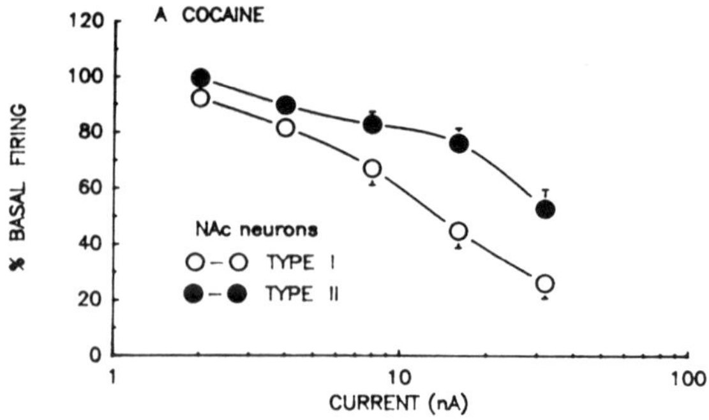

A

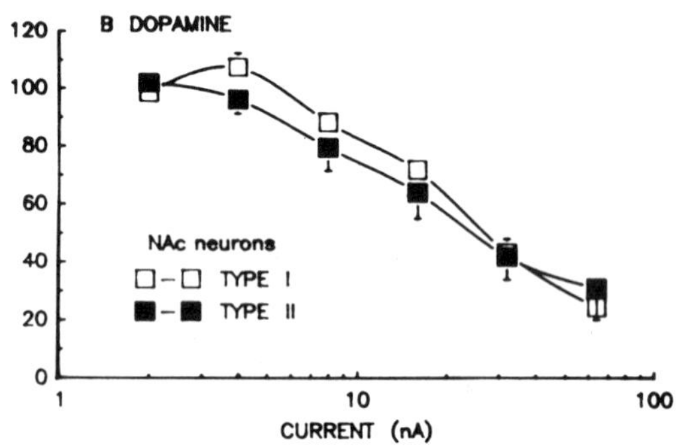

B

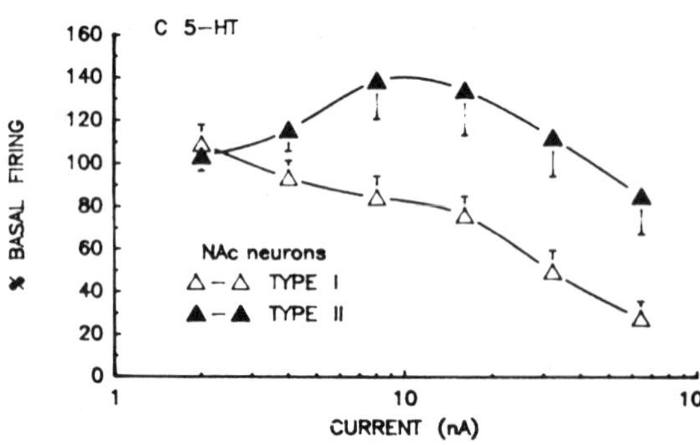

C

Figure 6. Comparison of the effects of cocaine (A), DA (B), and 5-HT (C) on the activity of type I and type II nucleus accumbens (NAc) neurons. These current-response curves demonstrate that cocaine (0.01 *M*) was more potent at inhibiting type I than II NAc cells, whereas DA (0.01 *M*) was equally potent on the two subtypes. Serotonin (5-HT) inhibited type I neurons, but excited most type II neurons.

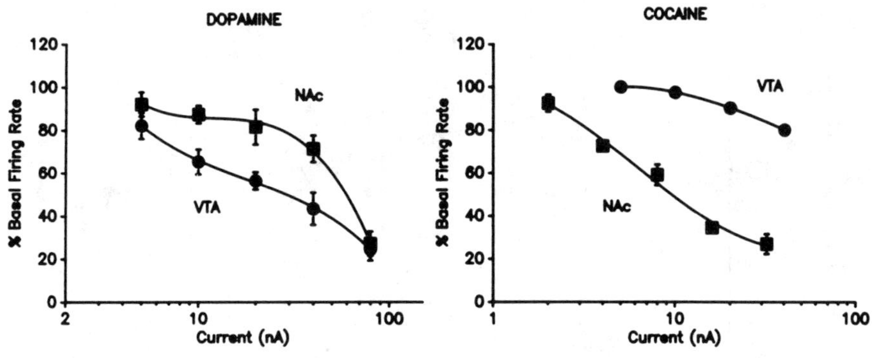

A B

Figure 7. Comparison of the effects of both DA (A) and cocaine (B), administered iontophoretically, on nucleus accumbens (NAc) neurons and A10 DA neurons in the rat ventral tegmental area (VTA). Both type I and type II neurons are included in the NAc data. Note that while DA is more potent on A10 DA neurons, cocaine is more potent on NAc neurons.

served to potentiate the inhibitory effects of DA on both type I and II neurons. On many cells, the effects of both DA and cocaine were readily blocked by (−)sulpiride. In addition, the D_1 selective DA antagonist SCH 23390 was also able to reduce the inhibitory effect of iontophoretic DA (White and Wang, 1986) and cocaine (unpublished findings) on NAc neurons. Interestingly, cocaine was more effective than DA on type I NAc cells (compare Figures 6A and B).

One of the most surprising findings obtained during these studies was that iontophoretic cocaine was significantly more effective at inhibiting NAc neurons than A10 DA neurons (Figure 7A). Previously, we (White and Wang, 1986) had demonstrated that D_2 autoreceptors were more sensitive to DA than were NAc D_2 receptors (Figure 7B). This surprising result might be explained by the fact that far greater concentrations of DA are released by nerve terminals than by DA dendrites (Kalivas *et al.*, 1989); therefore, inhibiting the re-uptake of synaptic DA within the NAc would have greater functional consequences than inhibiting the re-uptake of DA within the somatodendritic area of the VTA. Recent *in vivo* microdialysis studies support this hypothesis since extracellular DA concentrations in the NAc are increased by about 300% following i.v. cocaine administration, whereas extracellular levels of DA within the VTA increase by 180% (Bradberry and Roth, 1989). It is also possible that the uptake process in DA terminals and dendrites may differ. We have recently obtained preliminary evidence that cocaine is less effective as an inhibitor of [^{3}H]DA uptake into synaptosomes prepared from rat VTA as compared to NAc synaptosome (Figure 8). In fact, a considerable portion of DA uptake in the VTA was insensitive to cocaine, whereas 90% of DA uptake into the NAc was prevented by cocaine. These relative efficacies of cocaine at inhibiting NAc vs. VTA DA uptake parallel the contrasting electrophysiological effects which we observed in our iontophoretic experiments (compare Figures 7 and 8).

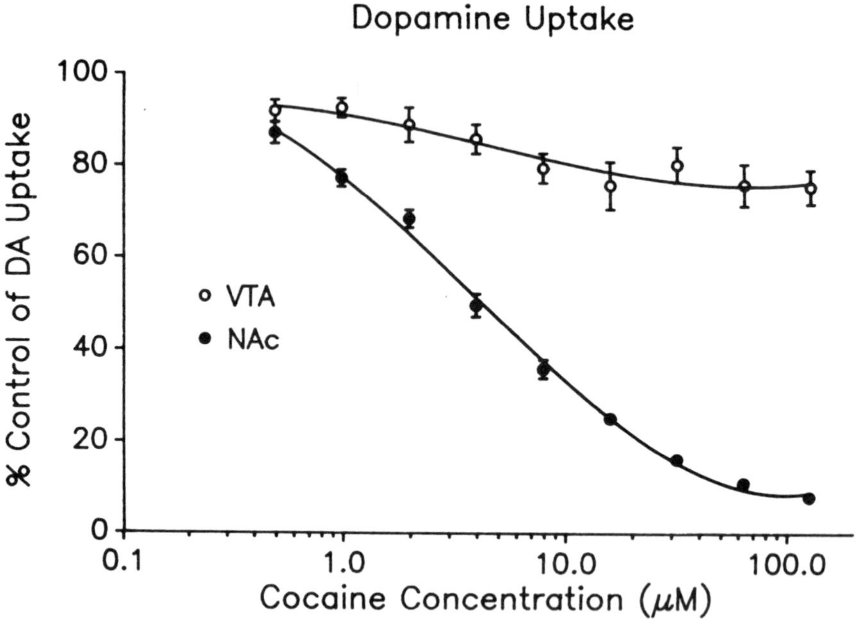

Figure 8. Comparison of the ability of cocaine to inhibit [^{3}H]DA (100 n*M*) into synaptosomal preparations (50 μl aliquots) of the rat nucleus accumbens (NAc) and ventral tegmental area (VTA). Tissue was typically pooled from four rats for the NAc preparations, and from seven rats for the VTA preparations. Tissues were incubated for 5 min at 37°C using nine concentrations of cocaine (0.48–125 μ*M*). Note the similarity of these curves to those obtained for the inhibition of VTA and NAc neurons during iontophoretic administration of cocaine (Figure 7).

Role for Both DA and 5-HT

Although our preliminary findings regarding DA uptake may account for differences between the relative efficacy of cocaine at inhibiting NAc vs. VTA neurons, they would not explain why type I NAc neurons were more sensitive to cocaine than to DA. Therefore, we hypothesized that cocaine might produce an effect on NAc neurons via mechanisms additional to that of DA re-uptake interference. In light of our previous study (White, 1986) indicating that most type I NAc neurons are inhibited by locally applied 5-HT as well as DA, whereas many type II neurons are typically excited by iontophoretic 5-HT (Figure 7C), and the fact that cocaine blocks the transporter for each of these amines, it seemed likely that the potent inhibitory effects of cocaine on type I NAc neurons may have been due to simultaneous potentiation of both DA- and 5-HT-induced inhibition. On the other hand, the weaker inhibition of type II neurons by cocaine (as compared to DA) may have resulted from competing inhibitory and excitatory influences of DA and 5-HT, respectively.

To test our hypothesis, we compared the effects of iontophoretically applied cocaine on NAc neurons to those of the selective 5-HT uptake blocker fluoxetine and the selective DA uptake blocker GBR 12909. When tested on type I NAc cells,

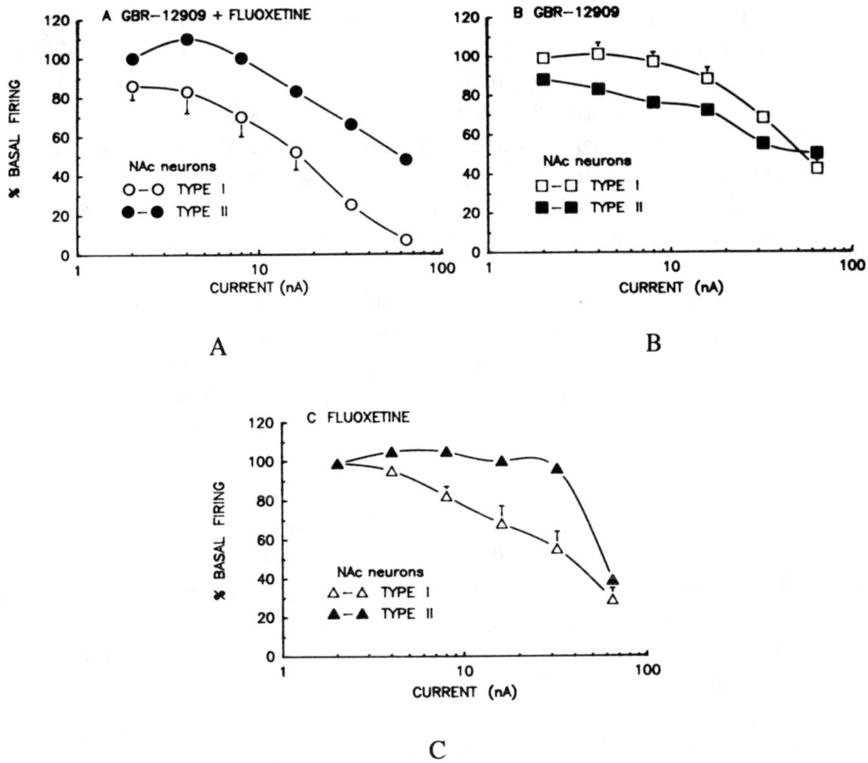

Figure 9. Comparison of the effects of GBR 12909 (B), fluoxetine (C), both at 0.01 *M*, and the combination of these agents (A) on the activity of type I and II nucleus accumbens (NAc) neurons. Sample sizes range from 9–12 cells per group, except for the type II cells tested with fluoxetine (n = 3). Because of this small sample size, error bars are not indicated.

each of these agents inhibited both spontaneous and glutamate-induced activity (Figure 9B and C). Moreover, when these two agents were co-administered, they produced a degree of inhibition which was quite similar to that produced by cocaine (Figure 9A). In contrast, fluoxetine (at low currents) tended to increase (or failed to affect) the activity of type II neurons, whereas GBR 12909 produced an inhibition (compare Figures 9B and C). Note that at higher currents, fluoxetine also inhibited firing. Therefore, these results support our hypothesis that the potent inhibitory effects of iontophoretic cocaine on type I NAc neurons are due to its ability to block the re-uptake of both DA and 5-HT, resulting in a simultaneous potentiation of both 5-HT and DA receptor-mediated inhibition.

Cocaine and the Role of the NAc-VTA Feedback Pathways

Background

Another of the interesting results obtained in our studies of cocaine on A10 DA cells was that systemic cocaine was more effective than iontophoretic cocaine at inhibiting DA cell firing. This finding cannot be explained by ineffective iontophoretic ejection of cocaine from the micropipettes since marked inhibition of NAc neurons by cocaine was observed under identical experimental conditions. Taken together with the finding that NAc cells were more sensitive to cocaine than A10 DA cells, we reasoned that the difference between the relative efficacies of iontophoretic and i.v. cocaine on A10 DA neurons may have been due to other factors and hypothesized that, in addition to DA autoreceptor activation, i.v. cocaine might also influence DA neurons via other mechanisms, in particular a long-loop inhibitory NAc-VTA feedback pathway.

Although considerably less extensive than the striatonigral pathway, a NAc-VTA GABAergic (inhibitory) feedback pathway has been indicated by a variety of anatomical (Nauta *et al.*, 1978; Phillipson, 1979), biochemical (Waddington and Cross, 1978; Walaas and Fonnum, 1980), and electrophysiological (German *et al.*, 1980; White and Wang, 1983a,b; Wolf *et al.*, 1978; Yim and Mogenson, 1980) studies. The latter two reports described inhibitory effects of electrical stimulation of the NAc on VTA neurons (both DA and unidentified non-DA cells) which are enhanced by iontophoretic application of the GABA re-uptake inhibitor nipecotic acid into the VTA and blocked by similar iontophoretic administration of the GABA antagonist picrotoxin (Wolf *et al.*, 1978; Yim and Mogenson, 1980). Biochemical studies have revealed that lesions of the NAc decrease levels of the GABA-synthesizing enzyme glutamic acid decarboxylase in the VTA (Waddington and Cross, 1978; Walaas and Fonnum, 1980).

Involvement of NAc-VTA Feedback Pathways in the Inhibition of Mesoaccumbens DA Neurons by Intravenous Cocaine

To determine whether a NAc-VTA feedback pathway might be involved in mediating the effects of i.v. cocaine on A10 DA neurons, we decided to eliminate such pathways in two different manners: (1) by lesioning cell bodies within the NAc with the excitotoxin ibotenic acid and (2) by cutting the pathway with acute hemitransections. As shown in Figure 10, each of these lesions significantly attenuated the inhibitory effects of i.v. cocaine on A10 DA neurons as compared to sham-operated animals, supporting a role for NAc-VTA pathways in the effects of cocaine on A10 DA neurons (Einhorn *et al.*, 1988). Specifically, these findings indicate that by blocking DA and 5-HT uptake mechanisms at nerve terminals within the NAc, cocaine increased synaptic concentrations of these transmitters and effectively inhibited NAc neurons. Because it is believed that the NAc-VTA pathway is primarily an inhibitory, probably GABAergic, projection, it is necessary to postulate that, like the striatonigral feedback pathway (Grace and Bunney, 1985), the primary target of NAc projections within the VTA would be an inhibitory neuron which makes synaptic contact with the A10 DA cells. This could be either an interneuron or an output neuron with collateral connections with A10 DA cells. Previous pharmacological evidence suggests the existence of such (inter)neurons, probably themselves GABAergic, within the VTA (Gysling and Wang, 1983; O'Brien and White, 1987; Waszczak and Walters, 1980). Thus, the net effect of inhibiting NAc neurons would be a compensatory inhibition of A10 DA neurons through a

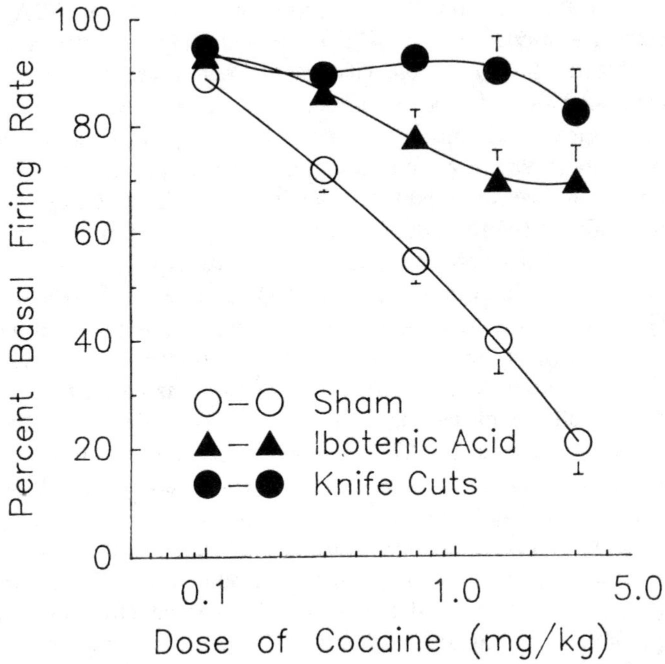

Figure 10. The effects of removing possible forebrain feedback pathways on the inhibitory potency of cocaine on A10 DA neurons within the rat ventral tegmental area (VTA). Both ibotenic acid-induced lesions of the NAc 1 week prior to recording ($n = 8$) and acute hemitransections of the brain rostral to the VTA ($n = 8$) significantly attenuated the inhibitory effects of i.v. cocaine on A10 DA neurons as compared to a sham-operated control group ($n = 9$). The data points for both of the lesion groups were significantly different from those of the sham group except at the 0.1-mg/kg dose of cocaine ($p < 0.01$). (Adapted from L.C. Einhorn, P.A. Johansen, and F.J. White. *J. Neurosci.* **8**: 100 (1988). With permission.)

GABA-GABA-DA feedback system, in which inhibition of the NAc GABA projection neurons would lead to a disinhibition of the VTA GABA (inter)neurons and thereby an inhibition of the VTA DA neuron.

If the presumed GABAergic NAc-VTA neurons terminate on both GABA (inter)neurons and A10 DA neurons, as has been suggested within the striatonigral pathway (Grace and Bunney, 1985), then the net effects of cocaine on a given A10 DA cell would result from the relative degree of innervation of that DA cell by the direct and indirect pathways. Such an arrangement could explain the occasional A10 DA cell which is excited by i.v. cocaine (Einhorn *et al.*, 1988), since excitation (disinhibition) would be predicted if certain A10 DA cells primarily receive direct NAc-VTA GABAergic inputs. Moreover, such a relationship might also help to explain the partial inhibition observed with cocaine on most A10 DA cells in that excitatory long-loop pathways may oppose inhibitory inputs on any given cell.

That cocaine-induced inhibition of A10 DA cell activity is attenuated by interruption of NAc-VTA feedback pathways is unique in that it represents the first report of an involvement of such a projection in the effects of a DA agonist on A10 DA cells. The only other report which indicated a role for this feedback pathway in regulating A10 DA neuronal activity was the demonstration that ibotenic acid lesions of the NAc prevented depolarization inactivation of A10 DA cells normally produced by repeated administration of the DA receptor antagonist haloperidol (White and Wang, 1983a). Several investigators have previously postulated that the mesolimbic (mesoaccumbens) DA system is under tighter autoregulatory control than the nigrostriatal (A9) DA system, whereas A9 DA neurons are influenced to a greater extent by long-loop feedback pathways than are A10 DA cells (e.g., see Andén *et al.*, 1983; Chiodo and Bunney, 1984; Wang, 1981c). One line of evidence for this hypothesis is that the inhibitory effects of the DA releasing agent D-amphetamine on A9 DA neurons are diminished by lesions placed along the striatonigral pathway (Bunney and Aghajanian, 1976, 1978), whereas such effects on A10 DA neurons are not altered by lesions of the NAc or by unilateral hemitransections (Wang, 1981c; White *et al.*, 1987). We have argued that the difference between amphetamine and cocaine with regard to the involvement of NAc-VTA feedback pathways relates to differences in their mode of action (Einhorn *et al.*, 1988). Enhancement of dendritic DA release by D-amphetamine would be expected to produce significantly greater extracellular concentrations of DA within the VTA (Kalivas *et al.*, 1989) than would blockade of DA re-uptake by cocaine (Bradberry and Roth, 1989). Given that the inhibitory effects of cocaine are much greater within the NAc (postsynaptic) than on A10 DA cells (presynaptic), we have hypothesized that the NAc-VTA feedback pathway is only observable (i.e., not redundant) when autoreceptor-induced inhibition of A10 DA neurons is insufficient to compensate for enhanced activation of postsynaptic DA receptors (Einhorn *et al.*, 1988; White *et al.*, 1987).

General Discussion

The results of our studies on the electrophysiological effects of cocaine have several implications for the hypothesis that enhanced activity within the mesoaccumbens DA system may be responsible for the rewarding effects of cocaine (and other drugs of abuse). First, although apparently contradictory, the decrease in mesoaccumbens A10 DA neuronal activity produced by cocaine is actually consistent with this hypothesis because this effect represents a compensatory decrease in impulse flow partially resulting from enhanced DA neurotransmission within the NAc. Other drugs of abuse have been shown to increase A10 unit activity, typically via nondopaminergic mechanisms. Among such drugs are the opioids morphine and heroin (Gysling and Wang, 1983; Matthews and German, 1984), ethanol (Gessa *et al.*, 1985), benzodiazepines (O'Brien and White, 1987), and nicotine (Mereu *et al.*, 1987). As we have previously argued (Einhorn *et al.*, 1988; White *et al.*, 1987), it is important to realize that the ability of all of these agents to increase A10 DA unit firing results *in* enhanced mesoaccumbens DA transmission, whereas the decrease in A10 DA neuronal impulse flow caused by DA agonists results *from* increased DA transmission leading to enhanced activation of DA receptors.

An important question remaining to be answered is why the increase in DA unit activity initially produced by other reinforcing drugs is not compensated by subsequent inhibition as a result of autoreceptor and/or NAc-VTA feedback inhibition. It is possible that autoregulatory mechanisms are not invoked because the drugs which stimulate DA transmission by working at non-DAergic receptors (opioid, benzodiazepine/GABA) in some way "override" this means of control, perhaps because dendritic release of DA occurs independently of impulse generation (Nissbrandt *et al.*, 1985) or, in some cases, may be inversely related to impulse flow (Chéramy *et al.*, 1981). Alternatively, it may be that the degree of excitation normally produced by these agents is, in fact, dampened by autoreceptor occupation by endogenous DA such that a preferential blockade of DA impulse-regulating DA autoreceptors might lead to even greater activation of the mesoaccumbens DA system by these various drugs. With respect to possible NAc-VTA feedback inhibition, it may be the case that this regulatory process is not activated following administration of the nondopaminergic agents because these drugs "short-circuit" long-loop control by directly affecting NAc neurons or the inhibitory (inter)neurons within the VTA (see further discussion below).

A final implication of our cocaine experiments relates to how the effects of cocaine on mesoaccumbens DA cells differ from those observed with other DA agonists. Because D_2 DA autoreceptors within the mesoaccumbens pathway are thought to be more sensitive than postsynaptic D_2 receptors within the NAc (White and Wang, 1986), DA agonists which possess D_2 receptor affinity not only inhibit NAc neurons by direct postsynaptic receptor stimulation, but also activate both somatodendritic autoreceptors which diminish DA impulse flow, and nerve terminal autoreceptors which reduce DA synthesis and release (Carlsson, 1975; Wolf and Roth, 1987). These compensatory mechanisms reduce DA activity within the mesoaccumbens system. In contrast to direct DA agonists, cocaine potently suppresses the firing of most NAc neurons, but only weakly activates somatodendritic autoreceptor mechanisms to diminish A10 DA neuronal impulse flow. In fact, during cocaine administration, additional long-loop (NAc-VTA) feedback processes are required to compensate, albeit still only partially, for enhanced postsynaptic inhibition. Therefore, the net effect of cocaine on the mesoaccumbens DA system would be an initial increase in synaptic levels of DA in the NAc (as a result of blocking the DA transporter), followed by continued impulse-dependent release of DA as a result of incomplete suppression of neuronal activity. As a result, cocaine may cause a greater enhancement of DA neurotransmission, at least initially, than other DA agonists. We have postulated (Einhorn *et al.*, 1988) that this unique spectrum of activity might account for the extremely potent rewarding effects of cocaine as compared to other drugs of abuse (Wise, 1984).

It is also interesting to draw possible parallels between the alterations in the mesoaccumbens DA system produced by three of the most important drug reinforcers, i.e., cocaine, amphetamine, and morphine (or heroin). Because amphetamine greatly enhances DA release by an impulse-independent mechanism (Carboni *et al.*, 1989), DA release in the NAc would continue unabated in spite of the complete suppression of A10 DA unit activity (Wang, 1981c; White *et al.*, 1987). Morphine (and heroin) may also possess an unusual ability to enhance mesoaccumbens DA neurotransmission because of poor compensatory activity. These agents stimulate the firing of A10 DA cells leading to enhanced release of DA within the NAc (Di Chiara and Imperato, 1988), which should inhibit NAc neuronal activity. In addition, these compounds may also directly inhibit NAc neurons by stimulating μ-type opioid receptors on those cells (Hakan and Henriksen, 1989). These two mechanisms would lead to a substantial inhibition of transmission through the NAc. According to the long-loop GABA-GABA-DA feedback hypothesis advanced above, this potent inhibition of NAc cells should activate (disinhibit) inhibitory GABAergic (inter)neurons within the VTA and thereby compensate for enhanced DA transmission within the NAc by shutting down A10 DA neurons. However, this may not occur because of direct inhibitory effects of these μ-opioid agonists on the non-DA VTA (inter)neurons (Gysling and Wang, 1983; Lacey *et al.*, 1989) which would override the disinhibition produced by the NAc-VTA pathway. Thus, DA neurons would remain activated. Although clearly speculative, this model should be viewed as a heuristic framework on which to base continued investigations and derive specific hypotheses. Future studies could test such hypotheses and, in doing so, should shed light on the relationship between alterations in reward-relevant neuronal circuits and the reinforcing efficacy of these important drugs of abuse.

EFFECTS OF REPEATED COCAINE ADMINISTRATION ON MESOACCUMBENS DA TRANSMISSION

Background

Because cocaine abuse is an addiction which is essentially chronic in nature, there has been considerable impetus on research into the neural mechanisms which might be altered by repeated cocaine administration. Because the re-uptake process is the primary means by which synaptic DA and 5-HT are inactivated, interruption of these monoamine transporters by acute administration of cocaine substantially increases synaptic concentrations of the transmitters, thereby prolonging stimulation of relevant receptors. The effects of repeated cocaine administration on DA function are less well understood. One of the most imporant behavioral consequences of repeated treatment with cocaine (or amphetamine) is a reverse tolerance or sensitization to certain of its behavioral effects, i.e., locomotor stimulation and stereotyped oral behaviors (see Chapter 7). However, the mechanisms underlying behavioral sensitization remain elusive (see Robinson and Becker, 1986, for review).

Following our studies of the acute effects of cocaine on neurons within the mesoaccumbens DA system, we next sought to determine the effects of repeated cocaine administration on neurotransmission within the mesoaccumbens DAergic pathway. In our initial experiments, we examined the extent to which repeated cocaine administration alters the sensitivity of both somatodendritic DA autoreceptors on A10 DA neurons and postsynaptic DA receptors located on target cells within the NAc (Henry *et al.*, 1989). In all of the studies described below, cocaine was injected (i.p.) in 10-mg/kg doses to rats either once or twice daily for a 14-d period. Animals were tested 12–24 h after the final injection, unless noted otherwise (see Henry *et al.*, 1989 for complete details).

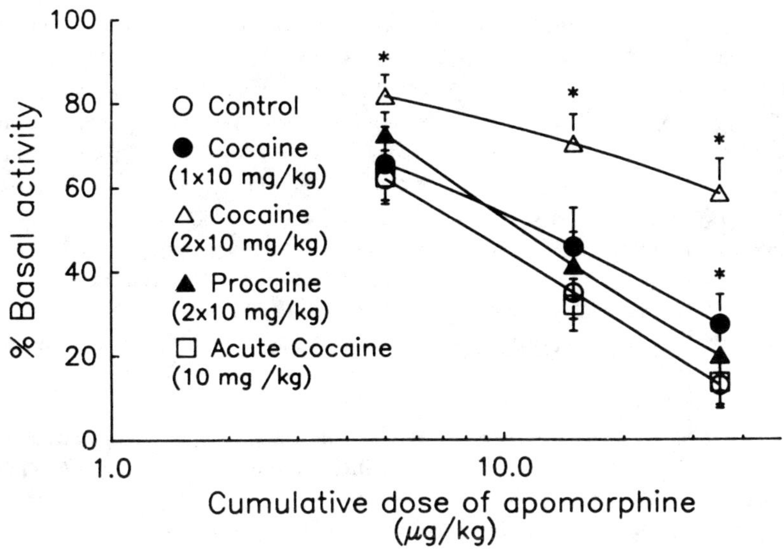

Figure 11. Repeated cocaine attenuated the inhibitory effects of i.v. apomorphine on A10 DA neurons. Dose-response curves for the inhibition of A10 DA neurons induced by i.v. administration of apomorphine (APO) in the various treatment groups (n = 8 cells per group). Asterisks represent values that were significantly different ($p < 0.01$) from control. Note that the rats which received once-daily cocaine injections differed from control only at the highest APO dose, and that repeated procaine failed to influence the sensitivity of A10 DA neurons to APO. (From Henry *et al.*, *J. Pharm. Exp. Ther.* **251**: 833 (1989). With permission.)

Effects of Repeated Cocaine on A10 DA Cells

Intravenous Challenge Experiments

In our first set of experiments, we determined the extent to which repeated cocaine administration alters the sensitivity of impulse-regulating somatodendritic autoreceptors on A10 DA neurons. Previous studies had demonstrated that these neurons were rendered subsensitive to the inhibitory effects of DA agonists following similar repeated administration with D-amphetamine (Kamata and Rebec, 1984; White and Wang, 1984a). As shown in Figure 11, twice-daily injections of cocaine greatly reduced the ability of low i.v. doses of apomorphine to inhibit A10 DA neurons (Henry *et al.*, 1989). However, A10 DA cells in rats that had received a single 10-mg/kg daily injection of cocaine for 14 d prior to testing exhibited normal sensitivity to apomorphine except at the highest dose (35 μg/kg) tested. Because the doses of apomorphine used in these experiments have been demonstrated to exert preferential agonist activity at highly sensitive DA autoreceptors (Skirboll *et al.*, 1979), our results suggest that impulse-regulating somatodendritic D_2 DA autoreceptors on A10 DA cells become subsensitive in response to repeated cocaine administration, presumably from enhanced stimulation by the elevated levels of endogenous neurotransmitter produced by the blockade of DA re-uptake by cocaine. This decrease in sensitivity was not due to an acute effect of cocaine, since a single injection of

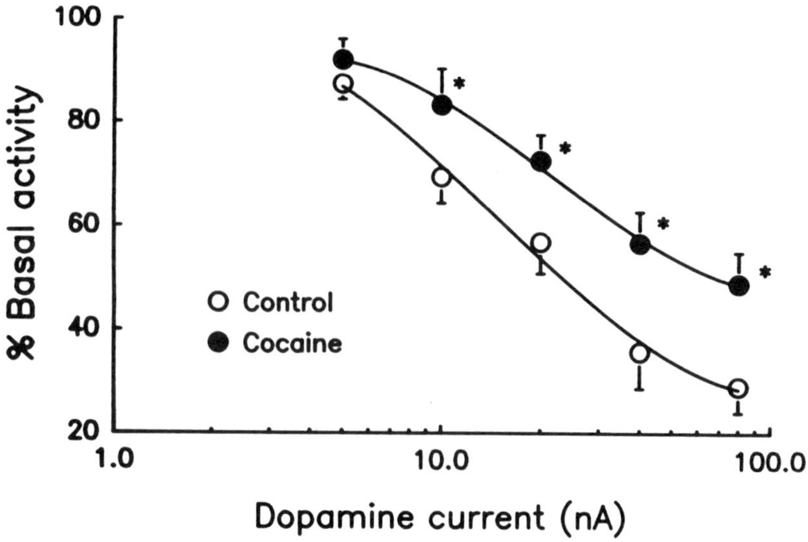

Figure 12. Repeated cocaine administration rendered A10 DA neurons subsensitive to iontophoretic DA. Current-response curves illustrating the effects of microiontophoretically applied DA on A10 DA neurons recorded from control rats and those receiving twice daily 10mg/kg injections of cocaine for 14 d (n = 10 cells per group). All animals were tested 12–24 h after the last saline or cocaine injection (*$p < 0.01$). (From Henry *et al., J. Pharm. Exp. Ther.* **251**: 833 (1989). With permission.)

cocaine 12–24 h prior to testing failed to alter the response of A10 DA neurons to apomorphine. In addition, since repeated procaine administration did not affect the inhibition of A10 DA cells induced by apomorphine, it is unlikely that local anesthetic effects are an important factor in the results observed following repeated cocaine administration (Henry *et al.*, 1989).

Microiontophoretic Studies

Because our acute studies had demonstrated that a major component of cocaine's effects on A10 DA neurons involved activation of a NAc-VTA feedback pathway from the NAc, it seemed possible that repeated cocaine administration might also alter the influence of this mechanism. In order to test directly the hypothesis that repeated cocaine administration reduced the responsiveness of A10 DA cells to i.v. apomorphine by producing subsensitive DA autoreceptors, microiontophoretic studies were conducted to rule out mechanisms outside the VTA. As shown in Figure 12, A10 DA neurons were also significantly subsensitive to iontophoretic DA following twice-daily injections of 10 mg/kg cocaine (Henry *et al.*, 1989). Therefore, despite the fact that cocaine only weakly enhances the stimulation of A10 DA autoreceptors by DA, this level of receptor activation by the endogenous agonist is apparently sufficient to desensitize autoreceptors following repeated cocaine administration. In contrast, repeated stimulation of DA autoreceptors with the low levels of the exogenous agonist apomorphine fails to down-regulate somatodendritic DA autoreceptors, perhaps due to a significant receptor reserve (Jeziorski and White, 1989).

Table 1
Effects of Repeated Cocaine Administration on the Spontaneous Activity of A10 DA Neurons

	N	Cells/track	Rate (spikes/s)
Control	6	1.45 ± 0.09	4.19 ± 0.22
Cocaine treated	5	2.08 ± 0.14[a]	5.58 ± 0.16[a]

[a] $p < 0.01$

In contrast to DA, there was no difference between control and cocaine-treated rats with respect to the sensitivity of A10 DA neurons to iontophoretic GABA. Since GABA-B receptors and D_2 DA autoreceptors may share the same K^+ conductance in midbrain DA neurons (Lacey *et al.*, 1988), the locus of cocaine-induced desensitization would appear to be restricted to D_2 DA autoreceptors.

Cell Population Analysis

Because the basal activity of A10 DA cells is dependent on the density or sensitivity of A10 DA autoreceptors (White and Wang, 1984c), autoreceptor subsensitivity caused by repeated cocaine administration might be expected to increase the firing rates of spontaneously active A10 DA neurons and activate DA neurons which, under normal conditions, are not firing due to autoinhibitory control (White and Wang, 1983a). As shown in Table 1, repeated cocaine administration significantly increased both the number and the mean basal firing rate of spontaneously active A10 DA cells (Henry *et al.*, 1989). These results are similar to previous reports of changes in A10 DA activity produced by repeated amphetamine administration (White and Wang, 1984a), and suggest that impulse-regulating DA autoreceptors on A10 DA neurons play an integral physiological role in regulating the activity of these neurons.

Lack of Persistence of Autoreceptor Subsensitivity

In our final set of experiments, we determined the extent to which autoreceptor subsensitivity induced by repeated cocaine administration persists following termination of cocaine injections. For these experiments, rats were tested either 1, 4, or 8 d following the last injection of the cocaine treatment regimen. In addition, a more complete dose-response relationship for the effects of apomorphine on A10 DA cells was established. As shown in Figure 13, we were able to replicate our original finding that repeated cocaine treatment renders A10 DA cells less sensitive to the inhibitory effects of apomorphine 12–24 h after the last injection (Ackerman and White, 1990). However, the extent of subsensitivity was not as dramatic as that which we originally observed. Whether this is due to the more gradual escalation in the apomorphine cumulative dose response curve (i.e., 1, 1, 2, 4, 8, etc. μg/kg as opposed to 5, 10, 20 μg/kg) is presently unknown. Following a 4-d withdrawal period, subsensitivity to apomorphine was still apparent; whereas after an 8-d withdrawal, sensitivity had returned nearly to control levels (Figure 13). These results indicate that autoreceptor subsensitivity induced by repeated cocaine admin-

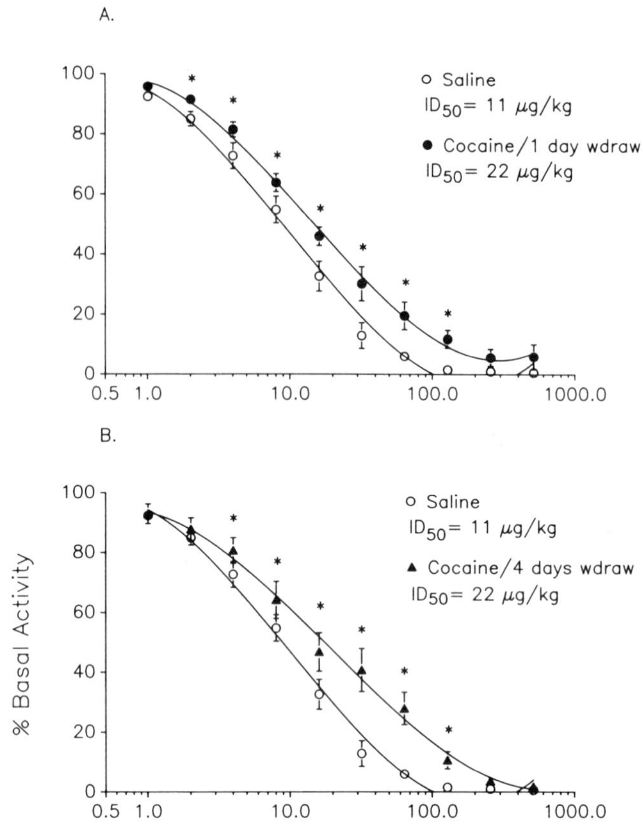

Figure 13. Inhibition of A10 DA neurons by apomorphine following withdrawal from repeated cocaine treatment. The dose-response curve for apomorphine is shown for rats given 10 mg/kg cocaine or 1 ml/kg saline i.p. twice daily for 14 d and tested 1 d (A), 4 d (B), or 8 d (C) after the last injection. The ID_{50} was calculated by linear regression analysis for individual cells, and statistical significance was determined by ANCOVA using basal firing rate as covariate. A significant difference between points in the dose-response curve was determined by Dunnett's test. An asterisk (*) indicates $p < 0.05$. (From J.M. Ackerman and F.J. White. *Neurosci. Lett.* **117**: 181 (1990). With permission.)

istration is not a persistent effect. Somewhat similar findings have been reported following repeated D-amphetamine administration (White and Wang, 1984a).

Effects of Repeated Cocaine on Nucleus Accumbens Neurons

Sensitization to Iontophoretic DA and 5-HT

Because our acute studies had demonstrated a profound inhibitory effect of cocaine in most NAc neurons, mediated by enhanced occupation of both DA and 5-HT receptors following uptake inhibition, we next determined the effects of repeated cocaine administration on the sensitivity of NAc neurons to iontophoretic administration of DA and 5-HT. Given that A10 DA neurons were hyperactive following

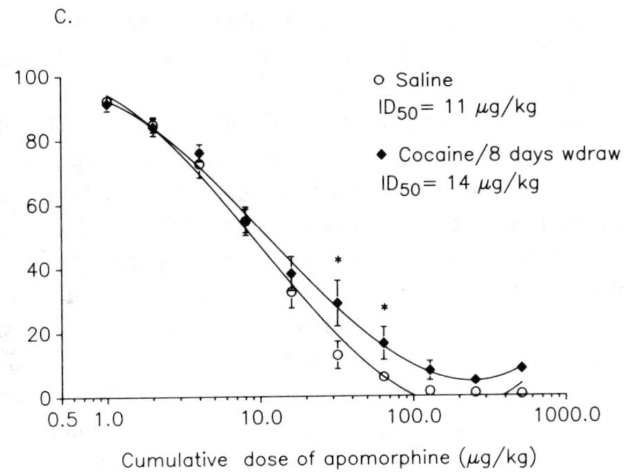

Figure 13C.

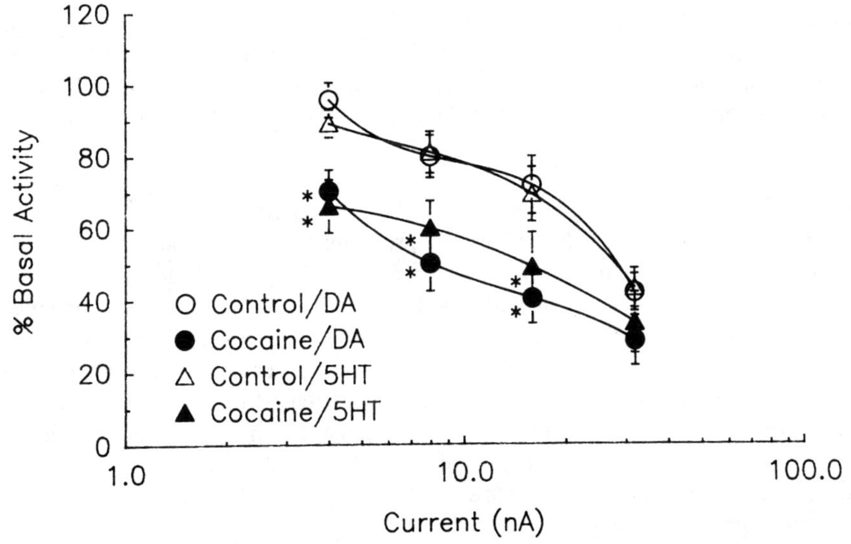

Figure 14. Nucleus accumbens (NAc) neurons in rats which pretreated with cocaine for 14 d were supersensitive to iontophoretic DA and 5-HT. Current-response curves illustrating the effects of microiontophoretically applied DA and 5-HT on NAc neurons recorded from control rats and those receiving twice daily 10-mg/kg injections of cocaine or saline for 14 d (n ≥ 12 cells for each group). All animals were tested 12–24 h after the last injection. Each point represents the mean ± SEM. (*$p < 0.01$).

repeated cocaine injections, we expected that postsynaptic DA receptors in the NAc might also be subsensitive to DA due to enhanced synaptic levels of DA. Surprisingly, our results (Henry *et al.*, 1989; Henry and White, 1991) indicated that DA receptors on NAc neurons were instead supersensitive to iontophoretic DA following repeated treatment with cocaine (Figure 14). In addition to the observed sensitization to DA, NAc neurons exhibited enhanced inhibitory responses to iontophoretic administration of 5-HT (Figure 14) following repeated cocaine administration (Henry and White, 1989).

When combined with the subsensitivity of somatodendritic D_2 autoreceptors on A10 DA cells, the supersensitivity of NAc neurons to DA and 5-HT suggests that repeated exposure to cocaine may produce a highly sensitized set of neurons within the NAc. Given acutely, cocaine blocks DA and 5-HT re-uptake into terminals of monoaminergic neurons innervating the NAc, thereby potently inhibiting NAc neurons while only weakly activating compensatory feedback mechanisms to reduce DA impulse flow and release. This unbalanced DA neurotransmission caused by acute cocaine administration is even further magnified by repeated cocaine treatment. Because impulse-regulating DA autoreceptors are desensitized, there is an overall increase in the number and firing rates of spontaneously active A10 DA neurons. At the same time, postsynaptic DA and 5-HT receptors in the NAc become more sensitive to their respective transmitters, leading to enhanced inhibition of NAc neuronal activity. Such alterations in the mesoaccumbens pathway could produce a system that is particularly sensitized to cocaine. Therefore, these alterations in receptor sensitivities may play a role in the progressive increases in stereotyped (Collins *et al.*, 1979; Post and Rose, 1976) and locomotor responses (Roy *et al.*, 1978) observed in rodents following repeated cocaine administration, and perhaps in the phenomenon of cocaine-induced psychosis (Post and Contel, 1983) observed in humans.

Role of D_1 and D_2 DA Receptors in NAc Sensitization

Previous studies have provided at least preliminary evidence indicating increases in both D_1 and D_2 DA receptors following repeated cocaine administration. Memo *et al.* (1981) found that repeated cocaine facilitated the responsiveness of adenylate cyclase in striatal membranes, which suggests that transmission through striatal D_1 DA receptors may be increased following repeated cocaine administration. More recently, Goeders and Kuhar (1987) found an increased density of [^{3}H]spiperone (D_2) binding sites in the NAc following repeated cocaine injections. In view of these findings, we investigated the involvement of DA receptor subtypes in the observed sensitization of NAc neurons following repeated cocaine using the same treatment protocol as described above.

As shown in Figure 15A, NAc neurons were significantly more sensitive to the inhibitory effects of the D_1 selective agonist SKF 38393 (administered iontophoretically) following 2 weeks of repeated cocaine injections (Henry and White, 1991). In contrast, there was no significant increase in the inhibitory effects of the selective D_2 agonist quinpirole (Figure 15A). These findings indicate that the sensitization of NAc neurons to iontophoretic DA observed in the previous studies was due primarily to enhanced sensitivity of the D_1 DA receptor. To our knowledge, there have been no published reports indicating that repeated cocaine administration increases NAc D_1 receptors, as measured by radioligand binding experiments. However, it is possible that repeated stimulation of D_1 receptors increases the functional sensitivity of D_1 receptors since Subramaniam *et al.*, (1988) reported

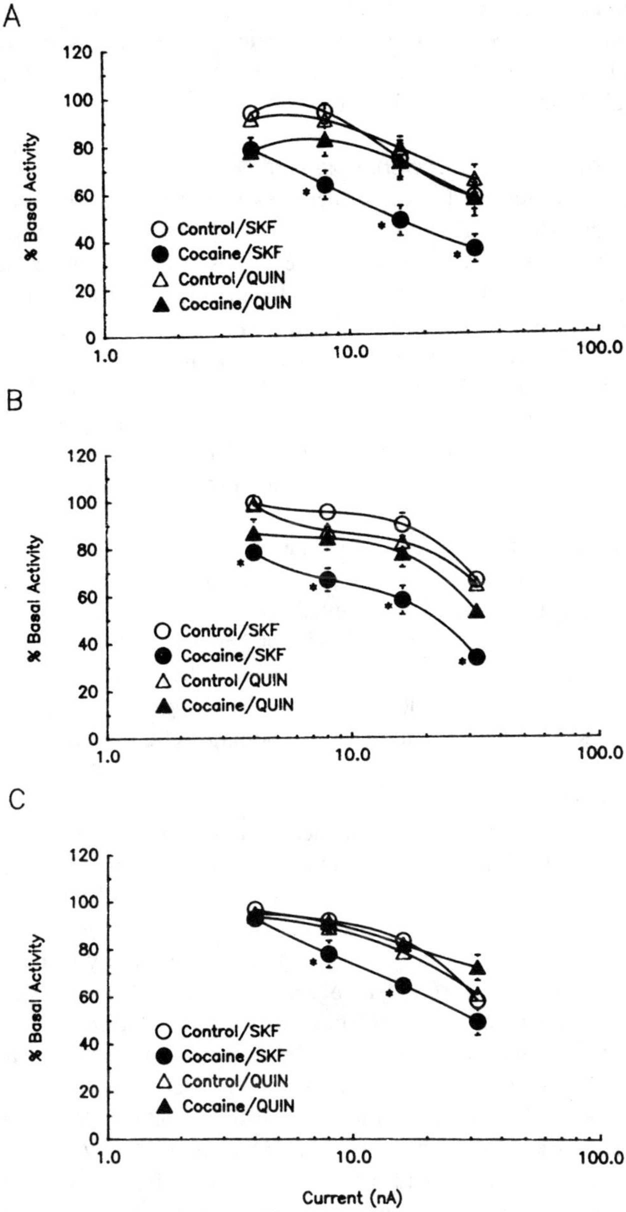

Figure 15. Following repeated cocaine treatment, nucleus accumbens (NAc) neurons were supersensitive to the D_1 agonist SKF 38393, but not the D_2 agonist quinpirole (both administered at 0.01 *M*). In all three plots, rats had received either repeated cocaine (10 mg/kg, i.p.) or saline for 14 d. A: Current-response curves illustrating the supersensitivity of NAc cells to iontophoretic administration of the D_1 agonist SKF 38393 12–24 h after the last cocaine injection. Following both a 1-week (B) and 1-month (C) withdrawal from cocaine treatment, NAc cells remained supersensitive to the effects of SKF 38393. (From D.J. Henry and F.J. White. *J. Pharmacol. Exp. Ther.* (in press). With permission.)

that repeated administration of the selective D_1 agonist SKF 38393 causes a slight up-regulation of D_1 receptors in the rat caudate-putamen and NAc, in agreement with recent behavioral (Braun and Chase, 1988; White *et al.*, 1989) and electrophysiological (White *et al.*, 1989) studies.

Persistence of D_1 Receptor Sensitization

The finding that repeated cocaine increases neurotransmission through D_1 DA receptors in the NAc suggests that this mechanism could be involved in the phenomenon of behavioral sensitization. However, since behavioral sensitization is a relatively persistent change (Robinson and Becker, 1986), it has been argued that the neural mechanisms underlying sensitization should also be persistent. As reviewed above, the subsensitivity of somatodendritic autoreceptors produced by repeated administration of cocaine or D-amphetamine (White and Wang, 1984a) was either not evident (cocaine) or less evident (amphetamine) following a 1-week withdrawal period. Our next set of experiments determined the extent to which D_1 receptor sensitization persisted following withdrawal from repeated cocaine administration.

Unlike autoreceptor down-regulation, the sensitization of NAc neurons to the inhibitory effects of SKF 38393 persisted for 8 d following the cessation of cocaine injections (Figure 15B). In addition, evidence for supersensitive D_1 receptor-mediated responses was still evident following a 1-month withdrawal period (Figure 15C). Therefore, unlike the subsensitivity/supersensitivity of D_2 DA autoreceptors, the of D_1 DA postsynaptic receptors within the NAc appears to be a relatively long-lasting adaptation produced by the repeated administration of cocaine.

General Discussion

Repeated cocaine administration resulted in an initial subsensitivity of impulse-regulating somatodendritic autoreceptors on A10 DA neurons, as well as enhanced sensitivities of postsynaptic D_1 DA and 5-HT (subtype undetermined) receptor-mediated responses within the nucleus accumbens. Although the autoreceptor subsensitivity did not persist for a week, the enhanced sensitivity of postsynaptic D_1 DA receptors was still observed 1 month after withdrawal from the cocaine treatment. Withdrawal studies with 5-HT have yet to be conducted.

One of the hallmark characteristics of behavioral sensitization to psychomotor stimulants is the enduring nature of this adaptation. Enhanced locomotor stimulant responses to cocaine have been reported to persist for months following the cessation of repeated administration (Post and Contel, 1983). Accordingly, it has been argued that any neural adaptation which is proposed to account for behavioral sensitization should also demonstrate persistence. In their thorough review of the literature regarding amphetamine sensitization, Robinson and Becker (1986) argued against both DA autoreceptor subsensitivity and postsynaptic DA receptor up-regulation as relevant neuronal substrates. One of the key arguments against autoreceptor down-regulation was the fact that it does not persist following withdrawal from repeated amphetamine administration (Antelman and Chiodo, 1981; White and Wang, 1984a). Because the great majority of radiolabeled ligand binding studies failed to provide evidence for postsynaptic DA receptor proliferation following repeated amphetamine, this mechanism was also dismissed as a possible mediator of behavioral sensitization (Robinson and Becker, 1986). These authors have argued that the enduring change in neuronal function accompanying amphetamine sensitization is an enhanced release of DA upon subsequent challenge.

In view of our own recent findings, we feel that the dismissal of changes in DA receptor-mediated effects as relevant neuronal adaptations for psychomotor stimulant sensitization may have been premature. With respect to ligand binding studies, it is essential to note that the studies which were reviewed never ascertained the effects of repeated amphetamine on D_1 receptor binding, as currently studied with selective ligands of high specific activity. Moreover, alterations in D_1 receptor-mediated responses may occur beyond the level of the recognition site, i.e., within transduction pathways. With regard to the failure of autoreceptor down-regulation to persist following withdrawal, it is possible that such subsensitivity is a necessary triggering event which might lead to other enduring neuronal adaptations such as enhanced release and D_1 receptor sensitization. Clearly, events occurring at the somatodendritic level are sufficient to produce sensitization since repeated administration of amphetamine into the VTA, *but not the NAc*, leads to sensitized locomotor responses to subsequent peripheral administration of amphetamine and cocaine (Kalivas and Weber, 1988).

In addition to the evidence presented herein, we have also recently demonstrated that repeated administration of the selective D_1 receptor agonist SKF 38393 results in *enhanced* inhibitory responses of striatal neurons to iontophoretic administration of SKF 38393 (White *et al.*, 1989). This effect could not be attributed to the partial agonist quality of SKF 38393 at D_1 receptors since the "full" D_1 agonist SKF 81297 (as defined by stimulation of adenylate cyclase) also produced sensitized responses following repeated administration. In addition, we and others have also observed enhanced behavioral responses to D_1 agonists (grooming) following repeated treatment with SKF 38393 (Braun and Chase, 1988; White *et al.*, 1989). Therefore, it appears that repeated stimulation of the D_1 receptor results in sensitized responses to subsequent administration of D_1 receptor agonists. However, unlike the effects of repeated cocaine administration, D_1 receptor sensitization following SKF 38393 administration did not persist for a month. Repeated administration of a selective D_1 agonist would not be expected to produce the initial autoreceptor subsensitivity observed with cocaine because DA autoreceptors appear to be exclusively of the D_2 subtype (White and Wang, 1984b) and function independently of D_1 receptor occupation (Wachtel *et al.*, 1989). Therefore, it is possible that subsensitivity of somatodendritic D_2 autoreceptors, although temporary, may be necessary for the induction of a persistent increase in D_1 receptor sensitivity.

SUMMARY AND CONCLUSIONS

Our experiments regarding the electrophysiological effects of cocaine within the rat mesoaccumbens DA system have discovered a profile of effects which distinguish cocaine from all other drugs which have been studied within this system. Unlike direct DA agonists and amphetamine, cocaine causes only a partial inhibition of A10 DA cells which is brought about not only by prolonged activation of impulse-regulating somatodendritic DA autoreceptors but also by activation of a NAc-VTA inhibitory feedback pathway. At the same time, the ability of cocaine to inhibit both the DA and 5-HT transporters within the NAc leads to marked inhibition of the majority of these DA target cells. Thus, the overall result of cocaine administration is a profound enhancement of neuronal inhibition within the NAc which is only partially offset by compensatory feedback mechanisms at the level of the DA neuron. Accordingly, impulse-dependent DA release continues while DA reuptake is blocked and synaptic levels of DA increase dramatically. Such poor compensatory responses lead to relatively unbalanced transmission within the mesoaccumbens system and may result in the potent rewarding effects of cocaine as compared to other drugs of abuse.

Repeated administration of cocaine may exaggerate this unbalanced transmission within the mesoaccumbens DA system. Immediately following withdrawal from 2 weeks of repeated cocaine administration, somatodendritic autoreceptors on A10 DA cells are subsensitive, leading to increases in the number of spontaneously active neurons and their average firing rates. In contrast, NAc neurons are rendered supersensitive to the inhibitory effects of both DA and 5-HT. Thus, cocaine administration would now lead to even more profound decreases in neuronal activity within the NAc, with considerably poorer compensatory decreases in DA unit activity. The DA supersensitivity is mediated by an enhanced responsiveness of D_1 DA receptors on NAc neurons and is a relatively enduring adaptation. In contrast, the subsensitivity of DA neurons to agonist-induced inhibition is mediated by reduced responsiveness of D_2 autoreceptors which fails to persist following cocaine withdrawal. Based upon other findings indicating that repeated stimulation of D_1 receptors with a direct selective D_1 agonist, which does not influence DA autoreceptors, causes D_1 receptor sensitization which does not persist, we have proposed that autoreceptor subsensitivity may be necessary to trigger a more persistent supersensitivity of postsynaptic D_1 receptor-mediated responses, which ultimately appears to be one of the underlying mechanisms of behavioral sensitization.

ACKNOWLEDGMENTS

This research was supported by USPHS Grants DA-04093 from the National Institute on Drug Abuse and MH-40832 from the National Institute of Mental Health. I thank Leslie C. Einhorn, Patricia A. Johansen, and Dr. Edward J. Roy for scientific contributions.

REFERENCES

J.M. Ackerman and F.J. White. A10 somatodendritic autoreceptor sensitivity following withdrawal from repeated cocaine treatment. *Neurosci. Lett.* **117**: 181 (1990).

G.K. Aghajanian and B.S. Bunney. Dopamine "Autoreceptors": pharmacological characterization by microiontophoretic single cell recording studies. *Naunyn-Schmiedeberg's Arch. Pharmacol.* **297**: 1 (1977).

N.E. Andén, M. Grabowska-Andén, S. Lindgren, and U. Thornström. Synthesis rate of dopamine: difference between corpus striatum and limbic system as possible explanation of variations in reactions to drugs. *Naunyn-Schmiedeberg's Arch. Pharmacol.* **323**: 193 (1983).

S.M. Antelman and L.A. Chiodo. Dopamine autoreceptor subsensitivity: a mechanism common to the treatment of depression and the induction of amphetamine psychosis? *Biol. Psychiat.* **16**: 717 (1981).

B.L. Baxter, M.I. Gluckman, L. Stein, and R.A. Scerni. Self-injection of apomorphine in the rat: positive reinforcement by a dopamine receptor stimulant. *Pharmacol. Biochem. Behav.* **2**: 387 (1974).

P.M. Beart and D. McDonald. Neurochemical studies of the mesolimbic dopaminergic pathway: somatodendritic mechanisms and GABAergic neurons in the rat ventral tegmentum. *J. Neurochem.* **34**. 1622 (1980).

A. Björklund and O. Lindvall. Dopamine in dendrites of substantia nigra neurons: suggestions for a role in dendritic terminals. *Brain Res.* **83**: 531 (1975).

M.A. Bozarth and R.A. Wise. Heroin reward is dependent on a dopaminergic substrate. *Life Sci.* **29**: 1881 (1981).

C.W. Bradberry and R.H. Roth. Cocaine increases extracellular dopamine in rat nucleus accumbens and ventral tegmental area as shown by *in vivo* microdialysis. *Neurosci. Lett.* **103**: 97 (1989).

A.R. Braun and T.N. Chase. Behavioral effects of chronic exposure to selective D-1 and D-2 dopamine receptor agonists. *Eur. J. Pharmacol.* **147**: 441 (1988).

B.S. Bunney and G.K. Aghajanian. D-Amphetamine-induced inhibition of central dopaminergic neurons: mediation by a striato-nigral feedback pathway. *Science* **192**: 391 (1976).

B.S. Bunney and G.K. Aghajanian. D-Amphetamine-induced depression of central dopamine neurons: evidence for mediation by both autoreceptors and a striatal-nigral feedback pathway. *Naunyn-Schmiedeberg's Arch. Pharmacol.* **304**: 255 (1978).

B.S. Bunney, J.R. Walters, R.H. Roth, and G.K. Aghajanian. Dopaminergic neurons: effect of antipsychotic drugs and amphetamine on single cell activity. *J. Pharmacol. Exp. Ther.* **135**: 560 (1973).

E. Carboni, A. Imperato, L. Perezzani, and G. Di Chiara. Amphetamine, cocaine, phencyclidine and nomifensine increase extracellular dopamine concentrations preferentially in the nucleus accumbens of freely moving rats. *Neuroscience* **28**: 653 (1989).

A. Carlsson. Receptor-mediated control of dopamine metabolism, in: *Pre- and Postsynaptic Receptors.* E. Usdin and W.E. Bunney, Jr., eds., Marcel Dekker, New York, p. 49 (1975).

A. Carlsson, M. Lindqvist, and T. Magnusson. 3,4-Dihydroxyphenylalanine and 5-hydroxytryptophan as reserpine antagonists. *Nature* **180**: 1200 (1957).

T.P. Carney. Alkaloids as local anesthetics, in: *The Alkaloids. Vol. V, Pharmacology.* R.N.F. Manske, ed., Academic Press, New York, p. 211 (1955).

A. Chéramy, V. Leviel and J. Glowinski. Dendritic release of dopamine in the substantia nigra. *Nature* **289**: 537 (1981).

L.A. Chiodo and B.S. Bunney. Effects of dopamine antagonists on midbrain dopamine cell activity, in: *Catecholamines: Neuropharmacology and Central Nervous System — Theoretical Aspects.* E. Usdin, A. Carlsson, A. Dahlström, and J. Engel, eds., Alan R. Liss, New York, p. 369 (1984).

D. Clark and F.J. White. The D_1 dopamine receptor: the search for a function. A critical evaluation of the D_1/D_2 dopamine receptor classification and its functional implications. *Synapse,* **1**: 347 (1987).

J.P. Collins, H. Lesse, and L.W. Lagan. Behavioral antecedents of cocaine-induced stereotypy. *Pharm. Biochem. Behav.* **11**: 683 (1979).

T.J. Crow. A map of the rat mesencephalon for electrical self-stimulation. *Brain Res.* **36**: 265 (1972).

H. de Wit and R.A. Wise. Blockade of cocaine reinforcement in rats with the dopamine receptor blocker pimozide, but not with the noradrenergic blockers phentolamine or phenoxybenzamine. *Can. J. Psychol.* **31**: 195 (1977).

G. Di Chiara and A. Imperato. Drugs abused by humans preferentially increase synaptic dopamine concentrations in the mesolimbic system of freely moving rats. *Proc. Natl. Acad. Sci. U.S.A.* **85**: 5274 (1988).

V.B. Domesick. Further observations of the anatomy of nucleus accumbens and caudatoputamen in the rat: similarities and contrasts, in: *The Neurobiology of the Nucleus Accumbens.* R.B. Chronister and J.F. DeFrance, eds., Haer Institute, Brunswick, ME, p. 147 (1981).

L.C. Einhorn, P.A. Johansen, and F.J. White. Electrophysiological effects of cocaine in the mesoaccumbens dopamine system. *J. Neurosci.* **8**: 100 (1988).

A. Ettenberg, H.O. Pettit, F.E. Bloom, and G.F. Koob. Heroin and cocaine intravenous self-administration in rats: mediation by separate neural systems. *Psychopharmacology* **78**: 204 (1982).

H.C. Fibiger. Drugs and reinforcement mechanisms: a critical review of the catecholamine theory. *Ann. Rev. Pharmacol. Toxicol.* **18**: 37 (1978).

M.W. Fischman, C.R. Schuster, and Y. Hatano. A comparison of the subjective and cardiovascular effects of cocaine and lidocaine in humans. *Pharmacol. Biochem. Behav.* **18**: 123 (1983).

D.C. German and D.M. Bowden. Catecholamine systems as the neural substrate for intracranial self-stimulation: a hypothesis. *Brain Res.* **73**: 381 (1974).

D.C. German, M. Dalsass, and R.S. Kiser. Electrophysiological examination of the ventral tegmental (A10) area in the rat. *Brain Res.* **181**: 191 (1980).

G.L. Gessa, F. Muntoni, M. Collu, L. Vargin, and G. Mereu. Low doses of ethanol activate dopaminergic neurons in the ventral tegmental area. *Brain Res.* **348**: 201 (1985).

N.E. Goeders, S.I. Dworkin, and J.E. Smith. Neuropharmacological assessment of cocaine self-administration into the medial prefrontal cortex. *Pharmacol. Biochem. Behav.* **24**: 1429 (1986).

N.E. Goeders and M.J. Kuhar. Chronic cocaine administration induces opposite changes in the striatum and nucleus accumbens. *Alcohol Drug Res.* **7**: 207 (1987).

N.E. Goeders and J.E. Smith. Cortical dopaminergic involvement in cocaine reinforcement. *Science* **221**: 773 (1983).

N.E. Goeders and J.E. Smith. Reinforcing properties of cocaine in the medial prefrontal cortex: primary action on presynaptic dopaminergic terminals. *Pharmacol. Biochem. Behav.* **25**: 191 (1986).

A.A. Grace and B.S. Bunney. Intracellular and extracellular electrophysiology of nigral dopaminergic neurons. 1. Identification and characterization. *Neuroscience* **10**: 301 (1983).

A.A. Grace and B.S. Bunney. Opposing effects of striatonigral feedback pathways on midbrain dopamine cell activity. *Brain Res.* **333**: 271 (1985).

P.M. Groves and J.C. Linder. Dendro-dendritic synapses in substantia nigra: descriptions based on analyses of serial section. *Exp. Brain Res.* **49**: 209 (1983).

K. Gysling and R.Y. Wang. Morphine-induced activation of A10 dopamine neurons in the rat. *Brain Res.* **277**: 119 (1983).

R.L. Hakan and S.J. Henriksen. Opiate influences on nucleus accumbens neuronal electrophysiology: dopamine and non-dopamine mechanisms. *J. Neurosci.* **9**: 3538 (1989).

D.J. Henry and F.J. White. Effects of repeated cocaine administration on the sensitivity of D_1 and D_2 receptors in the nucleus accumbens: electrophysiological studies. *Soc. Neurosci. Abstr.* **15**: 1013 (1989).

D.U. Henry and F.J. White. Repeated cocaine administration causes persistent enhancement of D, dopamine receptor sensitivity within the rat nucleus accumbens. *J. Pharmacol. Exp. Ther.*, in press, (1991).

D.J. Henry, M.A. Greene, and F.J. White. Electrophysiological effects of cocaine in the mesoaccumbens dopamine system: repeated administration. *J. Pharm. Exp. Ther.* **251**: 833 (1989).

B.G. Hoebel, A. Monaco, L. Hernandes, E. Aulisi, B.G. Stanley, and L. Lenard. Self-injection of amphetamine directly into the brain. *Psychopharmacology* **81**: 158 (1983).

R.B. Innis and G.K. Aghajanian. Pertussis toxin blocks autoreceptor-mediated inhibition of dopaminergic neurons in rat substantia nigra. *Brain Res.* **411**: 139 (1987).

M. Jeziorski and F.J. White. Dopamine agonists at repeated "autoreceptor-selective" doses: effects upon the sensitivity of A10 dopamine autoreceptors. *Synapse* **4**: 267 (1989).

P.W. Kalivas, A. Bourdelais, R. Abhold, and L. Abbott. Somatodendritic release of endogenous dopamine: *in vivo* dialysis in the A10 dopamine region. *Neurosci. Lett.* **100**: 215 (1989).

P.W. Kalivas and B. Weber. Amphetamine injection into the ventral mesencephalon sensitizes rats to peripheral amphetamine and cocaine. *J. Pharmacol. Exp. Ther.* **245** : 1095 (1988).

K. Kamata and G.V. Rebec. Long-term amphetamine treatment attenuates or reverses the depression of neuronal activity produced by dopamine agonists in the ventral tegmental area. *Life Sci.* **34**: 2419 (1984).

G.F. Koob, H.T. Le, and I. Creese. The D_1 dopamine receptor antagonist SCH 23390 increases cocaine self-administration in the rat. *Neurosci. Lett.* **79**: 315 (1987).

J. Korf, M. Zieleman, and G.H.C. Westerink. Dopamine release in the substantia nigra? *Nature* **260**: 257 (1976).

R. Kuczenski. Biochemical actions of amphetamine and other stimulants, in: *Stimulants: Neurochemical, Behavioral and Clinical Perspectives,* I. Creese, ed., Raven Press, New York, p. 31 (1983).

M.G. Lacey, N.B. Mercuri, and R.A. North. Dopamine acts on D_2 receptors to increase potassium conductance in neurons of the rat substantia nigra zona compacta. *J. Physiol. (London)* **392**: 197 (1987).

M.G. Lacey, N.B. Mercuri, and R.A. North. On the conductance increase activated by GABA-B and dopamine D_2 receptors in rat substantia nigra neurones. *J. Physiol.* **401**: 437 (1988).

M.G. Lacey, N.B. Mercuri, and R.A. North. Two cell types in rat substantia nigra zona compacta distinguished by membrane properties and the actions of dopamine and opioids. *J. Neurosci.* **9**: 1233 (1989).

W.H. Lyness, N.M. Friedle, and K.E. Moore. Destruction of dopaminergic nerve terminals in nucleus accumbens: effect on D-amphetamine self-administration. *Pharmacol. Biochem. Behav.* **11**: 553 (1979).

M.T. Martin-Iverson, C. Szostak, and H.C. Fibiger. 6- Hydroxydopamine lesions of the medial prefrontal cortex fail to influence intravenous self-administration of cocaine. *Psychopharmacology* **88**: 310 (1986).

R.T. Matthews and D.C. German. Electrophysiological evidence for excitation of rat ventral tegmental area dopamine neurons by morphine. *Neuroscience* **11**: 617 (1984).

M. Memo, S. Pradhan, and I. Hanbauer. Cocaine-induced supersensitivity of striatal dopamine receptors: role of endogenous calmodulin. *Neuropharmacology* **20**: 1145 (1981).

M. Mereu, K.-W.P. Yoon, V. Boi, G.L. Gessa, L. Naes, and T.C. Westfall. Preferential stimulation of ventral tegmental area dopaminergic neurons by nicotine. *Eur. J. Pharmacol.* **141**: 395 (1987).

W.J.H. Nauta, G.P. Smith, R.L.M. Faull, and V.B. Domesick. Efferent connections and nigral afferents of the nucleus accumbens septi in the rat. *Neuroscience* **3**: 385 (1978).

E.S. Nisenbaum, W.B. Orr, and T.W. Berger. Evidence for two functionally distinct subpopulations of neurons within the rat striatum. *J. Neurosci.* **8**: 4138 (1988).

H. Nissbrandt, E. Pileblad, and A. Carlsson. Evidence for dopamine release and metabolism beyond the control of nerve impulses and dopamine receptors in rat substantia nigra. *J. Pharmacol. Pharm.* **37**: 884 (1985).

D.P. O'Brien and F.J. White. Inhibition of non-dopamine cells in the ventral tegmental area by benzodiazepines: relationship to A10 dopamine cell activity. *Eur. J. Pharmacol.* **142**: 343 (1987).

J. Olds. Pleasure centers in the brain. *Sci. Am.* **195**: 105 (1956).

J. Olds and P.M. Milner. Positive reinforcement produced by electrical stimulation of septal area and other regions of rat brain. *J. Comp. Physiol. Psychol.* **47**: 419 (1954).

G. Paxinos and C. Watson. *The Rat Brain in Stereotaxic Coordinates.* Academic Press, New York (1982).

H.O. Pettit, A. Ettenberg, F.E. Bloom, and G.F. Koob. Destruction of dopamine in the nucleus accumbens selectively attenuates cocaine but not heroin self-administration in rats. *Psychopharmacology* **84**: 167 (1984).

A.G. Phillips, C.L.E. Broekkamp, and H.C. Fibiger. Strategies for studying the neurochemical substrates of drug reinforcement in rodents. *Prog. Neuropsychopharmacol. Biol. Psychiatry* **7**: 585 (1983).

O.T. Phillipson. Afferent projections to the ventral tegmental area of tsai and interfascicular nucleus: a horseradish peroxidase study in the rat. *J. Comp. Neurol.* **187**: 117 (1979).

R.M. Post, S.R.B. Weiss, and A. Pert. Sensitization and kindling effects of chronic cocaine administration. In: *Cocaine: Pharmacology, Physiology and Clinical Strategies.* J.M. Lakoski, M.P. Galloway, and F.J. White, eds., CRC Press, Boca Raton, FL, (1991).

R.M. Post and N.R. Contel. Human and animal studies of cocaine: implications for development of behavioral pathology, in: *Stimulants: Neurochemical, Behavioral and Clinical Perspectives.* I. Creese, ed., Raven Press, New York, p. 169 (1983).

R.M. Post and H. Rose. Increasing effects of repetitive cocaine administration in the rat. *Nature* **260**: 731 (1976).

M.C. Ritz, R.J. Lamb, S.R. Goldberg, and M.J. Kuhar. Cocaine receptors on dopamine transporters are related to self-administration. *Science* **237**: 1219 (1987).

D.C.S. Roberts. Neural substrates mediating cocaine reinforcement: the role of monoamine systems, in: *Cocaine: Pharmacology, Physiology and Clinical Strategies.* J.M. Lakoski, M.P. Galloway, and F.J. White, eds., CRC Press, Boca Raton, FL (1991).

D.C.S. Roberts, M.E. Corcoran, and H.C. Fibiger. On the role of ascending catecholaminergic systems in intravenous self-administration of cocaine. *Pharmacol. Biochem. Behav.* **6**: 615 (1977).

D.C.S. Roberts and G.F. Koob. Disruption of cocaine self-administration following 6-hydroxydopamine lesions of the ventral tegmental area in rats. *Pharmacol. Biochem. Behav.* **17**: 901 (1982).

D.C.S. Roberts, G.F. Koob, P. Klonoff, and H.C. Fibiger. Extinction and recovery of cocaine self-administration following 6-hydroxydopamine lesions of the nucleus accumbens. *Pharmacol. Biochem. Behav.* **12**: 781 (1980).

T.E. Robinson and J.B. Becker. Enduring changes in brain and behavior produced by chronic amphetamine administration: a review and evaluation of animal models of amphetamine psychosis. *Brain Res. Rev.* **11**: 157 (1986).

S.B. Ross and S.L. Renyi. Accumulation of tritiated 5-hydroxytryptamine in brain slices. *Life Sci.* **6**: 1407 (1967a).

S.B. Ross and A.L. Renyi. Inhibition of the uptake of tritiated catecholamines by antidepressant and related agents. *Eur. J. Pharmacol.* **2**: 181 (1967b).

S.N. Roy, A.K. Battacharyya, S. Pradhan, and S.N. Pradhan. Behavioral and neurochemical effects of repeated administration of cocaine in rats. *Neuropharmacology* **17**: 559 (1978).

L.R. Skirboll and B.S. Bunney. The effects of acute and chronic haloperidol treatment on spontaneously firing neurons in the caudate nucleus of the rat. *Life Sci.* **25**: 1419 (1979).

L.R. Skirboll, A.A. Grace, and B.S. Bunney. Dopamine auto and post-synaptic receptors: electrophysiological evidence for differential sensitivity to dopamine agonists. *Science* **206**: 80 (1979).

S.H. Snyder and J.T. Coyle. Regional differences in [^{3}H]-dopamine uptake into rat brain homogenates. *J. Pharmacol. Exp. Ther.* **165**: 78 (1969).

C. Spyraki, H.C. Fibiger, and A.G. Phillips. Attenuation of heroin reward in rats by disruption of the mesolimbic dopamine system. *Psychopharmacology* **79**: 278 (1983).

S. Subramaniam, I. Lucki, and P. McGonigle. Effects of chronic treatment with selective agonists on the subtypes of dopamine receptors. *Soc. Neurosci. Abstr.* **14**: 926 (1988).

P. Van der Zee, H.S. Koger, J. Gootjes, and W. Hespe. Aryl 1,4-dialk(en)ylpiperazines as selective and very potent inhibitors of dopamine uptake. *Eur. J. Med. Chem.* **15**: 363 (1980).

S.R. Wachtel, X.-T. Hu, M.P. Galloway, and F.J. White. D_1 dopamine receptor stimulation enables the post-synaptic, but not autoreceptor, effects of D_2 dopamine agonists in nigrostriatal and mesoaccumbens dopamine systems. *Synapse* **4**: 327 (1989).

J.L. Waddington and A.L. Cross. Neurochemical changes following kainic acid lesions of the nucleus accumbens: implications for a GABAergic accumbal-ventral tegmental pathway. *Life Sci.* **22**: 1011 (1978).

I. Walaas and F. Fonnum. Biochemical evidence for gamma-aminobutyrate containing fibers from the nucleus accumbens to the substantia nigra and ventral tegmental area in the rat. *Neuroscience* **5**: 63 (1980).

R.Y. Wang. Dopaminergic neurons in the rat ventral tegmental area. I. Identification and characterization. *Brain Res. Rev.* **3**: 123 (1981a).

R.Y. Wang. Dopaminergic neurons in the rat ventral tegmental area. II. Evidence for autoregulation. *Brain Res. Rev.* **3**: 141 (1981b).

R.Y. Wang. Dopaminergic neurons in the rat ventral tegmental area. III. Effects of D- and L-amphetamine. *Brain Res. Rev.* **3**: 153 (1981c).

M. Wassef, A. Berod, and C. Sotelo. Dopaminergic dendrites in the pars reticulata of the rat substantia nigra and their striatal input. Combined immmunocytochemical localization of tyrosine hydroxylase and anterograde degeneration. *Neuroscience* **6**: 2125 (1981).

B.L. Waszczak and J.R. Walters. Intravenous GABA agonist administration stimulates firing of A10 dopaminergic neurons. *Eur. J. Pharmacol.* **66**: 141 (1980).

F.J. White. Comparative effects of LSD and lisuride: clues to specific hallucinogenic drug actions. *Pharmacol. Biochem. Behav.* **24**: 365 (1986).

F.J. White, R.J. Brooderson, and X.-T. Hu. Repeated stimulation of D_1 dopamine receptors enhances grooming and neuronal responses to SKF-38393. *Soc. Neurosci. Abstr.* **15**: 1314 (1989).

F.J. White, L.C. Einhorn, P.A. Johansen, and S.R. Wachtel. Electrophysiological studies in the rat mesoaccumbens dopamine system: focus on dopamine receptor subtypes, interactions, and the effects of cocaine, in: *Neurophysiology of Dopaminergic Systems: Current Status and Clinical Perspectives.* L.A. Chiodo and A.S. Freeman, eds., Lakeshore Publishing, Detroit, p. 317 (1987).

F.J. White and R.Y. Wang. Comparison of the effects of chronic haloperidol treatment on A9 and A10 dopamine neurons in the rat. *Life Sci.* **32**: 983 (1983a).

F.J. White and R.Y. Wang. Comparison of the effects of LSD and lisuride on A10 dopamine neurons in the rat. *Neuropharmacology* **22**: 669 (1983b).

F.J. White and R.Y. Wang. Electrophysiological evidence for the existence of both D_1 and D_2 dopamine receptors in the rat nucleus accumbens. *J. Neurosci.* **6**: 274 (1986).

F.J. White and R.Y. Wang. Electrophysiological evidence for A10 dopamine autoreceptor subsensitivity following chronic D-amphetamine treatment. *Brain Res.* **309**: 283 (1984a).

F.J. White and R.Y. Wang. Pharmacological characterization of dopamine autoreceptors in rat ventral tegmental area: microiontophoretic studies. *J. Pharmacol. Exp. Ther.* **231**: 275 (1984b).

F.J.White and R.Y. Wang. A10 dopamine neurons: role of autoreceptors in determining firing rate and sensitivity to dopamine agonists. *Life Sci.* **34**: 1161 (1984c).

R.A. Wise. Action of drugs of abuse on brain reward systems. *Pharmacol. Biochem. Behav.* **13**: 213 (1980).

R.A. Wise. Neural mechanisms of the reinforcing actions of cocaine. *NIDA Res. Monographs* **50**: 15 (1984).

R.A. Wise. The role of reward pathways in the development of drug dependence. *Pharmacol. Therapeut.* **35**: 227 (1987).

R.A. Wise and M.A. Bozarth. Brain reward circuitry: four circuit elements ''wired'' in apparent series. *Brain Res. Bull.* **297**: 265 (1984).

R.A. Wise and P.-P. Rompre. Brain dopamine and reward. *Annu. Rev. Psychol.* **40**: 191 (1989).

M.E. Wolf and R.H. Roth. Dopamine autoreceptors, in: *Dopamine Receptors.* I. Creese and C. M. Fraser eds., Alan R. Liss, New York, p. 45 (1987).

P. Wolf, H.R. Olpe, D. Avrith and H.L. Haas. GABAergic inhibition of neurons in the ventral tegmental area. *Experimentia* **34**: 73 (1978).

D.T. Wong, J.S. Horng, F.P. Bymaster, K.L. Hauser and B.B. Molloy. A selective inhibitor of serotonin uptake: Lilly 110140, 3-(*p*-trifluoromethylphenoxy)-*N*- methyl-3-phenylpropylamine. *Life Sci.* **15**: 471 (1974).

W.L. Woolverton, L.I. Goldberg, and J. Ginos. Intravenous self-administration of dopamine receptor agonists by rhesus monkeys. *J. Pharmacol. Exp. Ther.* **230**: 678 (1984).

C.Y. Yim and G.J. Mogenson. Effect of picrotoxin and nipecotic acid on inhibitory response of dopaminergic neurons in the ventral tegmental area to stimulation of the nucleus accumbens. *Brain Res.* **199**: 466 (1980).

R.A. Yokel and R.A. Wise. Amphetamine-type reinforcement by dopamine agonists in the rat. *Psychopharmacology* **58**: 289 (1978).

K.A. Zito, G. Vickers, and D.C.S. Roberts. Disruption of cocaine and heroin self-administration following kainic acid lesions of the nucleus accumbens. *Pharmacol. Biochem. Behav.* **23**: 1029 (1985).

13

Electrophysiologic Effects of Cocaine on Serotonin Neuronal Systems

JOAN M. LAKOSKI, ERIC W. BLACK, AND HEATHER J. MODAY
Department of Pharmacology and Toxicology
University of Texas Medical Branch
Galveston, Texas

INTRODUCTION

The recognition of serotonin (5-hydroxytryptamine, 5-HT) as a neurotransmitter in both the central and peripheral nervous systems was facilitated by the now classical observations of Gaddum in the 1950s, which revealed multiple physiologic responses to this biogenic amine, and the identification of discrete serotonin-containing neuronal pathways by Dahlstrom and Fuxe in the 1960s. Today, 5-HT is recognized to have an important role in the regulation of numerous normal and human disease states, including pain and nociception, sleep, anxiety, depressive disorders, obesity, and feeding disorders, as well as cardiovascular and neuroendocrine function (Fuller, 1988; Lakoski and Haber, 1991). This indoleamine also mediates the psychopharmacology of abused substances including hallucinogens (Aghajanian *et al.*, 1987) and stimulants (Lakoski and Cunningham, 1988a; see Chapters 5 and 8). It is this interaction of the psychostimulant cocaine with serotonin neuronal function in the central nervous system (CNS) that will be the focus of this chapter. Specific neurochemical and behavioral actions of this stimulant will first be reviewed with respect to this neurotransmitter system, thereby providing a conceptual framework for the subsequent in-depth discussion of new insights provided by cellular eletrophysiologic approaches.

Neurochemical Profile of Cocaine on Serotonergic Function

A role for a potent interaction of cocaine with serotonergic neuronal systems emerged from studies investigating the ability of this stimulant to inhibit the re-uptake of biogenic amine neurotransmitters. Blackburn and colleagues (1967) and Ross and Renyi (1967) demonstrated that cocaine inhibited the re-uptake of 5-HT in neuronal tissue with an efficacy greater than the re-uptake inhibition of dopamine (DA) or norepinephrine (NE) (Ross and Renyi, 1969; Koe, 1976). This inhibitory effect on 5-HT transport processes results in an overall enhancement of 5-HT available at the synapse, with cocaine acting as an indirect 5-HT agonist. Thus, in addition to known actions of this stimulant at blocking Na^+ channels to mediate local anesthetic properties (Ritchie and Green, 1990), cocaine blocks the re-uptake of 5-HT into the presynaptic terminal. With respect to the binding site for cocaine on the transporter, Reith and colleagues (1983; 1985; Chapter 10) have demonstrated that cocaine binding sites are associated with DA and 5-HT transporter sites. The molecular mechanisms of such interactions at the re-uptake complex, however, remain to be addressed (Galloway, 1988).

Cocaine has also been demonstrated to modulate central serotonergic function via alteration of the synthesis of 5-HT. This stimulant has been identified to inhibit the re-uptake of the 5-HT precursor tryptophan, as well as enhance the activity of the rate-limiting enzyme, tryptophan hydroxylase (Knapp and Mandell, 1972), 1976; Schubert *et al.*, 1970). Decreases (Friedman *et al.*, 1975; Pradhan *et al.* 1978a, 1978b; Galloway, 1990), as well as no alterations (Taylor and Ho, 1977), have also been reported in the neuronal synthesis of 5-HT following acute administration of cocaine. These observations are consistent with the hypothesis that the potent neurochemical effects of this psychostimulant on brain function include a contribution of serotonergic neuronal systems.

Behavioral Profile of Cocaine on Serotonergic Function

Studies of the numerous behavioral effects of cocaine have also identified a role for serotonin-containing neuronal systems in mediating drug-induced behaviors. Potentiation of cocaine-induced enhancement of locomotor activity has been reported following either inhibition of 5-HT synthesis or selective 5-HT receptor blockade (Scheel-Kruger *et al.*, 1977). Additional studies have demonstrated the ability of the serotonin precursor 5-hydroxytryptophan to attenuate the locomotor enhancing effects of this stimulant (Pradhan *et al.*, 1978a).

Evidence is emerging from a variety of other behavioral paradigms that suggests that 5-HT may play an aversive role with respect to mediating the reward processes of psychoactive drugs (see Chapter 5). For example, lesions of ascending serotonergic pathways have been demonstrated to enhance the behavioral effects of systemic administration of amphetamine (Green and Harvey, 1974) and enhance morphine i.v. self-stimulation (Smith *et al.*, 1987). With respect to cocaine, serotonergic neuronal systems also have been identified to mediate both reinforcing (Lyness *et al.*, 1980; Porrino *et al.*, 1989; Carroll *et al.*, 1990a, 1990b) and nonreinforcing (Scheel-Kruger *et al.*, 1977; Pradhan *et al.*, 1978a) behavioral effects of this stimulant. Loh and Roberts (1990) have reported that discrete neurochemical lesions of ascending serotonergic pathways, via placement of 5,7-dihydroxytryptamine lesions in the medial forebrain bundle, enhance the apparent incentive value of cocaine. Similarly, 5-HT receptors in the dorsal raphe nucleus, but not the median raphe nucleus, mediate the action of the 5-HT_3 antagonist ondansetron to attenuate aversive behavior produced by cocaine administration.

Cellular Electrophysiologic Approaches to Assessment of Cocaine/Serotonin Interaction

The close scrutiny that the interaction of cocaine with serotonergic neuronal systems has received from both neurochemical and behavioral arenas must be viewed in the context of a pressing societal need to more fully understand the processes that underlie drug abuse and dependence (Koob and Bloom, 1988). From the vantage point that a widely abused stimulant, such as cocaine, must act on the molecules and cells of the nervous system in order to produce its numerous behavioral phenomena, it is important to address the pharmacologic and physiologic ramifications of such interactions. For example, with respect to a possible role of serotonin-containing neurons in mediating aversive effects of cocaine, do these effects involve neuronal pathways already implicated in reward circuitry (Koob and Bloom, 1988; see Chapters 4 and 5)? Do the serotonergic projections arising from the midbrain nuclei, in turn, selectively modify postsynaptic target regions in a specific and, perhaps, receptor selective manner? Can the new generation of agonists and antagonists selective for the multiplicity of serotonin receptors ultimately provide useful therapeutic agents in mediating long-term changes produced by chronic cocaine abuse? One approach to defining the cellular and molecular basis for the actions of cocaine in the CNS is the application of cellular electrophysiologic approaches to facilitate the direct assessment of 5-HT pharmacology and physiology. In addition to providing evidence of selective interaction of cocaine with this neurotransmitter system, this approach can help to provide a precise understanding of the functions of individual brain regions and, in turn, provide a framework for elucidating the integrative basis of nervous system function.

CELLULAR ELECTROPHYSIOLOGIC STUDIES: ACUTE COCAINE ADMINISTRATION

Utilizing the technology developed by Aghajanian and colleagues to record from identified monoamine-containing neurons in the CNS (Aghajanian *et al.*, 1970), extracellular recording studies of serotonin-containing neurons in the midbrain dorsal raphe nucleus (DRN) of the rat have been a useful neurobiological model for assessment of cocaine's action. These DRN neurons are advantageous as a model system for studies of the cellular and molecular physiological actions of cocaine, as they are well characterized with respect to their pharmacologic and physiologic properties (see Lakoski and Cunningham, 1988a, 1988b). 5-HT neurons in the DRN are inhibited by microionophoretic application of this agonist (Aghajanian *et al.*, 1970), hence the classification of this receptor as a 5-HT autoreceptor. Moreover, this serotonin somatodendritic receptor has been demonstrated to be of the 5-HT_{1A} subtype (Sprouse and Aghajanian, 1987). As illustrated in Figure 1, serotonin-containing neurons project from this nucleus to provide both stimulatory and inhibitory input to numerous forebrain regions. Several of these regions are associated with the "reward circuitry", including the nucleus accumbens and prefrontal cortex, and, therefore, brain areas that have merited consideration as sites of an interaction of 5-HT with cocaine.

Systemic and Microiontophoretic Administration of Cocaine on 5-HT DRN Neurons

The effects of cocaine on the spontaneous activity of 5-HT neurons in the DRN have been investigated in both the chloral hydrate (Cunningham and Lakoski, 1988,

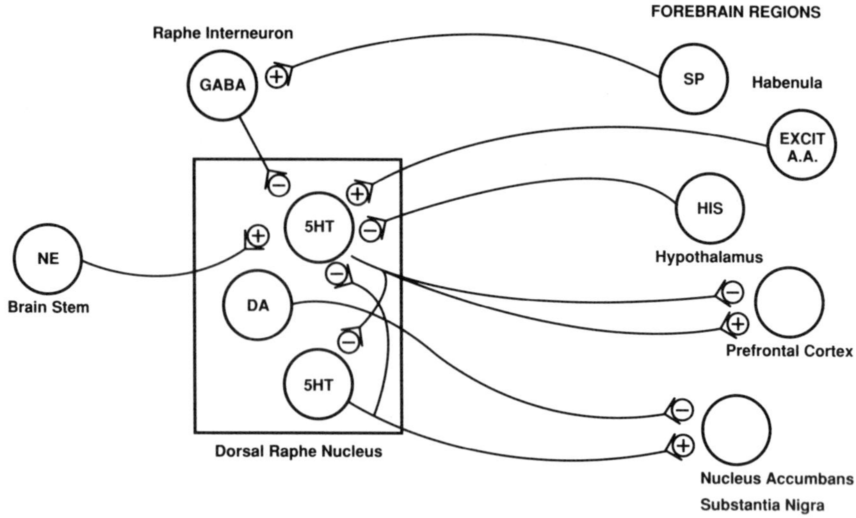

Figure 1. Diagram of the neurotransmitter afferent and efferent projections of the dorsal raphe nucleus. This schematic denotes the major identified neurotransmitter afferent and efferents of the midbrain dorsal raphe nucleus. Note that the terminal projections of this nucleus include the areas demonstrated to sustain self-stimulation behavior, including the nucleus accumbens and the prefrontal cortex. Serotonin (5-HT), dopamine (DA), norepinephrine (NE), γ-aminobutyric acid (GABA), substance P (SP), histamine (HIS), excitatory amino acids (EXCIT A.A.); (+) and (−) symbols denote stimulatory and inhibitory effects, respectively.

1990; Lakoski and Cunningham, 1988a, 1988b) and urethane (Pitts and Marwah, 1986, 1987) anesthetized *in vivo* male rat preparation using extracellular recording techniques. Intravenous administration of cocaine produces a potent and complete dose-dependent suppression of 5-HT cell firing (cumulative dose to produce 50% inhibition (ID_{50}) of 0.66 ± 0.11 mg/kg; Figure 2). As illustrated in Figure 2, systemic administration of the local anesthetic procaine, as well as lidocaine (Cunningham and Lakoski, 1990), does not significantly alter the activity of these neurons at cumulative doses up to 16 mg/kg. These data provide support for the hypothesis that these inhibitory effects of cocaine are not mediated via its known local anesthetic actions. While pharmacokinetic factors which control the distribution of these drugs to their sites of action might contribute to these observations, the doses of local anesthetic utilized have been previously demonstrated to have access to the brain (Riker *et al.*, 1981; Ritchie and Green, 1990) and produce changes in electroencephalographic activity (Tuttle and Elliott, 1969; Wagman *et al.*, 1967).

Microiontophoretic drug application techniques, used to complement these systemic studies in order to delineate the direct actions of cocaine on 5-HT DRN responses, have confirmed and extended these observations (Cunningham and Lakoski, 1988; Lakoski and Cunningham, 1988). Microiontophoretic application of cocaine (4 to 32 nA) consistently suppressed the spontaneous activity of these neurons in a current-dependent manner; while these responses recovered rapidly, continued application of cocaine over a several-minute period served to elicit an

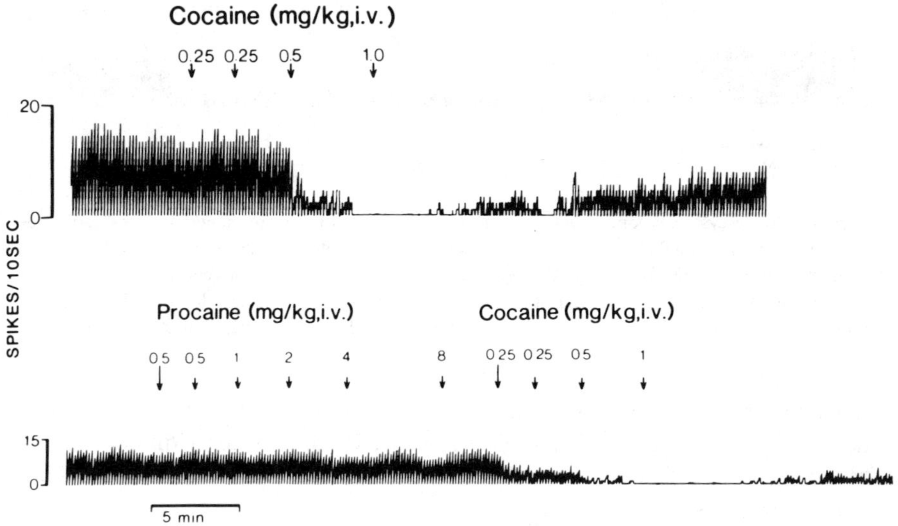

Figure 2. Effects of acute i.v. administration of cocaine and procaine on 5-HT-containing neurons recorded in the dorsal raphe nucleus of the rat. Data are frequency histograms obtained from extracellular single-unit recordings of spontaneously active neurons in the anesthetized preparation. Top panel: cocaine was administered in a cumulative dose manner (0.25, 0.25, 0.5, and 1.0 mg/kg, i.v.) as indicated at the arrows. This cell firing was inhibited with a cumulative dose of 2 mg/kg cocaine, i.v., and the effect began to reverse within a 5-min period. Bottom panel: Procaine was administered in a cumulative dose manner as indicated by the arrows to a total of 16 mg/kg, i.v., without any consistent alteration of basal activity. Cocaine was then subsequently administered to this cell in a cumulative dose manner and completely suppressed firing with a total dose of 2 mg/kg. Note: Both panels are data obtained from two separate recording preparations.

increasing magnitude of effect. As predicted from the known actions of cocaine as an indirect agonist, the co-microiontophoresis of cocaine significantly potentiated the inhibitory effects of 5-HT during the co-ejection with cocaine. Again, as observed with systemic administration of local anesthetics, the microiontophoretic application of procaine had little effect on the DRN cell firing rate. Thus, cocaine directly alters the activity of the midbrain 5-HT DRN neurons in a manner consistent with its action as a re-uptake inhibitor of biogenic amines.

Systemic administration of the levo- and detro-rotatory isomers of cocaine were also evaluated to establish the steroselectivity of the response of 5-HT DRN neurons to cocaine (Cunningham and Lakoski, 1990). The active isomer, (−)-cocaine completely suppressed cell firing with a potency similar to results obtained with commercial grade cocaine HCl; similar effects were not observed following i.v. of (+)-cocaine (Cunningham and Lakoski, 1990). The specificity of cocaine's inhibitory effects on serotonergic function was also assessed by the evaluation of the phenyltropane congener WIN 35428 and found to be a threefold more potent inhibitor of 5-HT neurons than cocaine (ID_{50} = 0.17 ± 0.03 mg/kg; Cunningham and Lakoski, 1990). While these data are in agreement with the observed greater potency of WIN 35428 to inhibit DA, NE, and 5-HT re-uptake mechanisms and to stimulate

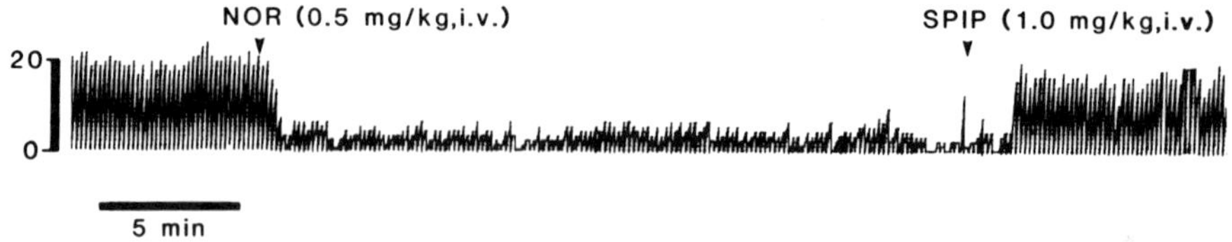

Figure 3. Acute systemic administration of norcocaine suppression of serotonergic dorsal raphe cell firing: attenuation of inhibitory effects with systemic administration of spiperone. In this frequency histogram of a spontaneously active 5-HT neuron, the i.v. administration of norcocaine (NOR; 0.5 mg/kg) significantly suppressed the basal cell firing rate within 1 min of drug administration. The i.v. administration of spiperone (SPIP; 1.0 mg/kg) rapidly restored the firing rate of this neuron to values observed before administration of norcocaine.

locomotor activity (Heikkila *et al.*, 1981; Reith *et al.*, 1986), the weaker local anesthetic profile of this analogue (Clarke *et al.*, 1973) provides additional support for the hypothesis that local anesthetic actions do not mediate this indoleamine-mediated response.

Utilizing parenteral drug administration in this *in vivo* model, metabolites of cocaine were evaluated for possible effects on 5-HT neuronal activity. The primary peripheral metabolite of cocaine, benzoylecgonine (Misra *et al,* 1974), and the primary central metabolite, norcocaine (Misra *et al.*, 1975; Nayak *et al.*, 1976), were evaluated following acute i.v. administration. While benzoylecgonine revealed little effect on DRN 5-HT cell firing (Cunningham and Lakoski, 1990), norcocaine proved to be as potent as cocaine in producing a reversible and complete suppression of neuronal activity (Figure 3; Lakoski and Paris, unpublished observations). While the reliability of the observations obtained with benzoylecgonine may be limited by its known restricted penetration of the blood brain barrier (Misra *et al.*, 1975), the direct interaction of norcocaine with central 5-HT neuronal responses is consistent with effects previously established at peripheral autonomic sites (Fozard *et al.*, 1983).

The specificity of this identified cocaine/serotonin interaction has been further addressed by the use of selective receptor antagonists. Systemic administration of antagonists selective for 5-HT$_2$ (ketanserin), terminal 5-HT autoreceptor (methiothepin), β-adrenergic (propranolol), dopamine D$_2$ (haloperidol), and γ-aminobutyric acid GABA$_A$ (picrotoxin) receptors were all found ineffective in attenuating cocaine's DRN effects. In contrast, the mixed 5-HT/D$_2$ antagonist spiperone, which has been shown to block 5-HT$_{1A}$ agonist-induced inhibition of DRN cell firing (Lum and Piercy, 1988), reversed the inhibition of 5-HT activity produced by cocaine both *in vivo* (Cunningham and Lakoski, 1990) and *in vitro* (Pan and Williams, 1989; Williams and Lacey, 1989). As illustrated in Figure 3, spiperone was also effective at reversing the inhibitory effects on cell firing in the DRN produced by i.v. administration of norcocaine (Lakoski and Paris, unpublished observations). Other pharmacologic antagonism of this response is produced by depletion of central neuronal stores of 5-HT with reserpine (Pitts and Marwah, 1987) or pretreatment with the serotonin synthesis inhibitor *p*-chlorophenylalanine (PCPA; Cunningham and Lakoski, 1990). Pretreatment with PCPA reduces the sensitivity of 5-HT DRN neurons to a subsequent systemic cocaine challenge as compared to saline-treated controls. These data support the hypothesis that a local inhibition of 5-HT re-uptake produced by the systemic administration of cocaine increases the 5-HT available for interaction at this autoreceptor. In turn, this indirect agonist action of cocaine modulates the cellular electrophysiologic responses in the DRN (see Chapter 14).

While these effects of cocaine on 5-HT DRN function may be mediated by a blockade of re-uptake processes which ultimately enhance serotonergic tone at the somatodendritic autoreceptor, this may perhaps be an oversimplification of the cellular mechanisms (see Pan and Williams, 1989). For example, cocaine has inhibitory effects on the norepinephrine-containing neurons recorded in the locus coeruleus (Pitts and Marwah 1987, 1989). As an α_1-noradrenergic-mediated excitatory input to the DRN has been established (Baraban and Aghajanian, 1980), the consequences of a pharmacologic alteration of this neuronal circuitry must be considered. Likewise, systemic administration of cocaine may be influencing other neuronal inputs to these DRN neurons, including those from GABA and histamine neurons (Lakoski and Aghajanian, 1983).

It is also important to recognize that the DRN includes non-5-HT-containing neurons, including DA neurons located in the rostral portion of this nucleus (Descarries *et al.*, 1986; Ochi and Shimizu, 1978). While dopamine has been shown to effect serotonergic function in the raphe nuclei (Chase, 1974; DeSimoni *et al.*, 1987; Lee and Geyer, 1984), DA is not co-localized within these 5-HT neurons (Geffard *et al.*, 1987). However, it is not yet known whether cocaine can effect the firing of DA and 5-HT neurons in this brain region in a differential or similar manner. Similarly, although as yet uninvestigated, an interaction of DA and 5-HT neuronal function in this midbrain nuclei may ensue from exposure to this psychostimulant.

Cocaine/Serotonin Interaction at DRN Terminal Projection Regions

The consequences of an interaction of cocaine with serotonergic function at postsynaptic terminal projection regions have only recently been investigated. In this regard, areas of the brain that have been recognized to mediate the reinforcing properties of cocaine (Koob and Bloom, 1988; see Chapter 5) have been the focus of current studies. One such forebrain region is the medial prefrontal cortex (PFC), which has been demonstrated to support self-administration behavior for this stimulant (Goeders and Smith, 1983) Not only does the PFC receive an extensive serotonergic innervation from the DRN, but numerous 5-HT receptor subtypes, including 5-HT_{1A}, 5-HT_2, and 5-HT_3 (Davies *et al.*, 1987; Ashby *et al.* 1989, 1990), have also been electrophysiologically characterized in this brain region. While marked inhibitory effects of microiontophoretically applied cocaine have been reported in the PFC (Qiao *et al.*, 1990; Lakoski and Moday, unpublished observations), the role of 5-HT in mediating these effects has yet to be established. Similarly, in the nucleus accumbens, another brain region critical to mediating the reinforcing effects of this stimulant, both *in vivo* (Qiao *et al.*, 1990; see White *et al.*, Chapter 12) and *in vitro* (Uchimura and North, 1990) electrophysiologic investigations have identified the ability of cocaine to modulate serotonergic-mediated responses.

CELLULAR ELECTROPHYSIOLOGIC STUDIES: CHRONIC COCAINE ADMINISTRATION

As the pharmacology of cocaine related to its abuse involves a pattern of repeated drug administration (Johanson and Fischman, 1989), the relevance of possible changes in serotonergic function produced by chronic cocaine administration merits consideration. With respect to the cellular electrophysiologic assessment of the consequences of long-term stimulant administration on this neurotransmitter system, preliminary investigations of repeated amphetamine exposure have identified both pharmacologic and physiologic changes in the receptor sensitivity at the 5-HT DRN autoreceptor (Heidenreich and Rebec, 1989). Again, the judicious application of cellular electrophysiologic recording techniques can provide much needed information with respect to effects produced by chronic cocaine on serotonergic neuronal systems, including both site- and receptor-specific interactions. Ultimately, such data will serve to enhance our understanding of the cellular and molecular mechanisms which may underlie the complex processes of tolerance and sensitization that are concomitant with repeated cocaine administration.

In addressing the physiologic consequences of long-term exposure to cocaine on 5-HT responses, a useful analogy is the results observed following chronic administration of tricyclic antidepressants, 5-HT$_{1A}$ agonists, and selective 5-HT re-uptake inhibitors (de Montigny and Aghajanian, 1978; Blier *et al.*, 1984; Chaput *et al.*, 1986; Blier and de Montigny, 1987; Blier *et al.*, 1988). This substantial body of work elegantly demonstrates numerous aspects of chronic drug administration that may be highly predictive of the consequences of a cocaine/serotonin interaction. As summarized in Figure 4, a region-specific modulation of serotonergic function is a consistent finding of this work. For example, changes in responsivity to 5-HT are divergent for presynaptic and postsynaptic locations following chronic administration of antidepressants or 5-HT$_{1A}$ agonists. However, any hypothesis of the long-term consequences of cocaine administration on 5-HT function must include the recognition of the neurochemical profile of cocaine as a potent blocker of biogenic amine re-uptake processes along with its lack of efficacy as an antidepressant. Therefore, perhaps the most appropriate pharmacologic analogy to the effects of chronic stimulant exposure is provided by studies of repeated administration of selective 5-HT re-uptake inhibitors, including zimelidine, indapaline, or fluoxetine (Chaput *et al.*, 1986; Blier and de Montigny, 1987; Blier *et al.*, 1988). In these studies, long-term drug pretreatment desensitizes the 5-HT-mediated responses recorded at both the somatodendritic autoreceptor in the DRN and at terminal projection sites in the hippocampus (Figure 4). Presumably, the desensitization observed by chronic 5-HT uptake inhibition is resultant from a greater concentration of 5-HT that is available in the synaptic cleft during the sustained re-uptake blockade. Indeed, support for such a mechanism has been provided by brain microdialysis techniques which have demonstrated acute increases in synaptic DA available in the striatum, ventral tegmental area, and PFC following administration of cocaine (DiChiara and Imperato, 1988; Akimoto *et al.*, 1989; Bradberry and Roth, 1989); the applicability of these microdialysis observations to the serotonergic system, however, remains to be definitively established.

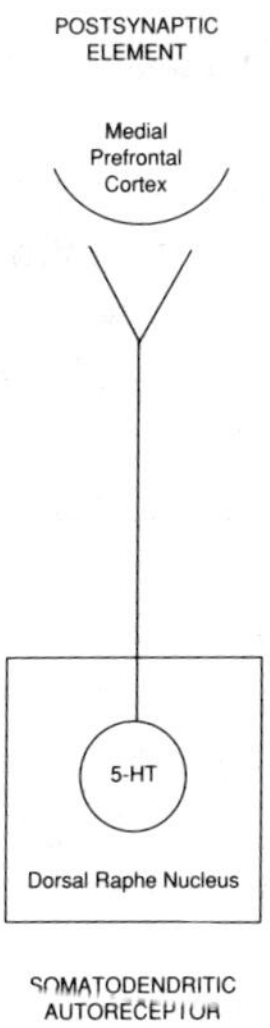

CHRONIC TREATMENT	BRAIN REGION	FOREBRAIN RESPONSE TO 5-HT
Tricyclic antidepressants (imipramine, chlorimipramine)	hippocampus, hypothalamus	↑ Sensitivity
$5\text{-}HT_{1A}$ agonist (gepirone)	hippocampus	No change
5-HT reuptake inhibitor (zimelidine, indapaline, fluoxetine)	hippocampus	↓ Sensitivity (tolerance)
Cocaine	**medial prefrontal cortex**	**HYPOTHESIS:**
		↓ Sensitivity (tolerance)

ACUTE TREATMENT	DRN RESPONSES
Tricyclic antidepressants	Inhibition of 5-HT cell firing
$5\text{-}HT_{1A}$ agonists	Inhibition of 5-HT cell firing
5-HT reuptake inhibitors	Inhibition of 5-HT cell firing
Cocaine	**Inhibition of 5-HT cell firing**

CHRONIC TREATMENT	DRN RESPONSES
Tricyclic antidepressants	No change to 5-HT
$5\text{-}HT_{1A}$ agonists	↓ Sensitivity to 5-HT, LSD
5-HT reuptake inhibitors	↓ Sensitivity to 5-HT
Cocaine	
Subchronic	**↓ Sensitivity to cocaine**
	No change to 5-HT
Chronic	**HYPOTHESIS:**
	↓ Sensitivity (tolerance) to cocaine
	↓ Sensitivity to 5-HT

Figure 4. Presynaptic and postsynaptic modulations of serotonergic receptor function: summary of results obtained from both acute and chronic drug administration of tricyclic antidepressants, 5-HT agonists, 5-HT re-uptake inhibitors, and stimulant (cocaine) administration on both somatodendritic autoreceptor and postsynaptic serotonin receptor function (see text for references to specific drug analysis). Highlighted in boldface type are both actual and predicted data for cellular electrophysiologic results obtained with cocaine.

Chronic Cocaine on 5-HT DRN Somatodentritic Autoreceptor Function

Our initial *in vivo* investigations of the effects of repeated or subchronic exposure to cocaine on 5-HT cell firing in the DRN have provided evidence of dose-dependent effects. In preliminary studies (Cunningham *et al.*, 1987), the pharmacologic profile of these neurons was investigated following two regimens of systemic cocaine administration: 10 mg/kg administered once daily for 10 d (i.p.) or 15 mg/kg administered twice daily for 7 d (i.p.) in male rats. At 24 h following completion of the once-daily cocaine treatment, the ability of an acute systemic cocaine challenge to inhibit 5-HT DRN activity was unchanged as compared to saline-treated controls. Similarly, following completion of the higher dose, twice-daily cocaine treatment, no significant differences were observed between treated and control animals in response to subsequent acute challenge with cocaine. However, in animals receiving subchronic cocaine twice daily for 7 d, at 24 h following termination of the treatment a significant reduction in the cumulative dose of the $5\text{-}HT_{1A}$ agonist 8-OH-DPAT [(8-hydroxy-2-(di-*n*-propyl-amino)-tetralin; i.v.)] needed to inhibit DRN cell firing was observed in cocaine vs. saline controls (IC_{50} = 1.33 ± 0.28 μg/

kg vs. 4.40 ± 1.31 μg/kg, respectively; data not shown). Additional neurochemical studies (Galloway and Lakoski, unpublished observations) measuring the levels of 5-HT and its metabolite 5-hydroxyindole acetic acid in the DRN of animals treated with this regimen of subchronic cocaine or saline failed to identify any alterations in this transmitter system. These data suggest that changes in the synthesis or turnover of 5-HT cannot solely account for the observed cocaine-induced change in the sensitivity of the 5-HT DRN autoreceptor.

This lack of effect on cocaine-mediated responses was somewhat puzzling as previous reports demonstrated that repeated administration of cocaine results in an increased sensitivity to the drug, as demonstrated by a variety of behavioral effects in numerous animal species (Stripling and Ellinwood, 1977; Robinson and Becker, 1986). Conversely, the continuous administration of cocaine via minipumps has been demonstrated to produce a tolerance to subsequent acute cocaine challenge in a time-dependent manner (Pitts and Marwah, 1989). Therefore, we evaluated responses in animals receiving a higher dose of cocaine (30 mg/kg, twice daily, i.p.) in a similar paradigm of subchronic drug administration (7 d) for subsequent *in vitro* electrophysiological and neurochemical studies of cocaine/5-HT interaction. Presumably in the earlier model the time period and/or pretreatment dosage were not sufficient to obtain the tolerance or sensitization effects induced by cocaine. In these studies, we also utilized a brain slice preparation for its numerous advantages, and these are discussed in detail in Chapter 14.

The repeated administration of cocaine produced a tolerance to the subsequent acute cocaine perfusion on 5-HT DRN neurons recorded *in vitro* as compared to saline-treated controls (Black and Lakoski, 1990). In the slice preparation continually exposed throughout the recording session to a low concentration of phenylephrine and 5-HT (1 μ*M*,) the concentration of cocaine required to inhibit the firing of 5-HT DRN neurons by 50% was greater for cocaine- vs. saline-treated controls (5.46 ± 0.50 μ*M* vs. 2.43 ± 0.35 μ*M*, respectively; Figures 5 and 6). Typically, cocaine concentrations of 10 μ*M* produced a complete inhibition of firing in the saline-treated controls, whereas cells recorded from the subchronic cocaine treatment required approximately 20 μ*M* to produce a complete cessation of firing. With this paradigm, no evidence for an acute development of tolerance from repeated perfusion with cocaine was evident in either treatment group (data not shown), as evaluated by repeated perfusion with identical concentrations of this stimulant. Interestingly, the careful evaluation of 5-HT-mediated responses revealed no differences in sensitivity produced by this subchronic cocaine regimen; equal concentrations of 5-HT produced identical inhibitory responses that were rapidly reversed by superfusion with the nonselective 5-HT antagonist spiperone.

These observations were complemented by radioligand observations of 5-HT_{1A} receptor function utilizing [^{3}H]8-OH-DPAT (Black and Lakoski, 1990). These studies addressed the question of whether the chronic cocaine-induced effects on serotonergic function (sensitization or tolerance) could be due to interactions at the 5-HT re-uptake site or the 5-HT_{1A} autoreceptor, or at both sites. Competitive radioligand binding studies carried out in membranes prepared from hippocampal tissue, which contains an enriched population of 5-HT_{1A} binding sites, demonstrated

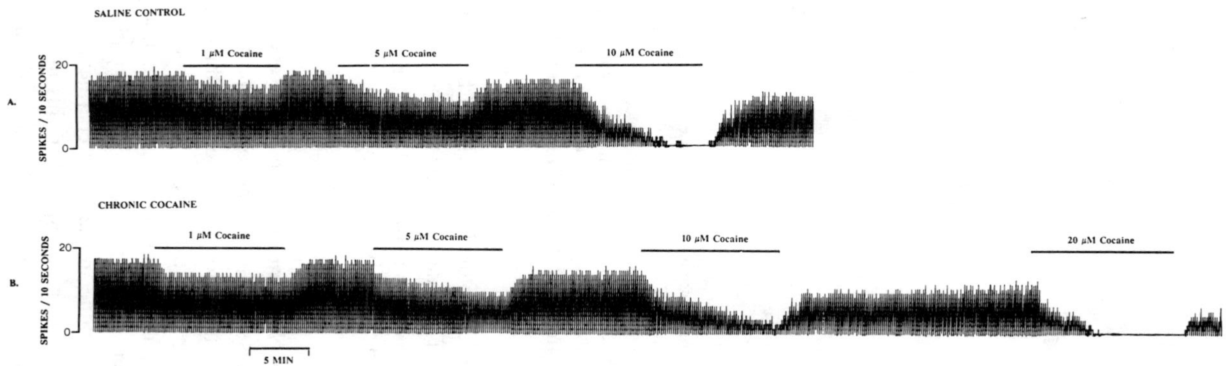

Figure 5. Effect of repeated cocaine administration of subsequent acute cocaine application to a serotonin-containing neuron recorded in an *in vitro* brain slice preparation of the rat dorsal raphe nucleus. The data are representative cumulative rate histograms which illustrate the effect of cocaine on neuronal cell firing rates determined using extracellular recording techniques. A: Cocaine administration to a 5-HT neuron from a saline-treated control. Typically, a concentration of 10 μ*M* cocaine was required to achieve complete cessation of cell firing. B: Cocaine administration to a 5-HT neuron from subchronic cocaine-treated group (30 mg/kg, twice daily for 7 d, i.p.). These neurons required acute perfusion of 20 μ*M* cocaine to achieve complete inhibition. (Data from E.W. Black and J.M. Lakoski. *Mol. Cell. Neurosci.* **1**: 84 (1990). With permission.)

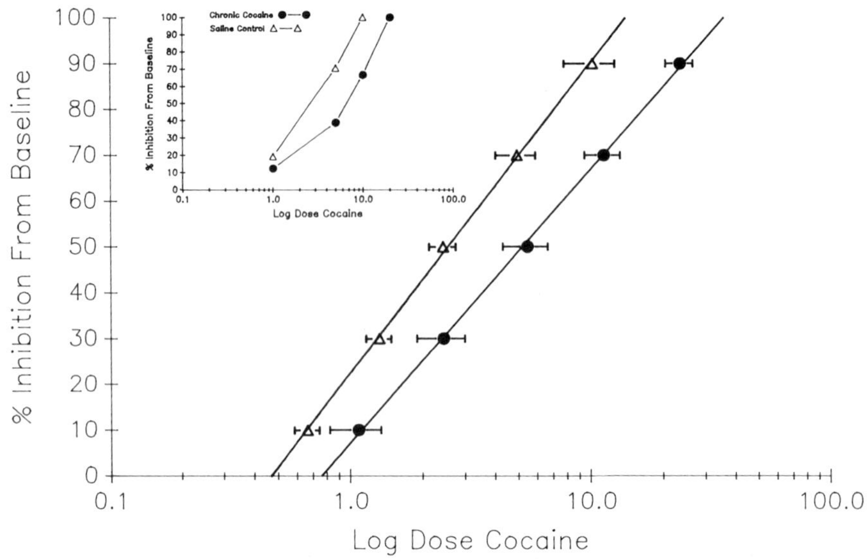

Figure 6. Summary of electrophysiologic results obtained from serotonin dorsal raphe neurons recorded *in vitro* of the acute cocaine challenge-induced inhibition observed following repeated cocaine administration. Data was recalculated from dose-response curves obtained for each neuron tested (see insert for an example) and graphed as mean values ± SEM. DRN serotonergic neurons from rats receiving repeated cocaine administration (closed circles) showed significant tolerance to acute cocaine application as compared to saline-treated controls (open triangles). (Data from E.W. Black and J.M. Lakoski. *Mol. Cell. Neurosci.* **1**: 84 (1990). With permission.)

no significant differences in the affinity (K_d) or receptor density (B_{max}) between subchronically cocaine- or saline-treated animals. These data indicate that the effects of chronic exposure to this stimulant are not due to a direct effect on the presynaptic autoreceptor, but are indirectly produced due to cocaine's inhibition of the re-uptake system.

Chronic Cocaine Administration on 5-HT DRN Terminal Projection Regions

Turning towards analysis of possible cocaine/serotonin interactions at postsynaptic sites, including brain regions where acute interactions have been established (e.g., nucleus accumbens or prefrontal cortex), the ramifications of chronic cocaine exposure on cellular physiological function have only recently received attention. Pitts and Marwah (1989) have established that repeated cocaine administration produces a functional tolerance to subsequent cocaine in the locus coeruleus, although the possible modulation of known 5-HT_2 receptor function in this nucleus was not investigated. Utilizing a model of long-term or chronic cocaine administration with 30 d of drug pretreatment (15 mg/kg, twice daily, i.p.), we have also identified a functional tolerance to the inhibitory effects produced by microiontophoretic application of cocaine in the prefrontal cortex of the rat (Lakoski and Moday, data not shown). However, the relevance of this shift in cocaine sensitivity observed in the PFC remains to be evaluated in light of known 5-HT_{1A}-, 5-HT_2- and 5-HT_3-mediated responses identified in this region. It remains to be seen if the pattern of monoamine receptor desensitization elucidated for the A10 DA pathway by White and colleagues (Henry *et al.*, 1989) will be a consistent feature of cocaine's interaction with other biogenic amines in the CNS.

SUMMARY

This review has focused on the body of literature obtained from electrophysiologic recording techniques that have examined the ability of acute and chronic cocaine to modulate central 5-HT function. A consensus is emerging from these preclinical studies that confirms an important interaction of this widely abused stimulant with indoleamine function at numerous neuronal sites. Indeed, as discussed by Galloway (1990), the importance of an interaction of dopamine-containing neuronal systems with serotonergic neurotransmitter function, each of which may be directly modulated by cocaine and, in turn, alter each other, deserves further elaboration. Likewise, the selective actions of cocaine on specific 5-HT, receptor subtypes and in a region-specific manner will continue to merit close attention.

It is important to emphasize that there are numerous clinical aspects of cocaine abuse that can gain urgently needed, new insights from such basic research studies. For example, the well-defined role of 5-HT in the etiology of several psychiatric disorders has prompted a concerted effort targeted at developing new drugs that are highly selective for 5-HT re-uptake sites and receptor subtypes (Charney *et al.*, 1990). As psychiatric complications are an all too common and severe problem concomitant with the use of cocaine, serotonergic compounds are now receiving careful consideration for their use in the treatment of this addiction (Gawin, 1988; Gawin and Ellinwood, 1988). As our collective understanding of the molecular and cellular basis of addiction continues to emerge (Koob and Bloom, 1988), the role of serotonergic neuronal systems in mediating such processes will be deserving of careful consideration.

ACKNOWLEDGMENTS

We wish to acknowledge the expert technical support of Ms. Kasha Wilhelm and Dr. J. Paris. Support for the preparation of this chapter was provided by the National Institute on Drug Abuse (DA 04296) and the John Sealy Memorial Endowment Fund at the University of Texas Medical Branch. JML is a recipient of a Research Career Development Award from the National Institute on Aging.

REFERENCES

G.K. Aghajanian, W.E. Foote, and M.H. Sheard. Action of psychotogenic drugs on single midbrain raphe neurons. *J. Pharmacol. Exp.* **171**: 178 (1970).

G.K. Aghajanian, J.S. Sprouse, and K. Rasmussen. Physiology of the midbrain serotonin system, in: *Psychopharmacology: The Third Generation of Progress*. H.Y. Meltzer, ed., Raven Press, New York. pp. 141–149 (1987).

K. Akimoto, T. Hamamura, and S. Otsuki. Subchronic cocaine treatment enhances cocaine-induced dopamine efflux, studied by *in vivo* intracerebral dialysis, *Brain Res.* **490**: 339 (1989).

C.R. Ashby, E. Edwards, K. Harkins, and R.Y. Wang. Characterization of 5-hydroxytryptamine$_3$ receptors in the medial prefrontal cortex: a microiontophoretic study. *Eur. J. Pharmacol.* **173**: 193 (1989).

C.R. Ashby, L.H. Jiang, R.J. Kasser, and R.Y. Wang. Electrophysiological characterization of 5-hydroxytryptamine$_2$ receptors in the rat medial prefrontal cortex, *J. Pharmacol. Exp. Ther.* **252**: 171 (1990).

J.W. Baraban and G.K. Aghajanian. Suppression of firing activity of 5-HT neurons in the dorsal raphe by alpha-adrenoceptor antagonists. *Neuropharmacology* **19**: 355 (1980).

E.W. Black and J.M. Lakoski. *In vitro* electrophysiology of dorsal raphe serotonergic neurons in subchronic cocaine treated rats: development of tolerance to acute cocaine administration. *Mol. Cell. Neurosci.* **1**: 84 (1990).

K.J. Blackburn, C.P. French, and R.J. Merrills. 5-hydroxytryptamine uptake by rat brain *in vitro*. *Life Sci.* **6**: 1653 (1967).

P. Blier, Y. Chaput, and C. de Montigny. Long-term 5-HT reuptake blockade, but not monoamine oxidase inhibition, decreases the function of terminal 5-HT autoreceptors: an electrophysiological study in rat brain. *Naunyn-Schmiedeberg's Arch. Pharmacol.* **337**: 246 (1988).

P. Blier, C. de Montigny, and D. Tardif. Effects of two antidepressant drugs mianserin and indalpine on the serotonergic system: single-cell studies in the rat. *Psychopharmacology* **84**: 242 (1984).

P. Blier and C. de Montigny. Modification of 5-HT neuron properties by sustained administration of the 5-HT$_{1A}$ agonist gepirone: electrophysiological studies in the rat brain. *Synapse* **1**: 470 (1987).

C.W. Bradberry and R.H. Roth. Cocaine increases extracellular dopamine in rat nucleus accumbens and ventral tegmental area as shown by *in vivo* microdialysis. *Neurosci. Lett.* **103**: 96 (1989).

M.E. Carroll, S.T. Lac, M. Asencio, and R. Kraugh. Fluoxetine reduced intravenous cocaine self-administration in rats. *Pharmacol. Biochem. Behav.* **35**: 237 (1990a).

M.E. Carroll, S.T. Lac, M. Asencio, and R. Kraugh. Intravenous cocaine self-administration in rats is reduced by dietary *l*-tryptophan. *Psychopharmacology* **100**: 293 (1990b).

Y. Chaput, C. de Montigny and R. Blier. Effects of a selective 5-HT reuptake blocker, citalopram, on the sensitivity of 5-HT autoreceptors: electrophysiological studies in the rat brain. *Naunyn-Schmiedeberg's Arch. Pharmacol.* **333**: 342 (1986).

D.S. Charney, J. H. Krystal, P.L. Delgado, and G.R. Heniger. Serotonin-specific drugs for anxiety and depressive disorders. *Annu. Rev. Med.* **41**: 437 (1990).

T.N. Chase. Serotonergic-dopaminergic interactions and extrapyramidal function. *Adv. Biochem. Psychopharmacol.* **11**: 377 (1974).

R.L. Clarke, S.J. Daum, A. J. Gambino, M.D. Aceto, J. Pearl, M. Levitt, W.R. Cumiskey and E.F. Bogado. Compounds affecting the central nervous system: 4.3 β-phenyltropane-2-carboxylic esters and analogs. *J. Med. Chem.* **16**: 1260 (1973).

K.A. Cunningham, E.K. Asprodini, N.A. Bernau, C.A. Richard, and J.M. Lakoski. Enhanced inhibitory responses of serotonin neurons in the dorsal raphe nucleus (DRN) after repeated cocaine exposure. *Soc. Neurosci. Abstr.* **13, Part 3**: 1651 (1987).

K.A. Cunningham and J.M. Lakoski. Electrophysiological effects of cocaine and procaine on dorsal raphe serotonin neurons. *Eur. J. Pharmacol.* **148**: 457 (1988).

K.A. Cunningham and J.M. Lakoski. The interaction of cocaine with serotonin dorsal raphe neurons: single-unit extracellular recording studies. *Neuropsychopharmacology* **3**: 41 (1990).

M.F. Davies, R.A. Deisz, D.A. Prince, and S.J. Peroutka. Two distinct effects of 5-hydroxytryptamine on single cortical neurons. *Brain Res.* **423**: 347 (1987).

C. de Montigny and G.K. Aghajanian. Tricyclic antidepressants: long-term treatment increases responsivity of rat forebrain neurons to serotonin. *Science* **202**: 1303 (1978).

L. Descarries, F. Berthelet, S. Garcia, and A. Beaudet. Dopaminergic projection from the nucleus raphe dorsalis to neostriatum in the rat. *J. Comp. Neurol.* **249**: 511 (1986).

M.G. DeSimoni, G. Dal Taso, F. Fodritto, A. Sokola, and S. Algeri. Modulation of striatal dopamine metabolism by the activity of dorsal raphe serotonergic afferents. *Brain Res.* **411**: 81 (1987).

G. DiChiara and A. Imperato. Drugs abused by humans preferentially increase synaptic dopamine concentrations in the mesolimbic systems of freely moving rats. *Proc. Natl. Acad. Sci., U.S.A.*, **85**: 5274 (1988).

T.V. Dunwiddie, J.T. William, D.H. Bobker, Z.Z. Pan, G.A. Gerhart, and M.S. Brodie. *In vitro* approaches to characterizing the electrophysiological effects of cocaine upon the central nervous system. In: *Cocaine: Pharmacology, Physiology and Clinical Strategies.* J.M. Lakoski, M.P. Galloway and F.J. White, eds., CRC Press, Boca Raton, FL, (1991).

J.R. Fozard. Differences between receptors for 5-hydroxytryptamine on autonomic neurones revealed by nor-(−)-cocaine. *J. Auton. Pharmacol.* **3**: 21 (1983).

E. Friedman, S. Gershon, and J. Rotrosen. Effects of acute cocaine treatment on the turnover of 5-hydroxytryptamine in the rat brain. *Br. J. Pharmacol.* **54**: 61 (1975).

R.W. Fuller. The pharmacology and therapeutic potential of serotonin receptor agonists and antagonists, in: *Advances in Drug Research.* Vol. 17. Academic Press, New York, p. 350 (1988).

M.P. Galloway. Neurochemical interactions of cocaine with dopaminergic systems. *Trend Pharmacol. Sci.* **9**: 451 (1988).

M.P. Galloway. Regulation of dopamine and serotonin synthesis by acute administration of cocaine. *Synapse* **6**: 63 (1990).

M.P. Galloway. Neuropharmacology of cocaine: effects on dopamine and serotonin systems, in: *Cocaine: Pharmacology, Physiology and Clinical Strategies.* J.M. Lakoski, M.P. Galloway, and F.J. White, eds., CRC Press, Boca Raton, FL (1991).

F.H. Gawin. Chronic neuropharmacology of cocaine: progress in pharmacotherapy. *J. Clin. Psychiatry* **49** (Suppl): 11 (1988).

F.H. Gawin and E.H. Ellinwood, Jr. Cocaine and other stimulants: actions, abuse and treatment. *N. Engl. J. Med.* **318**: 1173 (1988).

M. Geffard, S. Tuffet, N. Mons, and J.L. Chagnaud. Simultaneous detection of indoleamines and dopamine in rat dorsal raphe nuclei using specific antibodies. *Histochemistry* **88**: 61 (1987).

N.E. Goeders and J.E. Smith. Cortical dopaminergic involvement in cocaine reinforcement. *Science* **221:** 773 (1983).

T.K. Green and J.A. Harvey. Enhancement of amphetamine action after interruption of ascending serotonergic pathways. *J. Pharmacol. Exp. Ther.* **190**: 109 (1974).

B.A. Heidenreich and G.V. Rebec. Serotonergic dorsal raphe neurons: changes in spontaneous neuronal activity and responsiveness to 5-MeODMT following long-term amphetamine administration. *Neurosci. Lett.* **103**: 81 (1989).

R.E. Heikkila, F.S. Cabbat, and R.C. Duvoisin. Motor activity and rotational behavior after analogs of cocaine: correlations with dopamine uptake blockers. *Commun. Psychopharmacol.* **3**: 285 (1979).

D.J. Henry, M.A. Greene, and F.J. White. Electrophysiological effects of cocaine in the mesoaccumbens dopamine system: repeated administration. *J. Pharmacol. Exp. Ther.* **251**: 833 (1989).

C.-E. Johanson and M.W. Fischman. The pharmacology of cocaine related to its abuse. *Pharmacol. Rev.* **41**: 3 (1989).

S. Knapp and A.J. Mandell. Narcotic drugs: effects on the serotonin biosynthesis systems of the brain. *Science* **177**: 1209 (1972).

S. Knapp and A.J. Mandell. Cocaine and lithium: neurobiological antagonism in the serotonin biosynthetic system in the rat brain. *Life Sci.* **18:** 679 (1976).

B.K. Koe. Molecular geometry of inhibitors of the uptake of catecholamines and serotonin in synaptosomal preparations of rat brain. *J. Pharmacol. Exp. Ther.* **199**: 649 (1976).

G.F. Koob and F.E. Bloom. Cellular and molecular mechanisms of drug dependence. *Science* **242**: 715 (1988).

J.M. Lakoski and G.K. Aghajanian. Effects of histamine, H_1- and H_2-receptor antagonists on the activity of serotonergic neurons in the dorsal raphe nucleus. *J. Pharmacol. Exp. Ther.* **227**: 517 (1983).

J.M. Lakoski and K. Cunningham. Cocaine interaction with central monoaminergic systems: electrophysiological approaches. *Trend Pharmacol. Sci.* **9**: 177 (1988a).

J.M. Lakoski and K. Cunningham. The interaction of cocaine with central serotonergic neuronal systems: cellular electrophysiologic approaches. In: *Mechanisms of Cocaine Abuse and Toxicity, NIDA Research Monograph 88*. D. Clouet, K. Asghar, and R. Brown, eds., National Clearing House for Alcohol and Drug Information, Washington, DC, p. 78 (1988b).

J.M. Lakoski and B. Haber. Role of serotonin in the nervous system, in: *Encyclopedia of Human Biology*. R. Dulbecco, ed., Academic Press, San Diego, CA (1991).

E.H.Y. Lee and M.A. Geyer. Dopamine autoreceptor mediation of the effects of apomorphine on serotonin neurons. *Pharmacol. Biochem. Behav.* **21**: 301 (1984).

E.A. Loh and C.S. Roberts. Break-points on a progressive ratio schedule reinforced by intravenous cocaine increase following depletion of forebrain serotonin. *Psychopharmacology* **101**: 262 (1990).

J.T. Lum and M.F. Piercy. Electrophysiological evidence that spiperone is an antagonist of $5\text{-}HT_{1A}$ receptors in the dorsal raphe nucleus. *Eur. J. Pharmacol.* **149**: 9 (1988).

W.H. Lyness, N.M. Friedle, and K.E. Moore. Increased self-administration of D-amphetamine after destruction of 5-hydroxytryptamine neurons. *Pharmacol. Biochem. Behav.* **12**: 937 (1980).

A.L. Misra, P.K. Nayak, M.N. Patel, N.L. Vadlamani, and S.J. Mule. Identification of norcocaine as a metabolite of ^{3}H-cocaine in rat brain. *Experientia* **30**: 1312 (1974).

A.L. Misra, P.K. Nayak, R. Bloch, and S.J. Mule. Estimation and disposition of [^{3}H]benzoylecgonine and pharmacological activity of some cocaine metabolites. *J. Pharm. Pharmacol.* **27**: 784 (1975).

P.K. Nayak, A.L. Misra and S.J. Mule. Physiological disposition and biotransformation of [^{3}H]cocaine in acutely and chronically treated rats. *J. Pharmacol. Exp. Ther.* **196**: 556 (1976).

J. Ochi and K. Shimizu. Occurrence of dopamine-containing neurons in the midbrain raphe nuclei of the rat. *Neurosci. Lett.* **8:** 317 (1978).

Z.Z. Pan and J.T. Williams. Differential actions of cocaine and amphetamine on dorsal raphe neurons *in vitro*. *J. Pharmacol. Exp. Ther.* **251**: 56 (1989).

D.K. Pitts and J. Marwah. Electrophysiologic effects of cocaine on central monoaminergic neurons. *Eur. J. Pharmacol.* **131**: 95 (1986).

D.K. Pitts, and J. Marwah. Cocaine modulation of central monoaminergic neurotransmission. *Pharmacol. Biochem. Behav.* **26**: 453 (1987).

D.K. Pitts and J. Marwah. Chronic cocaine reduces α_2-adrenoreceptor elicited mydriasis and inhibition of locus coeruleus neurons. *Eur. J. Pharmacol.* **160**: 201 (1989).

L.J. Porrino, M.C. Ritz, N.L. Goodman, L.G. Sharpe, M.J. Kuhar, and S.R. Goldber. Differential effects of the pharmacologic manipulation of serotonin systems on cocaine and amphetamine self-administration in rats. *Life Sci.* **45**: 1529 (1989).

S.N. Pradhan, A.K. Bhattacharyya and S. Pradhan. Serotonergic manipulation of the behavioral effects of cocaine in rats. *Commun. Psychopharmacol.* **2**: 481 (1978a).

S.N. Pradhan, S.N. Roy, and S. Pradhan. Correlation of behavioral and neurochemical effects of acute administration of cocaine in rats. *Life Sci.* **22**: 1737 (1978b).

J.T. Qiao, P.M. Dougherty, R.C. Wiggins, and N. Dafny. Effects of microiontophoretic application of cocaine, alone and with receptor antagonists, upon the neurons of the medial prefrontal cortex, nucleus accumbens and caudate nucleus of rats. *Neuropharmacology* **29**: 370 (1990).

M.E.A. Reith, S.S. Kim, and A. Lajtha. Structural requirements for cocaine congeners to interact with ^{3}H-batrachotoxinin A 20-α-benzoate binding sites on sodium channels in mouse brain synaptosomes. *J. Biol Chem.* **261**: 7300 (1986).

M.E.A. Reith, B.E. Meisler, H. Sershen, and A. Lajtha. Sodium-independent binding of [^{3}H]cocaine in mouse striatum is serotonin related. *Brain Res.* **342**: 145 (1985).

M.E.A. Reith, H. Sershen, D.L. Allen and A. Lajtha. A portion of [^{3}H]cocaine binding in brain is associated with serotonergic neurons. *Mol. Pharmacol.* **23**: 600 (1983).

M.E.A. Reith. Central and peripheral cocaine binding sites, in: *Cocaine: Pharmacology, Physiology and Clinical Strategies*. J.M. Lakoski, M.P. Galloway, and F.J. White, eds., CRC Press, Boca Raton, FL (1991).

J.M. Ritchie and N.M. Greene. Local anesthetics, in: *Goodman and Gilman's The Pharmacologic Basis of Therapeutics* A.G. Gilman, T.W. Rall, A.S. Nies, and R. Taylor, eds., Pergamon press, Elmsford, NY, p. 311 (1990).

D.C.S. Roberts. Neural substrates mediating cocaine reinforcement: the role of monoamine systems, in: *Cocaine: Pharmacology, Physiology, and Clinical Strategies*. J.M. Lakoski, M.P. Galloway, and F.J. White, eds., CRC Press, Boca Raton, FL (1991).

T.E. Robinson and J.B. Becker. Enduring changes in brain and behavior produced by chronic amphetamine administration: a review and evaluation of animal models of amphetamine psychosis. *Brain Res. Rev.* **11**: 157 (1986).

D.K. Riker, R.A. Zito, and R.H. Roth. Lidocaine selectively diminishes meduallary serotonin metabolism in the rat. *Neuropharmacology* **20**: 1233 (1981).

S.B. Ross and A.L. Renyi. Inhibition of the uptake of tritiated catecholamines by antidepressants and related agents. *Eur. J. Pharmacol.* **2**: 181 (1967).

S.B. Ross and A.L. Renyi. Inhibition of the uptake of tritiated 5-hydroxytryptamine in brain tissue. *Eur. J. Pharmacol.* **7**: 270 (1969).

J. Scheel-Kruger, C. Braestrup, M. Nielson, K. Goembrowska, and E. Mogilnicka. Cocaine: discussion on the role of DA in the biochemical mechanism of action, in: *Cocaine and Other Stimulants*. E.H. Ellinwood and M.M. Kilbey, eds., Plenum Press, New York, p. 373 (1977).

J. Schubert, B. Fyro, K. Nyback, and G. Sedvall. Effects of cocaine and amphetamine on the metabolism of tryptophan and 5-hydroxytryptamine in mouse brain *in vivo*. *J. Pharm. Pharmacol.* **22**: 860 (1970).

J.E. Smith, K. Shultz, C. Co, N.E. Goeders, and S.I. Dworkin. Effects of 5,7-dihydroxytryptamine lesions of the nucleus accumbens on rat intravenous morphine self-administration. *Pharmacol. Biochem. Behav.* **26**: 607 (1987).

J.S. Sprouse and G.K. Aghajanian. Electrophysiologic responses of serotonergic dorsal raphe neurons to 5-HT_{1A} and 5-HT_{1B} agonists. *Synapse* **1**: 3 (1987).

J. Stripling and E. Ellinwood. Sensitization to cocaine following chronic administration in the rat, in: *Cocaine and Other Stimulants*. E.H. Ellinwood and M.M. Kilbey, eds. Plenum Press, New York, p. 327 (1977).

D. Taylor and B.T. Ho. Neurochemical effects of cocaine following acute and repeated injection. *J. Neurosci. Res.* **3**: 95 (1977).

W.W. Tuttle and H.W. Elliott. Electrographic and behavioral studies of convulsant in the cat. *Anesthesiology* **30**: 48 (1969).

N. Uchimura and R.A. North. Actions of cocaine on rat nucleus accumbens neurones *in vitro*. *Br. J. Pharmacol.* **99**: 736 (1990).

I.H. Wagman, R.H. de Jong, and D.A. Prince. Effects of lidocaine on the central nervous system. *Anesthesiology* **28**: 155 (1967).

F.J. White, D.J. Henry, X.T. Hu, M. Jeriorski, and J.M. Ackerman. Electrophysiological effects of cocaine in the mesoaccumbens dopamine system, in: *Cocaine: Pharmacology, Physiology and Clinical Strategies*. J.M. Lakoski, M.P. Galloway, and F.J. White, eds., CRC Press, Boca Raton, FL (1991).

J.T. Williams and M.G. Lacey. Actions of cocaine on central monoamine neurons: intracellular recordings *in vitro*. *Res. Mon. Natl. Inst. Drug Abuse* **50**: 15 (1989).

14

In Vitro Approaches to Characterizing the Electrophysiological Effects of Cocaine upon the Central Nervous System

THOMAS V. DUNWIDDIE[†], JOHN T. WILLIAMS*, DANIEL H. BOBKER*, ZHI-ZHONG PAN*, GREG A. GERHARDT[†], AND MARK S. BRODIE[†]

[†]*Departments of Pharmacology and Psychiatry*
University of Colorado Health Sciences Center
Denver, Colorado

**Vollum Institute*
Oregon Health Science University
Portland, Oregon

INTRODUCTION

Despite the fact that over a century has passed since the discovery that cocaine has significant pharmacological effects upon the central nervous system (CNS), the precise manner in which cocaine affects neural activity remains unclear. In general terms, there is already a great deal known concerning the mechanisms by which cocaine can act, but how actions at the cellular level translate into alterations in neural activity and ultimately behavior is not well understood.

Cellular Mechanisms of Cocaine Action

Behavioral neuropharmacological approaches have established that cocaine is a particularly potent reinforcer of behavior and, although the biological bases for these actions are not entirely resolved, it appears that dopamine-containing pathways to cortical and limbic structures are important elements in these reward pathways (Wise and Bozarth, 1987; Chapters 5 and 6). "Bottom-up" types of approaches have established a number of actions of cocaine at the cellular level. In particular, inhibition of high-affinity uptake of norepinephrine (NE), serotonin (5-HT), and dopamine (DA) can all occur at submicromolar concentrations of cocaine (Hertting *et al.*, 1961; Ross and Renyi, 1966; Koe, 1976; Hyttel, 1982). As is the case with the behavioral studies, the neurotransmitter that seems most likely to be directly involved in the reinforcing properties of cocaine is DA. Ritz *et al.* (1987) demonstrated that the potency of a variety of uptake inhibitors as behavioral reinforcers correlated most strongly with their ability to inhibit the uptake of DA; the ability to inhibit the uptake of NE and 5-HT appeared to have little relevance to their rewarding properties.

In addition to being an inhibitor of the re-uptake of monoamine transmitters, cocaine also has local anesthetic actions upon neural tissue, and continues to be used clinically for this purpose. The relative importance of local anesthetic activity to the behavioral aspects of cocaine action is unclear; although there is some evidence that animals will self-administer other local anesthetics (Ford and Balster, 1977), they are not particularly potent in this regard, and this is an action that is not common to all such drugs.

Until several years ago, what was lacking in our understanding of the actions of cocaine was the way in which this drug affected activity in the "middle ground", that is what types of cellular activity arise as a consequence of inhibition of monoamine uptake, and how these translate into cocaine-related behaviors. Prior to the 1980s, the only electrophysiological studies of cocaine action on the brain were studies that used field potential or EEG recording of cocaine-induced changes in neural activity (Ellinwood *et al.*, 1977; Herning *et al.*, 1985; Lesse and Collins, 1979; Scholfield and Harvey, 1975; Lesse and Harper, 1985; Stripling, 1982). Although these types of approaches may be useful in identifying brain regions that are affected by cocaine, they provide little if any information concerning the mechanisms by which such changes can occur. In addition, when drugs are injected systemically, even identifying the site of drug action can become difficult since there are few means of determining whether changes in a given region are direct or are mediated via remote loci of drug action.

For these reasons, our laboratories began several years ago to characterize the actions of cocaine on isolated, *in vitro* preparations of the rat brain. There are a variety of reasons why such preparations are advantageous for these kinds of studies (Dunwiddie *et al.*, 1983; Dunwiddie, 1986). One of the most important advantages of the brain slice preparation stems from the fact that many factors that are difficult to regulate under *in vivo* conditions can be brought under experimental control. The composition of the medium, temperature, oxygenation, etc., can all be manipulated. Moreover, the slice is isolated from indirect effects that might result from cocaine's cardiovascular effects. Another major advantage of this preparation is that it enables one to establish directly the site of drug action. In the intact animal, cocaine can modify neuronal activity either directly or indirectly via actions on afferents to the regions being studied, which might entail effects either upon the nerve terminals or upon the cell bodies that give rise to these afferents. Not only are the neurons of the slice isolated from the rest of the brain, but each neuron can be functionally isolated from other neurons in the slice by manipulating the ionic content of the medium. In addition, because of the stability and accessibility of the preparation, it is possible to record intracellularly and to use voltage clamp techniques to measure drug effects on the membrane properties of single neurons. Finally, it is possible in pharmacological studies to precisely control the concentrations of drug that give rise to specific pharmacological responses. Particularly in the case of drugs such as cocaine, which have multiple mechanisms of action within a relatively limited dose range, it is important to establish the concentration of drug that is producing the observed response. What is lacking, of course, from these kinds of studies is the overall view of what cocaine is doing in the brain; *in vitro* slices are inherently limited because the disruption of normal inputs from other parts of the brain limits the end response to only the direct actions of the drug. Thus, although in some cases we have characterized the effects of cocaine on both monoaminergic cell bodies (e.g., the locus coeruleus) and in projection areas (e.g., hippocampus), it is still difficult to predict what effects would be observed in an intact system. Nevertheless, these studies provide an important basis for understanding the way in which cocaine affects neuronal activity at the cellular level; once this background has been fully established, it may then be possible to extend these studies back into intact animals and, ultimately, to perhaps relate these electrophysiological changes to specific cocaine-induced behaviors.

ACTIONS OF COCAINE ON MONOAMINE-CONTAINING NEURONS

Ventral Tegmental Area and Substantia Nigra

Among the brain regions that seem most intimately involved in the rewarding properties of cocaine, the DA-containing neurons in the ventral tegmental area (VTA), and the mesolimbic and mesocortical regions that receive projections from this brain region, appear to be centrally involved (Wise 1987; Wise and Bozarth, 1987; Chapters 4 and 5 this volume). With this in mind, we initiated experiments in *in vitro* slices of the VTA, as well as the substantia nigra, that were intended to characterize the cellular actions of cocaine in these brain regions.

In initial studies, we characterized the basic physiology of putative DA-containing neurons in the VTA (A10 cell group) and the direct effects of DA on spontaneous activity in this brain region. Previous studies have demonstrated that the DA-containing cells of this brain region fire spontaneously with highly characteristic extracellular waveforms; the spikes themselves are unusually broad, often have an inflection on the rising phase of the waveform, and have a biphasic positive-negative shape when recorded extracellularly (Yim and Mogenson, 1980; Wang, 1981a; Wang, 1981b). Cells having these properties could be readily recorded from VTA brain slices, but unlike the situation *in vivo*, where the DA-containing cells often fire in with a characteristic bursting pattern, virtually all of the cells recorded in slices fired with a slow (1–5 Hz) and extremely regular pattern of firing (Brodie and Dunwiddie, 1987).

In the intact animal, DA released from dendritic sites acts upon D_2 receptors to inhibit the firing of the DA-containing cells of the VTA, much as it does in the substantia nigra (Aghajanian and Bunney, 1977; White and Wang, 1984). In our

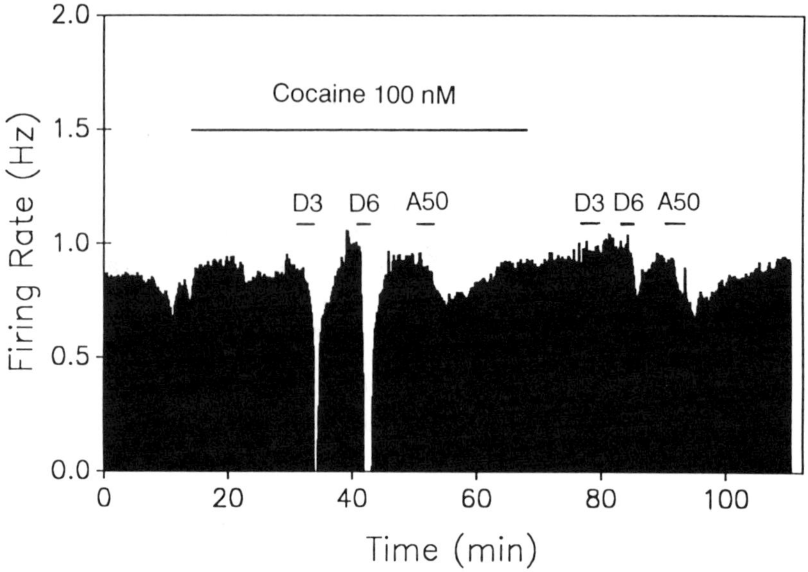

Figure 1. Electrophysiological responses recorded from a ventral tegmental area neuron *in vitro*. An extracellular recording of spontaneous activity was made from a single ventral tegmental area neuron in an *in vitro* slice preparation, and identified as a putative DA-containing neuron on the basis of its characteristic spike shape and duration, and sensitivity to inhibition by DA (Brodie and Dunwiddie, 1987). In the presence of 100 n*M* cocaine, superfusion with 3 and 6 μ*M* DA completely inhibited cell firing, whereas, under control conditions (following washout of the cocaine), 3μ*M* DA had no effect, and 6 μ*M* DA elicited about a 10% decrease in firing rate. Apomorphine (50 n*M*), a DA agonist that is not a substrate for re-uptake, elicited nearly identical responses under the two conditions.

in vitro experiments, DA also inhibited A10 neurons in a fashion that was dose dependent, readily reversible, and could be blocked by superfusion with the D_2 selective antagonist sulpiride (Brodie and Dunwiddie, 1987; Lacey *et al.*, 1987). Intracellular recording experiments demonstrated that this inhibition of firing occurred as a result of a hyperpolarization of the resting membrane potential that resulted from the activation of an outward current that could also be blocked by sulpiride (Lacey *et al.*, 1987).

Cocaine/DA Interactions

As might be expected, based upon cocaine's ability to inhibit the high-affinity uptake of DA, low doses of cocaine potentiated inhibitory responses of VTA neurons to superfusion with subthreshold concentrations of DA, but had no effect upon responses to apomorphine, which is not a substrate for re-uptake (Figure 1). In intracellular experiments, the hyperpolarization or outward current induced by DA (3 μ*M*) was increased more than 200% in the presence of 10 μ*M* cocaine. On the other hand, the hyperpolarization induced by 100 μ*M* DA was not affected by cocaine at all. This concentration of DA alone caused such a large hyperpolarization that blockade of re-uptake had no further effect, most probably because the response was already maximal.

Direct Effects of Cocaine

When VTA brain slices were superfused with low, micromolar concentrations of cocaine alone, the rate of spontaneous firing dropped markedly and could be completely inhibited in some cells. Several experiments suggest that this effect is due to a potentiation of the autoinhibitory effects of endogenous DA, rather than to a direct action of cocaine. These inhibitory effects could be readily blocked by concentrations of sulpiride that were sufficient to antagonize direct responses to DA (Figure 2; see also Lacey *et al.*, 1987). In addition, the mechanism underlying the direct response to cocaine was, like the direct DA effect, an activation of a potassium conductance. Finally, changes in rate were not accompanied by significant alterations in the height or shape of the extracellularly recorded action potential, as would be expected with a local anesthetic. In general, the responses to cocaine were quantitatively smaller than the responses to DA; the outward current induced by cocaine (10 μM) was about 33% of the maximum DA current.

Perhaps the most surprising aspect of these responses was the fact that similar actions occurred under conditions where synaptic transmission was blocked. In medium containing 0.25 m*M* calcium and 20 m*M* magnesium, regular spontaneous firing was maintained (albeit often at a somewhat higher rate), and the rate of firing could still be inhibited significantly by cocaine (Figure 3). The cocaine-induced hyperpolarization was also not changed when the slice was treated with tetrodotoxin, indicating that full amplitude action potentials were not necessary to support the release process (Lacey *et al.*, 1990). Several explanations might account for these observations. It seems likely that there is a pool of DA that is released spontaneously in a calcium-independent manner, and there is some biochemical evidence in support of this hypothesis. Woodward *et al.* (1988) have demonstrated both calcium-dependent and calcium-independent components of DA release from synaptosomes, and have hypothesized that the calcium-independent component may reflect a reversal of the DA transport system. A number of other transmitter systems also seem to have significant calcium-independent components to their release (e.g., Schwartz, 1987) which may play significant functional roles in regulating neuronal activity. However, another possible explanation that we have not been able to entirely discount is the possibility that enough DA "leaks" from damaged cells to provide a source of DA, even under conditions where synaptic release of transmitter cannot be supported.

Locus Coeruleus

The neurons in the locus coeruleus are a homogeneous population of noradrenergic cells that project widely throughout the CNS (Foote *et al.*, 1983). Included in the projection areas is the locus coeruleus itself. There is also histological evidence for dendro-dendritic connections (Groves and Wilson, 1980). Like the VTA, the neurons in the locus coeruleus have autoinhibitory receptors; in the case of the locus coeruleus, activation of α_2-adrenergic receptors inhibits the firing of these neurons by activating a potassium conductance (Egan *et al.*, 1983). From an electrophysiological standpoint, we have shown that electrical stimulation in the area of the locus coeruleus with focal stimulating electrodes evokes the release of NE that can

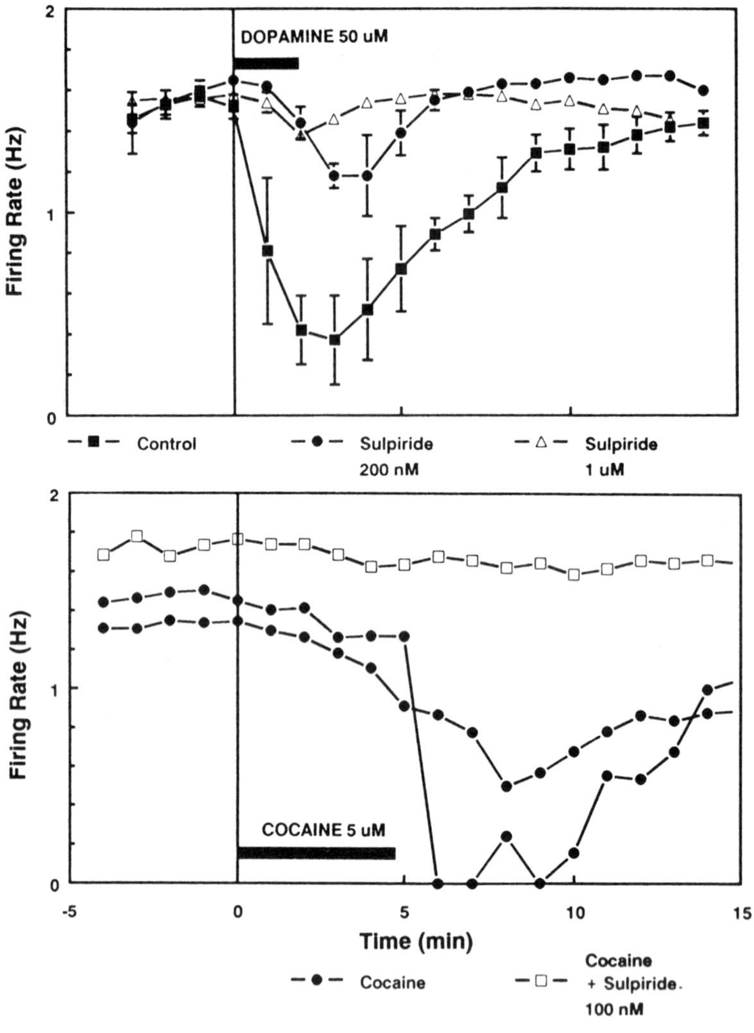

Figure 2. Effect of sulpiride on inhibitory responses to DA (upper) and cocaine (lower); all data were obtained from a single dopaminergic neuron in the ventral tegmental area during a continuous 2-h period of recording. Superfusion with DA (50 μM) elicited consistent inhibition of firing, which was reduced by pretreatment with the D_2 antagonist sulpiride. Each point in the upper figure is the mean ± SEM firing rate integrated over 1-min intervals before, during, and after DA perfusion (indicated by the bar). Two of the control responses were obtained prior to sulpiride infusion, and one following washout. It should be noted that the concentration of DA is the concentration in the superfusate, not the concentration in the chamber; with our perfusion system approximately 5–6 min of superfusion with drug would be required for the chamber to reach the indicated concentration, so that the actual concentration to which the slice is exposed is considerably less than 50 μM. In the bottom half of the figure, inhibition of the response to cocaine by sulpiride is illustrated. In cells such as this, which were quite sensitive to DA, cocaine by itself was able to elicit a marked inhibition of firing rate. Each line in this part of the figure denotes a single cocaine trial before, during, and after superfusion with 100 nM sulpiride. Note that the basal rate of firing of this cell was increased somewhat by sulpiride infusion.

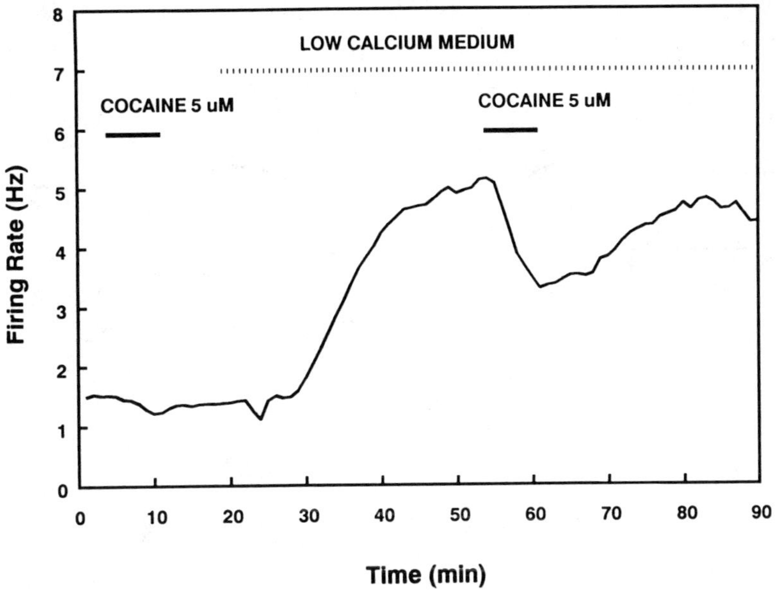

Figure 3. Effect of low calcium medium on the inhibition of ventral tegmental area neuron firing by cocaine. In this particular neuron, inhibition of firing induced by 5 μM cocaine was minimal during perfusion with normal medium. Superfusion with medium containing low calcium and sufficient magnesium to inhibit calcium-dependent release of transmitter increased the spontaneous firing rate, and enhanced the percent inhibition induced by superfusion with cocaine.

activate these receptors to inhibit firing (Egan *et al.*, 1983). The sensitivity of locus coeruleus neurons to NE and the presence of a noradrenergic synaptic potential was the basis for studying the action of cocaine on these cells.

Cocaine-Induced Hyperpolarization

In the slice preparation, locus coeruleus neurons fire action potentials spontaneously in a regular pattern (Williams *et al.*, 1984). Such spontaneous activity persists in medium containing tetrodotoxin, and is blocked by superfusion with agents such as $MgCl_2$ or $CoCl_2$. Superfusion of locus coeruleus slices with medium containing cocaine or other uptake blockers (e.g., desmethylimipramine; DMI) slowed the rate of spontaneous activity and often blocked firing completely. As with the inhibition by NE, the cocaine inhibition was blocked by the α_2-adrenoceptor antagonists idazoxan, yohimbine, or phentolamine (Egan *et al.*, 1983, Surprenant and Williams, 1987). In cells recorded under voltage clamp conditions, cocaine caused an outward current that was variable from cell to cell and ranged from 20–120 pA measured at -60 mV. The maximum outward current induced by α_2-adrenoceptor activation was about 250 pA (Christie *et al.*, 1987). In general, the outward current caused by cocaine (30 μM) was about 15–20% of the maximum current produced by NE.

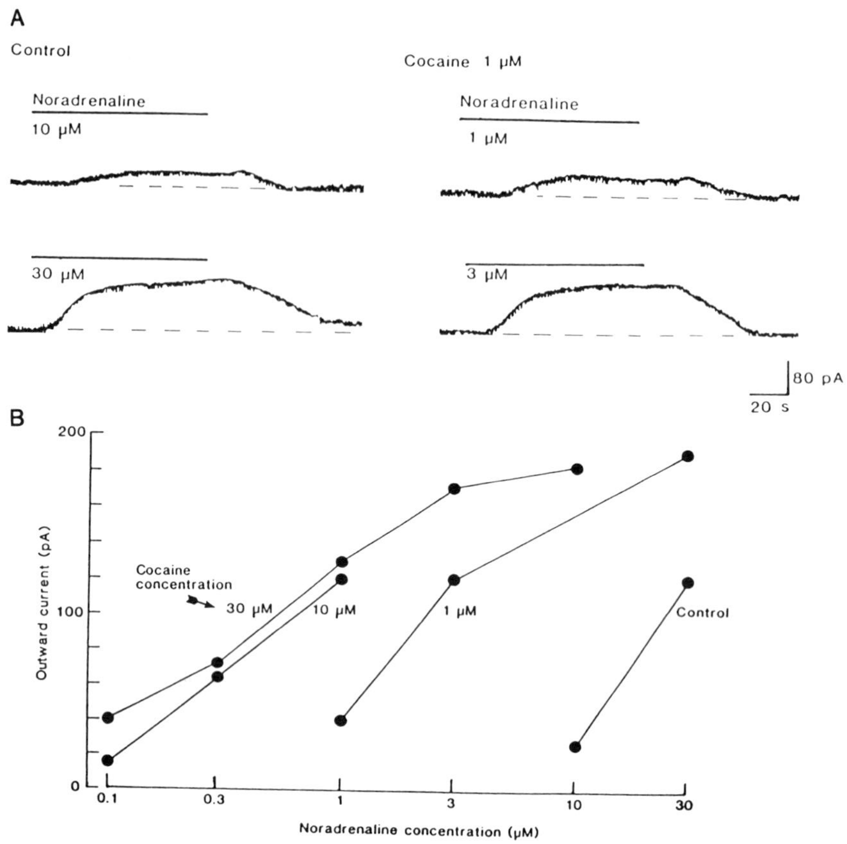

Figure 4. The upper part of the figure (**A**) shows the outwardly directed currents in a voltage-clamped locus coeruleus neuron that are activated by superfusing the slice with the indicated concentrations of NE alone (left column), and NE in the presence of 1 μ*M* cocaine (right). In the presence of cocaine equivalent responses were elicited by approximately tenfold lower concentrations of NE. **B** illustrates data obtained from a single neuron that was superfused with multiple concentrations of NE in the presence of increasing concentrations of cocaine. The concentration of cocaine is shown next to each curve, and the ordinate indicates the amount of outward current activated by superfusion with NE. (From A. Surprenant and J.T. Williams. *J. Physiol.* **382**: 87 (1987). With permission.)

Cocaine Augments the NE-Induced Hyperpolarization

In addition to exerting direct effects, cocaine produced a striking increase in the amplitude of the outward current or membrane hyperpolarization induced by NE. In the absence of cocaine, NE (10 μ*M*) caused a 2 to 4-mV hyperpolarization or an outward current of 20–40 pA. In the presence of 10 μ*M* cocaine, the same concentration of NE caused a maximal hyperpolarization (20–25 mV), which corresponded to an outward current of 200–250 pA. This potentiation reflected a leftward shift in the dose-response curve for NE in the presence of cocaine. This leftward shift was concentration dependent and reached a maximum at concentrations between 10 and 30 μ*M* cocaine (Figure 4). The average concentration of cocaine that caused a half-maximal shift in the NE concentration response curve was 4.0 ± 0.8 μM (Surprenant and Williams, 1987). This EC_{50} value was relatively consistent from neuron to neuron even though the overall magnitude of the leftward shift in the NE concentration response curve varied from 15- to 118-fold in different experiments. The differences in the magnitude of the shift presumably result from variations in the number of uptake sites between the outside of the slice and the neuron from which the recording was made. Neurons deep within the slice would be expected to be more affected by the blockade of NE uptake than neurons that are more directly exposed to the superfusion solution.

Cocaine Increased the Inhibitory Postsynaptic Potential

The amplitude and duration of the inhibitory postsynaptic potential (IPSP) elicited by focal stimulation of the locus coeruleus were markedly increased by cocaine (Figure 5). The duration of the IPSP under control conditions was about 2 s, and cocaine induced an approximately eightfold increase in the duration of the IPSP at the half-maximal amplitude. The EC_{50} for this action of cocaine was about 3 μM (Surprenant and Williams, 1987). At concentrations below 3 μM, cocaine increased the amplitude of the IPSP; but at higher concentrations, a *decrease* in the IPSP was observed. The action of cocaine on the amplitude of the IPSP probably results from a mixture of blockade of re-uptake, presynaptic inhibition of transmitter release, and postsynaptic occlusion of the α_2-adrenoceptor by endogenous NE. The interpretation of these results is therefore difficult (see section on dorsal raphe/prepositus and 5-HT mediated IPSPs).

Raphe Nucleus

In experiments on brain slices containing the serotonergic neurons of the dorsal raphe nucleus, cocaine was able to induce a hyperpolarization that appeared in many respects to be similar to that previously described in the VTA and the locus coeruleus. The amplitude of the hyperpolarization was dependent on the concentration of cocaine applied (EC_{50} ca. 4 μM). The maximum hyperpolarization averaged about 13 mV, which is approximately 80% of the maximum hyperpolarization that 5-HT agonists caused (Williams *et al.*, 1988; Pan and Williams, 1989). Spiperone, a 5-HT_{1A} receptor antagonist, completely blocked the hyperpolarization induced by cocaine; thus, it would appear that cocaine hyperpolarized neurons of the dorsal raphe indirectly by raising the concentration of 5-HT near the cell bodies, presumably by re-uptake blockade.

As with other monoamine-containing nuclei, the cocaine-induced hyperpolarization of raphe neurons appears to be an indirect response mediated via the ability of cocaine to potentiate the effects of spontaneously released endogenous 5-HT. However, in this context there are two unusual aspects to this response. First, dorsal raphe neurons in the slice preparation were not spontaneously active and most neurons maintain a stable membrane potential at about −65 mV (Williams *et al.*, 1988); in the absence of spontaneous activity, one would not expect release of 5-HT in the unstimulated slice if such release occurs in the classical calcium-dependent fashion by which most transmitters are released. Second, the hyperpolarization

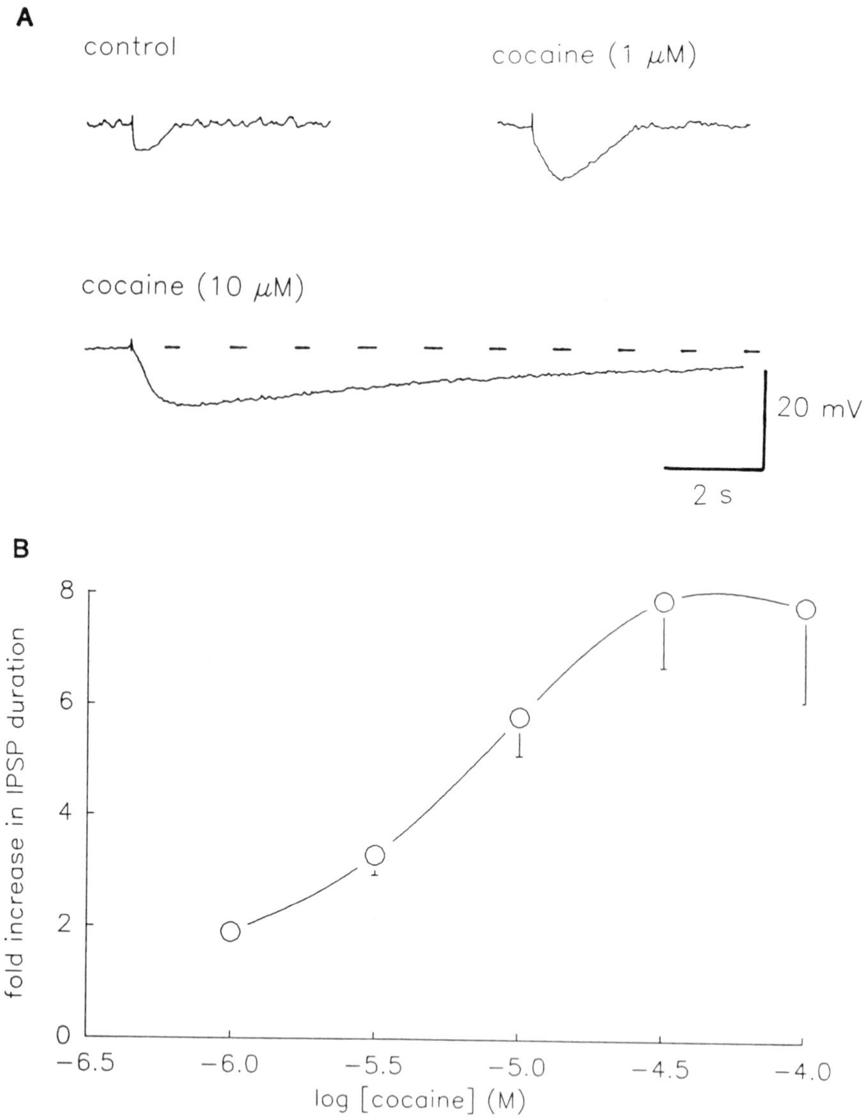

Figure 5. Voltage recordings of IPSPs induced in a single locus coeruleus neuron with and without superfusion with cocaine (**A**). IPSPs were evoked by a single supramaximal electrical stimulation pulse. Both the amplitude as well as the duration of the IPSP were increased by cocaine; the dose-response relation for the effect of cocaine on the duration of the IPSP is illustrated in **B**.

induced by cocaine was unaffected in animals that were pretreated with reserpine (5 mg/kg, 12 h), which should deplete most stores of 5-HT as well as NE and DA. Although the treatment was effective in blocking the vesicular release of 5-HT (as evidenced by the observation that no IPSPs could be evoked in the dorsal raphe by focal stimulation), immunohistochemical studies with antibodies directed at 5-HT demonstrated that the neurons still contained significant amounts of 5-HT. We hypothesize that the pool of 5-HT that mediates the cocaine-induced hyperpolarization may be a nonvesicular pool of 5-HT that is unaffected by reserpine.

Cocaine Augments the 5-HT-Induced Hyperpolarization

The 5-HT concentration response curve in the dorsal raphe was shifted to the left in the presence of cocaine (300 n*M* to 10 μ*M*). However, only a qualitative interpretation of these results was possible since cocaine itself caused a near maximal hyperpolarization that occluded the hyperpolarization induced by 5-HT. Cocaine (1 μ*M*) caused a 3-mV hyperpolarization and shifted the 5-HT concentration curve to the left about tenfold.

Cocaine Augments the 5-HT-Mediated IPSP

Focal stimulation in the area of the dorsal raphe produced several distinct post-synaptic potentials in dorsal raphe neurons. Within 1–5 ms following stimulation, a depolarizing synaptic potential occurred that reflects the summed action of an excitatory amino acid and GABA (Pan and Williams, 1989). These synaptic potentials occur in every cell, are 50–200 ms in duration, and are blocked by a combination of the excitatory amino acid antagonists CNQX (10 μ*M*) and APV (30 μ*M*), and the GABA antagonists bicuculline (30 μ*M*) and picrotoxin (100 μ*M*). An IPSP with a latency of 40–50 ms follows this initial response in about 75% of cells; this response has a total duration of about 2 s, reverses at the potassium equilibrium potential, and is blocked by spiperone (Yoshimura and Higashi, 1985; Pan *et al.*, 1989). This component of the response is most likely a serotonergic synaptic potential mediated via 5-HT_{1A} receptors. In about 30% of the cells, a slow depolarization follows the IPSP; this depolarization is 5–15 s in duration, 1–8 mV in amplitude, and is blocked by prazosin (100 n*M*). This component of the response is a noradrenergic, slow, excitatory postsynaptic potential mediated by activation of α_1 adrenoceptors (Yoshimura *et al.*, 1985; Pan and Williams, 1989). The action of cocaine on this noradrenergic input is an important consideration in *in vivo* experiments because this input is thought to mediate the repetitive activity in these neurons (Cunningham and Lakoski, 1988).

Effects upon Serotonergic IPSPs

We have focused on the slow, serotonergic component of the IPSP as a way of characterizing the actions of cocaine at a serotonergic synapse. Superfusion of cocaine (0.3–30 μ*M*) prolonged the duration of the 5-HT-mediated IPSP (Figure 6). The maximum increase in the IPSP was about eightfold with an EC_{50} of 1.5 μ*M*. At concentrations ≤3 μ*M*, cocaine generally increased the amplitude of the IPSP, whereas at higher concentrations the amplitude was depressed to less than control values. In order to characterize the action of cocaine on the IPSP in greater detail, experiments were carried out under conditions where all the other synaptic potentials were blocked pharmacologically. Under these conditions, the latency, time course of onset, amplitude, and time course of offset could be determined under control conditions and in different concentrations of cocaine.

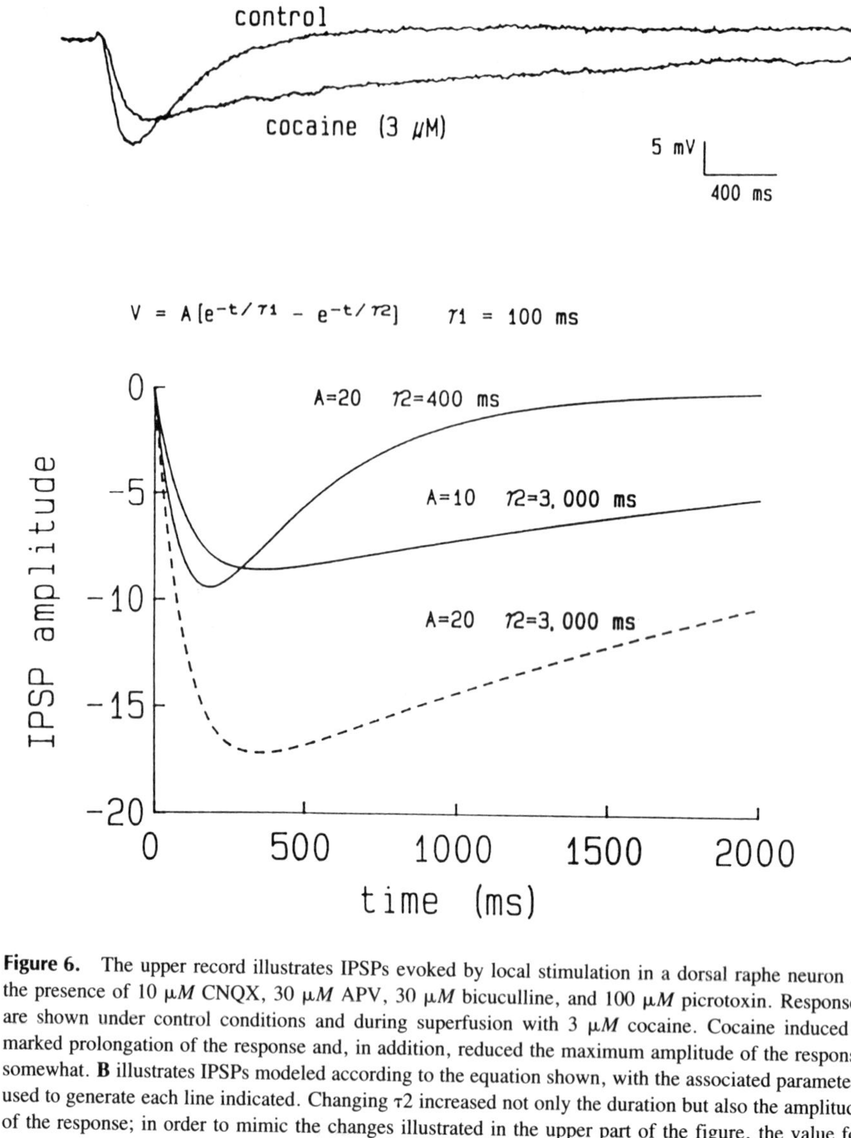

Figure 6. The upper record illustrates IPSPs evoked by local stimulation in a dorsal raphe neuron in the presence of 10 μM CNQX, 30 μM APV, 30 μM bicuculline, and 100 μM picrotoxin. Responses are shown under control conditions and during superfusion with 3 μM cocaine. Cocaine induced a marked prolongation of the response and, in addition, reduced the maximum amplitude of the response somewhat. **B** illustrates IPSPs modeled according to the equation shown, with the associated parameters used to generate each line indicated. Changing $\tau 2$ increased not only the duration but also the amplitude of the response; in order to mimic the changes illustrated in the upper part of the figure, the value for **A** had to be reduced by 50%. This change could represent either a decrease in postsynaptic sensitivity to the transmitter or, more likely, a reduction in serotonin release.

We found that when the other synaptic potentials were blocked, the IPSP could satisfactorily fit by the sum of two exponentials according to the following equation:

$$V = A[e^{-t/\tau 1} - e^{-t/\tau 2}]$$

where V is membrane voltage (mV), A is an amplitude constant, $\tau 1$ is the time constant for activation of the IPSP, and $\tau 2$ is the time constant for inactivation of the IPSP. The time constant of activation ($\tau 1$) was about 100 ms and the time constant of inactivation ($\tau 2$) was about 450 ms (Pan *et al.,* 1989). In terms of the kinetics of the response, only $\tau 2$ was affected by cocaine, increasing as the concentration of cocaine was increased. Neither the latency to onset nor $\tau 1$ were affected by cocaine. Such an observation suggests that the main effect of cocaine on the IPSP resulted from a decrease in the rate of removal of transmitter, which presumably reflects the inhibition of 5-HT uptake by cocaine.

In addition to modifying the time course of the response, cocaine exerted a biphasic action on the amplitude of the IPSP. At concentrations $\leq 3\ \mu M$, the amplitude was increased, whereas at higher concentrations the amplitude declined. Furthermore, although the rate of rise was unaffected by cocaine, the latency to the peak amplitude of the IPSP increased as the concentration of cocaine increased. To determine whether these changes could be explained solely on the basis of changes in $\tau 2$, we characterized the effect of changes in $\tau 2$ on the shape of the IPSP modeled according to the above equation. When increasing values for $\tau 2$ were used, the model synaptic potential was changed in three ways; the duration increased, the amplitude increased, and the time to reach the peak increased (Figure 6). Thus, the predictions from this simple model adequately fit the experimental observations at low concentrations of cocaine. However, the *decrease* in the observed synaptic potential amplitude at higher concentrations of cocaine could only be fit by decreasing the amplitude constant (A). From this analysis, it was clear that there were at least two actions of cocaine. The inhibition of transmitter uptake by cocaine could account for the increased amplitude, time to peak, and time constant of decay of the IPSP, but a second action had to be proposed to account for the decrease in the amplitude of the IPSP observed at higher concentrations of cocaine. The mechanism for this inhibition could not be predicted from the present experiments, but might involve presynaptic inhibition of transmitter release or postsynaptic occlusion of the response (however, see below, actions of cocaine in prepositus hypoglossi).

ACTIONS OF COCAINE IN MONOAMINE TARGET AREAS

Nucleus Prepositus Hypoglossi

Caudal to the dorsal raphe nucleus on the medullary border of the fourth ventricle is a large nucleus thought to be involved in vestibular/oculomotor function called the nucleus prepositus hypoglossi (PH) (Belknap and McCrea, 1988). There is a dense 5-HT innervation of an area of this nucleus that is within 300 μm of the fourth ventricle (Steinbusch, 1981). In brain slices containing this region, electrical stimulation near the fourth ventricle evoked an IPSP in neurons recorded within that area. This IPSP is similar to that found in the dorsal raphe in its transmitter (5-HT), time course, ionic mechanism, sensitivity to spiperone (it also appears to be mediated by activation of 5-HT_{1A} receptors), and its sensitivity to cocaine (Bobker and Williams, 1990). The important difference between the neurons in the PH and those in the dorsal raphe is that there are few or no 5-HT-containing cell bodies in the area of the PH. The IPSP found in PH cells results from release of 5-HT from nerve terminals, whereas 5-HT in the raphe may come from dendritic release as well as nerve terminals.

In marked contrast to the dorsal raphe nucleus, cocaine (100 n*M* to 10 μ*M*) by itself had no effect on the membrane potential of PH neurons. However, the concentration-response curve for exogenous 5-HT was shifted to the left by cocaine (10 μ*M*); the concentration response curve to 8OH-DPAT, a 5-HT_{1A} agonist that is *not* a substrate for uptake, was unaffected by cocaine (Bobker and Williams, in preparation).

The action of cocaine on the IPSP of PH neurons was almost identical to that described above in the dorsal raphe, in that the duration of the IPSP was prolonged in a concentration-dependent manner by cocaine. The amplitude of the IPSP was increased by low concentrations of cocaine (below 1 μ*M*), whereas at higher concentrations, the amplitude declined. Procaine (10 μ*M*) did not share this latter effect, suggesting that the local anesthetic actions of cocaine were not involved. In contrast, other 5-HT uptake inhibitors, such as fluoxetine, also reduced the IPSP. These experiments indicate that the inhibition of the amplitude of the IPSP is likely to result from presynaptic inhibition of 5-HT release, but the mechanism involved is unclear. Perhaps the most likely possibility is that when there is sufficient uptake inhibition, 5-HT accumulates near the terminal varicosities and inhibits the release of 5-HT through an 'autoreceptor'-type mechanism.

Nucleus Accumbens

Neurons of the nucleus accumbens are innervated by both DA (Bjorklund and Hokfelt, 1984) and 5-HT (Steinbusch, 1981) terminal fields that project from VTA and dorsal raphe, respectively. The DA input to the neurons in the nucleus accumbens is thought to be an important link in the brain circuitry underlying cocaine self-administration (Wise and Bozarth, 1987; Liebman and Cooper, 1989). On the other hand, the 5-HT input to nucleus accumbens appears to function in the opposite direction, that is, lesioning the 5-HT input increases the self-administration of cocaine (Loh and Roberts, 1988). DA has been reported to have both hyperpolarizing and depolarizing effects on these neurons. The hyperpolarization appears to be caused by an increase in potassium conductance, while the depolarization is blocked by sulpiride, and may involve activation of D_2-receptors (Uchimura *et al.*, 1986). In addition, 5-HT can depolarize accumbens neurons by activation of 5-HT_2 receptors (North and Uchimura, 1989).

Cocaine Augments the Effects of DA and 5-HT

In brain slices containing the nucleus accumbens, cocaine alone (300 n*M* to 30 μ*M*) had little or no effect on membrane potential (Uchimura and North, submitted). The effect of cocaine was tested on the depolarizations induced by exogenously applied 5-HT and DA, and the hyperpolarization induced by activation of D_1 receptors. Experiments studying the D_2-mediated depolarization were carried out in the presence of SCH23390 (1 μ*M*), to block the D_1 receptors; D_1-mediated hyperpolarizations were studied in the presence of sulpiride (1 μ*M*) to block the D_2

response. Cocaine caused a concentration-dependent leftward shift in the DA dose-response curve. The amount of the shift (expressed as the ratio of agonist concentrations giving an equivalent response) was the same whether the D_1 or D_2-receptor-mediated response was measured. Potentiation of responses to DA was calculated for a response to DA that was about half-maximal for both the depolarization and the hyperpolarization (4 mV). Cocaine (1 μ*M*) caused about a 3-fold increase in the sensitivity to DA, and a 50-fold shift at 30 μ*M*. The potentiation of the depolarizing response to 5-HT was larger and occurred at lower concentrations of cocaine. The maximum dose-ratio for 5-HT was about 120 occurring at a cocaine concentration of 30 μ*M*. The shift in the dose ratio induced by 1 μ*M* cocaine was 20 for 5-HT compared with 3 for DA.

These results indicate the presence of cocaine-sensitive uptake sites for both DA and 5-HT in the nucleus accumbens. It appears that these uptake sites are separate because the magnitude of the dose-response curve shift for each transmitter at a given concentration of cocaine was substantially different. The higher sensitivity for the 5-HT uptake site may have been expected based upon radioligand binding studies which have demonstrated that cocaine binds to 5-HT uptake sites with about a fourfold higher affinity than to DA uptake sites (Ritz *et al.*, 1987; but see Hyttel, 1982). On the other hand, this effect might also reflect the fact that there are perhaps more 5-HT uptake sites than DA uptake sites; hence, 5-HT uptake is more effective in reducing the amount of 5-HT that can penetrate the slice.

Relation to In Vivo Studies

It seems likely that the ability of cocaine to change the excitability of nucleus accumbens neurons results from interaction with both dopaminergic and serotonergic transmission. The same general conclusion was reached by White *et al.* (1987; see also Chapter 12). In this study, the inhibition of cell firing elicited by iontophoretic application of cocaine was examined in the anesthetized rat; the magnitude of this inhibition was found in many cells to be greater than that that could be achieved with either DA or 5-HT alone. On this basis, it was proposed that the effects of cocaine represented a potentiation of the effects of both DA and 5-HT. The situation will no doubt become even more complicated when the modulatory effects of 5-HT and DA are considered. Dopamine has been shown to depress a fast synaptic potential recorded in nucleus accumbens neurons (Higashi *et al.*, 1989), but only in animals that had been sensitized with prior treatments of amphetamine. Thus, cocaine may interact both pre- and postsynaptically with DA and 5-HT, and the specific response that is observed may also depend upon past exposure to drugs that interact with this system. Thus, although a general picture of the actions of cocaine on this system is developing, it may be some time before it is possible to predict what types of responses would be observed under various conditions in the intact animal.

ACTIONS OF COCAINE IN THE HIPPOCAMPUS

Previous studies regarding the effects of cocaine on hippocampal activity have focused primarily on the intact animal, and have demonstrated that cocaine exerts complex frequency-dependent effects on the threshold and duration of afterdischarges evoked by electrical stimulation of the hippocampus (Lesse and Collins, 1979; Lesse, 1980; Lesse and Collins, 1980; Lesse and Harper, 1985). While identifying the hippocampus as an important site of cocaine action, these types of investigations provide little insight into the cellular mechanism(s) by which cocaine affects neural activity. These types of effects might result from local anesthetic actions, from actions upon monoaminergic synapses, or from as yet unknown pharmacological actions of cocaine. For this reason, we have conducted experiments in *in vitro* hippocampal brain slices to characterize the actions of cocaine in this brain region from an electrophysiological standpoint.

Although the hippocampal slice affords a variety of advantages for studies of cocaine, the preparation of the slice severs all of the serotonergic and noradrenergic fibers innervating this brain region. Although the nerve terminals have been separated from their cell bodies, they nevertheless appear functional in the brain slice preparation, as suggested by certain findings. These nerve terminals can take up NE or 5-HT via high-affinity, energy-requiring transport mechanisms, and they can release these neurotransmitters upon stimulation with high K^+, or following superfusion of the slices with indirect releasing agents such as tyramine and amphetamine (Feuerstein *et al.*, 1985a, 1985b; Jackisch *et al.*, 1985; Jonzon and Fredholm, 1984; Mueller *et al.*, 1982). Thus, although it is not possible to study postsynaptic responses to endogenously released 5-HT or NE, as it is in the PH, we have been able to demonstrate a variety of electrophysiological responses that arise from actions of cocaine on these nerve terminals.

Direct Actions of Cocaine in the Hippocampus

In initial studies on the effects of cocaine in the hippocampal slice, we found that cocaine alone did not significantly affect evoked synaptic potentials at concentrations below 10 μM. At this concentration and above, effects similar to those produced by other local anesthetics were observed. In general, these were a decrease in the magnitude and an increase in the latency of the presynaptic fiber spike, the field EPSP response, and the population spike response (see section below on Local Anesthetic Actions in the Hippocampus). Thus, as in other monoamine target regions *in vitro*, cocaine by itself did not appear to elicit effects mediated via monoamine receptors.

Cocaine Potentiates Responses to NE in Hippocampus

Previous studies from our laboratory have demonstrated that superfusion of hippocampal slices with NE increases the excitability of pyramidal neurons in the CA1 region via an action at β-adrenergic receptors, and that this response can be easily measured using the extracellularly recorded population spike response to determine neuronal excitability (Mueller *et al.*, 1981; Mueller *et al.*, 1982a; Mueller *et al.*, 1982b). While superfusion of slices with cocaine alone had no effect on this potential, it markedly potentiated responses to a subthreshold concentration of NE (Figure 7), and shifted the NE dose-response curves to the left. A half-maximal response to cocaine was observed with a concentration of 400 nM, and its effects

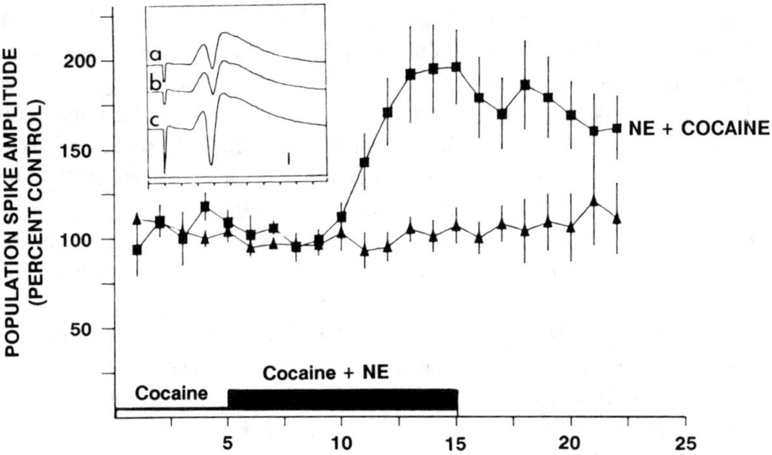

Figure 7. The effect on rat hippocampal slices of superfusion with 0.5 μ*M* NE is illustrated. The change induced by 0.5 μ*M* NE in the magnitude of the evoked population spike response is shown in the presence (squares) and absence (triangles) of 2.5 μ*M* cocaine. When pretreated with cocaine, slices were perfused with cocaine alone for 10 min, then co-perfused with NE + cocaine for 10 min as denoted by the bars along the abscissa. Values are shown as mean ± SEM (n = 7). Inset shows evoked population spike responses recorded from a single hippocampal slice before drug treatment (a), after 10 min of perfusion with 5 μ*M* cocaine (b), and after perfusion with 0.5 μ*M* NE and cocaine (c). Each averaged record represents three to four responses evoked at 1-min intervals. Marks along the horizontal axis are at 2-ms intervals; the vertical bar is 1 mV.

appeared to be maximal at 1 μ*M* (Yasuda *et al.*, 1984). In more recent biochemical experiments (Yasuda *et al.*, in preparation), we have shown directly that cocaine inhibits the uptake of ^{3}H-NE into hippocampal brain slices, with a K_1 value of 570 n*M*. Parallel experiments with DMI showed very similar effects, although DMI was significantly more potent in this regard, with an EC_{50} of 20 n*M* and 19 n*M* for the electrophysiological and uptake studies, respectively. Thus, there is a relatively good correspondence between the biochemical and electrophysiological effects of cocaine in this preparation and, at least as far as NE is concerned, they can be mimicked by another agent that blocks NE uptake.

To establish that the effects of cocaine were primarily due to the uptake inhibition, a number of experiments were conducted using other agents that shared some of the pharmacological effects of cocaine. These studies demonstrated that other NE uptake inhibitors such as imipramine also shared the ability to potentiate responses to low concentrations of NE, but that local anesthetics such as procaine and lidocaine did not (Yasuda *et al.*, 1984). As would be expected, cocaine did not potentiate responses to isoproterenol, a β-adrenergic agonist that is not a substrate for uptake. Finally, acute *in vitro* lesions of the noradrenergic nerve terminals with 6-hydroxydopamine (Yasuda *et al.*, 1986) markedly potentiated responses to NE (the concentration of NE required to elicit a maximal response was reduced from 5 μ*M* to 100 n*M*), and in such lesioned slices, cocaine no longer was able to potentiate the response to subthreshold concentrations of NE (Yasuda *et al.*, in preparation).

Thus, insofar as NE is concerned, cocaine's effects are consistent with its ability to inhibit NE uptake. The 50-fold shift in the dose-response curve for NE in 6-hydroxydopamine lesioned slices (Yasuda *et al.*, 1986) would suggest that active transport mechanisms are highly effective in reducing extracellular concentrations of exogenously applied NE before it can reach the appropriate receptors. Whether a similar shift in responsiveness would occur in terms of responses to endogenous NE is unclear. Compensatory mechanisms, such as the inhibitory autoreceptors found on noradrenergic nerve terminals (Langer, 1981), might limit release before such a marked potentiation could occur. Nevertheless, the concentrations of cocaine that are effective in inhibiting the uptake of ^{3}H-NE, and in potentiating electrophysiological responses to superfusion with NE, are within the range of concentrations that have been reported in cocaine abusers (maximum cocaine concentrations in blood have been reported in humans to range from 0.7-3 μM following various routes of administration; Javaid *et al.*, 1978; Van Dyke *et al.*, 1982; Paly *et al.*, 1982). Although we have not characterized the effects of cocaine on responses to 5-HT in the hippocampal slice preparation, it would seem likely that similar effects might be observed with this neurotransmittter as well.

ELECTROCHEMICAL MEASURES OF TRANSMITTER EFFLUX IN HIPPOCAMPAL SLICES

Another means by which to characterize the effects of cocaine on hippocampal slices takes advantage of the recent development of highly sensitive electrochemical techniques for measuring monoamine levels in intact tissue (Gerhardt *et al.*, 1984; Gerhardt *et al.*, 1986; Gerhardt *et al.*, 1987). Using this approach, we have been able to simultaneously record electrophysiological measures of neuronal activity and electrochemical measures of monoamine release from hippocampal slices *in vitro* (Su *et al.*, 1988, 1990). Figure 8 illustrates the results from one such experiment in which release was initiated with potassium. Superfusion of the slice with medium containing 50 mM potassium elicits a transient increase in the excitability of the pyramidal neurons, with a corresponding increase in the magnitude of the population spike response. Following this, evoked activity completely disappears; at this point, a marked increase in the extracellular concentration of electroactive species is observed at the electrochemical recording site in the dentate gyrus. Under these conditions, both NE and 5-HT contribute to the electrochemical "monoamine" signal, but the Nafion coating on the electrochemical electrode minimizes other electroactive agents (such as ascorbate) from being measured (Su *et al.*, 1990).

Cocaine Potentiates Potassium-Stimulated Monoamine Release

The magnitude of the electrochemical signals that are measured during these types of experiments can be significantly affected by cocaine. Figure 9 illustrates the results of four such experiments in which electrochemical releases were measured either in the presence or absence of cocaine. Cocaine superfusion alone did not produce detectable changes in the baseline electrochemical signals. However, co-

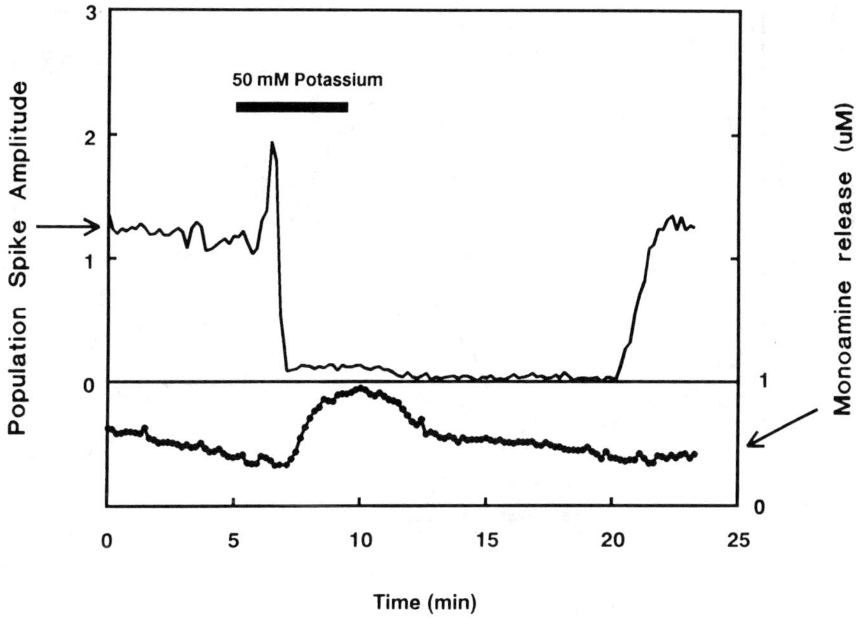

Figure 8. The effect on hippocampal slices of superfusion with 50 m*M* K^+ is shown. In this slice, alternating electrophysiological and electrochemical measurements were made at 5-s intervals before, during, and after superfusion with K^+. The amplitude of the population spike (upper part of figure) shows a transient increase, and then the synaptic response is completely lost; recovery of the evoked response usually took 10–20 min. The magnitude of the release (calibrated as μ*M* of NE) was approximately 0.6 μ*M*.

caine approximately doubled the extracellular concentration of monoamines that could be measured during superfusion with 50 m*M* K^+, but did not significantly affect the time course of the response. This latter observation has also been made using local application of cocaine to nerve terminals in the striatum (Gerhardt *et al.*, 1988). Thus, it would appear that uptake mechanisms in the slice cannot only prevent exogenous NE from reaching receptors within the slice, but can also prevent endogenously released monoamines from reaching an electrochemical electrode located on the surface of the slice.

LOCAL ANESTHETIC ACTIONS IN THE HIPPOCAMPUS

Although cocaine clearly can potentiate responses to exogenous NE in hippocampal slices, it also exerts local anesthetic effects at somewhat higher concentrations (Yasuda *et al.*, 1984). The threshold for significant effects on the population spike response is approximately 10 μ*M* (Yasuda *et al.*, 1984), and the concentrations of cocaine required to inhibit the presynaptic fiber spike, antidromic population spike, and synaptically evoked population spike by 50% ranged from 64 to 126 μ*M*. Thus, although significant effects can be observed at concentrations that

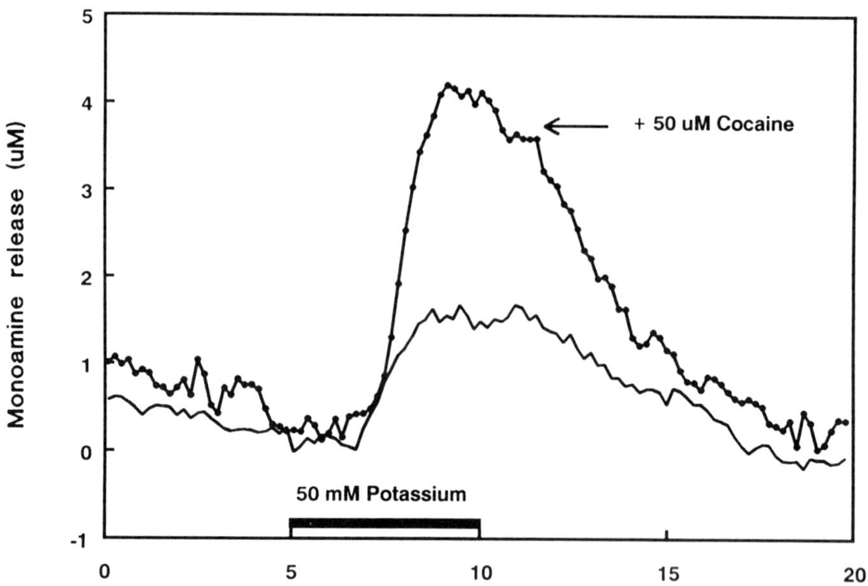

Figure 9. Effect of cocaine pretreatment on the magnitude of the electrochemical response to 50 m*M* K^+. Each record illustrates the mean electrochemical signal derived from four hippocampal slices treated with an identical protocol, except that the upper line represents slices continuously superfused with 50 μ*M* cocaine, and the lower line from controls. Although error bars are not shown for the two lines, the peak response under the two conditions was significantly different ($p < 0.05$). Similar effects were observed at lower concentrations of cocaine, but were not as robust.

approach the lower end of the range of drug concentrations at which cocaine toxicity appears, relatively high concentrations are required to effectively block transmission in this brain region.

Cellular Basis for the Proconvulsant Effects of Cocaine

Although most of the CNS effects of cocaine may be related to its effects upon monoamine uptake, we felt that a further investigation of the local anesthetic effects of cocaine was warranted because in many ways the CNS toxicity that is observed with high concentrations of this drug (e.g., seizures) resembles the clinical correlates of local anesthetic overdosage, rather than the actions of high doses of uptake inhibitors (although seizures can be observed with antidepressant overdose as well; Baldessarini, 1985). Because the hippocampal formation shows significant increases in metabolic activity at relatively modest concentrations of cocaine (Ingvar and Shapiro, 1981) and, because hippocampal afterdischarges are sensitive to the effect of cocaine (Lesse, 1980; Lesse and Collins, 1980), the hippocampus seemed a likely place to begin such studies. For these reasons, we characterized the local anesthetic effects of cocaine on hippocampal physiology in greater detail (Dunwiddie *et al.*, 1988).

A possible explanation for the proconvulsant effects of cocaine would be that it might interfere significantly with GABAergic inhibition. One protocol that has been

Table 1

Type of response	30 μM Cocaine % control	100 μM Cocaine % control
Field EPSP	87.5 ± 2.7[a] (11)	53.6 ± 4.6[b] (9)
Intracellular EPSP	81.2 ± 8.4 (6)	36.8 ± 5.2[b] (7)
$GABA_A$ IPSP	113 ± 13 (7)	40.6 ± 9.8[b] (8)
$GABA_B$ IPSP	94.3 ± 11.5 (8)	52.3 ± 3.9[b] (7)

[a] $p < 0.05$ vs. controls.
[b] $p < 0.01$ vs. controls.

Numbers in parentheses indicate the number of experiments. (From T.V. Dunwiddie, W.R. Proctor and J. Tyma. *Br. J. Pharmacol.* **95**: 1117 (1988). With permission.)

used to quantitate the degree of recurrent GABAergic inhibition in hippocampus is to activate the axons of the CA1 pyramidal cells antidromically, which then activate GABAergic interneurons via excitatory recurrent collaterals. Following such stimulation, one can assess the degree of GABAergic inhibition of the pyramidal cells by either stimulating them a second time via excitatory afferents during the period of inhibition, or by recording intracellularly and recording the inhibitory postsynaptic potentials directly. Both types of experiments point to the same conclusion; although cocaine in rather high concentrations can diminish synaptic inhibition, it is at least as effective in blocking excitatory synaptic events (Table 1). When evoked population responses are measured, the excitability of the pyramidal cell population is relatively unaffected at low doses (when neither EPSPs nor IPSPs are affected significantly) and is reduced at higher concentrations ($\geq$ 100 μM).

Cocaine Does Not Potentiate Interictal Spiking

Another and perhaps more sensitive way of testing for proconvulsant effects of drugs is to initiate interictal spiking by superfusing slices with GABAergic antagonists and/or increased K^+, superfuse with other drugs that might modify excitability, and monitor the change in the rate of interictal spiking. In the past, this approach has been used to characterize the actions of both convulsant and anticonvulsant drugs (Oliver *et al.*, 1977; Dunwiddie, 1980; Mueller and Dunwiddie, 1983). We have reported previously that direct application of both NE and isoproterenol increase the rate of interictal spiking by an action mediated by the β-adrenergic receptor (Mueller *et al.*, 1988; Dunwiddie *et al.*, 1988). Both of these agents decrease the calcium-dependent potassium conductance that may regulate the rate of burst discharge (Madison and Nicoll, 1982; Haas and Konnerth, 1983; Alger and Nicoll, 1980). However, in studies of the action of cocaine on interictal spiking, concentrations between 5 and 100 μM were never observed to do anything other than inhibit the rate of spiking (Dunwiddie *et al.*, 1988).

While these experiments cast doubt upon one of the major hypotheses concerning the proconvulsant effects of local anesthetic drugs, i.e., that local anesthetics have a selective action upon inhibitory circuitry, they give little indication as to how seizures might arise as a result of cocaine overdose. It is possible that local anesthetics induce seizures through actions on neural systems that do not involve the hippocampus, but there is no direct evidence that this is the case. Perhaps a more likely alternative is that cocaine modulates afferent activity to the hippocampus in such a way as to increase excitability. As discussed above, studies from our laboratory as well as others have clearly demonstrated that β-adrenergic effects in the hippocampus are proconvulsant. In the absence of ongoing noradrenergic transmission (i.e., in the hippocampus *in vitro*), there would be little or no ongoing release of neurotransmitter, and cocaine would have no "substrate" upon which to act to increase excitability. However, in the intact animal, cocaine might have a proconvulsant effect if it potentiated the β-adrenergic responses in the hippocampal formation. However, it is clear that this type of mechanism would only explain the proconvulsant effects of cocaine; other local anesthetics such as lidocaine, which are also proconvulsant, lack the ability of cocaine to inhibit the re-uptake of NE and hence could not act through a noradrenergic mechanism.

GENERAL CONCLUSIONS

Direct Effects on Monoaminergic Cell Bodies

Despite the disparate nature of the *in vitro* systems in which the effects of cocaine have been examined, there is remarkable consistency in terms of the general aspects of the responses which are observed in different systems. First, cocaine exerts a significant inhibitory effect by itself on monoamine-containing neurons within the locus coeruleus (NE), dorsal raphe (5-HT), and VTA/substantia nigra (DA), and in each case this appears to result from a potentiation of the actions of the endogenously released transmitter. There are also several indications that this endogenous transmitter may not come from a vesicular pool of transmitter that is released in a classical calcium-dependent fashion. In the VTA/substantia nigra, the release cannot be blocked with either low calcium medium or tetrodotoxin; in the raphe, spontaneous activity of the 5-HT-containing neurons is not required, nor is it blocked by pretreatment with reserpine, which depletes vesicular stores of transmitter.

Effects in Terminal Regions

A second common aspect of the effects of cocaine in all of the preparations tested is a failure of cocaine alone to induce monoaminergic responses in any of the target regions. This suggests that under basal conditions, the nerve terminals in these slices do not release significant amounts of transmitter; since they have been separated from their cell bodies in preparing the slices, this is perhaps not surprising. However, it is clear that cocaine can markedly potentiate the actions of exogenously applied catecholamines and indoleamines in terminal regions, and in areas containing monoaminergic cell bodies, and that it does so at concentrations that are entirely consistent with its ability to block the uptake of monoamines as measured with biochemical techniques. It is apparent that, in order for an exogenously applied transmitter to reach appropriate receptors, it must traverse areas in which high-affinity uptake mechanisms can markedly reduce the extracellular concentrations of transmitter. Likewise, the responses to *endogenous* transmitter released in response to stimulation are markedly potentiated by superfusion with cocaine. Which of these systems (if any) are directly related to the pharmacologically reinforcing properties of cocaine cannot be determined based upon these types of experiments. Nevertheless, these studies make it apparent that pharmacologically relevant concentrations of cocaine produce a variety of effects upon electrophysiological activity, and that any one of these might be a candidate for cellular bases of cocaine reinforcement.

Cocaine Does Not Induce Release of Monoamine

A conclusion from these studies that should be stressed is that there is little evidence that cocaine is able to evoke the release of monoamines in any of these brain regions. Pharmacologically, cocaine has many similarities to other psychomotor stimulants such as amphetamine, which have well-documented actions as neurotransmitter releasing agents. Nevertheless, neither the electrochemical studies of release in the hippocampus nor the electrophysiological studies in the hippocampus, nucleus prepositus, and nucleus accumbens provide any basis for proposing that cocaine exerts a similar action.

Functional Importance of Local Anesthetic Actions

One area that remains unresolved by our studies concerns the relevance of the local anesthetic actions of cocaine to the behavioral concomitants of cocaine use, and the ways in which cocaine (and other local anesthetics) induce CNS hyperexcitability. The concentration ranges in which we have observed local anesthetic effects of cocaine in the hippocampus (threshold $\geq 10\ \mu M$) are within the toxic range of serum cocaine levels, but it seems unlikely that they could contribute in a significant fashion to the actions of cocaine at levels normally attained by recreational users of cocaine. As to the mechanism underlying the proconvulsant effects of cocaine, the studies in the hippocampus have not suggested a mechanism that might underlie these effects. Although sodium channel blockers such as cocaine typically inhibit high-frequency firing, a pattern of activity typical of inhibitory interneurons in the hippocampus, we could not find any evidence that cocaine reduced the GABAergic inhibition of hippocampal pyramidal neurons that results from such activity.

Overall Effects of Cocaine on DA Systems

Clearly, the next step in terms of furthering our understanding of cocaine action is to relate the observations such as we have made in isolated preparations back to more global changes in electrophysiological activity in the intact animal. As has been pointed out by Einhorn *et al.* (1988), in determining the abuse potential of a drug such as cocaine, it is important to determine the net effect of the drug on the entire system, not just the effect on any one component. As but one example, the cocaine-induced reduction in firing rate of VTA neurons *in vitro* might lead one to predict that cocaine administration would result in *reduced* dopaminergic input to target regions, but this does not seem to be the case (White *et al.*, 1987); in fact, cocaine seems to inhibit firing in nucleus accumbens, which would be consistent with a potentiation of dopaminergic activity. Furthermore, there may also be differences in the responses that are observed *in vitro* and *in vivo* that reflect the loss of extrinsic inputs to the slice, the lack of anesthesia, and other factors as well. Einhorn *et al.* (1988) have shown in the intact rat that the apparent maximal response to i.v. injection of cocaine is an approximate 50% inhibition of cell firing, and the maximal response to locally applied cocaine, which would seem to be most comparable to our *in vitro* studies, is on the average a 20% inhibition of spontaneous firing. The apparently lower sensitivity of VTA neurons *in vivo* to cocaine might suggest that in the intact animal the firing rate of these neurons is influenced to a lesser degree by the autoinhibitory receptors. In any case, the major difference between the two preparations does not appear to be in the mechanism by which cocaine acts, but rather the magnitude of the response, which is dependent upon the integrity of other synaptic inputs, etc.

Overall Effects of Cocaine on Noradrenergic Systems

As for the effects of cocaine in the locus coeruleus, there are also differences between *in vitro* and *in vivo* preparations that parallel those seen in DA systems. Although cocaine can completely inhibit the firing of some neurons in the intact animal (Pitts and Marwah, 1987a), this was not true of the majority of cells. *In vitro*, however, the inhibition of firing rate was consistently greater than what was observed *in vivo*. As with the VTA, however, the differences in the two preparations are primarily ones of magnitude, rather than differences of a qualitative nature. One issue that has yet to be resolved is the net effect of cocaine on the noradrenergic system; it has been suggested that with the noradrenergic input to the cerebellar Purkinje cells, there is, if anything, a *decrease* in the net postsynaptic effect of NE at this synapse, which results from the increased inhibition of locus coeruleus neurons (Pitts and Marwah, 1987b). On the other hand, the effect of cocaine on noradrenergic IPSPs in the locus coeruleus suggests that potentiation should be observed, at least at relatively low concentrations. It will clearly require a more extensive analysis of these and other systems to completely resolve this issue.

The *in vitro* studies discussed in this review have contributed much to our understanding of the mechanisms by which cocaine can alter the activity of monoaminergic neurons, and the ways in which it can modify the effects of monoamines on target regions. These studies have shown that, in all of the monoaminergic regions tested, cocaine is able to inhibit the spontaneous firing of monoaminergic cells, and that this type of response is primarily due to its ability to inhibit uptake, rather than to a local anesthetic-type action. The *in vitro* studies have also shown that, as might be expected, cocaine potentiates the effects of exogenous transmitter (Yasuda *et al.*, 1984), but whether this effect is sufficient to offset the inhibitory effect of cocaine on the firing rate of monoaminergic neurons cannot be resolved with *in vitro* studies. As discussed above, *in vivo* studies suggest that in some regions the postsynaptic effects of monoamines are potentiated (e.g., nucleus accumbens), whereas in others (cerebellum) they are inhibited. The somewhat unique abilities of cocaine to serve as a powerful reinforcer of behavior may arise from a complex balance of increased *and* decreased postsynaptic actions of the monoamines; this being the case, further studies in monoaminergic target regions will clearly be required to determine which of these effects are directly related to pharmacological reinforcement, and which are not involved in drug self-administration. In summary, the present experiments all point towards the same general conclusion — that the pharmacological effects of cocaine are nearly all mediated via the inhibition of monoamine uptake. The difficulties involved in predicting or understanding the actions of cocaine largely reflect the complexities of monoaminergic systems and the ways in which inhibition of monoamine uptake translates into altered patterns of electrophysiological activity.

ACKNOWLEDGMENTS

This work was supported by grant DA 02702, DA 04523, and by the Veterans Administration Medical Research Service.

REFERENCES

G.K. Aghajanian and B.S. Bunney. Dopamine ''autoreceptors'': pharmacological characterization by microiontophoretic single cell recording studies. *Naunyn-Schmiedeberg's Arch. Pharmacol.* **297**: 1 (1977).

B.E. Alger and R.A. Nicoll. Epileptiform burst after hyperolarization: calcium-dependent potassium potential in hippocampal CA1 pyramidal cells. *Science* **210**: 1122 (1980).

R.J. Baldessarini. *The Pharmacological Basis of Therapeutics.* A.G. Gilman, L.S. Goodman, T.W. Rall, and F. Murad, eds., Macmillan, New York, p. 387 (1985).

D.B. Belknap and R.A. McCrea. Anatomical connections of the prepositus and abducens nuclei in the squirrel monkey. *J. Comp. Neurol.* **268**: 13 (1988).

A. Bjorklund and T. Hokfelt. *Handbook of Chemical Neuroanatomy.* Vol. 2. Elsevier, Amsterdam (1984).

D.H. Bobker and J.T. Williams. Serotonin-mediated inhibitory post-synaptic potential in guinea pig prepositus hypoglossi and feedback inhibition by serotonin. *J. Physiol.* **422**: 447 (1990).

M.S. Brodie and T.V. Dunwiddie. Cholecystokinin potentiates dopamine inhibition of mesencephalic dopamine neurons *in vitro. Brain Res.* **425**: 106 (1987).

M.J. Christie, J.T. Williams, and R.A. North. Cellular mechanisms of opioid tolerance: studies in single brain neurons. *Mol. Pharmacol.* **32**: 633 (1987).

J.T. Coyle and S.H. Snyder. Catecholamine uptake by synaptosomes in homogenates of rat brain: stereospecificity in different areas. *J. Pharmacol. Exp. Ther.* 221 (1969).

K.A. Cunningham and J.M. Lakoski. Electrophysiological effects of cocaine and procaine on dorsal raphe serotonin neurons. *Eur. J. Pharmacol.* **148**: 457 (1988).

T.V. Dunwiddie. Endogenously released adenosin regulates excitability in the *in vitro* hippocampus. *Epilepsia* **21**: 541 (1980).

T.V. Dunwiddie. The use of *in vitro* brain slices in neuropharmacology, in: *Electrophysiological Techniques in Pharmacology:* H.M. Geller, ed., p. 65 (1986).

T.V. Dunwiddie, A. Mueller, and A. Basile. The use of brain slices in central nervous system pharmacology. *Fed. Proc.* **42**: 2891 (1983).

T.V. Dunwiddie, W.R. Proctor, and J. Tyma. Local anaesthetic actions of cocaine: effects on excitatory and inhibitory synaptic responses in the hippocampus *in vitro. Br. J. Pharmacol.* **95**: 1117 (1988).

T.M. Egan, G. Henderson, R.A. North, and J.T. Williams. Noradrenaline-mediated synaptic inhibition in rat locus coeruleus neurones. *J. Physiol.* **345**: 477 (1983).

L.C. Einhorn, P.A. Johansen, and F.J. White. Electrophysiological effects of cocaine in the mesoaccumbens dopamine system: studies in the ventral tegmental area. *J. Neurosci.* **8**: 100 (1988).

E.H. Ellinwood, M.M. Kilbey, S. Catellani, and C. Khoury. Amygdala hyperspindling and seizures induced by cocaine. In *Cocaine and Other Stimulants.* (E.H. Ellinwood and M.M. Kilbey, eds., Plenum Press, New York, p. 303 (1977).

T.J. Feuerstein, G. Hertting, and R. Jackisch. Modulation of hippocampal serotonin (5-HT) release by endogenous adenosine. *Eur. J. Pharmacol.* **107**: 233 (1985a).

T.J. Feuerstein, G. Hertting, and R. Jackisch. Endogenous noradrenaline as modulator of hippocampal serotonin (5-HT) release. *Naunyn-Schmiedebergs Arch. Pharmacol.* **329**: 216 (1985b).

S.L. Foote, F.E. Bloom, and G. Aston-Jones. Nucleus locus coeruleus: new evidence of anatomical and physiological specificity. *Physiol. Rev.* **63**: 844 (1983).

R.D. Ford and R.L. Balster. Reinforcing properties of intravenous procaine in rhesus monkeys. *Pharmacol. Biochem. Behav.* **6**: 289 (1977).

G.A. Gerhardt, A. Gratton, and G.M. Rose. *In vivo* electrochemical studies of the effects of cocaine on dopamine nerve terminals in the rat striatum. *Physiol. Bohemoslovaca* **37**: 249 (1988).

G.A. Gerhardt, G.M. Rose, and B.J. Hoffer. Release of monamines from striatum of rat and mouse evoked by local application of potassium: evaluation of a new *in vivo* electrochemical technique. *J. Neurochem.* **46**: 842 (1986).

G.A. Gerhardt, G.M. Rose, and B.J. Hoffer. *In vivo* electrochemical demonstration of potassium-evoked monoamine release from rat cerebellum. *Brain Res.* **413**: 327 (1987).

G.A. Gerhardt, M.R. Palmer, A. Seiger, R.N. Adams, L. Olson, and B.J. Hoffer. Adrenergic transmission in hippocampus-locus coeruleus double grafts *in oculo:* demonstration by *in vivo* electrochemical detection. *Brain Res.* **306**: 319 (1984).

P.M. Groves and C.J. Wilson. Fine structure of the rat locus coeruleus. *J. Compar. Neurol.* **193**: 841 (1980).

H.L. Haas and A. Konnerth. Histamine and noradrenaline decrease calcium-activated potassium conductance in hippocampal pyramidal cells. *Nature* **302**: 432 (1983).

R.I. Herning, R.T. Jones, W.D. Hooker, J. Mendelson, and L. Blackwell. Cocaine increases EEG β: a replication and extension of Hans Berger's historic experiments. *Electroenceph. Clin. Neurophysiol.* **60**: 470 (1985).

G. Hertting, J. Axelrod, and L.G. Whitby. Effect of drugs on the uptake and metabolism of ^{3}H-norepinephrine. *J. Pharmacol. Exp. Ther.* **134**: 146 (1961).

H. Higashi, K. Inanaga, S. Nishi, and N. Uchimura. Enhancement of dopamine actions on rat nucleus accumbens neurones *in vitro* after metamphetamine treatment. *J. Physiol.* **408**: 587 (1989).

J. Hyttel. Citalopram — pharmacological profile of a specific serotonin uptake inhibitor with antidepressant activity. *Prog. Neuropsycholpharmacol. Biol. Psychiat.* **6**: 277 (1982).

M. Ingvar and H.M. Shapiro. Selective metabolic activation of the hippocampus during lidocaine-induced preseizure activity. *Anesthes.* **54**:33–37 (1981).

R. Jackisch, R. Fehr, and G. Hertting. Adenosine: an endogenous modulator of hippocampal noradrenaline release. *Neuropharmacology* **24**: 499 (1985).

J.E. Javaid, M.W. Fischman, C.R. Schuster, H. Dekirmenjian, and J.M. Davis. Cocaine plasma concentration: relation to physiological and subjective effects in humans. *Science* **202**: 227 (1978).

B. Jonzon and B.B. Fredholm. Adenosine receptor mediated inhibition of noradrenaline release from slices of rat hippocampus. *Life Sci.* **35**: 1971 (1984).

B.K. Koe. Molecular geometry of inhibitors of the uptake of catecholamines and serotonin in synaptosomal preparation of rat brain. *J. Pharmacol. Exp. Ther.* **199**: 649 (1976).

M.G. Lacey, N.B. Mercuri, and R.A. North. Dopamine acts at D_2 receptors to increase potassium conductance in neurons of the rat substantia nigra. *J. Physiol.* **392**: 397 (1987).

M.G. Lacey, N.B. Mercuri and R.A. North. Actions of cocaine on rat dopaminergic neurones *in vitro.* *Br. J. Pharmacol.* **97**: 731 (1990).

S.Z. Langer. Presynaptic regulation of the release of catecholamines. *Pharmacol. Rev.* **32**: 337 (1981).

H. Lesse and J.P. Collins. Effects of cocaine on propagation of limbic seizure activity. *Pharmacol. Biochem. Behav.* **11**: 689 (1979).

H. Lesse. Prolonged effects of cocaine on hippocampal activity. *Comm. Psychopharmacol.* **4**: 247 (1980).

H. Lesse and J.P. Collins. Differential effects of cocaine on limbic excitability. *Pharmacol. Biochem. Behav.* **13**: 695 (1980).

H. Lesse and R.K. Harper. Frequency-related, bidirectional limbic responses to cocaine: comparisons with amphetamine and lidocaine. *Brain Res.* **335**: 21 (1985).

J.M. Liebman and S.J. Cooper. *The Neuropharmacological Basis of Reward.* Clarendon, Oxford (1989).

E.A. Loh and D.C.S. Roberts. Increased motivation to self-administer cocaine following 5,7-dihydroxytryptamine lesion of the medial forebrain bundle in the rat. *Neurosci. Abstr.* **14**: 659 (1988).

D.V. Madison and R.A. Nicoll. Noradrenaline blocks accommodation of pyramidal cell discharge in the hippocampus. *Nature* **299**: 636 (1982).

A.L. Mueller, B.J. Hoffer, and T.V. Dunwiddie. Noradrenergic responses in rat hippocampus: evidence for medication by alpha and beta receptors in the *in vitro* slice. *Brain Res.* **214**: 113 (1981).

A.L. Mueller, K.L. Kirk, B.J. Hoffer, and T.V. Dunwiddie. Noradrenergic responses in rat hippocampus: electrophysiological actions of direct- and indirect-acting sympathomimetics in the *in vitro* slice. *J. Pharmacol. Exp. Ther.* **223**: 599 (1982).

A.L. Mueller, M.R. Palmer, B.J. Hoffer, and T.V. Dunwiddie. Hippocampal noradrenergic responses *in vivo* and *in vitro*. Characterization of alpha and beta components. *Naunyn-Schnmiedebergs Arch. Pharmacol.* **318(4)**: 259 (1982).

A.L. Mueller and T.V. Dunwiddie. Anticonvulsant and proconvulsant actions of alpha- and beta-noradrenergic agonists on epileptiform activity in rat hippocampus *in vitro. Epilepsia* **24**: 57 (1983).

R.A. North and N. Uchimura. 5-Hydroxytryptamine acts at 5-HT_2 receptors to decrease potassium conductance in rat nucleus accumbens neurones. *J. Physiol.* **417**: 1 (1989).

A.P. Oliver, B.J. Hoffer, and R.J. Wyatt. The hippocampal slice: a system for studying the pharmacology of seizures and for screening anticonvulsant drugs. *Epilepsia* **18**: 543 (1977).

D. Paly, P. Jatlow, C. Van Dyke, F.R. Jeri, and R. Byck. Plasma cocaine concentrations during cocaine paste smoking. *Life Sci.* **30**: 739 (1982).

Z.Z. Pan, and J.T. Williams. GABA- and glutamate-mediated synaptic potentials in rat dorsal raphe neurons *in vitro. J. Neurophysiol.* **61**: 719 (1989).

Z.Z. Pan, W.F. Colmers, and J.T. Williams, 5-HT-mediated synaptic potentials in the dorsal raphe nucleus: interactions with excitatory amino acid and GABA neurotransmission. *J. Neurophysiol.* **62**: 481 (1989).

Z.Z. Pan and J.T. Williams. Differential actions of cocaine and amphetamine on dorsal raphe neurons *in vitro. J. Pharmacol. Exp. Ther.* **251**: 56 (1989).

D.K. Pitts and J. Marwah. Electrophysiological actions of cocaine on noradrenergic neurons in rat locus ceruleus. *J. Pharmacol. Exp. Ther.* **240**: 345 (1987a).

D.K. Pitts and J. Marwah. Reciprocal pre- and postsynaptic actions of cocaine at a central noradrenergic synapse. *Exp. Neurol.* **98**: 518 (1987b).

M.C. Ritz, R.J. Lamb, S.R. Goldberg, and M.J. Kuhar. Cocaine receptors on dopamine transporters are related to self-administration of cocaine. *Science* **237**:1219–1223 (1987).

S.B. Ross and A.L. Renyi. Uptake of some tritiated sympathomimetic amines by mouse brain cortex *in vitro. Acta Pharmacol. Toxicol.* **24**: 297 (1966).

C.N. Scholfield and J.A. Harvey. Local anesthetics and barbiturates: effects on evoked potentials in isolated mammalian cortex. *J. Pharmacol. Exp. Ther.* **195**: 522 (1975).

E.A. Schwartz. Depolarization without calcium can release gamma-aminobutyric acid from a retinal neuron. *Science* **238**: 350 (1987).

H.W.M. Steinbusch. Distribution of serotonin-immunoreactivity in the central nervous system of the rat — cell bodies and terminals. *Neuroscience* **6:** 557 (1981).

J.S. Stripling. Origin of cocaine- and lidocaine-induced spindle activity within the olfactory forebrain of the rat. *Electroencephalogr. Clin. Neurophysiol.* **53**: 208 (1982).

M.-T. Su, T.V. Dunwiddie, and G.G. Gerhardt. Combined *in vivo* electrochemical and electrophysiological studies of monoamine overflow from rat hippocampal slices. *Soc. Neurosci. Abstr.* **14**: 213 (1988).

M.-T. Su, T.V. Dunwiddie, M. Mynlieff, and G. A. Gerhardt. Electrochemical characterization of stimulated norepinephrine release in locus coeruleus-hippocampus double brain grafts grown *in oculo, Neurosci. Lett.* **110**: 186 (1990).

A. Surprenant and J.T. Williams. Inhibitory synaptic potentials recorded from mammalian neurones prolonged by blockade of norepinephrine uptake. *J. Physiol.* **382:** 87 (1987).

N. Uchimura, H. Higashi, and S. Nishi. Hyperpolarizing and depolarizing actions of dopamine via D-1 receptors and D-2 receptors on nucleus accumbens neurons. *Brain Res.* **375**: 368 (1986).

N. Uchimura and R.A. North. Actions of cocaine on rat nucleus accumbens neurones *in vitro. Br. J. Pharmacol.* **99**: 736 (1990).

C. Van Dyke, J. Ungerer, P. Jatlow, P. Barash, and R. Byck. Intranasal cocaine: dose relationships of psychological effects and plasma levels. *Int. J. Psychiat. Med.* **12**: 1 (1982).

R.Y. Wang. Dopaminergic neurons in the rat ventral tegmental area. I. Identification and characterization. *Brain Res. Rev.* **3**: 123 (1981a).

R.Y. Wang. Dopaminergic neurons in the rat ventral tegmental area. II. Evidence for autoregulation. *Brain Res. Rev.* **3**: 141 (1981b).

F.J. White, S.R. Wachtel, P.A. Johansen, and L.C. Einhorn. Electrophysiological studies of the rat mesoaccumbens dopamine system: focus on dopamine receptor subtypes, interactions, and the effects of cocaine, in: *Neurophysiology of Dopaminergic Systems — Current Status and Clinical Perspectives.* L.A. Chiodo and A.S. Freeman, eds., Lakeshore Publishing, Detroit, p. 317 (1987).

F.J. White and R.Y. Wang. A10 dopamine neurons: role of autoreceptors in determining firing rate and sensitivity to dopamine agonists. *Life Sci.* **34**: 1161 (1984).

J.T. Williams, W.F. Colmers and Z.Z. Pan. Voltage- and ligand-activated inwardly rectifying currents in dorsal raphe neurons *in vitro. J. Neurosci.* **8**: 3499 (1988).

J.T. Williams, R.A. North, S.A. Shefner, S. Nishi, and T.M. Egan. Membrane properties of rat locus coeruleus neurones. *Neuroscience* **13**: 137 (1984).

R.A. Wise. The role of reward pathways in the development of drug dependence. *Pharmacol. Ther.* **35**: 227 (1987).

R.A Wise and M.A. Bozarth. A psychomotor stimulant theory of addiction. *Psychol. Rev.* **94:** 469 (1987).

J.J. Woodward, L.J. Chandler, and S.W. Leslie. Calcium-dependent and -independent release of endogenous dopamine from rat striatal synaptosomes. *Brain Res.* **473**: 91 (1988).

R.P. Yasuda, T.V. Dunwiddie, and N.R. Zahniser. The acute effects of 6-hydroxydopamine treatment on noradrenergic function in the rat hippocampus *in vitro. Br. Res.* **367**: 121 (1986).

R.P. Yasuda, N.R. Zahniser, and T.V. Dunwiddie. Electrophysiological effects of cocaine in the rat hippocampus *in vitro. Neurosci. Lett.* **45**: 199 (1984).

C.Y. Yim and G.J. Mogenson. Electrophysiological studies of neurons in the ventral tegmental area of Tsai. *Brain Res.* **181**: 301 (1980).

M. Yoshimura and H. Higashi. 5-Hydroxytryptamine mediates inhibitory postsynaptic potentials in rat dorsal raphe neurons. *Neurosci. Lett.* **53**: 69 (1985).

M. Yoshimura, H. Higashi, and S. Nishi. Noradrenaline mediates slow excitatory synaptic potentials in rat dorsal raphe neurons *in vitro. Neurosci. Lett.* **61**: 305 (1985).

15

Neuroendocrine Effects of Cocaine

LOUIS D. VAN DE KAR
Department of Pharmacology
Stritch School of Medicine
Loyola University of Chicago
Maywood, Illinois

INTRODUCTION

Cocaine affects the activity of dopaminergic, noradrenergic, and serotonergic neurons (see Chapters 8, 9, 11, 12, and 13). Thus, it is not surprising that cocaine would affect the secretion of hormones which are controlled by these neurotransmitters. Interest in the spectrum of effects which results from the use of cocaine, including neuroendocrine function, has only recently become urgent and studies on this topic are relatively scarce. There are major moral and legal barriers to studying the acute effect of cocaine in humans. Because of the highly addictive nature of cocaine, it is very difficult to conduct such a study, especially on naive humans. It is somewhat more acceptable to conduct acute experiments on humans who have been long-term cocaine abusers (Ambre *et al.*, 1988; Sherer, 1988). However, the possibility of tolerance and of neuroendocrine consequences of the long-term exposure to cocaine, which could interfere with the acute effects of cocaine, could make the data difficult to interpret. Consequently, there is very little information on the acute neuroendocrine effects of cocaine in humans. Most of the information on the acute neuroendocrine effects of cocaine was obtained in laboratory animals.

In order to understand how cocaine and other drugs of abuse affect the human brain, indirect approaches have to be found since direct approaches would be difficult if not impossible. One such approach is the determination of the effect of cocaine on the hormones which are controlled by known neurotransmitters in the brain. Thus, determining neuroendocrine function following this psychostimulant could be a "window into the brain". Studies in laboratory animals have led to the hypotheses that dopaminergic, noradrenergic, and serotonergic mechanisms mediate the behavioral effects of cocaine (see Chapters 5 and 6). Cocaine interferes with norepinephrine-, dopamine-, and serotonin-containing neurons in the brain, by inhibiting the uptake of the respective biogenic amines into their nerve terminals.

The effects of cocaine on these monoamine-containing neurons have been extensively documented and are reviewed in several chapters of this book (see Chapters 8, 9, 11, 12, and 13). The effects of cocaine on neuroendocrine function would appear to be mediated through its effect on dopaminergic, serotonergic, and (nor)adrenergic neurons, as these neurons play important roles in the regulation of the secretion of several hormones (see below).

This review will first briefly describe the components of the neuroendocrine system, followed by a brief depiction of the role of the cocaine-sensitive neurotransmitters, dopamine (DA), serotonin (5-HT), and norepinephrine (NE), in the regulation of neuroendocrine function. The acute effects of cocaine on neuroendocrine function in experimental animals and in humans will be addressed, followed by examination of the chronic effects of cocaine on neuroendocrine function in laboratory animals and in humans. An examination of the influence of endocrine status on the effects of cocaine on other brain mechanisms will conclude this review.

The Components of the Neuroendocrine System

The regulation of the secretion of hormones from the pituitary gland is controlled from two different sites.

Anterior Pituitary

Neurons in the hypothalamus secrete releasing and/or inhibiting factors into the pituitary portal vessels, where these substances are transported to the anterior lobe of the pituitary gland. When these factors reach the pituitary gland, they bind to cells in the anterior lobe to increase or decrease the secretion of the specific pituitary hormones into the circulation.

The anterior pituitary hormones reach their target glands to stimulate the synthesis and secretion of their hormones. The hormones from the target glands can then inhibit secretion of the respective pituitary hormone by negative feedback processes. This negative feedback occurs at the hypothalamic or at the pituitary level. For example, corticotropin releasing factor (CRF)-containing neurons in the hypothalamic paraventricular nucleus release CRF into the capillaries in the median eminence; CRF reaches the anterior pituitary through the portal vessels and then stimulates the release of corticotropin (ACTH) into the circulation. ACTH at the adrenal gland stimulates the synthesis and secretion of glucocorticoids and aldosterone. A feedback loop exists through which glucocorticoids bind to sites in the brain and to the corticotrophes in the anterior pituitary gland. Binding to these sites leads to inhibition of further secretion of CRF and, consequently, of ACTH (Tuomisto and Mannisto, 1985).

Posterior Pituitary

The hormones released from the hypophysial neural or posterior lobe, vasopressin and oxytocin, are synthesized in neurons in the hypothalamic supraoptic and paraventricular nuclei. These neurons send their axons to the neural lobe of the pituitary where vasopressin and oxytocin are released from the nerve terminals. Vasopressin controls water excretion from the kidneys and oxytocin stimulates milk ejection in

Table 1
Hypothalamic-Pituitary Hormones

Hypothalamic releasing factor	Gland or target	Hormone
LH-RH stimulates LH (FSH) (Luteinizing Hormone and Follicle Stimulating Hormone)	Gonads	(a) Female: (1) estrogen, (2) progesterone, (3) inhibin (b) Male: (1) testosterone (LH), (2) inhibin (FSH)
TRH stimulates TSH (Thyrotropin)	Thyroid	T_3, T_4
CRF stimulates ACTH (Adrenocorticotropin)	Adrenal	Cortisol
GH-RH (GRH) stimulates GH (Growth Hormone)	Liver	Somatomedin
PRH stimulates prolactin	Mammary glands	Milk
Somatostatin inhibits GH	Liver	Somatomedin
Dopamine inhibits prolactin	Mammary glands	Milk
Oxytocin	Uterus	Contracts uterus
	Mammary glands	Milk ejection
Vasopressin (ADH)	Kidney	Reabsorption of water in the distal nephron

lactating mothers. To date, there is no known hormonal feedback loop for the hormones of the neural lobe.

It is important to recognize that the pituitary gland does not regulate all endocrine glands. These glands are, instead, controlled by direct neural innervation and by other blood-borne substances. Examples of endocrine glands which are regulated independently of the pituitary gland include the C-cells in the thyroid gland which secrete calcitonin, the parathyroid gland which secretes parathyroid hormone (PTH), the pancreas which secretes insulin and glucagon, and the kidneys which secrete both renin and erythropoietin.

Outlined in Table 1 are the hypothalamic releasing/inhibiting hormones with their corresponding pituitary hormones and their target glands or tissues. The hypothalamic neurons containing releasing/inhibiting hormones are innervated by dopaminergic, serotonergic, and adrenergic neurons sensitive to the effect of cocaine. It is important, therefore, to examine the role of these aminergic neurons in the regulation of hypothalamic-pituitary function. The following is a survey of the role of DA-, 5-HT-, and NE-containing neurons in neuroendocrine function.

Dopaminergic Regulation of Neuroendocrine Function

Dopaminergic neurons in the arcuate nucleus release DA directly into the pituitary portal vessels. DA is then transported to the pituitary gland and binds to D_2 receptors to inhibit prolactin secretion (Ben-Jonathan *et al.*, 1977; MacLeod and Lehmeyer, 1974). Dopaminergic neurons also inhibit thyrotropin releasing hormone (TRH), and thus, thyrotropin (TSH) secretion, although this effect is not as important in the overall regulation of TSH secretion as is the case with prolactin (Vijayan *et al.*, 1978; Mannisto *et al.*, 1981). Dopaminergic neurons stimulate growth hormone (GH) secretion in normal humans (Bansal *et al.*, 1981; Lal *et al.*, 1977). Activation of brain dopaminergic D_2 receptors increases the secretion of ACTH (Fuller and Snoddy, 1984).

Serotonergic Regulation of Neuroendocrine Function

Serotonergic neurons originating in the dorsal raphe nucleus stimulate prolactin (Barofsky *et al.*, 1983; Van de Kar and Bethea, 1982; Van de Kar *et al.*, 1985a, 1985c), vasopressin (Iovino and Steardo, 1985; Brownfield *et al.*, 1988), and renin secretion (Gotoh *et al.*, 1987; Lorens and Van de Kar, 1987; Van de Kar and Richardson Morton, 1986; Van de Kar *et al.*, 1981, 1982, 1985b) via nerve terminals in the hypothalamus. Furthermore, serotonergic neurons outside the dorsal raphe nucleus stimulate ACTH (Feldman *et al.*, 1984, 1987; Fuller, 1981; Koenig *et al.*, 1987; Lorens and Van de Kar, 1987), and β-endorphin secretion (Koenig *et al.*, 1987).

Noradrenergic Regulation of Neuroendocrine Function

Adrenergic and noradrenergic neurons stimulate the secretion of ACTH (Feldman *et al.*, 1986, 1988; Spinedi *et al.*, 1988). Both basal and cold-induced TSH secretion are mediated by noradrenergic neurons which activate α_2-adrenoceptors (see Tuomisto and Mannisto, 1985, for review). Noradrenergic neurons also stimulate oxytocin (Crowley *et al.*, 1987) and luteinizing hormone (LH) secretion (Kalra and Kalra, 1983). Also, prolactin secretion is increased by a central noradrenergic mechanism, possibly through increased TRH secretion (Day *et al.*, 1982; Grosvenor and Mena, 1980). Activation of central α_2 receptors inhibits vasopressin (Reid *et al.*, 1979) and renin secretion (Nolan and Reid, 1978; Reid *et al.*, 1979).

NEUROENDOCRINE EFFECTS OF ACUTE COCAINE

This chapter will survey the evidence for effects of acute administration of cocaine on several hormones that have been evaluated to date.

Animal Studies

Administration of cocaine (20 mg/kg, i.p.) to conscious rats increases plasma ACTH, β-endorphin, and corticosterone concentrations. The effect of cocaine on plasma ACTH is maximal 10 min after administration while β-endorphin and corticosterone levels reach a peak at 20 min after cocaine injections (Moldow and Fishman, 1987). Similar findings were reported by Rivier and Vale (1987) who found that administration of cocaine i.v. to conscious rats produced a dose-dependent (1–10 mg/kg) elevation in plasma ACTH concentration. They demonstrated that the effect of cocaine on ACTH secretion is mediated by release of CRF, since co-administration of antiserum against rat CRF blocked the effect of cocaine (5 mg/kg, i.v.) on ACTH secretion. Cocaine, added *in vitro* to cultured pituitary cells, could not directly increase ACTH secretion.

Administration of cocaine at doses of 10–40 mg/kg (s.c.) decreased serum prolactin concentration in male rats. The effect was not very robust and only the highest dose (40 mg/kg) had a small but statistically significant effect (Ravitz and Moore, 1977). These authors also found that other psychostimulants which activate dopaminergic neurons, such as amphetamine and methylphenidate, produced a mild decrease in serum prolactin concentration. These data suggest that the suppression of prolactin secretion might be due to activation of DA receptors in the anterior pituitary. The inhibitory effect of cocaine on prolactin secretion was confirmed by Steger *et al.* (1981). In contrast, Pilotte and Sharpe (1988) did not detect a change in plasma prolactin levels in male rats which received an acute i.v. injection of 1 mg/kg of cocaine. The reason for the difference between these studies is unclear at the present time.

Steger *et al.* (1981) also found that administration of cocaine at doses of 10–20 mg/kg (i.p.) to ovariectomized female rats produced an increase in serum LH concentration, without affecting serum follicle stimulating hormone (FSH) concentration. However, a higher dose (40 mg/kg) of cocaine actually decreased serum LH levels (Steger *et al.*, 1981). If cocaine also increases LH secretion in males, a logical consequence would be that plasma testosterone levels would also be elevated. This was found to be the case. Administration of cocaine significantly increases plasma testosterone concentration in male rats at 90–120 min postinjection (Gordon *et al.*, 1980).

Lakoski *et al.* (1988) have recently investigated the effect of cocaine on renin secretion and found that the acute administration of cocaine (15 mg/kg, i.p.) decreases plasma renin concentration within 15 min. This effect could be a compensatory response to the blood pressure-elevating effect of cocaine and the subsequent inhibition of sympathetic output (Knuepfer *et al.*, 1988).

In conclusion, acute administration of cocaine can affect several hormones in laboratory animals. While this information is still incomplete, the data provide important new information which will be used as guidelines for future research on the neuroendocrine effects of cocaine in laboratory animals and in humans.

Human Studies

The ethical and practical problems mentioned above limit the evaluation of the acute neuroendocrine effects of cocaine in naive humans. For these reasons, very little is known about the acute effects of cocaine on neuroendocrine function in humans. Dackis and Gold (1985) found elevated plasma prolactin levels in cocaine abusers who were admitted with detectable cocaine concentrations in their blood. However, since these were probably chronic cocaine abusers, these effects of cocaine would be inappropriate in considering acute effect in naive subjects. No studies have been reported on the acute effect of cocaine on neuroendocrine function in either male or female naive human subjects.

Because of the paucity of human data, future studies of acute effects of cocaine must be obtained from appropriate animal models in the hope that these data can be extrapolated to humans. Such information would be useful to understand how cocaine influences neurotransmitter systems which regulate these hormones. Furthermore, this information could also explain the basis for the chronic neuroendocrine effects of cocaine.

NEUROENDOCRINE EFFECTS OF CHRONIC COCAINE

Chronic abuse of cocaine would be expected to lead to permanent or semipermanent modulation of the function of several amine-containing neurons (Galloway, 1988; Lakoski and Cunningham, 1988; Roy *et al.*, 1978). Neuroendocrine consequences of chronic cocaine abuse are to be expected, since these neurons are involved in the regulation of neuroendocrine function. For example, cocaine is believed to increase sexual performance. However, in male chronic cocaine abusers, there is a *decrease* in libido and sexual performance (Ashley, 1975; Spotts and Shontz, 1980; Siegel, 1982; Washton *et al.*, 1985). While more studies have been conducted on the chronic rather than the acute effects of cocaine, both in laboratory animals and in humans, the majority of these studies are fairly descriptive in nature, and do not provide insight into the mechanisms involved.

Animal Studies

Administration of cocaine (10 mg/kg, s.c., daily) to female rats disrupts their estrous cycle within 5–7 d. The main characteristic of the altered cycle was repetitive days of estrus and prolonged periods of diestrus (King *et al.*, 1988a). In addition, the *in vitro* increase of LH-RH release from hypothalamic tissue slices in response to NE and phorbol-12-myristate-13-acetate (PMA) was inhibited in these rats (King *et al.*, 1988b). Prenatal, chronic cocaine exposure of male rats (10 or 30 mg/kg, s.c., twice daily between days 15 and 20 of gestation) led to increased plasma LH levels and a decrease in sperm count when they mature (McGivern *et al.*, 1988). An increase in plasma LH would suggest that plasma testosterone would also be elevated. However, no alteration of plasma testosterone or FSH levels was observed (McGivern *et al.*, 1988). Basal prolactin concentrations were elevated in male rats which received daily injections of cocaine (1 mg/kg) for 4 consecutive days. Subsequent administration of cocaine to these rats produced a 66% decrease in plasma prolactin concentrations (Pilotte and Sharpe, 1988).

Chronic administration of 40 mg/kg (i.p.) cocaine to male rats for 21 d increases the size of their adrenal glands. The increased size could be due to increased ACTH secretion as acute cocaine injections increase ACTH secretion (Moldow and Fishman, 1987; Rivier and Vale, 1987). However, lower doses of cocaine (1, 10 mg/kg, i.p.) did not alter the size of the adrenal glands (Kirkby and Petchkovsky, 1973). In another study, administration of cocaine at doses of 2.5, 5.0, 10, or 25 mg/kg (i.p.) to male rats for 10 consecutive days led to a significant increase in plasma β-endorphin concentration (Forman and Estilow, 1988). This phenomenon did not seem to be dose dependent. β-Endorphin is often co-secreted with ACTH. Thus, it is not clear whether or not chronic cocaine consistently increases ACTH secretion. Finally, Lakoski *et al.* (1988) found that plasma renin concentration was elevated in rats which received twice-daily injection of cocaine (15 mg/kg, i.p.) for 7 d.

Human Studies

A few cautionary notes merit attention before discussing the neuroendocrine findings in human cocaine abusers. Most studies on cocaine abusers have methodological shortcomings which confuse the interpretation of the data. For example, the purity of the abused cocaine should be questioned and the period of cocaine abuse might be exaggerated upwards or downwards. The cocaine abuser could also have abused other drugs in addition to cocaine without informing the investigators. Other possible factors which contribute to the heterogeneity of the sample are (1) the state of health, (2) sleep habits, (3) nutritional status, and (4) the presence of stressors, etc. For these reasons, and despite serious difficulties, the results obtained in cocaine abusers will have to be verified and clinically replicated, preferably in diverse populations in different geographical locations.

Several studies indicate that plasma prolactin levels are elevated in cocaine abusers, especially if the samples were obtained within 72 h after the last administration of cocaine (Dackis and Gold, 1985; Dax *et al.*, 1988; Mendelson *et al.*, 1987, 1988; Wilkins *et al.*, 1988). Most investigators agree that the measurement of plasma prolactin concentrations in cocaine abusers may reflect changes in the activity of the dopaminergic tuberoinfundibular neurons (Dackis and Gold, 1985; Dax *et al.*, 1988; Mendelson *et al.*, 1988). In contrast, one study by Gawin and Kleber (1985) reported a *decrease* in plasma prolactin concentration in cocaine abusers who have been abstaining for 4–10 d. Thus, there appears to be a biphasic response to cocaine withdrawal, but additional studies are needed to evaluate the changes in prolactin secretion over the duration of abstinence from cocaine.

TSH, triiodothyronine (T3), and thyroxine (T4) concentrations in the blood of cocaine abusers were determined within 5 d after cessation of cocaine use and found to be similar to a control group. However, the increase of TSH observed following a challenge with TRH (500 μg) was lower in the cocaine abusers than in age-matched normal subjects (Dackis and Gold 1985; Dackis *et al.*, 1985). This finding could be interpreted as a down-regulation of pituitary TRH receptors, a phenomenon usually associated with increased availability of TRH in the pituitary. Since noradrenergic neurons stimulate TRH secretion (see Introduction), these findings could suggest that chronic cocaine exposure leads to noradrenergic hyperactivity which may be expressed in increased release of TRH from the hypothalamus.

After withdrawal from chronic high doses of cocaine, there is an increase in the size of adrenal glands in rats (Kirkby and Petchkovsky, 1973). However, in humans plasma cortisol levels were reported to remain within the normal range and no alteration in the circadian rhythm of cortisol was found (Dax *et al.*, 1988; Gawin and Kleber, 1985). These findings suggest that the increased size of the adrenal glands reflects a toxic effect of cocaine. Finally, GH was elevated in cocaine abusers who abstained from cocaine for 4–10 d (Gawin and Kleber, 1985). No additional studies have been reported to verify these findings.

In conclusion, the chronic effects of cocaine on neuroendocrine function are not well understood. Hyperprolactinemia seems to be the most prominent chronic effect of cocaine. This phenomenon could explain the decreased reproductive ability reported in many abusers of cocaine (Siegel, 1982; Washton *et al.*, 1985). The data from the human studies suffer from a lack of stringency in the conditions under which the data were collected. Although this is a problem that cannot easily be remedied, efforts should be made to standardize the conditions under which human data are collected. For example, a specific urinary or serum cocaine (or metabolite) titer could be set as a criterion for admission of cocaine abusers into a study.

Compatible, parallel studies in laboratory animals should be undertaken to verify and reproduce the important neuroendocrine findings.

INFLUENCE OF ENDOCRINE STATUS ON THE RESPONSE TO COCAINE

Thus far, this review has examined the acute and chronic effects of cocaine on hormone secretion. There is evidence from studies in laboratory animals that the status of some endocrine systems has an influence on the intake and behavioral or toxic effects of cocaine. For example, pretreatment with pregnenolone-16α-carbonitrile, spironolactone, or estradiol reduced the motor disturbances (dyskinesia) induced by cocaine (60 mg/kg, i.p.) in female rats, and reduced the concentration of cocaine in the blood (Rapp *et al.*, 1979). The liver-toxic effect of acute and chronic high doses of cocaine (50–100 mg/kg) was more pronounced in female than in male mice. This effect seems to be testosterone dependent since castration of male mice made them more vulnerable and supplements of testosterone reversed this effect (Thompson *et al.*, 1984). Similarly, ovariectomy of female mice followed by injection of testosterone propionate reversed their susceptibility to the hepatotoxic effect of cocaine.

There is also an apparent sex-dependent difference in the rate of self-administration of cocaine in rats trained to administer cocaine after injection of dopamine antagonists. Male rats self-administered less cocaine after an injection of haloperidol than female rats (Dalton *et al.*, 1986). These investigators subsequently extended their studies and observed haloperidol-stimulated cocaine self-administration over the estrous cycle. Rats in diestrus self-administered more cocaine than during the other days of the estrous cycle (Roberts *et al.*, 1987). While these findings could be related to the sensitivity of dopaminergic neurons to female sex hormones, this remains to be verified experimentally.

Diabetes seems to decrease the cocaine dose that is lethal in 50% (LD_{50}) of rats. Moreover, the chronic injection of cocaine was more toxic in diabetic rats than acute cocaine injections. Injection of insulin only partly restored the LD_{50} of cocaine towards the range found in nondiabetic rats (Hoskins *et al.*, 1988). These findings suggest that an investigation should be carried out to determine whether cocaine is more toxic in diabetic patients.

CONCLUDING REMARKS

This review has surveyed the evidence for acute and chronic effects of cocaine on the secretion of several hormones from the pituitary gland and from glands which are independent of the pituitary gland. Additionally, data have been reviewed suggesting that the effects of cocaine can be altered by the presence of certain hormones in the circulation. The current information regarding the effects of cocaine on neuroendocrine function is limited to only a few hormones and the results are at best preliminary. There are difficulties in conducting well-controlled studies in humans, and these difficulties will have to be acknowledged. Studies in laboratory animals will be vital for establishing a body of information on the neuroendocrine profile of cocaine. In turn, such studies will enable a better interpretation of the data which can be obtained in human studies.

ACKNOWLEDGMENTS

Supported by NIDA DA04865. The author wishes to thank Dr. Alexander H. Friedman and Rashel Piechowski for reviewing the manuscript.

REFERENCES

J.J. Ambre, S.M. Belknap, J. Nelson, T.I. Ruo, S.-G. Shin, and A.J. Atkinson. Acute tolerance to cocaine in humans. *Clin. Pharmacol. Ther.* **44**: 1 (1988).

R. Ashley. *Cocaine*. Warner Brooks, New York. (1975).

S.A. Bansal, L.A. Lee, and P.D. Woolf. Dopaminergic regulation of growth hormone (GH) secretion in normal man: correlation of L-dopa and dopamine levels with the GH response. *J. Clin. Endocrinol. Metab.* **53**: 301 (1981).

A.-L. Barofsky, J. Taylor, and V.J. Massari. Dorsal raphe-hypothalamic projections provide the stimulatory serotonergic input to suckling-induced prolactin release. *Endocrinology* **113**: 1894 (1983).

N. Ben-Jonathan, C. Oliver, H.J. Weiner, R. Mical, and J. Porter. Dopamine in hypophysial postal plasma of the rat during the estrous cycle and throughout pregnancy. *Endocrinology* **100**: 542 (1977).

M.S. Brownfield, J. Greathouse, S.A. Lorens, J. Armstrong, J.H. Urban, and L.D. Van de Kar. Neuropharmacological characterization of serotoninergic stimulation of vasopressin secretion in conscious rats. *Neuroendocrinology* **47**: 277 (1988).

W.R. Crowley, S.-W. Shyr, B. Kacsoh, and C.E. Grosvenor. Evidence for stimulatory noradrenergic and inhibitory dopaminergic regulation of oxytocin release in the lactating rat. *Endocrinology* **121**: 14 (1987).

C.A. Dackis, T.W. Estroff, D.R. Sweeney, A.L.C. Pottash, and M.S. Gold. Specificity of the TRH test for major depression in patients with serious cocaine abuse. *Am. J. Psychiatry* **142**: 1097 (1985).

C.A. Dackis and M.S. Gold. New concepts in cocaine addiction: the dopamine depletion hypothesis. *Neurosci. Biobehav. Rev.* **9**: 469 (1985).

J.C.H. Dalton, G.J. Vickers, and D.C.S. Roberts. Increased self-administration of cocaine following haloperidol: sex-dependent effects of the antiestrogen tamoxifen. *Pharmacol. Biochem. Behav.* **25**: 497 (1986).

E.M. Dax, W.W. Weddington, N.S. Pilotte, and J.J. Jaffe. Release of prolactin and cortisol following cocaine cessation in men. *Neurosci. Abstr.* **14**: 1051 (1988).

T.A. Day, P.M. Jervois, M.R. Menaude, and J.O. Willoughby. Catecholamine mechanisms in mediobasal hypothalamus influence prolactin but not growth hormone secretion. *Brain Res.* **253**: 213 (1982).

S. Feldman, N. Conforti, and E. Melamed. Norepinephrine depletion in the paraventricular nucleus inhibits the adrenocortical responses to neural stimuli. *Neurosci. Lett.* **64**: 191 (1986).

S. Feldman, N. Conforti, and E. Melamed. Paraventricular nucleus serotonin mediates neurally stimulated adrenocortical secretion. *Brain Res. Bull.* **18**: 165 (1987).

S. Feldman, N. Conforti, and E. Melamed. Involvement of ventral noradrenergic bundle in corticosterone secretion following neural stimuli. *Neuropharmacology* **27**: 129 (1988).

S. Feldman, E. Melamed, N. Conforti, and J. Seidenfeld. Effect of central serotonin depletion on adrenocortical responses to neural stimuli. *Exp. Neurol.* **85**: 661 (1984).

L.J. Forman and S. Estilow. Cocaine influences β-endorphin levels and release. *Life Sci.* **43**: 309 (1988).

R.W. Fuller. Serotonergic stimulation of pituitary-adrenocortical function in rats. *Neuroendocrinology* **32**: 118 (1981).

R.W. Fuller and H.D. Snoddy. Central dopamine receptors mediating pergolide-induced elevation of serum corticosterone in rats. Characterization by the use of antagonists. *Neuropharmacology* **23**: 1389 (1984).

M.P. Galloway. Neurochemical interactions of cocaine with dopaminergic systems. *Trends Pharmacol. Sci.* **9**: 451 (1988).

F.H. Gawin and H.D. Kleber. Neuroendocrine findings in chronic cocaine abusers: a preliminary report. *Br. J. Psychiatry* **147**: 569 (1985).

L.A. Gordon, D.I. Mostofsky, and G.G. Gordon. Changes in testosterone levels in the rat following intraperitoneal cocaine HCl. *Int. J. Neurosci.* **11**: 139 (1980).

E. Gotoh, K. Murakami, T.D. Bahnson, and W.F. Ganong. Role of brain serotonergic pathways and hpothalamus in regulation of renin secretion. *Am. J. Physiol.* **253**: R179 (1987).

C.E. Grosvenor and F. Mena. Evidence that thyrotropin-releasing hormone and a hypothalamic prolactin-releasing factor may function in the release of prolactin in the lactating rat. *Endocrinology* **107**: 863 (1980).

B. Hoskins, C.K. Burton, and I.K. Ho. Diabetes potentiation of cocaine toxicity. *Res. Commun. Substance Abuse* **9**: 117 (1988).

M. Iovino and L. Steardo. Effect of substances influencing brain serotonergic transmission on plasma vasopressin levels in the rat. *Eur. J. Pharmacol.* **113**: 99 (1985).

S.P. Kalra and P.S. Kalra. Neural regulation of luteinizing hormone secretion in the rat. *Endocrinol. Rev.* **4**: 311 (1983).

T.S. King, R.S. Schenken, I.S. Kang, D.K. Frasier, and R.M. Riehl. Effect of cocaine on GnRH release and estrous cyclicity in rats. *Endocrinology* **122**: 319 (1988a).

T.S. King, R.S. Schenken, and M.A. Javors. Effect of cocaine on *in vitro* noradrenergic and phorbol ester-stimulated hypothalamic GnRH release. *Soc. Neurosci. Abstr.* **14**: 661 (1988b).

R.J. Kirkby and L. Petchkovsky. Chronic administration of cocaine: effects on defecation and adrenal hypertrophy in the rat. *Neuropharmacology* **12**: 1001 (1973).

M.M. Knuepfer, D.M. Wehner, and T.L. Sellers. Hemodynamic response characteristics to cocaine (C) in rats. *Neurosci. Abstr.* **14**: 961 (388.4) (1988).

J.I. Koenig, G.A. Gudelsky, and H.Y. Meltzer. Stimulation of corticosterone and β-endorphin secretion in the rat by selective 5-HT receptor subtype activation. *Eur. J. Pharmacol.* **137**: 1 (1987).

J.M. Lakoski, R.E. Halpern, K.A. Cunningham, K. Kunimoto, and L.D. Van de Kar. Cocaine effects on renin secretion in conscious rats. *Neurosci. Abstr.* **14**: 1174 (469.1) (1988).

L.M. Lakoski and K.A. Cunningham. Cocaine interactions with central monoaminergic systems: electrophysiological approaches. *Trends Pharmacol. Sci.* **9**: 177 (1988).

S. Lal, H. Guyda, and S. Bikadoroff. Effect of methysergide and pimozide on apomorphine-induced growth hormone secretion in men. *J. Clin. Endocrinol. Metab.* **44**: 766 (1977).

S.A. Lorens and L.D. Van de Kar. Differential effects of serotonin (5-HT1A and 5-HT2) agonists and antagonists on renin and corticosterone secretion. *Neuroendocrinology* **45**: 305 (1987).

R.M. MacLeod and J.E. Lehmeyer. Studies on the mechanism of the dopamine-mediated inhibition of prolactin secretion. *Endocrinology* **94**: 1077 (1974).

P.T. Mannisto, J. Mattila, and S. Kaakkola. Possible involvement of nigrostriatal dopamine system in the inhibition of thyrotropin secretion in the rat. *Eur. J. Pharmacol.* **76**: 403 (1981).

R.F. McGivern, W.J. Raum, P. Peterson, and R.Z. Sokol. Long-term behavioral and endocrine effects of prenatal cocaine exposure in male rats. *Endocrinology* **122**: 219 (1988).

T.H. Mendelson, S.K. Teoh, U. Lange, N.K. Mello, R. Weiss, and A. Skupny. Hyperprolactinemia during cocaine withdrawal, in: *Problems in Drug Dependence, 1987, Proceedings of the 49th Annual Scientific Meeting of the Committee on the Problems of Drug Dependence.* L.S. Harris, ed., *NIDA Research Monograph #81,* Washington, DC (1987).

J.H. Mendelson, S.K. Teoh, U. Lange, N.K. Mello, R. Weiss, A. Skupny, and J. Ellingboe. Anterior pituitary, adrenal, and gonadal hormones during cocaine withdrawal. *Am. J. Psychiatry* **145**: 1094 (1988).

R.L. Moldow and A.J. Fishman. Cocaine induced secretion of ACTH, β-endorphin, and corticosterone. *Peptides* **8**: 819 (1987).

P.L. Nolan and I.A. Reid. Mechanism of suppression of renin secretion by clonidine in the dog. *Circ. Res.* **42**: 206 (1978).

N.S. Pilotte and L.G. Sharpe. Chronic cocaine (C) alters the release of prolactin (PRL) in male rats. *Neurosci. Abstr.* **14**: 73 (33.11) (1988).

U. Rapp, P. Kourounakis, and H. Selye. Effect of steroids and diethylstilbestrol on cocaine toxicity, plasma concentrations and urinary excretion. *Arzneim- Forsch/Drug Res.* **29**: 48 (1979).

A.J. Ravitz and K.E. Moore. Effects of amphetamine methylphenidate and cocaine on serum prolactin concentrations in the male rat. *Life Sci.* **21**: 267 (1977).

I.A. Reid, P.L. Nolan, J.A. Wolf, and L.C. Keil. Suppression of vasopressin secretion by clonidine: effect of α-adrenoceptor antagonists. *Endocrinology* **104**: 1403 (1979).

C. Rivier and W. Vale. Cocaine stimulates adrenocorticotropin (ACTH) secretion through a corticotropin-releasing factor (CRF)-mediated mechanism. *Brain Res.* **422**: 403 (1987).

D.C.S. Roberts, J.C.H. Dalton, and G.I. Vickers. Increased self-administration of cocaine following haloperidol: effect of ovariectomy, estrogen replacement, and estrous cycle. *Pharmacol. Biochem. Behav.* **26**: 37 (1987).

S.N. Roy, A.K. Bhattacharyya, S. Pradhan, and S.N. Pradhan. Behavioral and neurochemical effects of repeated administration of cocaine in rats. *Neuropharmacology* **17**: 559 (1978).

M.A. Sherer. Intravenous cocaine: psychiatric effects, biological mechanisms. *Biol. Psychiatry* **24**: 865 (1988).

R.K. Siegel. Cocaine and sexual dysfunction: the curse of mama coca. *J. Psychoact. Drugs* **14**: 71 (1982).

R.B. Smith and M.D. Pittsburgh. Cocaine and catecholamine interaction. *Arch. Otolaryngol.* **98**: 139 (1973).

E. Spinedi, C.A. Johnston, A. Chisari, and A. Negro-Vilar. Role of central epinephrine on the regulation of corticotropin-releasing factor and adrenocorticotropin secretion. *Endocrinology* **122**: 1977 (1988).

J.V. Spotts and F.C. Shontz. *Cocaine Users. A Representative Case Approach.* Free Press, New York (1980).

R.W. Steger, A.Y. Silverman, A. Johns, and R.H. Asch. Interactions of cocaine and delta-9-tetrahydrocannabinol with the hypothalamic-hypophysial axis of the female rat. *Fertility Sterility* **35**: 567 (1981).

M.L. Thompson, L. Shuster, E. Casey, and G.C. Kanel. Sex and strain differences in response to cocaine. *Biochem. Pharmacol.* **8**: 1299 (1984).

J. Tuomisto and P. Mannisto. Neurotransmitter regulation of anterior pituitary hormones. *Pharmacol. Rev.* **37**: 249 (1985).

L.D. Van de Kar and C.L. Bethea. Pharmacological evidence that serotonergic stimulation of prolactin secretion is mediated via the dorsal raphe nucleus. *Neuroendocrinology* **35**: 225 (1982).

L.D. Van de Kar, M. Karteszi, C.L. Bethea, and W.F. Ganong. Serotonergic stimulation of prolactin and corticosterone secretion is mediated by different pathways from the mediobasal hypothalamus. *Neuroendocrinology* **41**: 380 (1985a).

L.D. Van de Kar, S.A. Lorens, J.H. Urban, and C.L. Bethea. Effect of selective serotonin (5-HT) agonists and 5-HT2 antagonist on prolactin secretion. *Neuropharmacology* **28**: 299 (1989).

L.D. Van de Kar, K.D. Richardson, and J.H. Urban. Serotonin and norepinephrine-dependent effects of fenfluramine on plasma renin activity in conscious rats. *Neuropharmacology* **24**: 487 (1985b).

L.D. Van de Kar and K.D. Richardson-Morton. Serotonergic regulation of the release of renin is not mediated by the autonomic nervous system but involves beta adrenoceptors. *Neuropharmacology* **25**: 487 (1986).

L.D. Van de Kar, J.H. Urban, K.D. Richardson, and C.L. Bethea. Fenfluramine causes elevation in plasma prolactin levels via a serotonergic mechanism but causes elevation in plasma corticosterone levels via a mechanism that is independent of serotonin. *Neuroendocrinology* **41**: 283 (1985c).

L.D. Van de Kar, C.W. Wilkinson, and W.F. Ganong. Pharmacological evidence for a role of brain serotonin in the maintenance of plasma renin activity in unanesthetized rats. *J. Pharmacol. Exp. Ther.* **219**: 85 (1981).

L.D. Van de Kar, C.W. Wilkinson, Y. Skrobik, M.S. Brownfield, and W.F. Ganong. Evidence that serotonergic neurons in the dorsal raphe nucleus exert a stimulatory effect on the secretion of renin but not of corticosterone. *Brain Res.* **235**: 233 (1982).

E. Vijayan, L. Krulich, and S.M. McCann. Catecholaminergic regulation of TSH and growth hormone release in ovariectomized and ovariectomized, steroid-primed rats. *Neuroendocrinology* **26**: 174 (1978).

A.M. Washton, M.S. Gold, and A.C. Pottash. The 800-COCAINE helpline: survey of 500 callers, in: *Problems of Drug Dependence, NIDA Research Monograph 55*. L.S. Harris, ed., DHHS publication ADM 85-1393. U.S. Government Printing Office, Washington, DC (1985).

J.N. Wilkins, D.A. Gorelick, B.R. Bamshad, and D.Y. Setode. Elevated serum prolactin levels in newly abstinent male cocaine abusers. *Neurosci. Abstr.* **14**: 766 (309.12) (1988).

16

Effects of Cocaine and Other Drugs of Abuse on Immune Responses

HENRY U. BRYANT
Department of Immunology
Pulmonary and Leukotriene Research
Eli Lilly and Co.
Indianapolis, Indiana (HUB)

KATHRYN A. CUNNINGHAM AND THOMAS R. JERRELLS
Departments of Pharmacology and Toxicology (KAC) and Pathology (TRJ)
The University of Texas Medical Branch
Galveston, Texas

INTRODUCTION

The immune response consists of an orchestrated cascade of events that includes the recognition and processing of foreign antigen, proliferation and differentiation of the appropriate antigen-responsive cell types, and the production of autocrine and endocrine mediators for amplification and regulation of the response. While the immune system was at one time thought to act autonomously, it is now known that functionally relevant communicative networks exist among the nervous, endocrine, and immune systems. Felten *et al.* (1985) have demonstrated that major lymphoid tissue, such as the spleen and thymus, is innervated with both afferent and efferent nerve fibers. Furthermore, specific lesioning of the central nervous system (CNS) by chemical or physical manipulations results in alterations in immunity (Jankovic and Spector, 1986). Recently, endogenous neuropeptides and neurotransmitters have been shown to play an important regulatory role in the immune response (Besedovsky *et al.*, 1979), and a number of centrally active pharmacological agents have been shown to either directly or indirectly affect immune function (e.g., Pawlikowski *et al.*, 1987).

It is not surprising then that a number of drugs of abuse are now suspect for producing deleterious effects upon immune function. From an immunologic point of view, it is often difficult to look at the drug abuser per se and ascertain potential effects of drug abuse on the immune system, due to the immunotoxic health consequences often associated with drug addiction (e.g., sharing of potentially infected drug paraphernalia, lack of personal hygiene, and poor nutritional status). In fact, early epidemiological findings of increased rates of certain infections in narcotic abusers were originally attributed entirely to the sharing of contaminated needles and other factors. However, based on a number of preclinical studies in animal models and through some elegant work in human cell systems, it is now apparent that a number of addictive substances do possess the potential for deleterious effects on cells of the immune system. This review will specifically focus on three major drugs of abuse: cocaine, morphine, and marijuana. Effects of these narcotics will be discussed within the context of their effects on a variety of immune cell types and on infectivity.

EFFECTS OF DRUGS OF ABUSE ON LYMPHOCYTES

Cocaine

While the effects of cocaine on neurotransmission and behavior have been extensively studied, effects on immune parameters have received relatively little attention. Several investigators have examined cocaine-induced effects on lymphocyte function, both *in vitro* and *in vivo*. Lymphocytes represent a critical effector arm of the immune system and are responsible for both the cell-mediated responses (T-lymphocytes) and antibody production, or the humoral response (B-lymphocytes). Mitogens such as concanavalin A (Con A) or phytohemagglutinin (PHA) stimulate T-lymphocytes to proliferate and produce lymphokines in culture, while the bacterial lipopolysaccharide (LPS) stimulates murine and rat B-lymphocytes. Cocaine, when delivered *in vitro* to mitogen-stimulated lymphocytes, can be shown to suppress proliferative responses in cultures derived from both human and murine sources (Klein *et al.*, 1988). However, the concentrations of cocaine required to attain significant reductions *in vitro* are quite high (200 mcg/ml or approximately 0.6 m*M* for a reduction of 75%) relative to concentrations normally found in cocaine addicts (e.g., approximately 100 ng/ml in the plasma 2 h following a 2-mg/kg intranasal dose; Jacob *et al.*, 1984; or 2 h following a 0.6-mg/kg oral dose; Van Dyke *et al.*, 1986) or to the lethal blood concentration of cocaine (approximately 2–4 μg/ml in one case report; Sander *et al.*, 1985). Klein showed that these suppressive effects were not related to toxicity or reduced viability of the lymphocytes. Finally, T-lymphocyte responses appeared to be more sensitive to cocaine than B-lymphocytes.

In vivo administration of cocaine has also, in some instances, been associated with altered lymphocyte responses. Havas *et al.* (1987) demonstrated in mice that high doses of cocaine on an acute basis resulted in rather small increases in antibody production to a T-cell-independent antigen (pneumococcal polysaccharide type III) and larger elevations of antibody in response to a T-cell-dependent antigen (2,4-dinitrophenyl), although, again, doses at which these effects were observed are quite high (e.g., 50 mg/kg, five times in 24 h) when compared to doses at which other effects of cocaine are normally observed (e.g., 1 mg/kg is sufficient to significantly elevate heart rate and several other cardiovascular parameters; Billman and Hoskins, 1988). Chronic treatment of mice with lower doses of cocaine produce little or no effect on antibody production (Havas *et al.*, 1987). Conversely, findings by Watson *et al.* (1983) indicated that cocaine and several of its congeners suppress delayed-type hypersensitivity, a T-cell-mediated response. Work in our laboratory (Jerrells and Cunningham, unpublished) has shown that administration of cocaine to rats has little effect on the immune system. We found in these studies that animals given cocaine did not differ from controls in terms of thymus and spleen cell numbers. Splenic lymphocytes from the cocaine-treated animals were hyperresponsive to mitogenic stimulation in culture. Natural killer cell activity was also significantly greater from cocaine-treated rats when compared to the controls. These data are in contrast to the findings when experimental animals have been treated with morphine (discussed in this chapter) or ethanol (Jerrells *et al.*, 1986). It is apparent that cocaine has an effect on the immune system; however, in contrast to morphine and ethanol, the effect is to heighten, rather than blunt, immune responses. It remains to be determined if these heightened responses are pathological.

Evidence for potential effects of chronic cocaine exposure on lymphocyte function in man (e.g., cocaine addicts) is sparse. Donahoe *et al.* (1986) have utilized T-cell E-rosette formation as an indicator of T-lymphocyte function and demonstrated a suppression of this parameter in heroin abusers. Interestingly, in heroin addicts who also abuse cocaine, this suppression of E-rosette formation is reversed, suggesting some possible stimulatory effects on this T-cell parameter.

Cocaine is a local anesthetic and inhibits the re-uptake process for dopamine (DA), norepinephrine (NE), and serotonin (5-HT); thus, several mechanisms can be proposed for the potential effects of cocaine on lymphocyte function. Cocaine might directly affect lymphocyte responses via its local anesthetic properties and possible perturbation of calcium flux (calcium flux is essential for the induction of blastogenesis in lymphocytes) or to alterations in phosphatidyl inositol turnover within the lymphocyte (Allan and Michell, 1975). Alternatively, several indirect mechanisms might be proposed. These include (1) cocaine-induced stimulation of β-endorphin, and adrenocorticotrophin and corticosterone release (Gawin and Kleber, 1985; Moldow and Fischman, 1987; Rivier and Vale, 1987; Forman and Estilow, 1988), all of which are known immunomodulatory substances (Parrillo and Fauci, 1979; Holaday, 1985; Morley *et al.*, 1987); (2) altered adrenergic tone, since lymphocytes are known to contain β-adrenergic receptors and lymph organs possess adrenergic innervation and (Williams *et al.*, 1981); (3) possible hepatotoxicity (Kloss *et al.*, 1984), which is of particular relevance in the murine model since the liver is a major hematopoetic organ in the mouse. Potential interactions with DA systems appear to mediate some of the behavioral effects of cocaine, including its reinforcing properties. This provides one additional, interesting potential mechanism since prolactin, which is an important immunomodulatory hormone (Bernton *et al.*, 1988), is under tonic inhibition by DA. An analogous situation probably occurs with 5-HT systems. Cocaine administration potently diminishes central 5-HT activity, presumably as a result of its ability to inhibit the re-uptake of 5-HT (Cunningham and Lakoski, 1988; submitted). As is the case with DA, hormones derived from the hypothalamic-pituitary-adrenal axis in the rat are under serotonergic control (Koenig *et al.*, 1988); thus, perturbations of 5-HT tone could impact upon circulating levels of immunomodulatory hormones.

Morphine

The narcotic alkaloids derived from the opium poppy are perhaps the most heavily researched drugs of abuse with regard to effects on immune function. A large volume of literature is now available describing the effects of morphine and heroin in particular, as well as related compounds, on various immunologic parameters.

The discovery of opiate receptors on human peripheral lymphocytes was one of the first direct links between drugs of abuse and the immune system. It is now know that lymphocytes contain at least three of the traditional opiate receptor subtypes (μ, δ, and ϵ; Carr, 1988) and, in certain instance, respond to opiates following exposure *in vitro*. Morphine, as well as several of its endogenous opioid peptide counterparts, depresses the binding of sheep erythrocytes by T-lymphocytes (E-rosette formation; Wybran *et al.*, 1979). This suppression of a T-cell function by morphine is stereospecific and preventable by treatment with the opiate receptor antagonist, naloxone. Morphine administration *in vitro*, however, is not suppressive in all lymphocyte-based models. Human lymphocytes stimulated by mitogens such as PHA show an enhanced response when treated with either an agonist (morphine) or an antagonist (naloxone; Bocchini *et al.*, 1983). In mitogen-driven murine spleen cell proliferative assays, again, no suppression is observed with direct morphine administration *in vitro* (Bryant *et al.*, 1989). In fact, a slight stimulatory effect may be found (Bryant *et al.*, 1989). In contrast to the suppression of E-cell rosette formation, the effect on mitogen-induced proliferation is not preventable by naloxone. These examples are merely representative of the *in vitro* findings with opiates and immunologic systems; however, they do give some indication of the difficulties which exist in attempting to form a unified statement on direct effects of opiates on immune cells.

When immunologic parameters are evaluated following exposure to opiates *in vivo*, a more consistent picture emerges; that is, that the opiates generally appear to be suppressive when lymphocyte-dependent systems are considered. Lymphocyte proliferative responses to mitogens such as Con A or pokeweed mitogen are suppressed in heroin addicts (Brown *et al.*, 1974) and the total number of circulating lymphocytes are decreased in this population (Wybran *et al.*, 1979). In animal models, chronic exposure to morphine inhibits a number of important immune indices. For example, B-cell-related responses, such as antibody formation and serum hemolysin production in response to sheep erythrocytes, are inhibited by morphine. The suppression of the primary antibody response by morphine may be a T-cell-dependent effect, based on the observation made by Weber *et al.* (1987) that TNP-ovalbumin induced antibody production is preferentially suppressed by morphine. The suppression of antibody production by morphine is stereoselective and is prevented by naloxone (Weber *et al.*, 1987). Cellular-mediated immunity is also suppressed by morphine, as evidence by the inability of morphinized animals to produce a tuberculin reaction (Pellis *et al.*, 1986). The inflammatory response to carrageenan is also suppressed by morphine (Gyires *et al.*, 1985).

In a chronic exposure model that involves implantation of a morphine pellet (which allows for a continuous dosing regimen), we have documented a severe compromise in the performance of the immune system of mice. Morphine pellet implantation produces a marked atrophy and reduced cellularity of both the spleen and thymus within 48 h (Bryant *et al.*, 1988a). The splenic atrophy is associated with dramatic reductions in the proliferative responses of the splenocytes. Lymphocyte proliferation induced by T-cell mitogens (such as Con A) or B-cell mitogens (such as LPS; Bryant *et al.*, 1988a) or in an autologous mixed lymphocyte response (Bryant *et al.*, 1988b) are markedly suppressed within 2 to 3 d following implantation of amorphine pellet. Continuous exposure to an opiate receptor antagonist (naloxone) prevents the antiproliferative effects of chronic morphine exposure (Bryant *et al.*, 1987b).

The blastogenic response of lymphocytes depends upon the interplay of a complex cascade of events which include the recognition and processing of foreign antigen and the production of cytokine mediators for amplification of the response. The attenuated blastogenic responses to Con A in morphine-pelleted mice has been directly linked to a reduction in interleukin-2 production by the lymphocytes (Peritt *et al.*, 1988). The suppression of blastogenesis does not appear to be related to an effect at the interleukin-2 receptor because blast cells from mitogen-stimulated cultures of spleen cells from morphine-pelleted mice show an entirely normal response to exogenous interleukin-2 administration (Petitt *et al.*, 1988). Our most current findings suggest a reduction in interleukin-1, a critical monokine involved in the initiation of the proliferative cascade, may also be involved in this phenomenon.

The opiates are a class of drugs which possess a very complex pharmacology and are capable of producing a wide number of effects in the body. In addition to their well-characterized effects on pain perception and gastrointestinal motility, opiates affect various neuroendocrine parameters as well (Holaday, 1985). Thus, in many instances, effects of an opiate following *in vivo* administration may be secondary to actions in the CNS. This appears to be a likely case for at least some of the immunosuppressive effects attributed to chronic morphine exposure. With our chronic morphine pellet model, we have noted both adrenal hypertrophy and elevated levels of corticosterone (Bryant *et al.*, 1989). Since glucocorticoids are well-recognized immunosuppressants (Parrillo and Fauci, 1979) and since other immunosuppressive regimens are felt to be mediated through activation of glucocorticoids (e.g., Jerrells *et al.*, 1986), it seemed likely that morphine might be acting in such a manner. Therefore, we examined the immunosuppressive effects of morphine pellet implants in adrenalectomized animals and in mice treated with the glucocorticoid receptor antagonist, RU-486. As would be suspected in a glucocorticoid-mediated response, the immunosuppressive profile of morphine was blunted both by adrenalectomy and by glucocorticoid antagonist treatment (Bryant *et al.*, 1989). These studies indicate that in at least one model activation of the hypothalamo-pituitary-adrenal axis is a critical mechanism for the immunosuppressive effects of morphine. While these findings may not apply directly to all models of opiate-induced immunosuppression, they do provide for an important indirect mechanism which should be considered.

Marijuana

A variety of adverse effects (e.g., cardiovascular, psychologic, and genetic) have at one time or another been attributed to the widespread abuse of marijuana or related cannabinoids (e.g., hashish). However, relatively few pathological conditions have been unequivocally attributed to the use of marijuana for recreational purposes. Indeed, there is little agreement on either toxic or beneficial effects attributable to marijuana. Of concern to this review are the reports that have implicated marijuana and its major psychoactive constituent, Δ-9-tetrahydrocannabinol (THC), as immunosuppressive agents. As described here, there remains some debate as to the presence of real immunologic abnormalities in marijuana abusers.

In animal model of cellular- or humoral-mediated immunity, administration of THC or marijuana smoke consistently produces suppressive effects. Lymphocyte blastogenesis in response to T-lymphocyte mitogens (PHA or Con A) and B-lymphocyte mitogens (LPS) is suppressed by THC when delivered in a dose range of 50–200 mg/kg (Levy *et al.*, 1975) or by moderate to high exposure to marijuana smoke (Daul and Heath, 1975). The antiproliferative effect can be observed in as little as 3 h following exposure to THC and subsides within 48 h after THC administration (Levy *et al.*, 1975). Splenic weight (Petersen and Lemberger, 1976) and cellularity (Levy *et al.*, 1975) are also suppressed by THC. Other measures of T-lymphocyte-mediated immunity are suppressed as well following exposure to THC. For example, delayed-type hypersensitivity (Smith *et al.*, 1978; Watson *et al.*, 1983) and T-lymphocyte-dependent antibody responses (Petersen and Lemberger, 1976) are suppressed in THC-treated mice. An increase in skin graft survival has also been observed following chronic exposure to THC (Levy *et al.*, 1974). Antibody production by B-lymphocytes is also attenuated following *in vivo* exposure to THC or marijuana smoke in animals. Antibody formation in response to sheep erythrocytes in sensitized mice is suppressed by THC (Smith *et al.*, 1978; Watson *et al.*, 1983), as is the number of antibody-forming lymphocytes (Petersen and Lemberger, 1976). Additionally, immunoglobulin levels of mice chronically exposed to marijuana smoke are reduced relative to control animals (Daul and Heath, 1975).

Structure-activity relationships of structural modifications of THC indicate that the immunosuppressive effects of THC in mice are not related to effects on the CNS. Structural modifications devoid of activity in the CNS were observed as immunosuppressive by Smith *et al.* (1978). A direct effect of THC on lymphocyte responses is possible, as murine lymphocytes exposed to THC *in vitro* (concentration of 5–20 μM) display attenuated proliferative responses to PHA. Indirect mechanisms cannot be ruled out since THC significantly elevates plasma corticosterone levels (Jacobs *et al.*, 1979) and prostaglandin production (Burstein *et al.*, 1988), both of which are known to affect immune cell responses.

Effects of marijuana and its constituents on immune function in man are less clarified, and some degree of controversy exists as to whether there even is an effect. While the suppressive effects of marijuana smoking on T-cell rosette formation have been shown with consistency (Gupta *et al.*, 1974; Cushman and Kuhrana, 1975), reports on the effects of marijuana smoking on mitogen-induced blastogenesis are conflicting. In marijuana smokers, a reduction in the blastogenic response of circulating lymphocytes to PHA (Petersen *et al.*, 1974; Nahas *et al.*, 1974) or to allogenic cells in a mixed lymphocyte response (Nahas *et al.*, 1974a) has been reported by several groups. However, other labs using the same mitogenic stimuli (PHA) in marijuana smokers and making comparison to nonmarijuana smokers found no suppression in blastogenesis (White *et al.*, 1975; Lau *et al.*, 1976; Rachelefsky *et al.*, 1976). Administration of THC directly to test subjects was also found not to affect PHA-induced blastogenesis (Lau *et al.*, 1976). Skin sensitization with 2,4-dinitrochlorobenzene, a test of the integrity of T-lymphocyte-mediated immunity, is also unchanged in marijuana smokers (Silverstein and Lessin, 1974). Controversy also exists over the effects of marijuana on humoral immunity. Nahas *et al.* (1979) reported a significant reduction of IgG levels in marijuana smokers, while Rachelefsky *et al.* (1976) found no changes in IgG levels.

The controversy on the effects of marijuana *in vivo* may be attributable to any of a number of factors, including health status of subjects, age, gender, abuse of other drugs, and length and magnitude of exposure to marijuana. Following exposure of normal human lymphocytes to THC *in vitro*, effects of a suppressive nature are primarily observed. T-cell rosette formation is attenuated at THC concentrations in the micromolar range (Cushman and Khurana, 1975), as is the blastogenic response to PHA (Armand *et al.*, 1974). A direct effect is possible, as THC *in vitro* inhibits the biosynthesis of DNA, RNA, and protein in PHA-stimulated lymphocytes, although only at very high concentrations (e.g., 10 m*M*; Nahas *et al.*, 1977). Interestingly, some investigators have suggested that an unknown component of marijuana smoke may act as a direct antigenic stimulus (Shapiro *et al.*, 1976).

EFFECTS OF DRUGS OF ABUSE ON PHAGOCYTIC MONONUCLEAR CELLS

Cocaine

Macrophages and monocytes represent another key cellular component of the immune system. In addition to their role as phagocytic/scavenger-type cells, macrophages play a central role in the proliferative response of lymphocytes to antigens through their role as antigen-presenting cells and their production of monokine mediators such as interleukin-1. Surprisingly, relatively little has been reported on the effects of cocaine on polymorphonuclear cells. In human peripheral monocytes, Welch (1983) demonstrated a marked reduction of phagocytosis following *in vitro* administration of cocaine. A significant reduction of phagocytosis was noted with cocaine concentrations as low as 0.1 μg/ml and an exposure period of only 30 min. The suppression was dramatic (72%) at a concentration of 0.5 μg/ml and higher. Since these concentrations may be attained in the blood, this represents a potentially important pathophysiological consequence of cocaine abuse. These antiphagocytic effects of cocaine may be restricted to the more immature, circulating monocytes, as more mature tissue macrophages, such as those found in the lung, are relatively insensitive to cocaine (Davidson *et al.*, 1986).

Morphine

Several groups have studied the effects of morphine exposure on macrophage function. In man, morphine administration is associated with a depression of both the phagocytic and killing properties of monocytes (Tubaro *et al.*, 1985); the latter observation is most likely related to a reduced ability of monocytes in morphine-treated individuals to generate superoxide (Tubaro *et al.*, 1985).

Macrophages can be activated to become tumorcidal *in vitro* when stimulated with agents such as γ-interferon, or when they are obtained from animals treated with inducers like mycobacterium bovis (BCG). Chronic exposure of mice to morphine via morphine pellet implantation suppresses the tumorcidal activity of BCG-elicited macrophages (Bryant *et al.*, 1988b). Exogenous administration of γ-interferon, *in vitro*, to peritoneal macrophages obtained from morphine-pelleted, BCG-treated mice, restores tumorcidal activity. This suggests that suppression of γ-interferon production, as has been reported by Hung *et al.* (1973), is a likely contributor to this morphine-induced macrophage defect. Expression of class II major histocompatibility molecules (Ia) on peritoneal macrophages elicited by BCG was also reduced in the morphine-pelleted mice (Bryant *et al.*, 1988b).

Effects of morphine at the macrophage level may also be responsible for some of the antiproliferative effects of morphine which have been observed in lymphocyte systems. In one experiment aimed at assessing such a role, macrophages and T-lymphocytes were isolated from spleens of both control and morphine-pelleted mice. The macrophages and T-lymphocytes were then recombined in a cross-over design. With this approach, it was demonstrated that when lymphocytes from morphine-treated mice are incubated with macrophages from control mice, only a mild reduction of Con-A-induced proliferation is observed. However, when lymphocytes from control animals are co-cultured with macrophages from morphine-pelleted mice, the more familiar marked reduction of Con-A-induced proliferation is observed (Bryant *et al.*, 1988b). These findings suggest a critical role for the macrophage in morphine-induced anti-proliferative responses of T-lymphocytes.

Marijuana

Marijuana smoke and its major psychoactive component, THC, also produce a number of deleterious effects on macrophage function in both man and animals. Alveolar macrophages isolated from rats exposed to THS *in vivo* exhibit a depressed ability to kill *Staphylococcus aureas* in culture (Huber *et al.*, 1975). In work done with humans, histological and biochemical changes were found in pulmonary-derived macrophages from chronic marijuana smokers (Mann *et al.*, 1971). Petersen *et al.*, (1974) noted that one of the most striking immunologic aberrations in chronic marijuana smokers is a reduction of polymorphonuclear cells that were capable of phagocytosing yeast cells.

In vitro work with THC and its effects on macrophages suggest that direct effects of the cannabinoids are likely. Mouse peritoneal macrophages exposed to THC in culture exhibit marked histologic changes (Raz and Goldman, 1975). THC *in vitro* with murine macrophages also suppresses their ability to ingest yeast particles (Friedman *et al.*, 1986) or to spread on a glass plate, which is an assessment of macrophage activation (Lopez-Cepero *et al.*, 1986). The latter effects were observed in concentrations of 5–10 μg/ml. In human macrophage cell cultures, THC administration *in vitro* has been shown to be cytotoxic to pulmonary macrophages (Chari-Bitron *et al.*, 1971).

EFFECTS OF DRUGS OF ABUSE ON NATURAL KILLER CELL ACTIVITY

Cocaine

Natural killer cells represent a third type of immune cell known to be affected by various drugs of abuse. These granular lymphocytic-type cells are known to be lytic to certain tumor targets, and therefore work in the absence of antigen presentation (Tolmadge *et al.*, 1980). In addition to this well-known function in malignancy surveillance, natural killer cells may also activate in the presence of certain viral infections (Ching and Lopez, 1979) and appear to play a critical role as regulators of antibody production (Arai *et al.*, 1983.). Incubation of relatively high concentrations of cocaine (e.g., 0.5–1 μg/ml) with human natural killer cells *in vitro* results in a suppression of natural killer cell activity (Welch, 1983). However, lower concentrations of cocaine *in vitro*, which more closely resemble those observed following exposure to cocaine in humans, have no effect on natural killer cell activity (Van Dyke *et al.*, 1986). Perhaps the most interesting effects of cocaine on natural killer cell activity have been observed following exposure of human subjects to cocaine *in vivo*. An oral dose of 0.6 mg/kg in man has been associated with a significant *increase* in natural killer cell activity within 20 min of cocaine administration (Van Dyke *et al.*, 1986). This *in vivo* data, when considered with the *in vitro* data (which showed either a suppression or no effect at all), suggest that an indirect mechanism is likely for this effect of cocaine. Van Dyke *et al.* (1986) suggested that the stimulatory effect *in vivo* may be related to the stimulatory effects of cocaine on catecholamine release or to an inhibition of catecholamine reuptake by cocaine. This appears to be a plausible hypothesis in light of the known stimulatory effects of norepinephrine on natural killer cell activity (Tonnesen *et al.*, 1984). Since natural killer cells have been associated with an ability to impair antibody production by B-lymphocytes (Abruzzo and Rowley, 1983; Arai *et al.*, 1983), the stimulatory effects of cocaine on natural killer cell activity may ultimately lead to adverse effects on host defense (Van Dyke *et al.*, 1986).

Morphine

The effects of opiates on natural killer cell activity have been extensively examined in various animal models. Shavit *et al.* (1986) demonstrated a suppressive effect of morphine on natural killer cell activity. As with the effects of cocaine, the effect of morphine on natural killer cell activity does not appear to be a direct effect. In the case of morphine, the suppressive effect on natural killer cells may be mediated by activation of the adrenal gland and subsequent release of glucocorticoids (Shavit *et al.*, 1986). To further the notion that effects of morphine on natural killer cell function is not a direct effect, Weber *et al.* (1988) have performed very elegant studies demonstrating that the opiate is affecting a specific area of the brain, the periaqueductal gray matter of the mesencephalon, to produce this inhibitory effect. While the bulk of the literature in this area indicates a suppressive effect for morphine on natural killer cell activity, there are some contradictory reports. In mice, for example, Kraut and Greenberg (1986) reported a stimulatory action of morphine on natural killer cell function; however, this effect may be dose and time dependent.

Marijuana

Several reports indicate that natural killer cell activity, using both human and animal model systems, is suppressed by THC. THC concentrations of 5–15 μg/ml *in vitro* inhibit human natural killer cell activity in a manner that is reversible by addition of cytokines (such as interleukin-2) to the culture media (Specter *et al.*, 1988). In rats, natural killer cell activity is not affected by acute administration of THC *(in vivo)*. However, repeated administration (3 mg THC/kg daily for 25 d) resulted in a marked reduction of natural killer cell activity in rats (Patel *et al.*, 1985).

EFFECTS OF DRUGS OF ABUSE ON MODELS OF INFECTIVITY

Cocaine

Perhaps the most important and clinically relevant effect that drugs of abuse appear to have with regard to performance of the immune system is their ultimate effect on host resistance to infection. High rates of a number of infectious disorders are commonly observed in drug abusers. An important concern and potentially confounding variable when discussing drugs of abuse and incidence of infection is whether the phenomenon is related to an effect of the drug or simply to cross-contamination via the sharing of infected needles and drug paraphernalia. Knowledge of the effects of the drugs per se on the immune system (e.g., as described in the previous sections) and in animal models of infectivity would suggest that, with some of the abused substances, the drug does contribute directly to the higher rates of infection observed in drug addicts.

In man, cocaine abuse has been associated with an increased risk for endocarditis (Chambers *et al.*, 1987), hepatitis (Estroff *et al.*, 1987), and human immunodeficiency virus (HIV) seropositivity (Sterk, 1988). Endocarditis and hepatitis are primarily associated with i.v. cocaine administration, while the increase in exposure to HIV is observed in nonintravenous abuse of a refined form of cocaine (''crack''). There are, however, very little animal data to substantiate a direct role for cocaine in these infectious diseases. Havas *et al.* (1987) examined resistance to infection by *Streptococcus pneumoniae* and resistance to solid tumor injections in cocaine-treated mice. In this particular study, no reduction in resistance to infection (or tumor implantation) was observed in the cocaine-treated mice.

Secondary lines of host defense should also be considered when examining infectivity. Interestingly, suppressive effects of cocaine were demonstrated by Ukai *et al.* (1985) on mucociliary transport in the respiratory tract. Such an effect might conceivably result in an increased risk of pulmonary types of infection, although such a problem has not yet been identified in the cocaine abuser population.

Morphine

In reports as early as 1967, an increased incidence of a number of bacterial and viral related infections was described in heroin abusers (Briggs *et al.*, 1967; Cherubin, 1967; Louria *et al.*, 1967; Miller *et al.*, 1979). More recently, the concern of an increased incidence of acquired immunodeficiency syndrome (AIDS) in intravenous narcotic abusers has been reported (Fauci, 1988). While cocaine administration was not associated with discernable changes in the resistance of mice to an infection, chronic morphine exposure does appear to potentiate the lethality of various infections, thereby lending some relevance to the effects of morphine on the *in vitro* correlates already discussed. In mice given a sublethal inoculation of *Listeria monocytogenes*, morphine pellet implantation results in a marked increase in lethality associated with the bacterial infection (Bryant *et al.*, 1988b). Chronic exporsure to morphine also increases lethality following infection with *Candida albicans* and *Klebsiella pneumoniae* (Tubaro *et al.*, 1985). Recent work by Watson *et al.* (1988), using a murine leukemic retrovirus model (which closely resembles the HIV infection in man; Mosier *et al.*, 1985), has indicated deleterious effects of chronic morphine exposure in a viral model as well.

Marijuana

Consistent with its generally determined immunosuppressive effects, marijuana is also associated with an increased incidence and severity of infection. In mice infected with *Listeria monocytogenes,* THC lowered the 50% lethal dose of the pathogen, indicating a reduced ability to fight infection. Severity of viral-type infections is also increased in THC-treated animals. Both mice and guinea pigs infected with Herpes simplex virus type 2 presented a greater severity of vaginal herpes lesions and an increased number of lethalities when THC was given chronically. The increased severity of herpes infection in mice given THC has been related to reduced interferon levels.

CONCLUSIONS AND DISCUSSION

The deleterious effects of drugs of abuse have been discussed at length in both the scientific literature and the lay press. Abused substance control is a major concern of today's society. Cocaine is perhaps the drug of greatest national concern at the present time, due to the dramatic increases in its abuse in the past 10 years. Morphine and its related opiate narcotics represent perhaps one of the oldest abused substances, and marijuana remains one of the most common recreational drugs abused today. A number of pathological conditions have been associated or attributed to drug abuse. Effects of abused substances on immunologic parameters represent one important concern, particularly in light of the high rates of various infectious disorders observed in abusers of these controlled substances. A particularly important concern today is reflected in the findings that drug abusers have a much greater risk of exposure to HIV and the subsequent development of AIDS (Ginzberg *et al.*, 1985). Approximately 25% of reported AIDS cases occur in i.v. drug abusers, and 40–50% of i.v. drug abusers are seropositive for HIV (Robert-Guroff *et al.*, 1986). i.v. drug abusers, as a whole, now consititute the second largest group of persons to have developed AIDS and, in certain municipal areas of the country, the rate of AIDS infection in i.v. drug abusers has surpassed that of homosexual males (Schuster, 1988). i.v. drug abusers also constitute an important route of contact with the heterosexual population. While sharing of HIV-contaminated drug paraphernalia is most likely the primary culprit for this elevated rate of HIV infectivity, there is now some evidence suggesting the drugs of abuse themselves may also be important contributors (e.g., Watson *et al.*, 1988). Potential effects of drugs of abuse on conversion of HIV seropositive individuals to full-blown AIDS is another important consideration and potential point for an effect of drugs.

Since some of the abused substances are used for therapeutic purposes (e.g., opiates for pain management), the immunomodulatory effects of these drugs may have potentially important clinical relevance as well. This may be a particularly important concern in individuals at a high risk for infection or in immunocompromised patients (e.g., burn patients or patients on chemotherapy). In this regard, the formation of lymphokine-activated killer cells (LAK therapy), which has been advocated as a therapeutic approach in the treatment of certain cancers, is inhibited by exposure to morphine (Weber *et al.*, 1987). Therefore, it is possible that morphine may be contraindicated in patients undergoing LAK therapy. Further investigation in areas such as these is necessary.

Several other points should be highlighted when considering effects of abused substances on performance of the immune system. First, in some cases, the abused substance may not directly affect cells of the immune system; but rather, affect immune function indirectly. Therefore, it is important to consider *in vitro* data alongside *in vivo* data when evaluating the immunopharmacology of drugs of abuse. Second, the issues of tolerance and dependence associated with many drugs of abuse may impact the immune system — particularly in the habitual abuser, who often experiences self-induced withdrawal in order to reduce drug requirement to a less expensive, more affordable habit. Finally, the increased occurrence of polydrug use raises some potentially important concerns as to the combined effect of the substances on the abusers immune status.

In closing, it should also be noted that cocaine, morphine, and marijuana are not the only substances of abuse reported to affect immunologic end points. A number of other abused substances alter some facet of immunity, including alcohol (Jerrells *et al.*, 1986; MacGregor, 1986), tobacco smoke (Keast and Ayre, 1981), diazepam (Pawlikowski *et al.*, 1987), chlordiazepoxide (Nahas *et al.*, 1979), barbiturates (Puppo *et al.*, 1980), and nitrous oxide (Cullen and Duncan, 1976).

REFERENCES

L.V. Abruzzo and D.A. Rowley. Homeostasis of the antibody response: immunoregulation by natural killer cell. *Science* **222**: 581 (1983).

D. Allan and R.H. Michell. Enhanced synthesis *de novo* of phosphatidylinositol in lymphocytes treated with cationic amphiphilic drugs. *Biochem. J.* **148**: 471 (1975).

S. Arai, H. Yamamoto, K. Itohand, and K. Kumagai. Suppressive effect of human natural killer cell on pokeweed mitogen induced B-cell differentiation. *J. Immunol.* **131**: 581 (1983).

J.-P. Armand, J.T. Hsu, and G.G. Nahas. Inhibition of blastogenesis of T lymphocytes by Δ-9-THC. *Fed. Proc.* **33**: 539 (1974).

E.W. Bernton, M.S. Meltzer, and J.W. Holaday. Suppression of macrophage activation and T-lymphocyte function in hypoprolactinemic mice. *Science* **239**: 401 (1988).

H.O. Besedovsky, A. Del Rey, E. Sorkin, M. Da Prada, and H.H. Keller. Immunoregulation mediated by the sympathetic nervous system. *Cell. Immunol.* **48**: 346 (1979).

G.E. Billman and R.S. Hoskins. Cocaine-induced ventricullar fibrillation: protection afforded by the calcium antagonist verapamil. *FASEB J.* **2**: 2990 (1988).

G. Bocchini, G. Bobabbi, and A. Canevari. *Drug Alcohol Depend.* **11**: 233 (1983).

J.H. Briggs, C.G. McKerron, R.L. Souhami, D.J.E. Taylor, and H. Andrews. Severe systemic infections complicating "mainline" heroin addiction. *Lancet* **ii**: 1227 (1967).

S.M. Brown, B. Stimmel, R.N. Taub, S. Kochwa, and R.E. Rosenfeld. Immunologic dysfunction in heroin addicts. *Arch. Intern. Med.* **134**: 1001 (1974).

H.U. Bryant, E.W. Bernton, and J.W. Holaday. Immunosuppressive effects of chronic morphine treatment in mice. *Life Sci.* **41**: 1731 (1987a).

H.U. Bryant, E.W. Bernton, and J.W. Holaday. Morphine pellet induced immunomodulation in mice: temporal relationships. *J. Pharmacol. Exp. Ther.* **245**: 913 (1988a).

H.U. Bryant, E.W. Bernton, and J.W. Holaday. Morphine induced immunosuppression. Involvement of glucocorticoids and prolactin. In: *Proceedings of the 50th Annual Scientific Meeting of the Committee on Problems of Drug Dependence, NIDA Research Monograph.* in press (1989).

H.U. Bryant, E.W. Bernton, J.R. Kenner, T.K. Shakarjian, and J.W. Holaday. Suppression of macrophage function and increased lethality to bacterial infection in mice given morphine pellet implants. *FASEB J.* **2**: A1260 (1988b).

H.U. Bryant, T.K. Shakarjian, E.W. Bernton, and J.W. Holaday. Morphine induced immunosuppression in mice: effects of naloxone and morphine tolerance. *Pharmacologist* **29**: 106 (1987b).

S.H. Burstein, K. Hull, S.A. Hunter, and V. Latham. Cannabinoids and pain responses: a possible role for prostaglandins. *FASEB J.* **2**: 3022 (1988).

D.J. Carr. Opioid receptors on cells of the immune system. *Prog. Neuroendocrinimmunol.* **1** **(2)**: 8, (1988).

H.F. Chambers, D.L. Morris, M.G. Tauber, and G. Modin. Cocaine use and the risk for endocarditis in intravenous drug users. *Ann. Intern. Med.* **106**: 833 (1987).

A. Chari-Bitron. Stabilization of rat erythrocyte membrane by delta-1-tetrahydrocannabinol. *Life Sci.* **10**: 1273 (1971).

C.E. Cherubin. The medical sequelae of narcotic addiction. *Ann. Intern. Med.* **67**: 23 (1967).

C. Ching and C. Lopez. Natural killing of herpes simplex virus type-1 infected target cells: normal human responses and influence of antiviral antibody. *Infect. Immunol.* **26**: 49 (1979).

B.F. Cullen and P.G. Duncan. Inhibition of cell mediated cytotoxicity by halothane and nitrous oxide. *Anesthesiology* **44**: 386 (1976).

P. Cushman and R. Khurana. Effects of marijuana smoking and tetrahydrocannabinol on T-cell rosettes. *Fed. Proc.* **34**: 783 (1975).

K.A. Cunningham and J.M. Lakoski. Electrophysiological effects of cocaine and procaine on dorsal raphe serotonin neurons. *Eur. J. Pharmacol.* **148**: 457 (1988).

K.S. Cunningham and J.M. Lakoski. The interaction of cocaine with serotonin: single unit extracellular recording studies. *Neuropsychopharmacology* (submitted, 1989).

C.B. Daul and R.G. Heath. The effect of chronic marihuana usage on the immunological status of rhesus monkeys. *Life Sci.* **17**: 875 (1975).

A.C. Davidson, I. Barbosa, J.S. Kooner, and A.S. Hamblin. The influence of topical anaesthetic agents at bronchoaveolar lavage upon cellular yield and identification and upon macrophage function. *Br. J. Dis. Chest* **80**: 385 (1986).

R.M. Donahoe, J.K.A. Nicholson, J. J. Madden, F. Hollingsworth, D.A. Shafer, D. Gordon, P. Bokos, and A. Falek. Coordinate and independent effects of heroin, cocaine and alcohol abuse on T-cell E-rosette formation and antigenic marker expression. *Clin. Immunol. Immunopathol.* **41**: 254 (1986).

T.V. Estroff, I.L. Extein, D. Malaspina and M.S. Gold. Hepatitis in 101 consecutive suburban cocaine and opiate abusers. *Int. J. Psychiatry Med.* **16**: 237 (1987)

A.S. Fauci. The human immunodefficiency virus: infectivity and mechanisms of pathogenesis. *Science* **239**: 617 (1988).

D.L. Felten, S.Y. Felten, S.L. Carlson, J.A. Olschowka, and S. Livnat. Noradrenergic and peptidergic innervation of lymphoid tissue. *J. Immunol.* **135**: 755s (1985).

L.J. Forman and S. Estilow. Cocaine influences β-endorphin levels and release. *Life Sci.* **43**: 309 (1988).

M. Friedman, M. Lopez-Cepero, T. Klein, and H. Friedman. Suppressive effect of Δ-9-tetrahydrocannabinol *in vitro* on phagocytosis by murine macrophages. *Proc. Soc. Exp. Biol. Med.* **182**: 225 (1986).

F.H. Gawin and H.D. Kleber. Neuroendocrine findings in chronic cocaine abusers: a preliminary report. *Br. J. Psychiatry* **147**: 569 (1985).

H.M. Ginzburg, S.H. Weiss, M.G. MacDonald, and R.L. Hubbard. HTLV-III exposure among drug users. *Cancer Res.* **45** (Supply): 46055 (1985).

S. Gupta, M.H. Grieco, and P. Cushman. Impairment of rosette-forming T lymphocytes in chronic marihuana smokers. *N. Engl. J. Med.* **291**: 874 (1974).

K.Gyires, I. Budavari, S. Furst, and I. Mulnar. Morphine inhibits the carrageenan induced oedema and the chemoluminescence of leucocytes stimulated by zymosan. *J. Pharm. Pharmacol.* **37**: 100 (1985).

H.F. Havas, M. Dellaria, G. Schiffman, E.B. Geller, and M.W. Adler. Effect of cocaine on the immune response and host resistance in BALB/c mice. *Int. Arch. Allergy Appl. Immunol.* **83**: 377 (1987).

J.W. Holaday. Endogenous opioids and their receptors. In: *Current Concepts*. Scope Publications, Upjohn, Kalamazoo, MI. p. 26 (1985).

G.L. Huber, G.A. Simmons, C.R. McCarthy, M.B. Cutting, R. Lagurda, and W. Pereira. Depressant effect of marihuana smoke on antibactericidal activity of pulmonary alveolar macrophages. *Chest* **68**: 6 (1975).

C.Y. Hung, S.S. Lefkowitz, and W.F. Geber. Interferon inhibition by narcotic analgesics. *Proc. Soc. Exp. Biol. Med.* **142**: 106 (1973).

P. Jacob, B.A. Elias-Baker, R.T. Jones, and N.L. Benowitz. Determination of cocaine in plasma by automated gas chromatography. *J. Chromatogr.* **306**: 173 (1984).

J.A. Jacobs, A.J. Dellarco, R.A. Mannfred, and J. Harclerode. The effect of Δ-9-tetrahydrocannabinol, cannabidiol and shock on plasma corticosterone concentrations in rats. *J. Pharm. Pharmacol.* **31**: 341 (1979).

B.D. Jankovic and N.H. Spector. Effects on the immune system of lesioning and stimulation of the nervous system: neuroimmunomodulation. In: *Enkephalins and Endorphins: Stress and the Immune System*. N.P. Plotnikoff, R.E. Faith, A.J. Murgo, and R.A. Good, eds., Plenum Press, New York, p. 189 (1986).

T.R. Jerrels, C.A. Marietta, M.J. Eckardt, E. Majchrowicz, and F.F. Weight. Effects of ethanol administration on parameters of immunocompetency in rats. *J. Leuk. Biol.* **39**: 499 (1986).

D. Keast and D.J. Ayre. Effects of chronic tobacco smoke exposure on immune response in aged mice. *Arch. Environ. Health* **36**: 201 (1981).

J.I. Koenig, H.Y. Meltzer, and G.A. Gudelsky. Hormone responses to selective serotonin receptor stimulation in the rat, in: *Serotonin Agonists as Psychoactive Drugs*. N.P.P. Books, R.H. Rech, and G.A. Gudelsky, eds., Ann Arbor, p. 283 (1988).

T.W. Klein, C.A. Newton, and H. Friedman. Suppression of human and mouse lymphocyte proliferation by cocaine, in: *Psychological, Neuropschiatric, and Substance Abuse Aspects of AIDS*. T.P. Bridge *et al.*, eds., Raven Press, New York, p. 139 (1988).

M.W. Kloss, G.M. Rosen, and E.J. Rauckman. Cocaine-mediated hepatotoxicity. A critical review. *Biochem. Pharmacol.* **33**: 169 (1984).

R.P. Kraut and A.H. Greenberg. Effects of endogenous and exogenous opioids on splenic natural killer cell activity. *Natl. Immunol. Cell Growth Regul.* **5**: 28 (1986).

R.J. Lau, D.G. Tubergen, M. Barr, E.F. Domino, N. Benowitz, and R.T. Jones. Phytohemagglutinin-induced lymphocyte transformation in humans receiving Δ-9-thetrahydrocannabinol. *Science* **192**: 805 (1976).

J.A. Levy, A.E. Munson, L.S. Harris, and W.L. Dewey. Effects of Δ-8 and Δ-9-tetrahydrocannabinol of the immune response in mice. *Pharmacologist* **16**: 259 (1974).

J.A. Levy, A.E. Munson, L.S. Harris, and W.L. Dewey. Effects of Δ-9-THC on the immune response of mice. *Fed. Proc.* **34**: 782 (1975).

M. Lopez-Cepero, M. Friedman, T. Klein, and H. Friedman. Tetrahydrocannabinol-induced suppression of macrophage spreading and phagocytic activity *in vitro*. *J. Leukocyte Biol.* **39**: 679, (1986).

D.B. Louria, T. Hensle, and J. Rose. The major medical complications of heroin addiction. *Ann. Intern. Med.* **67**: 1 (1967).

R.R. MacGregor. Alcohol and immune defense. *JAMA* **256**: 1474 (1986).

P.E.G. Mann, A.B. Cohen, T.N. Finley, and A.J. Ladman. Alveolar macrophages: structural and functional differences between nonsmokers and smokers of marihuana and tobacco. *Lab. Invest.* **25**: 111 (1971).

D.J. Miller, H. Kleber, and J.R. Bloomer. Chronic hepatitis associated with drug abuse: significance of hepatitis B virus. *Yale J. Biol. Med.* **52**: 135 (1979).

R.L. Moldow and A.J. Fischman. Cocaine induced secretion of ACTH, β-endorphin and corticosterone. *Peptides* **8**: 819 (1987).

J.E. Morley, N.E. Kay, G.F. Solomon, and N.P. Polotnikoff. Neuropeptides: conductors of the immune orchestra. *Life Sci.* **41**: 527 (1987).

D.E. Mosier, R.A. Yetter, and H.C. Morse. Retroviral induction of acute lymphoproliferative disease and profound immunosuppression in adult C57BL/6 mice. *J. Exp. Med.* **161**: 766 (1985).

G.G. Nahas, M. Davies, and E.F. Osserman. Serum immunoglobulin in chronic marihjuana smokers. *Fed. Proc.* **38**: 591 (1979).

G.G. Nahas, B. Desoize, and C. Legar. Effects of psychotropic drugs on DNA synthesis in cultured lymphocytes. *Proc. Soc. Exp. Biol. Med.* **160**: 344 (1979).

G.G. Nahas, A. Morishima, and B. Desoize. Effects of cannabinoids on macromolecular synthesis and replication of cultured lymphocytes. *Fed. Proc.* **36**: 1748 (1977).

G.G. Nahas, N. Sucia-Foca, J.-P. Armand, and A. Morishima. Inhibition of cellular mediated immunity in marijuana smokers. *Science* **189**: 419 (1974).

G.G. Nahas, N. Sucia-Foca, J.-P. Armand, and A. Morishima. Marijuana smoking: impairment of cellular mediated immunity and cell division in man. *Fed. Proc.* **33**: 539 (1974a).

J.E. Parrillo and A.S. Fauci. Mechanisms of glucocorticoid action on immune processes. *Annu. Rev. Pharmacol. Toxicol.* **19**: 129 (1979).

V. Patel, M. Borysenko, M.S.A. Kumar, and W.J. Millard. Effects of acute and subchronic Δ-9-tetrahydrocannabinol administration on the plasma catecholamine, β-endorphin and corticosterone levels and splenic natural killer cell activity in rats. *Proc. Soc. Exp. Biol. Med.* **180**: 400 (1985).

M. Pawlikowski, H. Stepien, Z. Mroz-Wasilewska, and A. Pawlikowski. Effect of diazepam on cell proliferation in cerebral cortex, anterior pituitary and thymus of developing rats. *Life Sci.* **40**: 1131 (1987).

N.R. Pellis, C. Harper, and N. Dafny. Suppression of the induction of delayed hypersensitivity in rats by repetitive morphine treatments. *Exp. Neurol.* **93**: 92 (1986).

D. Peritt, T.J. Jerrells, J.W. Holaday, and H.U. Bryant. Suppression of interleukin-2 production with chronic morphine treatment. *FASEB J.* **2**: A485 (1988).

B.H. Petersen, J. Graham, L. Lemberger, and B. Dalton. Studies of the immune response in chronic marijuana smokers. *Pharmacologist* **16**: 259 (1974).

B.H. Petersen and L. Lemberger. Effect of Δ-9-tetrahydrocannabinol administration on antibody production in mice. *Fed. Proc.* **35**: 333 (1976).

F. Puppo, G.F. Adami, G. Corsinig, G.M. Zavariss, and J. Zattoni. Effect of a single oral dose of phenobarbitone on lymphocyte blastogenesis response in man. *Br. J. Anaesth.* **52**: 1205 (1980).

G.S. Rachelefsky, G. Opelz, R. Mickey, P. Lessin, M. Kiuchi, M. Silverstein, and E.R. Stiehm. Intact humoral and cell-mediated immunity in chronic marijuana smoking. *J. Allergy Clin. Immunol.* **58**: 483 (1976).

A. Raz and R. Goldman. Effect of hashish compounds on mouse peritoneal macrophages. *Isr. J. Med. Sci.* **11**: 1177 (1975).

C. Rivier and W. Vale. Cocaine stimulates adrenocorticotropin secretion through a corticotropin-releasing factor mediated mechanism. *Brain Res.* **422**: 403 (1987).

M. Robert-Guroff, S.H. Weiss, J.A. Giron, A.M. Jennings, H.M. Ginzburg, I.B. Margolis, W.A. Blattner, and R.C. Gallo. Prevalence of antibodies to HTLV-I, -II, and -III in intravenous drug abusers from an AIDS endemic region. *JAMA* **255**: 3133 (1986).

R. Sander, M.A. Ryser, T.C. Lamoreaux, and K. Raleigh. An epidemic of cocaine associated deaths in Utah. *J. Forensic Sci.* **307**: 478 (1985).

C.R. Schuster. Keynote Address: NIDA Research Program on Treatment: Drugs related to intravenous drug abuse and HIV infection. From: Current chemical and pharmacological advances on drugs of abuse which alter immune function and their impact upon HIV infection. *NIDA Res. Monogr.* in press (1988).

C.M. Shapiro, A.R. Orlina, P.J. Unger, M. Telfer, and A.A. Billings. Marihuana-induced antibody response. *J. Lab. Clin. Med.* **88**: 194 (1976).

Y. Shavit, G.W. Terman, J.W. Lewis, C.J. Zane, R.P. Gale, and J.C. Liebeskind. Effects of footshock stress and morphine on natural killer lymphocytes in rats: studies of tolerance and cross-tolerance. *Brain Res.* **372**: 382 (1986).

M.J. Silverstein and P.J. Lessin. Normal skin test responses in chronic marijuana users. *Science* **186**: 740 (1974).

S.H. Smith, L.S. Harris, I.M. Uwaydah, and A.E. Munson. Structure-activity relationships of natural and synthetic cannabinoids in suppression of humoral and cell-mediated immunity. *J. Pharmacol. Exp. Ther.* **207**: 165 (1978).

S. Specter, M. Riverbark, G. Lancz, T. Klein, and H. Friedman. Interleukin 2 mediated abrogation of tetrahydro-cannabinol induced inhibition of human natural killer cell activity. *Int. J. Immunopharmacol.* **10** (Suppl. 1): 58 (1988).

C. Sterk. Cocaine and HIV seropositivity. *Lancet* **i**: 1052 (1988).

J.E. Tolmadge, K.M. Meyers, D.J. Prieur, and J.R. Starkey. Role of natural killer cells in tumor growth and metastasis in beige mice. *Nature* **284**: 623 (1980).

E. Tonnesen, J. Tonnesen, and N.J. Christensen. Augmentation of cytotoxicity by natural killler cells after adrenalin administration in man. *Acta Pathol. Microbiol. Immunol. Scand.* **92**: 81 (1984).

E. Tubaro, U. Avico, C. Santiangeli, P. Zuccaro, G. Cavallo, R. Pacifici, C. Croce, and G. Borelli. Morphine and methadone impact on human phagocytic physiology. *Int. J. Immunopharmacol.* **7**: 865 (1985).

E. Tubaro, G. Borelli, C. Croce, G. Cavallo, and C. Santiangeli. Effect of morphine on resistance to infection. *J. Infect. Dis.* **148**: 656 (1985).

K. Ukai, Y. Sakakura, and S. Saida. Interaction between mucociliary transport and the ciliary beat of chicken nasal mucosa. *Arch. Otorhinolaryngol.* **242**: 225 (1985).

C. Van Dyke, A. Stesin, R. Jones, A. Chuntharapai, and W. Seaman. Cocaine increases natural killer cell activity. *J. Clin. Invest.* **77**: 1387 (1986).

E.S. Watson, J.C. Murphy, H.N. El Sohly, M.A. El Sohly, and C.E. Turner. Effects of the administration of coca alkaloids on the primary immune responses of mice: interaction with Δ-9-tetrahydrocannabinol and ethanol. *Toxicol. Appl. Pharmacol.* **71**: 1 (1983).

R.R. Watson, R.H. Prabhala, H.R. Draban, M.D. Yahya, and T.L. Smith. Changes in lymphocyte and macrophage subsets due to morphine and ethanol treatment during a retrovirus infection causing murine AIDS. *Life Sci.* **43**: v (1988).

R.J. Weber, B. Ikejiri, K.C. Rice, A. Pert and A.A. Hagan. Opiate receptor mediated regulation of the immune response *in vivo,* in: *Problems of Drug Dependence, 1986. Proceedings of the 48th Annual Scientific Meeting, the Committee on Problems of Drug Dependence, NIDA Research Monograph 76.* 1068 (1987).

R.J. Weber. Immunologic effects of drugs of abuse. In: *Problems of Drug Dependence, 1988.* L.S. Harris, ed., NIDA Research Monograph **90**. 99 (1988).

W.D. Welch. Effect of cocaine on non-specific human cellular host defenses. *Res. Commun. Subst. Abuse* **4**: 1 (1983).

S.White, B.C. Brin, and B.W. Janicki. Mitogen-induced blastogenic response of lymphocytes from marihuana smokers. *Science* **188**: 71 (1975).

J.M. Williams, R.G. Peterson, P.A. Shea, J.F. Schmedtke, D.C. Bauer, and D.L. Felten. Sympathetic innervation of murine thymus and spleen: evidence for a functional link between the nervous and immune systems. *Brain Res. Bull.* **6**: 83 (1981).

J. Wybran, T. Appleboom, J.P. Famey, and A. Govaerts. Suggestive evidence for receptors for morphine and methionine enkephalin on normal human blood T-lymphocytes. *J.Immunol.* **123**: 1068 (1979).

17

Mechanisms Responsible for Cardiovascular Disorders Associated with Cocaine Abuse

RICHARD A. GILLIS AND JOHN A. QUEST
Department of Pharmacology
Georgetown University School of Medicine
Washington, District of Columbia

R. DOUGLAS WILKERSON
Department of Pharmacology
Medical College of Ohio
Toledo, Ohio

INTRODUCTION

Since cocaine became available in relatively pure form and began to flourish in use among the general populace over a century ago, it has been recognized as an agent of abuse with the added detriment of provoking organ system toxicity. Most of the adverse effects associated with cocaine abuse have been attributed primarily to actions on the central nervous system (CNS) and include delirium, hyperpyrexia, seizures followed by respiratory arrest or coma, and then death (Wetli and Wright, 1979). In contrast, with the exception of anecdotal reports, there have been few historical descriptions of potential adverse effects of cocaine on the cardiovascular system. One of the earliest of these was made by Freud in his *Uber Coca* in 1884 in which he described a slight slackening of his pulse, followed by a subsequent moderate increase in pulse rate following the oral administration of cocaine (Freud, 1884). A few years later, more ominous cardiovascular symptoms were reported by none other than the presiding Surgeon General of the U.S., Dr. William A. Hammond, who personally examined cocaine's effects by self-injecting the drug under the skin of the forearm in gradually increasing doses on successive evenings. Increases in the action of the heart and irregularities in cardiac rhythm and force were observed to such a great extent that he became apprehensive of serious results and concluded that a dose of cocaine sufficient to produce death would do so by its action on the heart (Hammond, 1887). The scarcity of such early reports relating to cardiovascular toxicity with cocaine presumably relates, in part, to the diminished use of the chemical at the turn of the 20th century as a result of its removal from popular tonics and beverages such as Coca-Cola and its regulation by the Pure Food and Drug Act of 1906 and the Harrison Narcotics Act of 1914.

With the renewed interest in the use of cocaine as a recreational drug over the past few decades there has been an accelerating number of clinically documented episodes of cardiovascular complications related to cocaine use. These include acute myocardial infarction, angina pectoris, cardiac arrhythmias, heart failure, pulmonary edema, subarachnoid hemorrhage, cerebral infarction, myocarditis, aortic rupture, intestinal ischemia, and obstetrical complications secondary to vasoconstriction (Cregler and Mark, 1986). These effects probably do not result from accidental overdose of cocaine, but rather appear to arise due to unexpected clinical reactions from recreational doses of the drug. Evidence for this is that an overdose of cocaine results primarily in convulsions and respiratory arrest (Finkle and McCloskey, 1977); the above-described cardiovascular complications are usually not associated with either of these events (Isner *et al.*, 1986). Further suggestive evidence against an overdose situation is that the amounts of cocaine consumed in association with cardiovascular complications are generally much less than those reported in cases of cocaine overdose (Isner *et al.*, 1986).

In this chapter, an attempt has been made to focus on the possible mechanism(s) through which recreational doses of cocaine produce myocardial infarction, angina pectoris, and cardiac arrhythmias, and to discuss factors which might predispose individuals to these cardiovascular complications of cocaine abuse. We also review pertinent laboratory animal experiments that seek to explain the actions of cocaine on myocardial and vascular tissue, particularly those performed using doses of the drug that are in the range of those encountered with recreational use. Finally, we attempt to indicate therapeutic approaches for treating cocaine-induced cardiovascular disorders based upon our evaluation of the clinical data and the animal studies.

MAJOR CARDIOVASCULAR COMPLICATIONS ASSOCIATED WITH COCAINE ABUSE

Cocaine-Induced Myocardial Infarction

The most frequently reported cardiovascular complication associated with cocaine abuse is myocardial infarction. There have been at least 46 cases described since 1982 (Coleman *et al.*, 1982; Kossowsky and Lyon, 1984; Schnachne *et al.*, 1984; Pasternak *et al;* 1985; Howard *et al.*, 1985; Cregler and Mark, 1986; Gould *et al.*, 1985; Wilkens *et al.*, 1985; Chiu *et al.*, 1986; Weiss, 1986; Rollingher *et al.*, 1986; Simpson and Edwards, 1986; Isner *et al.*, 1986; Rod and Zucker, 1987; Wehbie *et al.*, 1987; Zimmerman *et al.*, 1987; Altieri *et al.*, 1987; Smith *et al.*, 1987; Lam and Goldschlager, 1988; Ascher *et al.*, 1988; Hadjimiltiades *et al.*, 1988; Wang *et al.*, 1988). The subjects ranged in age from 19–47 years (mean age of approximately 34 years). Of interest is the fact that at least 36 of the patients were males, thus raising the question of whether males either use the drug more frequently or whether they might be more susceptible to cocaine's effects on the coronary circulation. The route of administration of cocaine was intranasal in 27 patients and intravenous in 10 patients. The dose of cocaine used was clearly reported in only five instances: 0.15 g (intranasal, two subjects), 0.25 g (intranasal, one subject) 1 g (intranasal, one subject), and 1.5 g (intranasal and over a 5-h period, one subject). Four of the patients died, presumably, from cocaine-induced myocardial infarction. The time interval between cocaine administration and symptoms of cardiovascular complications was immediate in four patients, and within 45 min in seven patients. This interval for the rest of the patients was 1 h or more, or data were not presented.

In terms of site of infarction, anterior infarctions occurred about twice as often as inferior infarctions, raising the possibility that the left anterior descending coronary artery is more susceptible to the deleterious effects of cocaine than the right or left circumflex coronary artery.

In patients evaluated by coronary arteriography, approximately half had angiographically normal coronary arteries. It should be pointed out, however, that the presence of angiographically normal coronary arteries does not insure that the vessel is indeed "normal". A case in point is the patient described by Zimmerman and colleagues (1987). Their patient had angiographically normal coronary arteries, but an iatrogenic intimal dissection of the left anterior descending coronary necessitated that coronary artery bypass be performed. Close examination of the infarct-related vessel at the time of surgery revealed an atherosclerotic plague at the site of spasm.

Cocaine-Induced Angina Pectoris

Several case reports indicate that angina pectoris has occurred following cocaine use. The majority of these have been reviewed by Mathias (1986). One of the first reports was by Coleman and colleagues (1982) of a 38-year-old male smoker who administered approximately 0.5 g cocaine by the intranasal route. Angina occurred that was relieved by administering nitroglycerin.

Similar cases of chest pain have been reported in other subjects following the administration of unspecified doses of cocaine by the intranasal and i.v. routes (Pasternak *et al.*, 1985; Smith *et al.*, 1987; Lam and Goldschlager, 1988). The episodes of angina have occurred in the presence of stenosed coronary arteries as well as in the presence of angiographically normal coronary vessels. It should be noted, however, that chest pain in cocaine abusers is not always the result of cocaine-induced myocardial ischemia. Rubin and Neurgartern (1989) recently reported on a group of patients with chest pain and elevated serum creatine phosphokinase levels who were hospitalized for evaluation of possible cocaine-induced mycardial infarction. In each case, myocardial infarction was ruled out by serial electrocardiograms and absence of the cardiac specific MB isoenzyme of creatine phosphokinase. It was concluded that the elevated serum creatine phosphoskinase was the result of cocaine-induced rhabdomyolysis.

Cocaine-Induced Arrhythmias

Primary arrhythmias due to cocaine have not been clearly documented and appear to occur only rarely as an initial complication of cocaine abuse. Most arrhythmias associated with cocaine have been secondary to myocardial ischemia or acute myocardial infarction. In addition, cocaine-induced convulsions, respiratory arrest, and acidosis may also result in cardiac arrhythmias (Young and Glauber, 1947; Jonsson *et al*; 1983; Nanji and Filipenko, 1984). There is one report by Benchimol and colleagues (1978) describing a patient, with no history of cardiovascular disease, chest pain, or dyspnea, who exhibited intermittent episodes of a wide QRS-rhythm at a rate of 94 beats per minute. The arrhythmia had a left bundle-branch block configuration and was occasionally preceded by a fusion beat, suggesting a ventricular origin. The patient had self-administered at least 5 i.v. injections of cocaine over a 24-h period with an estimated total dose of 140 mg.

EVIDENCE RELATING COCAINE USE TO THE OCCURRENCE OF CARDIOVASCULAR COMPLICATIONS

Unfortunately, much of the evidence indicating that cocaine administration can cause acute myocardial infarction, angina pectoris, and cardiac arrhythmias is based on case reports where there was either a poorly documented or questionable relationship between the time at which cocaine was administered and the onset of cardiovascular complications. For example, there are case reports where the onset of symptoms of an acute myocardial infarction occurred several hours after peak cardiovascular effects and blood levels of cocaine had been achieved (Schachne *et al.*, 1984; Isner *et al.*, 1986; Smith *et al.*, 1987; Zimmerman *et al.*, 1987). Similarly, the onset of symptoms of angina pectoris has been reported to be delayed relative to the time that peak cardiovascular effects and blood levels of cocaine had been achieved (Lam and Goldschlager, 1988).

In the cases with acute myocardial infarction, the most common route of administration was intranasal. According to Van Dyck and colleagues (1977), intranasal administration of cocaine results in a rapid rise in plasma concentration over 15–20 min, a peak concentration at 15–60 min, and a gradual decrease in plasma levels over the next 3–5 h. According to Javaid and colleagues (1978), the maximum high occurs at 15–20 min after intranasal administration, and a return to predrug physiological and subjective levels occurs within 60–90 min after intranasal administration. Hence, the time course of action of cocaine given by the intranasal route does not match the time course for the occurrence of acute myocardial infarction as described in many of the case reports (Schachne *et al.*, 1984; Isner *et al.*, 1986; Smith *et al.*, 1987; Zimmerman *et al.*, 1987; Ascher *et al.*, 1988). In the cases involving angina pectoris (Lam and Goldschlager, 1988), cocaine was administered either i.v. or by smoking as the free base. With i.v. administration, peak blood levels occur within 5 min of administration, and the half-life ranges from 16–87 min (Javaid *et al.*, 1978). The peak heart rate increase occurs within 10 min and disappears within 30–40 min (Javaid *et al.*, 1978; Perez-Reyes *et al.*, 1982). Inhalation of the free base results in an almost identical time course of action for the development of sinus tachycardia and hypertension (Perez-Reyes *et al.*, 1982).

Again, the reported onset of 45 min to 2 h for angina symptoms (Lam and Goldschlager, 1988) would have occurred after the hemodynamic effects of cocaine had worn off.

The most compelling evidence that cocaine may precipitate cardiovascular toxicity is that in at least ten instances (Coleman *et al.*, 1982; Kossowsky and Lyon, 1984; Weiss, 1986; Smith *et al.*, 1987; Zimmerman *et al.*, 1987; Lam and Goldschlager, 1988), patients who had an initial myocardial ischemic event after cocaine administration developed a second ischemic event when rechallenged with the drug at a later date. For example, Weiss (1986) describes a patient who developed an acute inferior infarction while abusing cocaine. This patient reported having brief episodes of chest pain after most prior episodes of intranasal cocaine use. Approximately 1 month after developing the initial myocardial infarction, he had a new inferior infarction immediately after cocaine use.

HEMODYNAMIC MECHANISMS INVOLVED IN MYOCARDIAL ISCHEMIC EVENTS

The major adverse cardiovascular effects of cocaine, acute myocardial infarction, and angina pectoris appear to occur because of a mismatch between myocardial oxygen damand and myocardial oxygen delivery. Cocaine could cause a disparity in the relationship of these two parameters by (1) increasing myocardial oxygen demand out of proportion to myocardial oxygen delivery, (2) decreasing myocardial oxygen delivery, or (3) producing both of these effects. In addition, the cardiac arrhythmias reported after cocaine use could also occur secondarily to myocardial ischemia and/or occlusion of a major coronary vessel. Cardiac arrhythmias could also occur because of indirect effects of cocaine on cardiac tissue mediated by the sympathetic nervous system and/or by direct local anesthetic effects of the drug on the myocardium.

Effects on Myocardial Oxygen Demand

A key question in focusing on the mismatch between oxygen demand and oxygen delivery is whether cocaine is capable of influencing cardiovascular function in a manner that leads to an increase in the oxygen consumption of the heart. In this regard, clinical data involving i.v. and intranasally administered cocaine HCl and the smoking of cocaine free base (''crack'') indicate that recreational doses of the drug can elevate heart rate and blood pressure, changes known to result in an increase in myocardial oxygen demand.

In studies involving i.v. injections of cocaine, Fischman and colleagues (1985) administered doses of 16–48 mg to human subjects and observed tachycardia responses ranging from 18–42 beats per minute, with peak responses occurring in 8–12 min. The cocaine plasma levels associated with these responses ranged from approximately 125–440 ng/ml and peaked at about 5 min after injection. According to the subjects who received these i.v. doses, the subjective effects they experienced (e.g., mood changes and euphoria) were similar to those produced by the doses of cocaine they usually took on a recreational basis. Similar i.v. doses of cocaine (10 and 25 mg) in humans were found by Resnick and co-workers (1977) to increase systolic blood pressure by 13 and 25 mm Hg, respectively, and diastolic blood pressure by 5 and 14 mm Hg, respectively.

An intranasally administered dose of cocaine that corresponds pharmacologically to the i.v. dose of 16 mg discussed above is 96 mg (Javaid *et al.*, 1978). This inference can be made on the basis of plasma levels of cocaine observed after each route of administration. For example, an i.v. dose of 16 mg and an intranasal dose of 96 mg both yield peak plasma levels on the order of 200 ng/ml. Peak plasma levels occur about 30 min after intranasal administration. A dose of 96 mg cocaine administered by the intranasal route increases heart rate about 25% (e.g., by approximately 20 beats per minute) and peak changes in heart rate occur in about 20–30 min (Javaid *et al.*, 1978).

A dose of 50 mg cocaine free base administered by smoking corresponds to an i.v. dose of approximately 20 mg of cocaine HCl (Perez-Reyes *et al.*, 1982). This inference can be made on the basis of subjective and cardiovascular responses obtained by each route of administration. For example, 20 mg i.v. increases heart rate, systolic pressure, and diastolic pressure 34, 20, and 15%, respectively, while 50 mg free base increases heart rate, systolic pressure, and diastolic pressure by 42, 27, and 18%, respectively. Peak effects of free base occurred at about 6–12 min after smoking. This time-action curve for free base was similar to that observed with i.v. administration. In general, the cardiovascular effects noted above with each of the three routes of administration are over in 30 min to 1 h.

Data from animal studies also indicate that cocaine can produce hemodynamic changes that result in an increase in myocardial oxygen consumption. However, only a few such studies have utilized cocaine in doses that correspond to those that are generally used on a recreational basis in man. Gonzalez and Byrd (1977) adminsitered cocaine hydrochloride i.v. in doses of 0.1–1.0 mg/kg to conscious squirrel monkeys and observed dose-related increases in arterial blood pressure and heart rate. Arterial blood pressure was a more sensitive indicator of a response than heart rate. More recently, Wilkerson (1988) demonstrated that i.v. cocaine produced dose-related increases in arterial blood pressure in conscious dogs (Figure 1). Most importantly, doses of cocaine of 0.25 mg/kg and higher resulted in significant dose-dependent elevations in the rate-pressure product (Figure 2). This product represents an index of myocardial oxygen demand (Gobel *el al.*, 1978) which must be met by a corresponding increase in myocardial oxygen supply (i.e., an increase in coronary blood flow) to avoid creating a myocardial oxygen debt. The ability of cocaine to increase the rate-pressure product has also been demonstrated in anesthetized dogs following i.v. administration of 1 mg/kg of cocaine (Wilkerson, 1989), and preliminary data of a similar nature have been obtained in sedated dogs (Kuhn *et al.*, 1989). In general, upon i.v. administration, rate-pressure product increased up to fourfold with the duration of the effect ranging from approximately 15–30 min (Wilkerson 1988; Kuhn *et al.*, 1989). In most other studies involving animals, only large doses of cocaine (e.g., 5–10 mg/kg), considerably above those used in humans (approximately 0.2–0.5 mg/kg), have been evaluated. Data from these

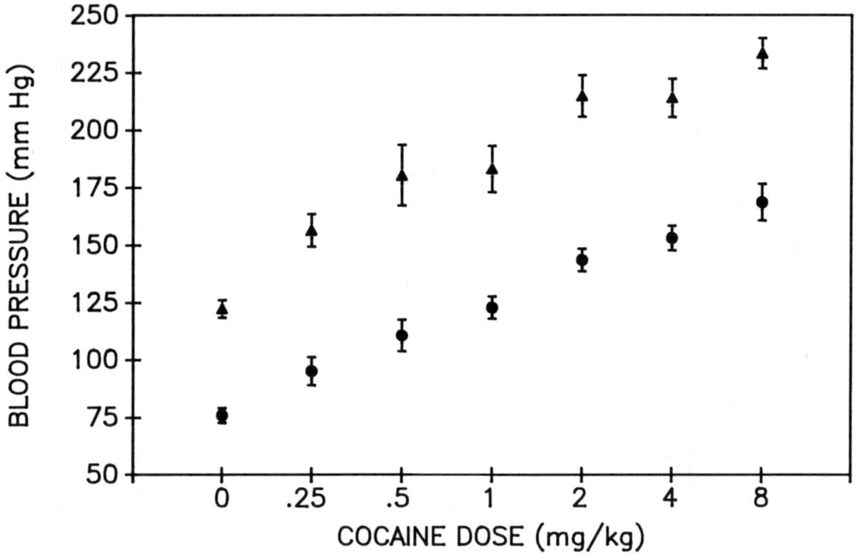

Figure 1. Changes in systolic (▲) and diastolic (●) blood pressure produced by intravenously administered cocaine in conscious dogs. The interval between cocaine doses was at least 24 h. All data are expressed as the mean ± S.E. of the animals studied.

animal studies using such high doses may not be useful for determining the mechanisms involved for the cardiovascular complications that occur in man. Future studies in animals should employ lower doses of cocaine for this purpose.

Effects on Myocardial Oxygen Delivery

Based upon cocaine's effect to increase myocardial oxygen demand, it would be expected that a corresponding increase in myocardial oxygen delivery would result from autoregulatory mechansims in the coronary vasculature. Data addressing this point can be found in both clinical and experimental studies.

Patients with no evidence of coronary artery disease have been described as suffering an acute myocardial infarction after cocaine use due to coronary spasm (Simpson and Edwards, 1986; Hadjimiltiades *et al.*, 1988; Cregler and Mark, 1986; Altieri *et al.*, 1987; Ascher *et al.*, 1988). The evidence for the occurrence of coronary spasm is indirect and consists of severe chest pain at rest with ST segment elevation (Altieri *et al.*, 1987; Zimmerman *et al.*, 1987). Stasis may also occur at the site of spasm and could result in platelet aggregation, thrombus formation, and an ensuing myocardial infarction (Vincent *et al.*, 1983). Coronary spasm induced by cocaine could also contribute to the occurrence of acute myocardial infarction in patients with coronary artery disease by disrupting the endothelium at the site of an atherosclerotic plaque, thereby resulting in platelet aggregation and thrombus formation (Gertz *et al.*, 1981).

Although coronary artery spasm has been suggested as an important mechanism for cocaine-induced myocardial infarction in clinical studies, there is currently no

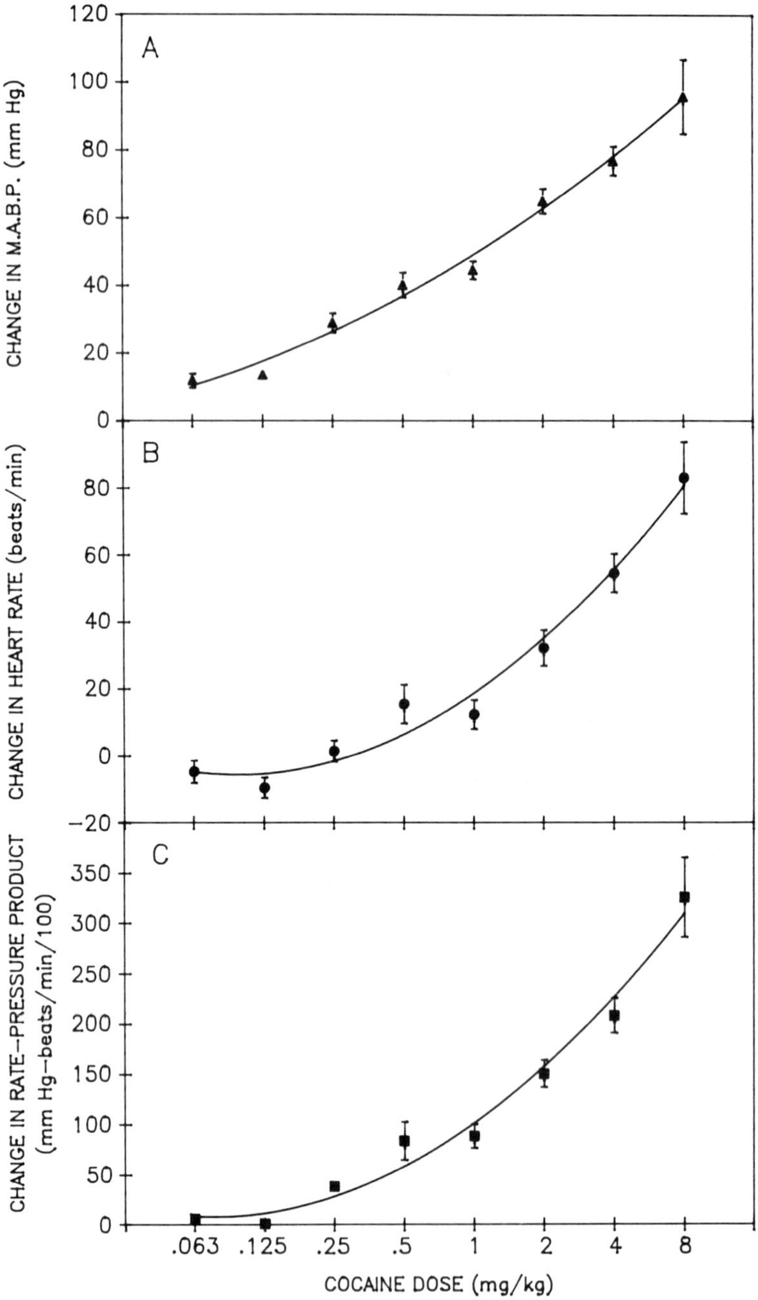

Figure 2. Dose-response relationships for the effects of intravenously administered cocaine on mean arterial blood pressure (M.A.B.P.) (panel A), heart rate (panel B), and the product of systolic blood pressure and heart rate (panel C) in conscious dogs. The interval between cocaine doses was at least 24 h. All data are expressed as the mean ± S.E. of the animals studied. (From R.D. Wilkerson, *J. Pharmacol. Exp. Ther.* **246**: 466 (1988). With permission.)

direct evidence from either the scientific literature or from ongoing studies that coronary spasm occurs after cocaine administration. Only a few studies have been performed where the effect of cocaine on the coronary vasculature in animals has been studied. In two of these, exceedingly high doses were evaluated by the i.v. route (approximately 5–10 mg/kg). In one case, 200 mg was given to dogs and resulted in an increase in coronary vascular resistance (Pierre *et al.*, 1985). In another study where 5 mg/kg cocaine was given to dogs, coronary blood flow after stimulation of the sympathetic nervous system was augmented compared to responses obtained in treated dogs not receiving cocaine (Levy and Blattberg, 1976). In our preliminary studies, no evidence of an increase in coronary vascular resistance was observed after 0.25–1.0 mg/kg cocaine was given to cats (Jain *et al.*, 1987). In fact, coronary vascular resistance was unchanged, while a significant increase in arterial pressure developed. This latter finding fits with the idea that cocaine decreases the metabolic coronary dilation that would follow in response to its effects to raise arterial blood pressure. We recently obtained more direct experimental evidence for this idea from studies in anesthetized dogs (Wilkerson, 1989). The increase in myocardial oxygen consumption (i.e., increase in rate-pressure product) produced by an i.v. injection of 1 mg/kg of cocaine was found to be accompanied by an increase in left circumflex coronary artery blood flow. The expected response of the coronary circulation to such an increase in the metabolic demand of the heart would be vasodilation; however, calculated coronary vascular resistance actually increased after cocaine administration. The observed increase in coronary blood flow after cocaine thus resulted from an increase in coronary perfusion pressure due to the peripheral effect of cocaine to raise arterial blood pressure, while the normal metabolic autoregulatory coronary vasodilatory response to the drug-induced increased myocardial oxygen demand appeared to be blocked. In confirmation of these findings, Kuhn and colleagues (1989) demonstrated that i.v. doses of 0.5–2 mg/kg of cocaine in sedated dogs produced widespread constriction of both epicardial and coronary resistance vessels at the time that marked augmentation of myocardial oxygen demand was observed, as manifested by increases in heart rate, blood pressure, and cardiac output. Taken together, the findings obtained with relatively low doses of cocaine in experimental animals suggest that diffuse constriction of coronary resistance vessels, rather than focal vasospasm in large coronary vessels, is the predominant effect of the drug. Indeed, we have even looked for evidence of coronary spasm under conditions where the endothelial layer of the coronary artery has been disrupted, and have observed coronary constriction, but not spasm (Kuhn *et al.*, 1989). Finally, constriction of the coronary blood vessels has been reported with cocaine in studies conducted *in vitro* on porcine and bovine coronary arteries (Vargas *et al.*, 1988; Foy *et al.*, 1989), but the only study to date using human coronary vessels *in vitro* reported that cocaine did not cause vasoconstriction. In fact, cocaine caused vasodilation in human coronary arteries whose tone was elevated with KCl (Perrault *et al.*, 1989). In summary, cocaine has a deleterious effect on both myocardial oxygen demand and delivery; the drug increases demand while simultaneously restricting oxygen delivery to the myocardium.

Since coronary vasospasm does not appear to be a predominant effect of cocaine, the mechanism whereby the drug might induce an acute myocardial infarction in patients with coronary artery disease may be due to generalized coronary vasoconstriction. As a result of this effect, vessel shear rate would increase since this effect is known to occur with decreasing vessel diameter (Falk, 1983). Shear rate is a measure of the difference in blood flow velocity between the center of the vessel and along the vessel wall. An increase in shear rate can result in platelet deposition at the site of an atherometous plaque and may also predispose a plaque to disrupt (Fuster *et al.*, 1988). Plaque disruption is now known to be the underlying event that frequently precedes acute myocardial infarction (Falk, 1983). Plaque disruption may result in further luminal narrowing and enhanced platelet activation and deposition. Activated platelets may result in additional vasoconstriction and luminal compromise by releasing thromboxane A_2, serotonin, and platelet-derived growth factor (Lam *et al.*, 1987). These events occurring in tandem could cause complete thrombotic occlusion and acute myocardial infarction in patients with coronary artery disease.

Evidence of the occurrence of thrombotic occlusion in a major coronary artery during cocaine abuse has been obtained by several investigators. Simpson and Edwards (1986) describe autopsy findings where acute coronary obstruction resulted from platelet thrombosis of the left main coronary artery along an area of intimal irregularity. This patient showed chronic coronary obstruction due to nonatherosclerotic intimal proliferation of smooth muscle cells. Kossowsky and Lyon (1984) describe autopsy findings of a thrombotic occlusion of the proximal left anterior descending coronary artery associated with mild atherosclerosis. Smith and colleagues (1987) present evidence from coronary angiography studies indicating that a thrombus was present in five of their nine patients. Similar evidence has been presented for two additional patients (Zimmerman *et al.*, 1987; Hadjimiltiades *et al.*, 1988).

Smith and colleagues (1987) raise the interesting point that cocaine could increase platelet aggregability (presumably by a direct action) and this could contribute to the formation of a thrombus in patients and subsequent acute myocardial infarction. While very little is known about cocaine and platelet function, Togna and colleagues (1985) reported that cocaine increased platelet aggregability and thromboxane production and decreased endothelial prostacylin release *in vitro*. However, the concentration of cocaine required for these effects was too high to be encountered with recreational use of the drug. Kramer and colleagues (1988) reported that cocaine *in vitro*, in a concentration range that could be encountered with recreational use, inhibited platelet aggregation induced by epinephrine, ADP, and collagen, and shifted the ratio of prostacycline:thromboxane A_2 toward prostacycline excess. Hence, it is unlikely that acute myocardial infarction is due to a deleterious effect of cocaine per se on platelet and hemostatic mechanisms.

We have no explanation at the present time for why acute myocardial infarction occurs in patients with normal coronary arteries. It is possible that coronary vasospasm might be an underlying event in this situation. However, it is more likely that the coronary vessels in these patients are not normal, although angiographically described as being so (Simpson and Edwards, 1986).

MECHANISMS FOR COCAINE-INDUCED INCREASES IN HEART RATE, ARTERIAL BLOOD PRESSURE, AND CORONARY VASOMOTOR TONE

The above-described clinical and animal studies indicate that an imbalance between myocardial oxygen demand and supply may play a role in the cardiovascular complications associated with cocaine abuse. Cocaine has at least three actions on the cardiocascular system that contribute to this effect; the drug increases heart rate and blood pressure, and increases tone in both the epicardial and resistance vessels of the coronary circulation. The mechanisms for these effects are described below.

Tachycardia produced by cocaine appears to be due primarily to effects mediated by activation of the sympathetic nervous system at the level of the cardiac β1 receptor, as studies in experimental animals have shown that this response is blocked by propranolol (Tessel *et al.*, 1978; Kuhn *et al.*, submitted for publication). A role for catecholamines in the increase in heart rate evoked by cocaine was also suggested early on from the studies of Trendelenburg (1968), in which tachycardia produced by the drug was blocked by reserpine pretreatment in isolated guinea pig hearts. The findings of Trendelenburg (1968) in isolated reserpinized hearts also rule out a direct myocardial effect of cocaine to increase heart rate; under these experimental conditions, the only response he observed was bradycardia, presumably resulting form the local anesthetic action of cocaine. In a recent study, Sharkey and colleagues (1988) found that the (−) enantiomer of cocaine, known to be present in extracts of the coca plant, produced competitive inhibition of M_2 muscarinic receptors in rat heart and brain tissue and also antagonized methacholine-induced reductions in heart rate in isolated, spontaneously beating guinea pig right atria. They speculated that the cardiac antimuscarinic effect of cocaine, particularly at higher, toxic doses of the drug, might play a role in the cardiotoxic actions of cocaine; that is, cocaine ingestion has been shown to produce a reflex increase in vagal tone secondary to its hypertensive effect (Gay, 1982). At low doses of cocaine where the antimuscarnic effects are not predominant, this reflex vagal response may serve to partially offset tachycardia normally seen with the drug. However, at higher dose levels of cocaine, the M_2 cardiac receptor blocking action could prevent the reflex vagal cardioinhibitory effect and thus leave the primary sympathomimetic effect of cocaine on heart rate unopposed.

Several studies have demonstrated that the increase in arterial blood pressure produced by cocaine is abolished or markedly diminished by α-adrenergic blocking drugs or by reserpine (Teeters *et al.*, 1963; Kuhn *et al.*, 1989). The vascular response to cocaine is therefore largely due to an action of cocaine to increase α_1-adrenergic receptor activation and, thus, to increase total peripheral resistance. The effect of cocaine on vascular resistance acts in concert with the above-described tachycardic effect to result in the overall increase in arterial blood pressure seen with the drug. An increase in stroke volume does not appear to contribute to the rise of the blood pressure, as this parameter has been shown to decline or remain relatively unchanged after cocaine administration (Catravas and Waters, 1978; Bedotto *et al.*, 1988). The absence of a role for increased stroke volume in the pressor effect of cocaine fits with data indicating that the drug does not produce a significant increase in myocardial contractile force (O'Keefe *et al.*, 1977; Wilkerson *et al.*, 1988). In fact, it has been recently shown that regional ejection fraction is deceased after i.v. administration of 4 mg/kg to conscious dogs (Fraker *et al.*, 1989).

Vasoconstriction medicated by α-adrenergic receptors also accounts for the adverse effects of cocaine on the coronary circulation. This was indicated by Kuhn *et al.*, (1989) who demonstrated that phentolamine prevented the decrease in the diameter of the left anterior descending coronary artery and the associated increase in coronary vascular resistance to i.v. cocaine in lightly sedated dogs. Similar results were reported by Vargas and colleagues (1988) for prazosin on cocaine-induced contractions of isolated pig coronary artery strips.

Consistent with the evidence that cocaine can increase heart rate, blood pressure, and coronary artery tone through a catecholaminergic mechanism are reports indicating that serum levels of catecholamines are elevated after cocaine treatment in animals (Chieuh and Kopin, 1978) and humans (Karch, 1987) exposed to the drug. In the report of Karch (1987), six subjects hospitalized with cardiac symptoms (e.g., sinus tachycardia and premature ventricular contractions) immediately after using "crack" displayed markedly elevated circulating levels of norepinephrine and epinephrine. Cocaine appeared to cause a release of the catecholamines from sympathetic nerve terminals and the adrenal medulla rather than by influencing catecholamine synthesis, since circulating levels of dopamine remained essentially unchanged in these patients.

PROPOSED PHARMACOLOGICAL ACTIONS OF COCAINE THAT RESULT IN DELETERIOUS CARDIOVASCULAR EFFECTS

At least four mechanisms have been proposed to account for the sympathomimetic effects of cocaine: (1) inhibition of norepinephrine uptake into the peripheral sympathetic nerve terminal, (2) stimulation of brain centers that influence sympathetic outflow to the heart and blood vessels, (3) a direct action on cardiac and vascular tissue to cause the release of catecholamines, and (4) sensitization of postsynaptic adrenergic receptors to heighten their responsiveness to released neurotransmitter. We have focused upon the first two mechanisms since they are considered to be the most important in the actions of cocaine. The reason for this is that much higher doses of cocaine are required to demonstrate the direct release of catecholamines from tissue and also to alter the sensitivity of postsynaptic receptors to catecholamines (Van Dyke and Byck, 1977). We also briefly discuss an additional mechanism of action of cocaine that may be involved in its deleterious cardiovascular actions, the well-known local anesthetic effect of the drug.

The re-uptake of monoaminergic amines by sympathetic nerve terminals is the major process for terminating the actions of neurally released, exogenously administered, and circulating catecholamines (Weiner and Taylor, 1985). Cocaine, by virtue of its ability to inhibit peripheral norepinephrine uptake, has been found to potentiate and prolong the effects of exogenously administered norepinephrine (Trendelenburg, 1959; Whitby *et al.*, 1960, Koerker and Moran, 1971, Jain *et al.*,

1987). Surprisingly, however, most investigators report that cocaine either does not potentiate the cardiovascular effects produced by sympathetic nerve stimulation (Pappas *et al.*, 1965; Koerker and Moran, 1971; Levy and Blattberg, 1976; Matsuda *et al.*, 1979), or that the potentiation that does occur is less than that seen with exogenously administered catecholamines (Bayorh *et al.*, 1983). The doses of cocaine that have been tested for potentiating effects on both neurally released and injected catecholamines in these studies usually ranged from 1 to 10 mg/kg. This dose is relatively high and would probably not be encountered with recreational abuse of cocaine.

We recently re-explored this issue in anesthetized cats and obtained several findings that might help to explain why the previous studies did not demonstrate potentiation of cardiac responses to sympathetic nerve stimulation (Jain *et al.*, 1990). First, and most importantly, we found that low i.v. doses of cocaine (i.e., 0.0625–0.25 mg/kg) consistently potentiated the cardioaccelerator response to sympathetic nerve stimulation in a linear fashion, whereas administration of larger doses of cocaine (i.e., 0.5 and 2.0 mg/kg) led to a decline in the degree of potentiation. Thus, it appears that the doses of cocaine used in previous studies (i.e., 1–10 mg/kg) were too high to demonstrate potentiation. Second, we observed that the potentiating effects of the lower i.v. doses of cocaine were maximal at about 1 min after injection and began to dissipate after 10–12 min. This suggests that time is a critical factor for observing the response and that previous investigators who did not test for potentiation immediately after administering cocaine may have missed the response. Third, tachyphylaxis occurred to cocaine-evoked potentiation of sympathetic neural effects on the heart in our studies upon repeated injections of the drug. Interestingly, tachyphylaxis to cocaine's effect seemed to be dose related, as it did not occur with repeated doses of 0.0625 mg/kg but was consistently seen with repeated administration of a 0.25-mg/kg dose given at approximately 60-min intervals. It is possible that tachyphylaxis may have negated the efforts of others to demonstrate potentiation, particularly those who administered cocaine by continuous i.v. infusion over several or more minutes. Fourth, we evaluated the ability of cocaine to potentiate sympathetically induced cardiac responses using submaximal rather than maximal neural stimulation. This factor is important since, as shown by Trendelenburg (1959), cocaine does not potentiate sympathetic response to supramaximal electrical stimulation. In some studies, the inability of investigators to observe potentiation can be attributed to the fact that they tested cocaine against near-maximal cardiac changes produced by nerve stimulation (e.g., Inoue and Zipes, 1988).

Two interesting questions raised by these results are why potentiation is not seen with larger doses of cocaine, and why does tachyphylaxis occur to the potentiating effect of lower doses (i.e., 0.25 mg/kg) of cocaine? In reference to the first question, a local anesthetic effect appears to be responsible for the fall-off in potentiation (Jain *et al.*, submitted for publication). For the second question, the reason appears to be stimulation of presynaptic α_2-adrenoceptors by norepinephrine accumulating in the synapse after cocaine administration. This was shown by demonstrating that yohimbine pretreatment can prevent tachyphylaxis to cocaine-induced potentiation of nerve stimulation (Jain *et al.*, submitted for publication).

The idea that cocaine augments sympathetic nervous function by acting in the CNS is also considered to be an important mechanism of action for the cardiovascular effects of cocaine. The evidence that favors a CNS site of action of cocaine is as follows: (1) administration of cocaine (0.4–10 mg/kg, i.v.) results in an increase in plasma epinephrine level that is prevented by splanchnic nerve section; in contrast, another amine uptake inhibitor, desipramine, did not alter plasma epinephrine levels (Chieuh and Kopin, 1978); (2) sectioning or pithing the spinal cord reduces the sympathomimetic actions of cocaine (MacGregor, 1939; Pals *et al.*, 1968); (3) dogs under anesthesia do not exhibit the pronounced cardiovascular effects that are seen in conscious dogs after cocaine administration (Wilkerson, 1988); and (4) pretreatment of conscious dogs with the ganglionic blocker hexamethonium prevents the arterial blood pressure increase evoked by i.v. administration of cocaine (Wilkerson, 1988).

The evidence against a CNS site for the sympathomimetic effects of cocaine is (1) vertebral artery injections of cocaine in cats does not produce a rise in arterial blood pressure; indeed, a decrease in blood pressure is observed (Jolly and Steinhaus, 1955; Erzouki *et al.*, submitted for publication). (2) Systemic administration of the quaternary analogue of cocaine, cocaine methiodide, an agent that presumably does not cross the blood brain barrier, produces the same qualitative cardiovascular effects normally seen with cocaine after i.v. administration to sedated dogs (Kuhn *et al.*, 1988). (3) Preliminary data involving recordings of spontaneous sympathetic nerve activity in anesthetized cats indicate that i.v. administration of cocaine does not cause an increase in neural discharge; instead, a decrease in sympathetic firing is noted (Erzouki *et al.*, submitted for publication).

It is obvious from the above information that the role of the CNS in the cardiovascular actions of cocaine is uncertain and needs to be addressed in more detail in future experiments.

The doses of cocaine required to produce local anesthetic effects are generally larger (approximately tenfold higher) than those required to produce inhibition of catecholamine uptake (Yasuda *et al.*, 1984). Nevertheless, cardiovascular consequences resulting from local anesthetic activity do appear to be a feature of the drug and include cardiac arrhythmias and myocardial depression. The powerful local anesthetic effect of cocaine is due to its ability to block sodium entry into neural tissue. This effect also occurs in cardiac tissue and results in a decrease in the rate of rise of phase zero of the transmembrane action potential (Weidmann, 1955) and impaired conduction in the heart, as reflected by a widening of the QRS duration and an increase in the P-R interval. Newer data of Tracy and colleagues (1989) also indicate that cocaine may also block calcium entry into A-V nodal cells.

In contrast to the sympathomimetic action of cocaine, tachyphylaxis does not occur to the local anesthetic effect (Bachenheimer *et al.*, 1988). This may have relevance to situations such as that described by Benchimol *et al.* (1978), where subjects developed cardiac arrhythmias after self-administration of repeated doses of cocaine over a 24-h period; that is, with repeated or "binge" use of cocaine, the direct local anesthetic effect on the myocardium might be more important for eliciting cardiotoxicity than the sympathomimetic action because of the occurrence of acute tolerance to the latter effect.

FACTORS PREDISPOSING TO COCAINE CARDIOTOXICITY

Clinical studies involving cocaine-induced cardiotoxicity have identified two factors, tobacco smoking and the presence of coronary artery disease, as being present in a high percentage of patients, and it is possible that these might enhance the propensity of abusers of the drug to develop acute myocardial infarction or angina pectoris (Coleman *et al.*, 1982; Kossowsky and Lyon, 1984; Schachne *et al.*, 1984; Pasternak *et al.*, 1985; Howard *et al.*, 1985; Cregler and Mark, 1986; Gould *et al.*, 1985; Wilkens *et al.*, 1985; Chiu *et al.*, 1986; Weiss, 1986; Rollingher *et al.*, 1986; Simpson and Edwards, 1986; Isner *et al.*, 1986; Rod and Zucker, 1987; Wehbie *et al.*, 1987; Zimmerman *et al.*, 1987; Altieri *et al.*, 1987; Smith *et al.*, 1987; Lam and Goldschlager, 1988; Ascher *et al.*, 1988; Hadjimiltiades *et al.*, 1988; Wang *et al.*, 1988). In these studies, a total of 46 patients was reported to experience myocardial infarction or angina after cocaine use. Data were provided on tobacco use for 33 of these patients, and 29 individuals (88%) were reported to be smokers. Similarly, data were reported on the condition of the coronary arteries in 39 of these patients, and 18 individuals (44%) were reported to have diseased coronary arteries. It is possible that tobacco could predispose toward cocaine-induced cardiotoxicity by enhancing platelet activation (Fuster *et al.*, 1988). Furthermore, nicotine has been reported to produce stimulation of parasympathetic nerves that innervate the coronary blood vessels leading to coronary vasoconstriction (Young *et al.*, 1988). This vagal mechanism for coronary constriction could act in concert with cocaine's sympathomimetic vasoconstrictor action on the coronary arteries (Wilkens *et al.*, 1985) and result in synergistic or additive effects on coronary flow. In subjects with underlying coronary artery disease, cocaine could evoke acute myocardial infarction, because of plaque disruption as discussed above, and angina pectoris due to its ability to increase myocardial oxygen demand in the face of the fixed coronary resistance.

TREATMENT OF COCAINE-INDUCED CARDIOVASCULAR COMPLICATIONS

Current experimental evidence indicates that the two most important pharmacological actions of cocaine that are associated with life-threatening cardiac complications are inhibition of norepinephrine uptake at the sympathetic nerve terminal and local anesthesia. As such, the most logical therapy for treating patients with cardiovascular disorders would involve drugs that either oppose the indirect sympathomimetic effects of cocaine or counteract the direct local anesthetic effect of excessive doses of cocaine encountered during "binging". In the case of the sympathomimetic effects of cocaine, changes in cardiovascular function are produced by activation of both α- and β-adrenergic receptors, and a rational drug to use would be one such as labetabol that blocks both of these receptors. In one preliminary clinical study, this drug was tested against cocaine-induced cardiotoxicity and was found to be efficacious (Gay and Loper, 1988). The use of a drug that blocks only β receptors would not appear to be beneficial as a therapeutic agent since it may elicit a hypertensive crisis, presumably due to unopposed α-adrenergic effects (Ramoska and Sacchetti, 1985). The use of nonspecific-acting drugs such a nitroglycerin and calcium-entry blockers might also be useful for antagonizing cocaine's coronary vasoconstrictor effects, as well as its action to increase the oxygen demand of the heart. Rollingher and co-workers (1986) have indeed obtained evidence that both types of nonspecific agents can ameliorate cocaine-induced myocardial ischemia in humans, and Trouve and Nahas (1986) have shown that nitrendipine can offset cardiovascular toxicity produced by large doses of cocaine in rats. In the case of cardiotoxic effects produced by cocaine's local anesthetic effect, such as prolongation of the QRS duration and impaired cardiac conduction, successful therapy might involve the administration of sodium i.v., either in the form of hypertonic saline, sodium lactate, or sodium bicarbonate. Bajai and colleagues (1986) have recently obtained favorable results in treating this type of problem using hypertonic saline.

SUMMARY

A number of cardiovascular disorders have been associated with cocaine abuse. These include acute myocardial infarction, angina pectoris, cardiac arrhythmias, heart failure, pulmonary edema, subarachnoid hemorrhage, cerebral infarction, myocarditis, aortic rupture, intestinal ischemia, and obstetrical obstructions. This chapter focuses upon the first three aforementioned disorders. Acute myocardial infarction is the most prevalent disorder observed. The site of infarction appears to occur mainly in the anterior descending branch of the left coronary artery, and occurs in patients with both diseased and normal coronary arteries. Angina pectoris also occurs with cocaine use in patients with both stenosed and normal coronary arteries. Primary arrhythmias due to cocaine have not been clearly documented and appear to occur secondary to myocardial ischemia and infarction or convulsions and respiratory arrest. The best clinical evidence to associate cocaine with cardiovascular disorders is derived from challenge-rechallenge studies.

Myocardial ischemia associated with cocaine results from a deleterious effect on both myocardial oxygen demand and delivery; the drug increases demand (by increasing heart rate and arterial blood pressure), while simultaneously restricting oxygen delivery (by constricting both epicardial and resistance vessels). Coronary spasm does not appear to be a predominant effect of cocaine; instead, generalized coronary vasoconstriction and the resulting increase in vessel shear rate accompanied by platelet deposition and/or plaque disruption appears to be the cause of acute myocardial infarction.

The mechanisms responsible for cocaine-induced tachycardia, pressor and coronary vasoconstriction involve activation of the sympathetic nervous system. This sympathomimetic effect is primarily due to inhibition of catecholamine uptake into the sympathetic nerve terminal. Local anesthetic effects of cocaine may by important for cardiovascular complications that occur with high doses of cocaine, that is, with accumulation of the drug in plasma after binging. Local anesthetic effects counteract the sympathomimetic effects of cocaine and can result in disturbances in cardiac conduction. Interestingly, tachyphylaxis develops to the sympathomimetic effects of cocaine, but not to the local anesthetic effects of the drug.

Factors that could predispose to cocaine-induced cardiotoxicity include the presence of coronary artery disease and tobacco smoking. In the case of the latter, nicotine enhances platelet activation and can produce coronary vasoconstriction. Finally, cocaine-induced cardiotoxicity appears to be treatable by first determining if it is due to sympathomimetic effects or to local anesthetic effects. In the former case, labetalol is a rational drug since both α- and β-receptors mediate the sympathomimetic effects of cocaine. In the latter case, administration of sodium i.v. could counteract the local anesthetic effects of the drug.

REFERENCES

P.I. Altieri, J.M. Toro, H. Banch, and E.H. Carrion. Coronary artery spasm in patients with cocaine-induced acute myocardial infarction. *J. Am. Coll. Cardiol.* **9**: 172A (1987).

E.K. Ascher, J.C.E. Stauffer, and W.A. Gaasch. Coronary artery spasm, cardiac arrest, transient electrocardiographic Q waves and stunned myocardium in cocaine-associated myocardial infarction. *Am. J. Cardiol.* **61**: 939 (1988).

A.K. Bajai, R.L. Woosley, and D.M. Doden. Reversal of the QRS prolonging effects of *O*-desmethyl encainide by hypertonic saline. *Clin. Res.* **34**: 281A (1986).

M.A. Bayorh, Z. Zukowska-Grojec, and I.J. Kopin. Effect of desipramine and cocaine on plasma norepinephrine and pressor responses to adrenergic stimulation in pithed rats. *J. Clin. Pharmacol.* **23**: 24 (1983).

J.B. Bedotto, R.W. Lee, L.D. Lancaster, M. Olajos, and S. Goldman. Cocaine and cardiovascular function in dogs: effects on heart and peripheral circulation. *J. Am. Coll. Cardiol.* **11**: 1337 (1988).

A. Benchimol, H. Bartall, and K.B. Desser. Accelerated ventricular rhythm and cocaine abuse. *Annu. Intern. Med.* **88**: 519 (1978).

J.D. Catravas and I.W. Waters. Acute cocaine intoxication in the conscious dog: studies on the mechanism of lethality. *J. Pharmacol. Exp. Ther.* **217**: 350 (1981).

C.C. Chieuh and I.J. Kopin. Centrally mediated release by cocaine of endogenous norepinephrine and norepinephrine from the sympathoadrenal medullary system of unanesthetized rats. *J. Pharmacol. Exp. Ther.* **205**: 148 (1978).

Y.C. Chiu, K. Brecht, D.S. DasGupta, and E. Mhoon. Myocardial infarction with topical cocaine anesthesia for nasal surgery. *Arch. Otolaryngol. Head Neck Surg.* **112**: 988 (1986).

D.L. Coleman, T.F. Ross, and J.L. Naughton. Myocardial ischemia and infarction related to recreational cocaine use. *West J. Med.* **136**: 444 (1982).

L.L. Cregler and H. Mark. Medical complications of cocaine abuse. *N. Engl. J. Med.* **315**: 1495 (1986).

H.K. Erzouki, P.T. Abrahams, K.L. Dretchen, N.W. Fleming, R.K. Jain, P. Hamosh, and R.A. Gillis. Cocaine acts in the CNS to inhibit sympathetic centers. *Circulation* **80**: II-554 (1989).

E. Falk. Plaque rupture with severe pre-existing stenosis precipitating coronary thromboses. Characteristics of coronary atherosclerotic plaques underlying fatal occlusive thrombi. *Br. Heart J.* **50**: 127 (1983).

B.S. Finkle and K.L. McCloskey. The forensic toxicology of cocaine. *Natl. Inst. Drug Abuse Res. Monogr. Ser.* **13**: 153 (1977).

M.W. Fishman, C.R. Schuster, J. Javaid, Y. Hatano, and J. Davis. Acute tolerance development to the cardiovascular and subjective effects of cocaine. *J. Pharmacol. Exp. Ther.* **235**: 677 (1985).

R.A. Foy, M.M. Levine, and R.D. Wilkerson. Cocaine causes constriction of bovine coronary arteries *in vitro. FASEB J.* **3**: A896, 1989.

T. Fraker, P. Temsey-Armos, and R. Wilkerson. Cocaine causes transient myocardial depression in conscious dogs. *J. Mol. Cells Cardiol.* **21** (Suppl. II): S-196 (1989).

S. Freud. Uber Coca. *Zentbl. Ther.* **2**: 289 (1884).

V. Fuster, L. Badimon, M. Cohen, J.A. Ambrose, J.J. Badimon, and J. Chesebro. Insights into the pathogenesis of acute ischemic syndromes. *Circulation* **77**: 1213 (1988).

G.R. Gay. Clinical management of acute and chronic cocaine poisoning. *Annu. Emerg. Med.* **11**: 562 (1982).

G.R. Gay and K.A. Loper. Control of cocaine-induced hypertension with labetalol. *Anesth. Analg. (Cleveland)* **67**: 92 (1988).

S.D. Gertz, G. Uretsky, R.S. Wajnberg, N. Lavot, and M.S. Gotsman. Endothehal cell damage and thrombus formation after partial artenal constriction: relevance to the role of coronary artery spasm in the pathogenesis of myocardial infarction. *Circulation* **63**: 476 (1981).

F.L. Gobel, L.A. Nordstrom, R.R. Nelson, C.R. Jorgensen, and Y. Wang. The rate-pressure product as an index of myocardial oxygen consumption during exercise in patients with angina. *Circulation* **57**: 549 (1978).

F.A. Gonzalez and L.D. Byrd. Physiological effects of cocaine in the squirrel monkey. *Life Sci.* **21**: 1417 (1977).

L. Gould, C. Gopalaswamy, C. Patel, and R. Betzu. Cocaine induced myocardial infarction. *NY State J. Med.* **85**: 660 (1985).

S. Hadjimiltiades, V. Covalesky, B.V. Manno, W.S. Haaz, and G.S. Mintz. Coronary arteriographic findings in cocaine abuse-induced myocardial infarction. *Cath. Cardiovasc. Diag.* **14**: 33 (1988).

W.A. Hammond. Coca: its preparations and their therapeutic qualities, with some remarks on the so-called "cocaine habit". *Trans. Med. Soc. Virginia* pp. 212, November (1887).

R.E. Howard, D.C. Hueter, and G.J. Davis. Acute myocardial infarction following cocaine abuse in a young woman with normal coronary arteries. *JAMA* **254**: 95 (1985).

J.M. Isner, N.A.M. Estes, P.D. Thompson, M.R. Costanzo-Nordin, R. Subramanian, G. Miller, G. Katsas, K. Sweeney, and W.Q. Sturner. Acute cardiac events temporally related to cocaine abuse. *N. Engl. J. Med.* **315**: 1438 (1986).

H. Inoue and D.P. Zipes. Cocaine-induced supersensitivity and arrhythmogenesis. *J. Am. Coll. Cardiol.* **11**: 867 (1988).

R.K. Jain, P.J. Gatti, M. Visner, M.G. Moront, C.E. Rackley, and R.A. Gillis. Dose-dependent effects of cocaine on cardiovascular function of anesthetized cats. *Clin. Res.* **35**: 289A (1987).

R.K. Jain, M. Jain, L. Bachenheimer, C. Tracy, M.S. Visner, and R.A. Gillis. Factors influencing the ability of cocaine to potentiate cardiac responses elicited by sympathetic nerve stimulation. *J. Pharmacol. Exp. Ther.* **252**: 147 (1990).

J.I. Javaid, M.W. Fischman, C.R. Schuster, H. Dekirmenjian, and J.M. Davis. Cocaine plasma concentration: relation to physiological and subjective effects in humans. *Science* **202**: 227 (1978).

E.R. Jolly and J.E. Steinhaus. The effects of drugs injected into limited portions of the cerebral circulation. *J. Pharmacol. Exp. Ther.* **116**: 272 (1955).

S. Jonsson, M. O'Meara, and J.B. Young. Acute cocaine poisoning. *Am. J. Med.* **75**: 1061 (1983).

S.B. Karch. Serum catecholamines in cocaine-intoxicated patients with cardiac symptoms. *Ann. Emerg. Med.* **16**: 481 (1987).

R.L. Koerker and N.C. Moran. An evaluation of the inability of cocaine to potentiate the responses to cardiac sympathetic nerve stimulation in the dog. *J. Pharmacol. Exp. Ther.* **178**: 482 (1971).

W.A. Kossowsky and A.F. Lyon. Cocaine and acute myocardial infarction. *Chest* **86**: 729 (1984).

T.L. Kramer, A.R. Silvers, P.A. Recht, W. Lovenvirth, J. Belanoff, G. Weissman, and R.I. Levin. Cocaine alters platelet and endothebal cell eicosanoid synthesis. *Clin. Res.* **36**: 365A (1988).

F.E. Kuhn, R.A. Gillis, S.K. Wahlstrom, R.E. Tessel, M.S. Visner, and G.L. Schaer. Cocaine-induced cardiovascular effects: lack of evidence for a central nervous system site of action. *Circulation* **78**: II-78 (1988).

F.E. Kuhn, R.A. Gillis, M.S. Visner, and G.L. Schaer. Cocaine induces alpha receptor mediated epicardial and resistance coronary artery vasoconstriction in dogs. *Circulation* (submitted).

F.E. Kuhn, R.A. Gillis, S.K. Wahlstrom, M.S. Visner, and G.L. Schaer. Cocaine induced deleterious effects on the canine coronary circulation. *J. Am. Coll. Cardiol.* **13**: 79A (1989).

J.Y.T. Lam, J.H. Chesebro, P.M. Steele, L. Badimon, and V. Fuster. Is vasospasm related to platelet deposition? Relationship in a porcine preparation of arterial injury *in vivo*. *Circulation* **75**: 243 (1987).

D. Lam and N. Goldschlager. Myocardial injury associated with polysubstance abuse. *Am. Heart J.* **115**: 675 (1988).

M.N. Levy and B. Blattberg. Progressive reduction in norepinephrine overflow during cardiac sympathetic nerve stimulation in the anesthetized dog. *Cardiovasc. Res.* **10**: 549 (1976).

D.F. MacGregor. The relation of cocaine and procaine to the sympathetic system. *J. Pharmacol. Exp. Ther.* **66**: 393 (1939).

D.W. Mathias. Cocaine associated myocardial ischemia: review of clinical and angiographic findings. *Am. J. Med.* **81**: 675 (1986).

Y. Matsuda, Y. Matsuda, and M.N. Levy. The effects of cocaine and metanephrine on the cardiac responses to sympathetic nerve stimulation in dogs. *Circ. Res.* **45**: 180 (1979).

A.A. Nanji and J.D. Filipenko. Asystole and ventricular fibrillation associated with cocaine intoxication. *Chest* **85**: 132 (1984).

D.D. O'Keefe, R.N. Grantham, E.A. Beirhold, G.Z. Kapellakis, M.A. Vasu, R.A. Guyton, and W.M. Daggett. Cocaine and the contractile response to catecholamines in right ventricular failure. *Am. J. Physiol.* **233**: H399 (1977).

N. Pappas, H.S. Margolius, and T.E. Gaffney. The influence of cocaine on the chronotropic effects of endogenously released and injected norepinephrine. *Pharmacologist* **7**: 184 (1965).

P.F. Pasternak, S.B. Colvin, and F.G. Baumann. Cocaine-induced angina pectoris and acute myocardial infarction in patients younger than 40 years. *Am. J. Cardiol.* **55**: 847 (1985).

M. Perez-Reyes, S. DiGuiseppi, G. Ondrusek, A.R. Jeffcoat, and C.E. Cook. Free-base cocaine smoking. *Clin. Pharmacol. Ther.* **32**: 459 (1982).

L. Perrault, K. Egashira, M. Briggs, K.G. Morgan, and J.G. Morgan. Effects of cocaine on human epicardial coronary arteries. *J. Mol. Cell. Cardiol.* **21** (Suppl. II): S-197 (1989).

A.Pierre, W. Kossowsky, S.T. Chow, and A.R. Abadir. Coronary and systemic hemodynamics after intravenous injection of cocaine. *Anesthesiology* **63**: A28 (1985).

E. Ramoska and A.D. Sacchetti. Propranolol-induced hypertension in treatment of cocaine intoxication. *Ann. Emerg. Med.* **14**: 1112 (1985).

R.B. Resnick, R.S. Kestenbaum, and L.K. Schwartz. Acute systemic effects of cocaine in man: a controlled study by intranasal and intravenous routes. *Science* **195**: 696 (1977).

J.L. Rod and R.P. Zucker. Acute myocardial infarction shortly after cocaine inhalation. *Am. J. Cardiol.* **59**: 161 (1987).

I.M. Rollingher, A.S. Belzberg, and I.L. Macdonald. Cocaine-induced myocardial infarction. *Can. Med. Assoc. J.* **135**: 45 (1986).

R.B. Rubin and Neugarten. Cocaine-induced rhabdomyolysis masquerading as myocardial ischemia. *Am. J. Med.* **86**: 551 (1989).

J.S. Schachne, B.H. Roberts, and P.D. Thompson. Coronary-artery spasm and myocardial infarction associated with cocaine use. *N. Engl. J. Med.* **310**: 1665 (1984).

J.Sharkey, M.C. Ritz, J.A. Schender, R.C. Hansen, and M.J. Kuhar. Cocaine inhibits muscarinic cholinergic receptors in heart and brain. *J. Pharmacol. Exp. Ther.* **246**: 1048 (1988).

R.W. Simpson and W.D. Edwards. Pathogenesis of cocaine-induced ischemic heart disease. *Arch. Pathol. Lab. Med.* **110**: 479 (1986).

H.W.B. Smith, H.A. Liberman, S.L. Brody, L.L. Batty, B.C. Donohue, and D.C. Morris. Acute myocardial infarction temporally related to cocaine use. *Ann. Intern. Med.* **107**: 13 (1987).

W.R. Teeters, T. Koppanyi, and F.F. Cowan. Cocaine tachyphylaxis. *Life Sci.* **7**: 509 (1963).

R.E. Tessel, C.B. Smith, D.N. Russ, and L.B. Hough. The effects of cocaine HCl and a quaternary derivative of cocaine, cocaine methiodide, on isolated guinea pig and rat atria. *J. Pharmacol. Exp. Ther.* **205**: 569 (1978).

G. Togna, E. Tempesta, A.R. Togna, N. Dolci, B. Cebo, and L. Caprino. Platelet responsiveness and biosynthesis of thromboxane and prostacylin in response to in vitro cocaine treatment. *Hemostasis* **15**: 100 (1985).

U. Trendelenberg. The supersensitivity caused by cocaine. *J. Pharmacol. Exp. Ther.* **125**: 55 (1959).

U.Trendelenberg. The effect of cocaine on the pacemaker of isolated guinea pig atria. *J. Pharmacol. Exp. Ther.* **161**: 222 (1968).

R. Trouve and G. Nahas. Nitrendipine: an antidote to cardiac and lethal toxicity of cocaine. *Proc. Soc. Exp. Biol. Med.* **183**: 392 (1986).

C. Van Dyke, P.G. Barash, P. Jatlow, and R. Byck. Cocaine: plasma concentrations after intranasal application in man. *Science* **191**: 859 (1976).

C. Van Dyke and R. Byck. Cocaine 1884–1974, in: *Cocaine and Other Stimulants: Advances in Behavioral Biology*. Vol. 21. E.H. Ellengwood, Jr. and M.M. Kilbey, eds., Plenum Press, New York, p. 1 (1977).

R. Vargas, Z. Zukowska-Grojec, R.A. Gillis, and P.W. Ramwell. Cocaine and coronary artery reactivity. *Circulation* **78**: II-223 (1988).

G.M. Vincent, J.L. Anderson, and H.W. Marshall. Coronary spasm producing coronary thrombosis and myocardial infarction. *N. Engl. J. Med.* **309**: 220 (1983).

T. Wang, F. Hadidi, F. Triana, and M. Bargout. Myocardial infarction associated with the use of cocaine. *Am J. Med. Sci.* **295**: 569 (1988).

C.S. Wehbie, H.J. Vidallet, Jr., F.I. Navetta, and R.H. Peter. Acute myocardial infarction associated with initial cocaine use. *So. Med. J.* **80**: 933 (1987).

S. Weidmann. Effects of calicum ions and local anaesthetics on electrical properties of purkinje fibers. *J. Physiol.* **129**: 568 (1955).

N. Weiner and P. Taylor. Drugs acting at synaptic and neuroeffector junctional sites, in: *Goodman and Gilman's The Pharmacological basis of Therapeutics.* 7th ed., A.G. Gilman, L.S. Goodman, T.W. Rall, and F. Murad, eds., MacMillian, New York, p. 66 (1985).

R.J. Weiss. Recurrent myocardial infarction caused by cocaine abuse. *Am. Heart J.* **111**: 793 (1986).

C.V. Wetli and R.K. Wright. Death caused by recreational cocaine use. *JAMA* **241**: 2519 (1979).

L.G. Whitby, G. Hertting and J. Axelrod. Effect of cocaine on the disposition of noradrenaline labelled with tritium. *Nature* **187**: 604 (1960).

R.D. Wilkerson. Cardiovascular effects of cocaine in conscious dogs: importance of fully functional autonomic and central nervous systems. *J. Pharmacol. Exp. Ther.* **246**: 466 (1988).

R.D. Wilkerson. Cardiovascular effects of cocaine: enhancement by yohimbine and atropine. *J. Pharmacol. Exp. Ther.* **248**: 57 (1989).

C.E. Wilkens, V.S. Mathur, R.C. Ty, and R.J. Hall. Myocardial infarction associated with cocaine abuse. *Texas Heart Inst. J.* **12**: 385 (1985).

R.P. Yasuda, N.R. Zahniser, and T.V. Dunwiddie. Electrophysiological effects of cocaine in the rat hippocampus *in vitro*. *Neurosci. Lett.* **45**: 199 (1984).

D. Young and J.J. Glauber. Electrocardiographic changes resulting from acute cocaine intoxication. *Am. Heart J.* **34**: 272 (1947).

M.A. Young, D.R. Knight, and S.V. Vatner. Parasympathetic coronary vasoconstriction induced by nicotine in conscious calves. *Circ. Res.* **62**: 891 (1988).

F.H. Zimmerman, G.M. Gustafson, and H.G. Kemp. Recurrent myocardial infarction associated with cocaine abuse in a young man with normal coronary arteries: evidence for coronary artery spasm culminating in thrombosis. *J. Am. Coll. Cardiol.* **9**: 964 (1987).

18

Perinatal Exposure to Cocaine: Human Studies

LORETTA P. FINNEGAN, JONEL M. MELLOTT, LYNN R. WILLIAMS, AND RONALD J. WAPNER
Department of Pediatrics
Jefferson Medical College
Thomas Jefferson University
Philadelphia, Pennsylvania

INTRODUCTION

The rising incidence of cocaine abuse in recent years indicates a health-care problem of major proportions. It is now estimated that up to 30 million Americans have used cocaine at least once (Abelson and Miller, 1985), that up to 8 million use it regularly (Stone *et al.*, 1984), and another 5000 Americans try cocaine for the first time each day (Schnoll *et al.*, 1984). Cocaine use among pregnant women has paralleled the increased use in the general population (Chasnoff *et al.*, 1989; Fantel and Macphail, 1982). The number of pregnant women using cocaine has grown with the drug's increasing availability, its highly addictive quality, and its persistent image as a socially acceptable and harmless agent. Although the destructive effects of cocaine on the adult user are well known, much less has been reported regarding maternal cocaine use and its effect upon the perinatal period. Over the past 6 years, though, research regarding cocaine use, pregnancy, and the neonate began to emerge. In addition to the hazards associated with cocaine use in the general population, these investigations have revealed a wide variety of risks specific to women and children.

UTEROPLACENTAL EFFECTS

A common misconception of pregnancy is that the placenta can protect the developing fetus from the effects of toxic substances taken during pregnancy. Regarding cocaine, however, its low molecular weight and high solubility in both water and lipids allows it to readily cross the placenta as well as the fetal blood brain barrier. When the mother employs i.v. or inhalation routes, the resulting high serum concentrations may not undergo first-pass metabolism and may therefore cross the placenta in large quantities. Additionally, because of the relatively low pH of fetal blood (cocaine is a weak base) and the low fetal level of plasma esterases which metabolize this drug, cocaine may easily accumulate in the fetus. Furthermore, the binge pattern commonly associated with cocaine use to regain the euphoric state can possibly lead to very high levels of fetal exposure (LeBlanc *et al.*, 1987).

The extent of drug transfer depends in part upon the gestational age of the placenta. Transfer is greatest during the first and third trimesters and slowest during the second. During the first trimester, the placenta is rudimentary and not efficient in regulating drug transfer. Since this is the period of organogenesis, cocaine exposure during the first trimester may be manifested in a number of ways. Several studies have reported varying teratogenic potential for cocaine. By the second trimester, the placenta is relatively thick with a small surface area and greater density, thereby hampering diffusion. However, by the third trimester, the placenta has significantly thinned, resulting in an increased surface area which enhances diffusion of all psychoactive substances and readily exposes the growing fetus to maternally ingested cocaine (Flandermeyer, 1987).

One of cocaine's inherent properties may somewhat serve to decrease fetal cocaine exposure. The placental transfer of cocaine is inhibited by the catecholamine-mediated vasoconstrictive properties of this drug. When uterine, placental, and umbilical blood vessels constrict, fetal blood flow is restricted and the amount of cocaine delivered to the fetus reduced. However, a deleterious effect of this vasoconstriction is that the fetus is also deprived of essential gas and nutrient exchange so that fetal hypoxia simultaneously occurs (Flandermeyer, 1987). This decreased blood flow and resultant decrease in gaseous and nutritional exchange has been suggested as a contributing factor for the intrauterine growth retardation often associated with chronic cocaine use during pregnancy. A possible increase in congenital malformations in infants born to cocaine-using pregnant women may also be associated with this decrease in fetal blood supply during critical periods of morphogenesis and growth (Chasnoff *et al.*, 1989).

Several animal studies have demonstrated the effect of cocaine on the pregnant uterus. Woods (1987) showed that maternal cocaine administration led to a dose-dependent increase in maternal blood pressure and decrease in uterine blood flow. Catecholamine-mediated vasoconstriction was thought to cause the increased uterine vascular resistance. The resulting fetal hypoxemia, tachycardia, and hypertension prompted the release of even more fetal catecholamines, thereby exaggerating the cocaine-related effect. When he administered cocaine directly to the fetus, the elevations in blood pressure and heart rate were less than was seen with maternal cocaine administration. (Although Woods' experiments were conducted in sheep, doses were selected to approximate the usual blood levels seen after recreational use in humans.)

Moore (1986) demonstrated in sheep that doses of cocaine equivalent to levels observed in humans resulted in a significant reduction in uterine blood flow for at least 15 min. He found a doubling of circulating norepinephrine concentration after cocaine administration, along with fetal and maternal hypertension. The cocaine-induced vasoconstriction was attributed, at least partially, to catecholamine potentiation, especially blockade of norepinephrine re-uptake.

Rosenfeld's (1977) investigations, using pregnant ewes, revealed that reproductive tissues (placenta, endometrium, and myometrium) were far more sensitive to the vasoconstrictive effect of norepinephrine than were other body tissues. Similar studies in mice showed higher concentrations in the uterus and placenta at 15 min and 1, 3, and 6-h intervals than in other body tissues (Shah *et al.*, 1980). In addition, they found that fetal brain and liver accumulate far less cocaine (one third and one fourth of the concentrations in those maternal organs, respectively), despite high levels in the uterus and placenta. Moore (1986) also found that fetal cocaine levels 5 min after cocaine administration to pregnant ewes were only 12% of the maternal levels. The fetal concentration would be expected to be higher because of the lower pH of fetal blood. Higher concentration of ionized/cationic forms of local anesthetics usually appear on the acidotic side of a permeable membrane such as the placenta (Shah *et al.*, 1980). The observed results may be due to either a decreased affinity for immature tissues or a reduction in placental penetration secondary to cocaine's vasoconstrictive effect on the uteroplacental complex. It may also account for the higher rate of obstetrical rather than teratogenic complications.

The course of labor may also be affected by maternal cocaine use. Intravenous administration of local anesthetics appears to cause a direct increase in uterine muscle tone and placental vasoconstriction. The effect does not seem to be adrenergic, because it is not inhibited by α-adrenergic blockade (McGaughey *et al.*, 1962). Crack has also been shown to have a direct action in increasing uterine contractility, and may precipitate the onset of premature labor (Baxi and Petrie, 1987).

A higher rate of placental abruptions is considered to be one of the complications of maternal cocaine use. Its incidence is especially increased during the third trimester (Acker *et al.*, 1983). Bingol (1987) reported a significantly higher rate of stillbirths among cocaine-using women, with all stillbirths related to abruptio placentae occurring shortly after smoked, i.v., or intranasal cocaine use. Acker (1983) reported two cases of placental abruption associated with i.v. and intranasal cocaine use. Both women demonstrated acute elevations in blood pressure at the time of their abruptions. Like other investigators, Acker noted the association between cocaine and immediate hypertension and tachycardia due to central stimulation and blockade of catecholamine re-uptake.

Pritchard (1970) studied the etiology of severe placental abruption in 201 cases which led to fetal death. Hypertension was identified as the cause in nearly half of the cases. A diastolic blood pressure of greater than 90 mm for over 6 h was seen in 47% of abruptions which led to fetal demise, compared to an overall incidence of intrapartum hypertension of 10% at the same hospital. Although pharmacologic agents theoretically should be able to block cocaine-related pressor responses, studies have shown that phentolamine infusion before and during cocaine administration has no significant effect on the duration or magnitude of compromised uterine blood flow (Moore *et al.*, 1986).

Another effect of cocaine on the placenta has been reported by Wang (1987) who found that the placentas of women who used cocaine during pregnancy had a decrease in the total number of β-adrenergic receptor binding sites, μ-opiate receptor binding sites, and δ-opiate receptor binding sites. The decreases in receptor binding sites were not associated with a decrease in receptor binding affinity, therefore indicating a true down-regulation of receptor binding sites. Wang attributes this down-regulation to cocaine's effects on the peripheral nervous system, whereby norepinephrine re-uptake is inhibited at presynaptic nerve terminals. The resultant increased synaptic catecholamine concentrations lead to the decreased number of β-adrenergic receptor binding sites. Wang suggests that this receptor down-regulation may have significant implications in view of the transplacental transfer of cocaine. A transient elevation in fetal catecholamine level may cause a similar down-regulation in the adrenergic receptors of the developing fetal nervous system and result in derangement of synaptic development. Wang concedes that the significance of down-regulation of the opiate receptors among cocaine users is unclear. He suggests the possibility that cocaine may trigger the release of an endogenous opiate peptide and the transient increase in opiate agonist may result in the observed down-regulation of both μ- and δ-opiate receptors.

EFFECTS ON PREGNANCY

Church (1988) has shown the deleterious effects of cocaine on pregnancy in animals. He found cocaine administration to Long-Evans rats on gestation days 7 through 19 inclusive led to dose-dependent decreases in maternal weight gain and food and water consumption. Additionally, increased doses of cocaine were correlated with increased maternal and fetal mortality, fetal edema, abruptio placentae, lack of fetal weight gain, and fetal cephalic hemorrhage related to vasodilation of the head and face. However, no significant increase in congenital malformations was found. On examination at autopsy, one rat fetus was reported as microcephalic and another was described as anopthalmic.

Human data collection is often clouded by many confounding variables such as maternal age and undernutrition, varying concentrations and methods of cocaine usage, multiple drug use, poor pre- and post-natal care, and the possible presence of toxic adulterants that are mixed with or used to process cocaine. A controlled animal study such as the one by Church can limit the number of variables so that cocaine's effects alone may be studied. Although projections to humans is risky, this is consistent with human observations that maternal cocaine use may place the human fetus at risk for low birth weight, fetal edema, placental abruption, infarctions, and death (Church *et al.*, 1988).

The toxicity of abused chemicals is compounded by other drug-related problems such as maternal appetite suppression contributing to poor fetal outcome. Chronic amphetamine users often develop severe vitamin deficiencies, particularly of the water-soluble B-complex and C vitamins (Wesson and Smith, 1978). It is reasonable to assume that the increased sympathetic stimulation and/or anorectic effect associated with chronic cocaine abuse may lead to similar deficits. Many of these vitamins (e.g., riboflavin, thiamine, and niacin) are essential for normal neuronal function and biosynthesis of dopamine, norepinephrine, acetylcholine, and epinephrine.

One study found that approximately 73% of cocaine users admitted for treatment had at least one vitamin deficiency (40% were deficient in B_6, 28% in B_1, and 23% in Vitamin C) (Estroff *et al.*, 1983). The correction of these deficiencies is particularly necessary in pregnant women undergoing treatment in order to reestablish neurochemical balance.

Women who use cocaine during pregnancy are more likely than nonusers to have spontaneous first trimester abortions, a finding consistent with the implication of placental vasoconstriction induced by cocaine use. (Chasnoff *et al.*, 1988). Third trimester use of cocaine by pregnant women has been reported to induce a sudden onset of uterine contractions, fetal tachycardia, and excessive fetal activity within hours or even minutes following maternal cocaine ingestion. Also, in a report by Chasnoff, 52 cocaine-using women had a higher rate of complications of labor and delivery compared to a matched group of 73 non-cocaine-using, methadone-maintained women whose incidence of complications of labor and delivery was similar to that of the general population.

Due to cocaine-mediated increases in norepinephrine levels, an increase in uterine contractility and in the incidence of preterm and precipitous labor occurs during pregnancy. Additional complications following delivery among cocaine-using women include a higher incidence of maternal hemorrhage and post-partum anemia. These maternal problems during and following delivery are paralleled in the high rate of fetal distress noted in cocaine-exposed infants. Increased fetal distress is indicated by the higher incidence of fetal monitor abnormalities, fetal meconium staining, and lower Apgar scores in the cocaine-exposed infants as compared to those infants whose mothers were maintained on methadone but did not use cocaine during pregnancy. Chasnoff has reported two infants who suffered perinatal cerebral infarction probably related to their mother's cocaine use in the 48–72 h prior to delivery. This finding tends to parallel reports of acute cerebral and myocardial infarction in young adults shortly after cocaine use.

MacGregor (1987) also reported lower birth weight, lower gestational age at delivery, and increased frequency of placental abruptions (7.1 vs. 1.4%) with cocaine use. There was a higher incidence of preterm labor (21.4 vs. 1.4%) and premature delivery (24.3 vs. 2.9%) in the cocaine-exposed group compared to an appropriately matched control group; this is clinically relevant in that up to 75% of perinatal mortality and morbidity occurs in preterm infants. MacGregor also found that only 33% of the women eliminated cocaine use during their pregnancies (based on random urine screening), which was significantly less than the success of interventions with other substances of abuse.

In one of the largest human studies, Livesay (1987) studied 239 pregnant women divided into 3 groups: (1) 93 cocaine-using, drug-dependent women, (2) 83 non-cocaine-using, drug-dependent women, and (3) 63 non-drug-dependent women. The women in the first two groups were enrolled in a comprehensive program for pregnant, addicted women in Philadelphia and the third consisted of those who attended the same hospital prenatal clinic as the cocaine-exposed women (Table 1).

Table 1
Study Groups (N = 239)

Group I: (n = 93)	Cocaine-using, drug-dependent women
Group II: (n = 83)	Non-cocaine-using, drug-dependent women
Group III: (n = 63)	Non-drug-dependent women

Table 2
A Comparison of Intrapartum and Maternal Outcome for the Three Study Groups

	Group 1 (Cocaine DDW) n = 93	Group II (Non-cocaine DDW) n = 83	Group III (Drug free) n = 63
Stat C-section, %[a]	10.8	3.6	3.2
Abruptio placentae, %	8.7	3.6	1.6
Meconium staining, %[b]	31.7	23.1	10.3

[a] Stat C-section: emergency cesarean sections for fetal distress.
[b] Meconium staining: staining of amniotic fluid.

The three groups were similar in maternal age, parity, nicotine use, and socioeconomic level. There was no significant difference between groups 1 and 2 for daily dose of methadone in milligrams (38.4 vs. 42.4 mg). The cocaine-using, drug-dependent women had significantly fewer prenatal visits than either of the other two groups. Sixty-four percent of the cocaine group used the drug at least once daily and could therefore be considered heavy users.

Maternal outcome corroborated previous studies. The cocaine group had more frequent emergency cesarean sections (10.8%) than the non-cocaine, drug-dependent group (3.6%) or the drug-free group (3.2%). Abruptio placentae occurred in 8.7% of cocaine-using, drug-dependent women; 3.6% of the non-cocaine drug-dependent women; and 1.6% of drug-free women. There was a significant difference in the three groups in the occurrence of meconium staining, with the cocaine-exposed group having the highest incidence (Table 2).

The cocaine group had three spontaneous abortions and four fetal deaths. The narcotic-exposed group had no spontaneous abortions and five fetal deaths. The comparison group had neither. The cocaine-exposed infants had significant increases in premature birth and small increases for gestational-age infants. No significant differences were seen in the incidence of SIDS or death due to congenital malformations (Table 3). (See additional information about SIDS under Neonatal Effects.)

Three additional reports imply that cocaine use during pregnancy may be related to an increase in both maternal and fetal fatality. Henderson (1988) reports rupture of an intracranial aneurysm after smoking crack in a 24-year-old woman who was 34 weeks pregnant. Woods reports that the cardiovascular circulation may be more sensitive in the pregnant state due to the interaction between progesterone and cocaine. Cocaine-induced hypertension and tachycardia further strain the already overworked maternal heart, indicating at least the potential for other vascular effects as well. In another report, Critchley (1988) describes the death of a fetus *in utero*

Table 3
Condition of Fetus Neonate

	Group I (Cocaine DDW) n = 93 freq. %	Group II (Non-cocaine DDW) n = 83 freq. %	Group III (Drug free) n = 63 freq. %
Spontaneous abortions	3 (3)	0 (0)	0 (0)
Fetal deaths	4 (4)	5 (6)	0 (0)
Live births	83 (92)	72 (92)	63 (100)
Premature	19 (20)	9 (11)	63 (100)
Postmature	0 (0)	0 (0)	3 (5)
Gestational age[a,b]			
SGA	13 (15)	6 (8)	3 (5)
AGA	71 (84)	67 (89)	48 (91)
LGA	1 (1)	2 (3)	2 (4)
Unclassified	8	7	10
SIDS	2	2	0
Death due to congenital anomalies	1 (3 d)	0	0

[a] SGA: small for gestational age. AGA: appropriate. LGA: large.
[b] Percentage calculated on classified infants only.

following maternal cocaine use. Toxicological screening of fetal whole blood revealed cocaine concentrations of 1 mg/l, an amount often associated with fatalities in adults. Thus, infants *in utero* may suffer cocaine intoxication even when the mother herself fails to manifest the clinical signs of acute cocaine intoxication. Most drugs taken by pregnant women have a longer half-life in the fetus than in the adult because the fetal liver is incompletely developed and connot excrete drugs quickly. Furthermore, fetuses and infants have a relative deficiency of plasma cholinesterases (the enzymes that break down cocaine), thereby allowing cocaine to accumulate and thus increase the possibility of fetal toxicity.

TERATOGENIC EFFECTS

Several studies in both animals and humans have revealed an increased frequency of developmental abnormalities associated with maternal cocaine use. An investigation in mice by Mahalik (1980) found a higher incidence of soft tissue abnormalities (especially skeletal anomalies) in the animals exposed to cocaine than in controls. They postulated that cocaine-induced vasoconstriction and hypoxia resulted in these anomalies, based on evidence that hypoxic conditions elicited similar anomalies at similar critical periods of organogenesis. In addition, they found a higher resorption ratio among cocaine-exposed litters, yet did not find differences in maternal weights nor correlation between fetal growth retardation and increased susceptibility to teratogenesis.

Other researchers, however, reported no increase in congenital anomalies with maternal cocaine use. Fantel and MacPhail (1982) noted a significant decrease in maternal and fetal weights in rats given high-dose i.p. cocaine, as well as fetal edema and higher resorption frequencies. They found no increase in congenital malformations, which they attribute to cocaine-induced vasoconstriction causing reduced fetal cocaine concentrations and, therefore, lower teratogenic potential.

Cocaine's teratogenic effect on the developing human fetus remains controversial at this time. There have been some isolated reports and small pilot studies that suggest cocaine might be responsible for congenital anomalies. Bingol *et al.* (1987) found a higher rate of congenital malformations among cocaine users (10%) vs. polydrug users (4.5%) and controls (2%); the malformations seen included cardiac, skeletal, and skull abnormalities. Although the mechanism is unknown, they speculate that cocaine induces vasoconstriction, sudden hypertension, and cardiac arrhythmias, all of which disrupt the blood supply to the fetus and thereby disrupt morphogenesis. Hypoxia and hyperpyrexia may also contribute to these effects.

Following the deliveries of 1 infant with prune belly syndrome (a complex of signs including major malformations of the genitourinary tract, bilateral hydronephrosis, and bilateral cryptorchidism) and another infant with hypospadias from cocaine-abusing women, Chasnoff compared the incidence of genitourinary malformations in 50 infants born to cocaine-abusing women compared to 30 infants born to polydrug, non-cocaine-using women. Of the 50 cocaine-exposed infants, 9 demonstrated some form of an anomaly whereas only 1 infant with Fetal Alcohol Syndrome was noted among the 30 polydrug, non-cocaine-exposed infants. Of the nine cocaine-exposed infants with malformations, two had ileal atresia and seven had a malformation of the genitourinary tract, including two infants with multiple congenital malformations similar to the previously described prune belly syndrome, three infants with hydronephrosis, and two infants with second degree hypospadias. Chasnoff offered cocaine's pharmacologic effects as a possible explanation leading to these anomalies. Infarctions or decreases in fetal blood flow early in gestation may disrupt organ system formation, resulting in defects such as ileal atresia. Additionally, inadequate blood flow and resultant hypoxia may cause deformity or disruption of morphogenesis at other times during pregnancy. Chasnoff suggests that the prune belly syndrome may be a result of fetal vascular compromise, such as might occur with maternal cocaine use at or around the 12th week of gestation.

Recently a unifying hypothesis linking cocaine to congenital anomalies has been suggested by Hoyme *et al.* (1990). The authors present a range of malformations under an umbrella of the term "fetal vascular disruption"; these anomalies include growth deficiency, nonduodenal intestinal atresia, limb reduction defects, renal anomalies and aplasia cutis congenita.

Hyperthermia is another adverse effect of cocaine. Body temperature may be elevated from 3–5°C above normal following cocaine ingestion. This hyperthermia may also be a contributing factor for cocaine's possible teratogenic effect (Flandermeyer, 1987).

In contrast to these studies, however, Hadeed and Seigel (1989) noted no congenital malformations in either 32 cocaine-exposed infants or 29 control infants. Zuckerman *et al.* (1989), in a larger study, concluded that a positive urine assay for cocaine metabolites was not significantly associated with an increased malformation rate.

In addition, Mellott (1989) in Philadelphia examined the incidence and type of congenital anomalies noted in the neonate at birth following maternal cocaine use during pregnancy. Subjects included 136 pregnant women and their 139 offspring (3 sets of twins are included). The mothers were divided into two groups based on whether or not they were opiate addicted and maintained on methadone during pregnancy. All women used cocaine during pregnancy. The first group comprised 44 women who were maintained on methadone and used cocaine during pregnancy, and 92 women, who used cocaine as their primary drug of abuse while pregnant, comprised the second group (all 3 sets of twins were in this group). All women were enrolled in a drug treatment program providing both pre- and postnatal services. Although the women maintained on methadone were older, the two groups were similar in mean gravidity and parity. The methadone-maintained group had a higher percentage of fetal loss; however, there were no statistical differences in birth weight, length, head circumference, prematurity, or Apgar scores between the two groups. In both groups, the percentage of congenital anomalies was similar and approximately equal to nationally reported figures. Congenital anomalies were observed in four infants. Two infants had heart defects and two had minor limb malformations. These abnormalities more closely agree with the findings of the Bingol study than the one by Chasnoff. All of these defects are commonly seen and cannot be solely attributed to the maternal use of cocaine during pregnancy. Recently, one of the largest clinical studies regarding the issue of urogenital abnormalities as a result of cocaine was reported by Rajegowda. From 1987 to 1990, 18,056 deliveries were evaluated; cocaine use was deteremined in 1,320. The incidence of congenital urogenital abnormalities was lower in the cocaine exposed infants (although no significant difference was noted).

Obviously, more research in this area needs to be conducted. Several factors, however, make this a difficult area to study. First, drug and alcohol dependence are often one of the most frequently missed diagnoses. Second, only naturally occurring anomalies can be reported, and numerous extraneous variables make it difficult to ascertain a definite cause-effect relationship. In the population at large, there is a baseline rate of birth defects. When studying a small population of women who abuse cocaine, it is difficult to determine if a particular anomaly was due to the cocaine ingested during pregnancy or if it would have occurred independent of drug abuse. Furthermore, because many studies are retrospective in nature, they may be misleading, depending on the accuracy of detail recalled by the pregnant subjects or the accuracy of records used in compiling data for the study. Prospective studies are complicated by the intervention strategies that attempt to persuade women to discontinue or modify their drug use to minimize the risk to the pregnancy and the unborn child. Finally, research in this field is also complicated by the ethical constraints which prevent extensive experimentation on fetal and human development.

Table 4
Outcome of the Neonate

Signs	Group I (Cocaine DDW) n = 93	Group II (Non-cocaine) n = 83	Group III (Drug free) n = 63
Birth weight, g[a,b]	2653	2860	3095
Birth length, cm[a,b]	46.9	48.5	49.7
Head circumference, cm[b]	32.5	33.0	33.8
Gestational age, weeks[b]	36.8	37.9	39.1
Apgar — 1 min[b]	7.3	7.6	7.9
Apgar — 5 min	8.5	8.7	8.8
Nuchal cord, %[c,d]	15.7	21.5	37.9
Pharmacotherapy, %[e]	36.0	43.0	—
Congenital anomalies, %	4.3	6.0	—

[a] Cocaine group differs from non-cocaine, drug-dependent group at the $p \leq 0.05$ level.
[b] Cocaine group differs from drug-free group at the $p \leq 0.01$ level.
[c] Nuchal cord: umbilical cord entangled around infant's neck one or more times.
[d] Groups differed significantly at the $p \leq 0.01$ level.
[e] Pharmacotherapy: treatment for neonatal abstinence according to Finnegan protocol.

NEONATAL EFFECTS

The effects of maternal cocaine abuse do not cease at birth. Many studies report abnormal findings in infants who were exposed to cocaine *in utero*. Ryan (1987) and Livesay (1987) report that infants born to drug-dependent women have a poorer general outcome than do those infants born to drug-free women. In the Livesay study, infants born to women who used cocaine and methadone or other drugs during pregnancy had decreased birth weight, head circumference, length, and Apgar scores compared to infants born to methadone-maintained women who did not use cocaine during pregnancy and non-drug-exposed comparison infants (Table 4).

Bingol (1987) also reported lower birth weights with cocaine exposure. Some investigators attribute this variance in birth weight to the poor nutritional status and the anorectic effects of cocaine use causing poor maternal weight gain, rather than to the drug itself. However, multiple regression analysis of Oro's (1987) data indicated that the drug itself contributed to the birth weight, length, head circumference, and gestational age, independent of maternal factors.

Although Chasnoff (1985) originally reported no decrease in head circumference in cocaine-exposed newborns compared to controls, his subsequent data (1989) did reveal smaller head size if cocaine was used throughout the pregnancy. Decreases in neonatal head circumference in cocaine-exposed newborns have been confirmed by Hadeed (1989), Zuckerman (1989), Cherukuri (1988), and Fulroth (1989). These various data sets suggest, therefore, that maternal cocaine use is associated with a symmetric growth retardation in offspring, with reductions to both birth weight and head circumference. This pattern appears similar to that described in opiate-exposed newborns by Doberczak and Kandall (1987), although the mechanism in the latter case appears due to reduced organ cell number.

Cocaine-related fetal growth retardation may be the result of an altered placental or fetal hormonal milieu. To date, no human studies have been conducted regarding the effect of cocaine on reproductive hormones, but it is likely that the initial effect of cocaine would be to increase secretion of luteinizing hormone (LH) and suppress prolactin (PRL) secretion. Cocaine's action at the nerve terminal, whereby the re-uptake of norepinephrine and dopamine is blocked, would lead to enhanced neuronal stimulation. Since dopamine suppresses prolactin secretion, PRL levels would drop. LH is inhibited by PRL, so LH concentration would increase. Chronic cocaine use could, however, result in temporary depletion of hypothalamic neurotransmitters leading to decreased dopamine levels and a loss of PRL inhibition. The elevated prolactin levels would then inhibit the release of gonadotropins such as LH, thus possibly compromising placental development and fetal growth. Studies in ovariectomized rats show that moderate doses (10–20 mg/kg of cocaine) increase, and high doses (40 mg/kg cocaine) decrease, serum luteinizing hormone. Follicle stimulating hormone was unaffected at any dosage, whereas serum PRL was decreased by cocaine at either dosage (Smith and Asch, 1987). Additionally, Scher (1982) reports that a single 2-mg/kg dose of cocaine by i.v. bolus injection administered to nonhuman female primates led to an increase in LH levels within 30 min and a decrease in PRL to a nondetectable level within 1–2 h which remained depressed for up to 24 h. These studies suggest the possibility that cocaine may play a role in reproductive hormone release. Additional research on human subjects is necessary to draw any further conclusions (Smith and Asch, 1987).

An interesting finding in the Livesay study (1987) (Table 4) was the significantly higher occurrence of nuchal cords in the drug-free group. In this case, the trend went from the highest incidence in the comparison group to the lowest in the cocaine exposed group. Silver (1987) postulated two possible etiologies for his finding of fewer nuchal cords in infants of drug-dependent women than in infants of drug-free women: (1) the fetuses are less active secondary to *in utero* drug exposure and (2) the fetuses have shorter umbilical cords. Shorter umbilical cords have been associated with any situation that limits or suppresses fetal movement (Miller *et al.*, 1981; Moessinger *et al.*, 1982). The depressant nature of the drugs used by the methadone exposed group could lead to decreased fetal movement, shorter umbilical cords, and fewer nuchal cords than in the non-drug-exposed infants. The even fewer number of nuchal cords in the cocaine-exposed infants can perhaps be explained as follows. Cocaine causes a transient hyperaroused state which is followed by the "cocaine crash". This crash, in an adult withdrawing from cocaine, has been characterized by drug craving, irritability, appetite changes, irregular sleep patterns, anergia, psychomotor retardation, amotivation, and depression (Gold *et al.*, 1986). These have all been related to the depletion of DA in the brain that occurs with the regular use of cocaine. The transmitter cannot be synthesized at a fast enough rate to replace what is lost into the synapse due to cocaine's blockage of the re-uptake mechanism. Perhaps there is also depletion of DA stores in the fetal brain with fetal counterparts to the adult signs and symptoms that result in decreased fetal movement.

Table 5
Average Frequency of Abstinence Scores for Infants of Drug-Dependent Women (Cocaine vs. Non-cocaine Exposed)

Symptoms	Cocaine exposed (N = 75)	Non-cocaine exposed (X = 68)
High-pitched cry	4.22[a]	10.50
Decreased sleep	10.42	15.35
Hyperactive Moro	6.11	10.37
Disturbed tremors	66.41	108.01
Undisturbed tremors	15.25	24.79
Increased muscle tone	59.05[a]	102.15
Excoriations	4.15	8.88
Myoclonic jerks	0.48	0.57
Convulsions	0.02	0.01
Sweating	3.28	5.59
Fever	16.51[a]	30.82
Yawning	1.25	2.44
Mottling	10.71[a]	26.87
Nasal stuffiness	17.37	25.51
Sneezing	5.77	10.43
Nasal flaring	0.43	0.75
Increased respirations	22.23	31.96
Increased suck	7.24	10.87
Poor feeding	4.00	6.66
Vomiting	3.61	5.26
Loose stools	7.05[a]	13.57

[a] Cocaine group differs significantly from non-cocaine at the $p < 0.05$ level.

In the Livesay study, no difference existed in the occurrence of congenital anomalies, intracranial hemorrhage, or need for pharmacotherapy to treat neonatal abstinence (Table 4). Neonatal abstinence was assessed for the infants of the drug-dependent mothers utilizing the score of Finnegan (1986). Mean neonatal abstinence scores were lower in the cocaine-exposed infants for all 21 signs. Differences were significant ($p < 0.05$) with respect to cry, disturbed tremors, increased muscle tone, excoriations, fever, mottling, and loose stools. (Differences for undisturbed tremors approached significance.) It must be remembered that both groups of infants were exposed to methadone and that the major symptomatology is that of opiate withdrawal but this does suggest that cocaine may reduce opioid withdrawal severity (Table 5). Kosten has reported a similar response in her studies in rats and adult males when she evaluated the effects of chronic cocaine dependence on the severity of naloxone-precipitated opioid withdrawal in opioid addicts. Kosten postulates that this drug-drug interaction may occur through alpha-2 adrenergic neurons in the locus coeruleus, similar to the effect of clonidine on opioid withdrawal.

Madden's (1986) study of eight cocaine-exposed infants described one premature birth, two small for gestational age neonates, and one infant who had withdrawal symptoms. Anecdotal reports from parents and caregivers suggested that cocaine-exposed infants seemed to have prolonged periods of deep sleep. This may be due to cocaine-induced suppression of REM sleep and rebound after withdrawal (Fischman *et al.*, 1976). However, Chasnoff reports an increased degree of irritability, tremulousness, and state lability on the Brazelton Neonatal Behavioral Assessment Scale in 52 cocaine-exposed infants when compared to 73 infants born to non-cocaine-using, methadone-maintained women. As in the Ryan and Livesay studies, these women were matched for maternal age, socioeconomic status, gravidity, nicotine, marijuana, and alcohol use. Additionally, cluster analysis in Chasnoff's study, revealed that cocaine-exposed infants have a greater deficiency in state control than do non-cocaine, methadone-exposed infants. Chasnoff found that the irritability and tremulousness characteristic of the cocaine-exposed infant interferes with the ability of the infant to interact with and respond to the caretaker, thereby making it difficult for the caretaker to establish an appropriate relationship with the infant. The lack of proper infant response induces increasing passivity on the part of the caretaker for interaction, thus setting up a cycle of increasing passivity in both participants in the relationship.

Oro's (1987) studies on prenatal drug abuse are consistent with other researchers. She found significantly lower birth weight, length, head circumference, and gestational age among infants exposed to cocaine *in utero*. Significant neurologic and physiologic alterations were seen in most cocaine- or amphetamine-exposed neonates, including abnormal sleep patterns (81%), tremors (71%), poor feeding (58%), hypotonia, vomiting, and fever. These babies tended to spend long periods of time in a dull-alert state with eyes open, and demonstrated poor visual processing of faces and objects. (Cocaine was also associated with increased rates of fetal distress, placental abruptions, intrauterine growth retardation, and prematurity.) The greatest morbidity was seen with cocaine plus narcotic use, despite the theoretical possibility of antagonism between the CNS stimulant and depressive effects of these drugs.

Additional signs of prior cocaine exposure in the neonate include transient signs of CNS irritability, including increased muscular rigidity and startle response, increased tone and tremors, increased respiratory and heart rates, poor tolerance for oral feedings, diarrhea, and disturbed sleep patterns (Smith, 1988). Clinicians must determine if these symptoms are due to the frequent depressant use by the cocaine-abusing woman before implicating cocaine. These symptoms usually do not persist beyond the neonatal period. In one study, only muscular rigidity persisted beyond this time and normal muscle tone returned by the 4th month of life (LeBlanc *et al.*, 1987).

In addition to neurobehavioral abnormalities, Doberczak and Kandall studied serial electroencephalographic tracings in cocaine-exposed neonates (1988). Electroencephalograms were abnormal in 17 of 38 infants during the first week of life, characterized as showing cerebral irritation with bursts of sharp waves and spikes and features of discontinuity. These abnormalities were unpredictable and did not correlate with variables such as maternal drug use, neonatal characteristics, or severity of neurologic dysfunction. In that study the authors demonstrated that all abnormal electroencephalograms reverted to normal when followed over a 3- to 12-month period, but the longer term impact of these changes is not known. Despite these abnormalities, no clinical seizures were observed during a 7- to 10-day observation period. Other workers have also noted the absence of clinical seizures in cocaine-exposed neonates. Chasnoff *et al.* (1989) did note seizures, however, in 6 of 52 infants born to mothers after cocaine use throughout the pregnancy; two of those infants were reported to have abnormal electroencephalograms. In addition, Kramer *et al.*, (1990) reported 16 infants born to cocaine-using mothers who developed seizures after birth. All 16 seizure incidents occurred within 36 hours after delivery and were either subtle (10), focal (1), tonic (2) or clonic (3). Many of the seizures were poorly responsive to therapeutic levels of anticonvulsant medications.

In addition to clinical encephalopathic and electroencephalographic changes, a study of echoencephalographic abnormalities in stimulant-exposed neonates has recently been published by Dixon and Bejar (1989). This study assesses a mixed intrauterine exposure to cocaine and methamphetamine in addition to some opiates. Thirty-five percent of the neonates had abnormal ECHO studies, with the highest incidence in the stimulant (cocaine-methamphetamine) subgroup. Lesions suggestive of prior hemorrhagic or ischemic injury with cavitation, located anterior and inferior to the lateral ventricles, in the frontal lobes and basal ganglia, were found in 8% of the infants. Intraventricular hemorrhage was found in 12% of those drug-exposed babies, subependymal hemorrhage in 11% and subarachnoid hemorrhage in 14%. In addition, ventricular dilatation suggesting diffuse atrophy was seen in 10% of the study infants. Finally, cerebral infarction was evident in two of the cocaine-exposed infants. Despite these impressive ECHO studies, infants with abnormal findings did not display identifiable neurobehavioral abnormalities during the newborn period. This is not totally surprising since the location of the lesions may indicate damage which would be detectable when the infant is older and challenged with more complicated visual-motor and cognitive tasks.

Isenberg (1987) reports another possible sign of cocaine exposure in the neonate. When compared to 36 non-cocaine-exposed infants, he found a greater incidence of dilated or tortuous iris vessels in 13 cocaine-exposed infants. Like the previously mentioned signs, this abnormality of iris vasculature gradually resolved without apparent visual sequelae. Isenberg acknowledges that other conditions may yield similar findings; however, the presence of this sign should alert the physician to rule out cocaine intoxication when an infant presents with dilated and tortuous iris vasculature.

Maternal prenatal cocaine use may place the infant at a potentially increased risk for Sudden Infant Death Syndrome (SIDS) although the risk of SIDS in cocaine-exposed infants is not known. The incidence of SIDS in the general population of the U.S. is about 0.5%. In Chasnoff's retrospective study on cocaine and pregnancy, he found the SIDS rate for cocaine-exposed infants to reach nearly 15% as 10 of the 66 cocaine-exposed infants reviewed died of SIDS. Autopsies of all infants revealed no other causes of death. In another study of infants born to 50 mothers maintained on methadone and not using cocaine during pregnancy, the rate of SIDS was 4%. This is comparable to the previously reported five- to tenfold increase in SIDS among infants born to opiate-abusing women (Ryan *et al.*, 1987).

In contrast to these findings, Bauchner *et al.* (1988) found no increased incidence of SIDS in cocaine-exposed infants (1/174; 5.6/10,000) compared to controls (4/

821; 4.9/1000), although both incidences were increased over nationwide rates in their low socioeconomic Boston population. Recently Kandall *et al.* (1991) reviewed SIDS in drug-exposed infants in a population of about 1.21 million births in New York City between 1979 and 1989. Once other high-risk variables (ethnicity, low maternal age, socioeconomic status, maternal cigarette smoking, low birth weight) were controlled, cocaine exposure resulted in only a very modest increase in the rate of SIDS. Many factors may play a role in the etiology of SIDS and therefore the direct relationship between cocaine and SIDS remains uncertain.

Large-scale epidemiologic studies are needed to separate the effect of cocaine itself from other drug-associated factors, e.g., low socioeconomic status, race, cigarette smoking, alcohol, and polydrug abuse, etc., which may impact on the SIDS rate.

Another new source of cocaine exposure for the neonate has been reported. Chasnoff detected cocaine in maternal breast milk following cocaine ingestion. This is a reasonable finding given cocaine's lipophilic properties which make it easily able to cross biologic membranes. This finding was discovered following presentation of a 2-week-old infant with signs of acute cocaine intoxication. The child was markedly irritable, vomiting, and having diarrhea. The mother noted that the infant's pupils were dilated, unfocused, and poorly responsive to light. The child had increased sucking reflex, hyperactive Moro reflex, increased symmetrical deep tendon reflexes, and a marked lability of mood (Abelson and Miller, 1985). Following admission, the infant's blood pressure, heart, and respiratory rates were noted to be elevated. These symptoms gradually waned over the course of the next 72 h. Maternal breast milk samples were found to be negative for cocaine and its metabolites by 36 h after the mother's last cocaine use, and the infant's urine sample was negative by 60 h after the last breast feeding. In light of this report, it is advisable to warn all drug-abusing women of the dangers of cocaine exposure when breast feeding the newborn (Chasnoff *et al.*, 1987).

Additionally, Chaney (1988) reported an infant who developed severe apnea and seizures following acute cocaine exposure while breast feeding. Apparently, the mother applied cocaine to her nipples for relief of soreness and then proceeded to breast feed without washing off her breasts. Three hours after breast feeding, the mother found her infant gasping and choking with limbs flexed and stiff. No other causes for the seizure could be found upon hospital admission. It is assumed that the infant both inhaled and ingested the cocaine, resulting in the convulsions. This incident again indicates the dangers of neonatal cocaine exposure.

In summary, the results of human studies suggest that maternal and infant outcome are generally poorer when women are drug dependent and using cocaine. A consistent trend is seen in the parameters that define fetal/neonatal and maternal outcome. In addition to the effects on dopaminergic neurotransmission, cocaine also acts on the CNS to increase release and prevent re-uptake of norepinephrine at nerve terminals (Ryan *et al.*, 1987). This results in increased norepinephrine levels resulting in tachycardia, vasoconstriction, and elevated blood pressure (Ryan *et al.*, 1987). Because of placental vasoconstriction, there is also decreased blood flow to the fetus (Schnoll *et al.*, 1984). Increased uterine contractility has also been shown

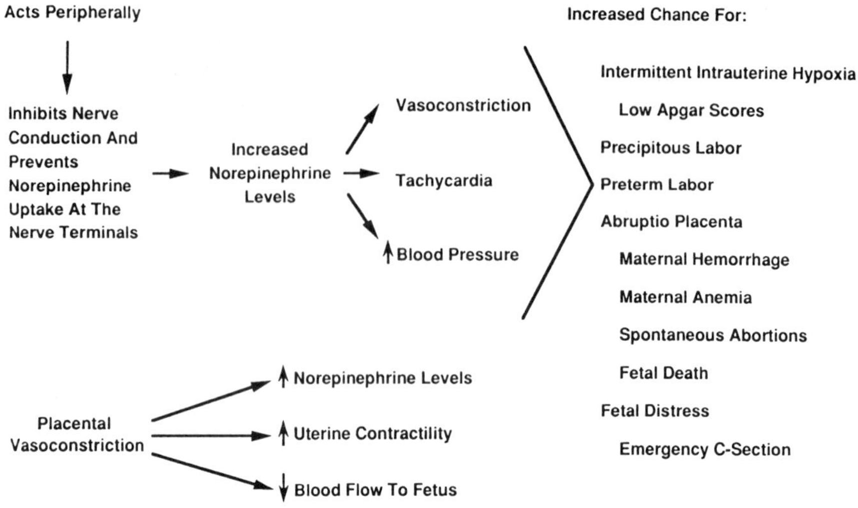

Figure 1. Actions of cocaine in pregnancy.

with elevated norepinephrine levels (Smith, 1988) and may be one of the factors predisposing the women to premature labor. Furthermore, the placental vasoconstriction may cause acute and chronic fetal hypoxia and result in fetal distress (possibly leading to emergency cesarean section), meconium staining of amniotic fluid (Stone *et al.*, 1984), low Apgar Scores, and small for gestational age infants. Abruptio placentae may occur with cocaine-induced hypertension (Wang and Schnoll, 1987), as suggested by the association between hypertension and premature separation of the placenta (Blair, 1973; Prichard *et al.*, 1970). Spontaneous abortions, fetal death, and premature labor have been related to placental abruption in several studies (see Figure 1) Chasnoff *et al.*; Bingol *et al.* (1987); Maguire (1986); Lifschitz *et al.*, 1986).

The lifestyle of drug-dependent women is erratic and hardly conducive to achieving an optimal outcome in pregnancy. The use of cocaine appears to adversely affect this already poor situation. Moreover, cocaine has been found to effectively suppress appetite and decrease the user's interest in fulfilling other bodily needs (Gold *et al.*, 1986). Because of the drug's highly addictive quality, the search for and use of the drug becomes the most important goal in the user's life. The ''high'' it produces is rapid and intense, but it is followed by a deep ''crash''. This leads the user to seek other drugs to come down from the intense stimulation of cocaine and soften the crash (Wilbur, 1986). Therefore, pregnant, drug-dependent women who abuse cocaine are at high risk because their lifestyles are dominated by drug-seeking behavior and use throughout pregnancy, frequent polydrug abuse, poor nutrition, and inadequate prenatal care. This combination of factors probably accounts for lower birth weight and length, head circumference, and gestational age found in the cocaine-exposed infants and the general overall increased perinatal morbidity and mortality.

In order to meet the needs of the pregnant, cocaine-dependent woman and her child, it is essential to further define treatment modalities. As in the opiate-dependent, pregnant woman and her child, the medical, psychological, and sociological issues must be addressed. Unfortunately, due to the lack of any pharmacological agent appropriate for the pregnant, cocaine-dependent woman, compliance in outpatient drug treatment programs is negligible. In an effort to decrease the tremendous monies that are currently being expended as a result of perinatal morbidity, child abuse, and the many socioeconomic problems in this population, it is essential to develop comprehensive residential facilities for the women and their children. Although, on the surface, these recommendations appear to be costly and difficult operationally, a comprehensive residential facility is our only option at this time in order to eliminate the vast problems and the growing epidemic in cocaine drug-dependent, pregnant women and their children.

REFERENCES

H.I. Abelson and J.D. Miller. A decade of trends in cocaine use in the household population. *NIDA Research Monograph Series.* **61**: 35 (1985).

D. Acker, B. Sachs, K. Tracy, and W. Wise. Abruptio placentae associated with cocaine use. *Am. J. Obstet. Gynecol.* **146(2)**: 220 (1983).

H. Bauchner, B. Zuckerman, M. McClain, D. Frank, L.E. Fried, and H. Kayne. Risk of sudden infant death syndrome among infants with *in utero* exposure to cocaine. *J. Pediatr.* **11**: 831 (1988).

L.V. Baxi and R. Petrie. Pharmacologic effects on labor: effect of drugs on dystocia, labor and uterine activity. *Clin. Obstet. Gynecol.* **30(1)**: 19 (1987).

N. Bingol, M. Fuchs, V. Diaz, R.K. Stone, and D.S. Gromisch. Teratogenicity of cocaine in humans. *J. Pediatrics* **119(1)**: 93 (1987).

R.G. Blair. Abruption of the placenta: a review of 189 cases occurring between 1965 and 1969. *J. Obstet. Gynecol.,* (Br. Commonwealth). **80**: 242 (1973).

N.E. Chaney, J. Frank, and W.B. Wadlington. Cocaine convulsions in a breast-feeding baby. *J. Pediatrics* **112(1)**: 134 (1988).

I.J. Chasnoff, G.M. Chisum, and W.E. Kaplan. Maternal cocaine use and genitourinary malformations. *Teratology* **37**: 201 (1988).

I.J. Chasnoff, K.A. Burns, and W.J. Burns. Cocaine use in pregnancy: perinatal morbidity and mortality. *Neurotoxicol. Teratol.* **9**: 291 (1987).

I.J. Chasnoff, D.R. Griffith, S. MacGregor, S.K. Dirkes, and K.A. Burns. Temporal patterns of cocaine use in pregnancy. *JAMA* **261**: 1741 (1989).

I.J. Chasnoff, D.E. Lewis, and L. Squires. Cocaine intoxication in a breast fed baby. *Pediatrics* **680(6)**: 836 (1987).

I.J. Chasnoff. Perinatal effects of cocaine. *Contemporary Obstet. Gynecol.* **29**: 163 (1987).

I.J. Chasnoff, W.J. Burns, S.H. Schnoll, and K.A. Burns. Cocaine use in pregnancy. *N. Engl. J. Med.* **313(11)**: 666 (1985).

I.J. Chasnoff, M.E. Bussey, R. Savich, and C.M. Stack. Perinatal cerebral infarction and maternal cocaine use. *J. Pediatrics* **108(3)**: 456 (1986).

P. Cherukuri, H. Minkoff, J. Feldman, A. Parekh, and L. Glass. A cohort study of alkaloidal cocaine ("crack") in pregnancy. *Obstet. Gynecol.* **72**: 147 (1988).

M.W. Church, B.A. Dintcheff, and P.K. Gessner. Dose dependent consequences on pregnancy outcome in the long-evans rat. *Neurotoxicol. Teratol.* **10**: 51 (1988).

H.O. Critchley, S.M. Woods, A.J. Barson, T. Richardson, and B.A. Lieberman. Fetal death in utero and cocaine abuse. Case report. *Br. J. Obstet. Gynecol.* **95**: 195 (1988).

S.D. Dixon and R. Bejar. Echoencephalographic findings in neonates associated with maternal cocaine and methamphetamine use: incidence and clinical correlates. *J. Pediatr.* **115**: 770 (1989).

T.M. Doberczak, S. Shanzer, R.T. Senie, and S.R. Kandall. Neonatal neurologic and encephalographic effects of intrauterine cocaine exposure. *J. Pediatr.* **113**: 354 (1988).

T.M. Doberczak, J.C. Thornton, J. Bernstein, and S.R. Kandall. Impact of maternal drug dependency on birth weight and head circumference of offspring. *Am. J. Dis. Child.* **141**: 1163 (1987).

T.W. Estroff, C.A. Dackis, D.E. Sweeney, M.S. Gold, and A.L.C. Potash. The vitamin deficiencies of cocaine abuse, unpublished data (1983).

A.G. Fantel and B.J. MacPhail. The teratogenicity of cocaine. *Teratology* **26**: 17 (1982).

L. Finnegan. Neonatal abstinence syndrome: assessment and pharmacotherapy. *Neonatal Therapy: An Update.* F.F. Rubaltelli and B. Granati, eds., Elsvier, New York, p. 122 (1986).

M.W. Fischman, C.R. Schuster, L. Resnekov, J.F.E. Schick, N.A. Krasnegor, W. Fennell, and D.X. Freedman: Cardiovascular and subjective effects of intravenous cocaine administration in humans. *Arch. Gen. Psych.* **33**: 983 (1976).

A.A. Flandermeyer. A Comparison of the Effects of Heroin and Cocaine Abuse Upon the Neonate. *Neonatal Network.* 42-8, December (1987).

R. Fulroth, B. Phillips, and D.J. Durand. Perinatal outcome of infants exposed to cocaine and/or heroin *in utero*. *Am. J. Dis. Child.* **143**: 905 (1989).

M.S. Gold, A.M. Washton, C.A. Dackis, and J.C. Chatlos. New treatments for opiate and cocaine users: but what about marijuana? *Psychiatr. Ann.* **16**: 206 (1986).

A.G. Goodman, L.S. Goodman, and A. Gilman, eds., *The Pharmacologic Basis of Therapeutics.* Macmillan, New York. (1980).

A.J. Hadeed and S.R. Seigel. Maternal cocaine use during pregnancy: effect on the newborn infant. *Pediatrics* **84**: 205 (1989).

C.E. Henderson and M. Torbay. Rupture of Intracranial Aneurysm Associated with Cocaine Use During Pregnancy. *Am. J. Perinatal.* **5(2)**: 142 (1988).

H.E. Hoyme, K.L. Jones, S.D. Dixon, T. Jewett, J.W. Hanson, L.K. Robinson, M.E. Msall, and J.E. Allanson. Prenatal cocaine exposure and fetal vascular disruption. *Pediatrics* **85**: 753 (1990).

S.T. Isenberg, A. Spicer, and S.H. Inkelis. Ocular signs of cocaine intoxication in neonates. *Am. J. Opthamal.* **103**: 211 (1987).

S.R. Kandall and J. Gaines. Maternal substance use and subsequent sudden infant death syndrome (SIDS) in offspring. *Neurotox. Teratol.* **13**: 235 (1991).

T.A. Kosten. Cocaine attentuates the severity of naloxone-precipitated opioid withdrawal. *Life Sci.* **47**: 1617 (1990).

L.D. Kramer, G.E. Locke, A. Ogunyemi, and L. Nelson. Neonatal cocaine-related seizures. *J. Child Neurol.* **5**: 60 (1990).

P.E. LeBlanc, A.J. Parekh, B. Naso, and L. Glass. Effects of intrauterine exposure to alkoloidal cocaine (crack). *AJDC* **141**: 937 (1987).

M.H. Lifschitz, F. Walters, and G.S. Wilson. Perinatal effects of cocaine abuse. *Pediatric Res.* **20**: 206 (1986).

S. Livesay, S. Ehrlich, and L. Finnegan. Cocaine and pregnancy: maternal and infant outcome. *Pediatric Res.* **21**: 387 (1987).

S.N. MacGregor, L.G. Keith, I.J. Chasnoff, M.A. Rosner, G.M. Chisum, P. Shaw, and J.P. Minogue. Cocaine use during pregnancy: adverse perinatal outcome. *Am. J. Obstet. Gynecol.* **157(3)**: 686 (1987).

J.D. Madden, T.F. Payne, and S. Miller. Maternal cocaine abuse and effect on the newborn. *Pediatrics* **77(2)**: 209 (1986).

M.P. Mahalik, R.F. Gautein and D.E. Mann. Teratogenic potential of cocaine hydrochloride in CF-1 mice. *J. Pharm. Sci.* **69(6)**: 703 (1980).

R. Maquire. Document coke's harm to fetus. *Medical Tribune* **27**: 17 (1986).

H.S. McGaughey, F.I. Corey, D. Westwood, and W.N. Thornton. Effect of synthetic anesthetics on spontaneous mobility of human uterine muscle *in vitro*. *Obstet. Gynecol.* **19**: 233 (1962).

J. Mellott, K. Kaltenbach, and L. Finnegan. Evaluaiton of neonatal outcome after in-utero exposure to cocaine. *Pediatric Res.* **23**: (1989).

M.E. Miller, M. Higgenbottom, and D.W. Smith. Short umbilical cord: its origin and relevance. *Pediatrics* **67**: 618 (1981).

A.C. Moessinger, W.A. Blanc, P.A. Marone, and D.C. Polsen. Umbilical cord length as an index of fetal activity: experimental study and clinical implications. *Pediatr. Res.* **16**: 109 (1982).

T.R. Moore, J. Sorg, L. Miller, T.C. Key, and R. Resnik. Hemodynamic effects of intravenous cocaine on the pregnant ewe and fetus. *Am. J. Obstet. Gynec.* **155(4)**: 833 (1986).

A.S. Oro and S.D. Dixon. Perinatal cocaine and methamphetamine exposure: maternal and neonatal correlates. *J. Pediatrics* **111(4)**: 571 (1987).

M.A. Plessinger and J.R. Woods. The cardiovascular effects of cocaine use in pregnancy. *Reprod. Toxicol.* **5(2)**: 99 (1991).

"Perinatal Toxicity of Cocaine", The Medical Letter, **30(767)**: 59 (1988).

J.A. Prichard, R. Mason, M. Corley, and S. Pritchard. Genesis of severe placental abruption. *Am. J. Obstet. Gynecol.* **108(1)**: 22 (1970).

C.R. Rosenfeld and J. West. Circulatory response to systemic infusion of norepinephrine in the pregnant ewe. *Am. J. Obstet. Gynecol.* **127(4)**: 376 (1977).

L. Ryan, S. Ehrlich, and L. Finnegan. Cocain abuse in pregnancy: effects on fetus and newborn. *Neurotoxicol. Teratol.* **9**: 295 (1987).

P.M. Scher, R.G. Almirez, R.W. Steger, and C.G. Smith. The effects of cocaine on reproductive hormones in the primate (abs). *Pharmacologist* **24**: 185 (1982).

S.H. Schnoll, A.N. Daghestani, and T.R. Hanses. Cocaine dependence. Research Staff Physician **30**: 24 (1984).

N.S. Shah, D.A. May, and J.D. Yates. Disposition of levo-^{3}H cocaine in pregnant and non-pregnant mice. *Toxicol. Appl. Pharmacol.* **53**: 279 (1980).

H. Silver, R. Wapner, M. Lorez-Vega, and L. Finnegan. Addiction in pregnancy: high risk intrapartuym management and outcome. *J. Perinatol.* Aug/Sept. (1987).

C.G. Smith and R.H. Asch. Drug abuse and reproduction. *Fertility Sterility* **48(3)**: 355 (1987).

J. Smith. The dangers of prenatal cocaine use. *MCN.* **13**: 174, May/June, (1988).

N. Stone, M. Fromme, and D. Kagan. *Cocaine: Seduction and Solution.* Clarkson, N. Plotter, New York. (1984).

C.H. Wang and S.H. Schnoll. Prenatal cocaine use associated with down regulation of receptors in human placenta. *Neurotoxicol. Teratol.* **9**: 301 (1987).

D.R. Wesson and D.E. Smith. CNS Stimulants. Treatment Aspects of Drug Dependence, A. Schecter, ed., CRC Press, Boca Raton, FL (1978).

R.A. Wilbur. A drug to fight cocaine. *Science 86* **7**: 42 (1986).

J.R. Woods, M.A. Plessinger, and K.E. Clark. Effect of cocaine on uterine blood flow and fetal oxygenation. *JAMA* **257(7)**: 957 (1987).

B. Zuckerman, D. Frank, R. Hingson, H. Amaro, S.M. Levenson, H. Kayne, S. Parker, R. Vinci, K. Aboagye, L.E. Fried, H. Cabral, R. Timperi, and H. Bauchner. Effects of maternal marijuana and cocaine use on fetal growth. *N. Engl. J. Med.* **320**: 762 (1989).

19

Treatment Strategies for Cocaine Detoxification

CHARLES A. DACKIS
Medical Director
Hampton Hospital
Rancocas, New Jersey
and
Instructor in Psychiatry
College of Physicians and Surgeons
Columbia University
New York, New York

MARK S. GOLD
Director of Research
Fair Oaks Hospital
Summit, New Jersey

INTRODUCTION

Cocaine addiction can be viewed from societal, psychological, behavioral, and biological perspectives in an attempt to better understand its nature and contain its spread. The biological forces and dynamics that drive compulsive cocaine use include the acute pleasurable effects of intoxication, and protracted neurochemical disruptions resulting from chronic cocaine exposure. Addicts attempting to become abstinent are often subject to severe craving, withdrawal symptoms, and obsessions about cocaine euphoria. These symptoms result in part from abnormal brain function acquired from chronic cocaine use. For instance, cocaine euphoria results from the acute pharmacological action of cocaine on the brain's pleasure centers, and cocaine withdrawal and craving appear to result from catecholamine depletion. Strategies that normalize cocaine-induced brain disruptions are gaining increasing attention as these abnormalities and their clinical ramifications are further identified and understood. Effective pharmacological treatments would greatly enhance behavioral and psychological approaches to the cocaine addict and improve recovery rates in this potentially fatal disorder. This chapter will discuss current and future detoxification strategies for cocaine addicts with a focus on the normalization of clinically relevant neurochemical alterations.

COCAINE EUPHORIA AND DOPAMINE ACTIVATION

Cocaine euphoria is a logical starting point to discuss the biological aspects of cocaine addiction. The pharmacological action of cocaine on the brain produces an immediate feeling of euphoria in naive and chronic users. Cocaine addicts consistently describe acute intoxication as a ''rush'' of pleasure lasting for several seconds, followed by a lower level but more sustained feeling of euphoria. The initial pleasure is so intense that spontaneous ejaculation without genital stimulation has even been described after i.v. cocaine administration (Dimijian, 1974). Cocaine intoxication involves pleasure that is far outside the normal range of human expeience, and its reinforcing properties exceed those of other pleasurable activities. This explains why abuse continues in spite of exorbitant cost, psychosocial deterioration, and associated medical and psychiatric hazards.

Other effects of intoxication include elevated energy and libido, hyperkinesis, decreased appetite and sleep requirement, increased self-confidence, and enhancement of alertness and sensory experience. Physiological effects of hypertension, tachycardia, diaphoresis, and mydriasis coincide with the subjective state, and with blood levels of the drug. The initial pleasure of cocaine intoxication is fleeting, and euphoria gives way to irritability, ''crashing'' mood, and craving for more cocaine as the positive subjective effects wear off. In addition to craving, the memory of euphoria reinforces subsequent drug use.

Stimulant addicts often find they derive less pleasure from cocaine as their addiction progresses. This phenomenon may lead addicts to increase their frequency of use or change the route of administration (injecting or smoking) in an attempt to recapture previously attained euphoria (Siegel, 1985). The loss of pleasure suggests that some degree of disruption in the pleasure centers probably occurs to explain the diminished potency of stimulants over time. Also, since stimulant addicts frequently complain of generalized boredom and disinterest, their overall capacity to experience enjoyment might be impaired as a result of cocaine exposure.

Because of their intimate involvement in cocaine reward and euphoria, the discussion on cocaine euphoria centers on dopamine (DA) systems in the brain. Localized ''pleasure centers'' were first described as anatomically discrete brain areas that support electrical self-stimulation in animals (Olds, 1956). Central stimulants such as cocaine and amphetamine were also found to stimulate these ''pleasure centers'' and the concept of endogenous reward systems in the brain was developed and researched. Early studies found that cocaine is vigorously self-administered by animals to the point of serious toxicity or death (Deneau *et al.*, 1969) and preferred over food, water, and mating (Pickens and Harris, 1968). The ability of cocaine to override these survival drives is a dramatic indication of its power over behavior. Animals are more likely to administer cocaine than attend to basic survival drives because the stimulation of endogenous reward centers by cocaine "tricks" the organism into thinking a survival drive is being satiated.

We now know that the brain's reward centers are rich in DA-containing neurons, and that DA activation forms the neurochemical basis of natural and cocaine-induced reward (Wise, 1978). The DA theory of reward is based on a number of findings that are relevant to this discussion. Selective DA receptor antagonists eliminate cocaine self-administration by animals (DeWit and Wise, 1977), and attenuate the euphoria of central stimulants in human subjects (Gunne *et al.*, 1972). Direct DA agonists such as apomorphine and piribedil are self-administered by animals, although not nearly to the degree seen with direct DA agonists (such as cocaine, amphetamine, or normifensine) that block the synaptic re-uptake of DA. The increased reward potency of indirect DA agonists is consistent with the role of DA in stimulant reward, but inconsistent with the simple notion that DA neurotransmission alone forms its basis.

Anatomical evidence is also consistent with the role of DA circuits in the mediation of cocaine reward. Reward pathways have been extensively delineated and include descending fibers from the lateral hypothalamus that project to the ventral tegmentum via the medial forebrain bundle (Wise, 1985). These myelinated fibers descend and terminate directly on DA neurons in the ventral tegmentum, which is precisely the location of major DA cell groups (Corbett and Wise, 1980) with which reward fibers make direct synaptic contact (Fouriezos and Wise, 1976). The ventral tegmentum, in turn, projects to several other DA areas in the striatum, limbic system, and cortex. Specific lesions of these pathways disconnect the reward circuitry. Neurotoxin lesions of the DA-containing nucleus accumbens attenuates the self-administration of amphetamine (Lyness *et al.*, 1979), as does lesioning the tract between the ventral tegmentum and its limbic or cortical projections (Roberts and Koob, 1977). In addition, animals selectively self-administer cocaine into the nucleus accumbens (Goeders and Smith, 1983). Central stimulant reward therefore depends on the functional integrity and activation of discrete and localized DA structures in the brain that have evolved to insure survival behaviors.

Ascribing behavioral and subjective phenomena to the activation or disruption of specific neurotransmitter systems is an important step in identifying effective pharmacotherapies for cocaine addicts. For instance, we have outlined how DA activation by cocaine is largely responsible for the euphoria associated with cocaine intoxication. This finding suggests that DA antagonists might block cocaine euphoria and have clinical utility similar to that of naltrexone with opiate addicts (Gold *et al.*, 1982). Other examples of DA-related symptoms seen during cocaine intoxication are hyperkinesis, psychotic symptoms (Post, 1975), and hypersexuality (Cocores *et al.*, 1978). The next section will review DA alterations found during cocaine withdrawal. It is implicit that pharmacological agents capable of reversing these alterations might be effective in the management of cocaine withdrawal symptoms.

COCAINE CRAVING AND DOPAMINE DEPLETION

Cocaine addicts prefer cocaine euphoria over other pleasurable activities and

Table 1
DA Disruptions Caused by Cocaine

Acute Effects
DA re-uptake blockade (acute DA stimulation)
Increased synaptic DA metabolism (acute DA depletion)
Intraneuronal DA metabolism (acute DA depletion)
Chronic Effects
Decreased brain DA levels
Increased DA binding sites (compensated depletion)
Increased tyrosine hydroxylase activity (increased synthesis)
Hyperprolactinemia (decreased DA functional tone)
Decreased DA metabolites (HVA)

acquire a craving for the drug that resembles but surpasses the intensity of natural cravings for food, sex, and thirst alleviation. Interestingly, DA circuits are also involved in these natural drive states (Wise and Bozarth, 1987). During cocaine binges, the craving for more cocaine (often referred to as ''cocaine hunger'') develops rapidly as the drug's effects wear off. This acute craving is often the first withdrawal symptom encountered by addicts. This section will outline the evidence that chronic cocaine exposure leads to DA depletion, which in turn gives rise to cocaine craving. In this way, cocaine craving might be viewed as an acquired natural drive. Similarly, the subjective experience of craving represents a neurochemical imbalance that is susceptible to pharmacotherapy and detoxification with DA agonists.

Cocaine acutely blocks the synaptic re-uptake of DA (Ross and Renyi, 1966), which results in the flooding and stimulation of the postsynaptic DA receptors. This acute effect is consistent with the rewarding quality of cocaine intoxication. Conversely, several lines of evidence indicate that chronic cocaine exposure depletes brain DA, leading to the functional inhibition of dopamine circuits (see Table 1). Brain DA in the prefrontal cortex is persistently reduced (23%) in rats exposed to repeated cocaine administrations (Wyatt *et al.*, 1988). This direct measurement is supported by a number of indirect findings. DA binding sites are increased in the caudate nucleus after chronic exposure to cocaine (Taylor *et al.*, 1979), which is a compensatory response to decreased DA availability at the postsynaptic receptor. Increased D_2 DA receptors were also found in the accumbans after cocaine exposure (Goeders and Kuhar, 1987). Another compensatory response to DA depletion is an increase in tyrosine hydroxylase activity, leading to increased DA synthesis after cocaine administration (Patrick and Barchas, 1977). In human subjects exposed to cocaine, hyperprolactinemia (Dackis *et al.*, 1984; Mendelson *et al.*, 1987) and decreased homovanillic acid (HVA) (Extein *et al.*, 1987) have been reported and are consistent with functional DA depletion. Hyperprolactinemia may lead to sexual dysfunction with infertility and amenorrhea in females and impotence in male cocaine abusers (Corcores *et al.*, 1978). Hyperprolactinemia and HVA reductions gradually worsen for several weeks after cocaine abstinence, suggesting that DA depletion evolves and persists over this period of time.

DA depletion by cocaine has been theorized to result from the blockade of DA re-uptake from the synapse (Dackis *et al.*, 1985). After re-uptake, DA is normally stored in secretory vesicles and reutilized by the presynaptic neuron. DA re-uptake blockade by cocaine prevents this ''recycling'' by trapping DA in the synapse where it is exposed to metabolism. In fact, high levels of the synaptic metabolite 3-methoxytyramine have been reported after cocaine administration (DiGiulio *et al.*, 1971). Cocaine also inhibits vesicle binding of DA, leading to intraneuronal metabolism (Carlsson *et al.*, 1963). The step from neurochemical alterations to clinical symptomatology is speculative but necessary in order to achieve practical treatments. Similar to the ascribing of cocaine reward to DA activation (Wise, 1978), we hypothesized that DA depletion forms the neurochemical basis of cocaine craving (Dackis and Gold, 1985). The anticraving efficacy of bromocriptine, a DA receptor agonist, further supports the DA depletion hypothesis of stimulant craving (Dackis and Gold, 1985a) and will be discussed later.

Craving alternates with euphoria to form a cycle of addiction that becomes increasingly entrenched and uncontrollable. Developing gradually over time, craving is a crucial clinical phenomenon because it leads directly to continued use or recidivism through negative reinforcement. Cocaine craving can be separated into at least three subtypes. Acute craving begins minutes after the intoxication state abates. The user experiences a rapid drop in euphoria that gives way to irritability, depression, and a desperate urge to avert the evolving dysphoria through the use of more drug. Acute craving perpetuates cocaine binges as addicts typically consume their entire available supply to postpone the unavoidable ''crash''.

Intermediate craving begins after the cocaine ''crash'' has resolved and decreases irregularly over a period of several weeks. Most cocaine addicts are unable or unwilling to remain abstinent during this period of diminution. Intermediate craving may correlate with the time course of biological abnormalities such as hyperprolactinemia (Mendelson *et al.*, 1987), DA depletion (Wyatt *et al.*, 1988), and depressed HVA levels (Extein *et al.*, 1987), and might therefore be most amenable to DA agonist treatment. The final subtype is provoked craving, which is triggered by conditioned cues of the drug environment such as freebase pipes, mirrors, or the actual sight of cocaine. These cues are capable of producing intense craving even after long periods of abstinence (Dackis *et al.*, 1985).

The role of DA in cocaine euphoria and craving is an exciting example of how neurochemical effects can underlie clinical symptoms and identify potential pharmacological treatments. The effects of cocaine on other neurotransmitter systems are much less correlated with clinical syndromes and less researched at the present time. In addition, there are obviously many effects of this substance that have yet to be identified. Since cocaine's effects on norepinephrine (NE) and serotonin (5-HT) systems may contribute to cocaine withdrawal, these will be briefly reviewed in the next sections.

COCAINE ACTIONS ON NOREPINEPHRINE AND SEROTONIN

Norepinephrine

Cocaine blocks norepinephrine (NE) re-uptake from the synapse (Ross, 1966),

Table 2
NE Disruptions Caused by Cocaine

Acute Effects
NE re-uptake blockade (acute NE stimulation)
Increased synaptic NE metabolism (acute NE depletion)
Intraneuronal NE metabolism (acute NE depletion)
Chronic Effects
Increased β-adrenergic binding sites (compensated depletion)
Decreased brain NE levels
Increased tyrosine hydroxylase activity (increased synthesis)
Decreased NE metabolites (MHPG)

analogous to its major acute DA effect. This action leads to an acute potentiation of NE neurotransmission. Although neurotransmission is potentiated, the firing rates of presynaptic NE neurons decrease with acute cocaine administration (Pitts and Marwah, 1986). This appears to result from α_2-adrenergic receptor feedback inhibition and possibly the activation of inhibitory epinephrine neurons by cocaine (Hokfelt *et al.*, 1973). Acute cocaine administration therefore appears to inhibit presynaptic NE neurons, while indirectly stimulating their postsynaptic receptor sites.

Several studies suggest that NE depletion occurs after cocaine exposure. After acute administration of cocaine, brain NE levels increase but then fall below their baseline (Pradhan *et al.*, 1978). In human cocaine addicts, the NE metabolite 3-methoxy-4-hydroxy-phenylglycol (MHPG) is decreased (Tennant, 1985). Compensatory indications of NE depletion have also been reported. These include postsynaptic supersensitivity (increased populations of β-adrenergic receptors), increased cyclic-AMP levels after cocaine exposure (Banerjee *et al.*, 1979), and increases in tyrosine hydroxylase activity with chronic cocaine exposure (Ho *et al.*, 1977). Based on these data, it would appear that cocaine acutely stimulates but chronically inhibits NE circuits, analogous to its actions on DA systems (see Table 2).

As seen with DA, acute activation of NE circuits might underlie several effects of cocaine intoxication. These include hypertension, tachycardia, mydriasis, diaphoresis, vasoconstriction, and tremor. In fact, these noradrenergic symptoms can be replicated in monkeys by the direct electrical stimulation of NE nuclei (Redmond *et al.*, 1976). Chronic NE depletion in cocaine abusers might explain the frequent occurrence of depressive symptoms such as sleep disturbance, anergia, decreased libido, suicidality, anhedonia, and poor concentration (Gawin *et al.*, 1988). NE depletion by cocaine implies that direct NE agonists might have efficacy in the treatment of cocaine abstinence symptoms.

Serotonin

Cocaine's acute actions on serotonin (5-HT) neurons are largely inhibitory (see Table 3). Sudden reductions in brain 5-HT levels occur after the administration of cocaine, which is opposite to the acute increases seen with NE and DA (Pradhan

Table 3
Cocaine Actions on Serotonin (5-HT) Neurons

Acute Effects	
	5-HT re-uptake blockade
	Reduced brain 5-HT levels
	Reduced 5-HT synthesis (decreased precursor availability)
	Antagonism of peripheral 5-HT receptors
Chronic Effects	
	Reduced 5-HT metabolites (5-HIAA, 5-hydroxytryptamine)
	Decreased tryptophan hydroxylase activity (reduced 5-HT synthesis)

et al., 1978). Levels of 5-hydroxyindoleacetic acid (5-HIAA), a 5-HT metabolite, are also reduced after acute cocaine administration (Taylor and Ho, 1977). Reductions of 5-HT by cocaine may result from the inhibition of 5-HT synthesis. Cocaine interferes with the active uptake of tryptophan (Knapp and Mandell, 1972), leading to decreased 5-HT precursor supply. In addition, the tryptophan hydroxylase is directly inhibited by cocaine, further reducing the rate of 5-HT synthesis (Taylor and Ho, 1977). 5-hydroxytryptamine is also reduced by cocaine, indicating decreased 5-HT turnover (Friedman *et al.*, 1975). Inhibition of 5-HT synthesis by cocaine is opposite to the activation of NE and DA synthesis induced by cocaine.

The direct and indirect effects of cocaine on 5-HT postsynaptic receptors are somewhat contradictory. Cocaine blocks 5-HT re-uptake in animals (Friedman *et al.*, 1975) and from human platelets (Dackis *et al.*, 1985c) which would potentiate 5-HT neurotransmission. However, direct antagonism of peripheral 5-HT receptors has also been reported (Gyermek and Bindler, 1962). It is likely that the blockade of 5-HT re-uptake is negated by the many inhibitory effects of cocaine on the 5-HT system.

There are numerous acute and chronic disruptions of the 5-HT system by cocaine. Furthermore, 5-HT alterations alter behavioral changes in animals (Pradhan *et al.*, 1978a) and may produce symptomatology in cocaine users. The acute insomnia seen in cocaine intoxication may result in part from the acute inhibition of 5-HT neurons. Enhancement of sensory experience by cocaine is found with other 5-HT agents such as LSD-25. Sexual effects of cocaine may also result from 5-HT disruptions. Further research should clarify whether serotonergic agents have pharmacological roles in the treatment of cocaine addicts.

NEUROENDOCRINE EFFECTS: PROLACTIN AND THYROID AXIS

Before discussing pharmacological treatments of cocaine addiction, the thyroid and prolactin axes will be briefly reviewed with regard to cocaine. As in previous sections pertaining to neurotransmitters, this discussion will focus on endocrine imbalances that may be responsible for clinically significant symptoms. Delineating the acute and chronic effects of cocaine on these neuroendocrine systems should enhance our understanding of cocaine's actions on the brain, and point toward possible pharmacotherapies. A clear example of this process was how the finding of hyperprolactinemia in cocaine addicts led to the DA depletion hypothesis of cocaine craving, and the successful trials of bromocriptine for this condition (Dackis and Gold, 1985).

Prolactin

As previously discussed in relation to DA depletion, cocaine addicts often have hyperprolactinemia during cocaine withdrawal (Dackis *et al.*, 1984) which appears to persist for at least 4 weeks (Mendelson *et al.*, 1987). Since prolactin-secreting cells are tonically inhibited by tuberoinfundibular DA neurons (MacLeod, 1976), hypeprolactinemia is consistent with DA depletion by cocaine. The time course of hyperprolactinemia may correspond with the period of maximal craving and highest risk of recidivism. Patients with hyperprolactinemia may benefit the most from bromocriptine treatment.

Elevated prolactin has been reported to cause gynecomastia, galactorrhea, and contribute to the sexual dysfunction that frequently occurs in chronic cocaine addicts (Cocores *et al.*, 1978). Sexual problems described in cocaine addicts include infertility, impotence, and amenorrhea (Seigel, 1982). Sexual dysfunction in cocaine addicts might also be due directly to DA depletion, since DA is involved in sexual reward (Wise and Bozarth, 1987). Although cocaine initially enhances sexual drive (Dimijian, 1974), chronic addicts typically describe declining sexual performance over time, especially as their addiction progresses. This pattern is consistent with cocaine's acute activation but chronic inhibition of DA circuits in the brain's pleasure center.

Sexual dysfunction will be overlooked by clinicians unless a proper sexual interview is conducted. The prevalence of sexual problems in cocaine addicts is probably grossly underestimated. Clinicians should maintain a high level of suspicion for sexual dysfunction in these patients, particularly during the acute detoxification phase. Bromocriptine rapidly normalizes hyperprolactinemia and has been reported to be effective in treating sexual dysfunction in cocaine abusers (Cocores *et al.*, 1978). Since hyperprolactinemia produced by cocaine might contribute to impotence and infertility in cocaine addicts, prolactin levels should be obtained in symptomatic patients and bromocriptine treatment considered.

Thyroid Axis

Animal studies report thyroid axis activation by central stimulants. Monkeys treated with dextroamphetamine have elevated levels of thyroxine (T_4) and thyroid-stimulating hormone (TSH), and cocaine addicts have been found to have elevated T_4 levels during intoxication (Morley *et al.*, 1980). These findings are interesting because many clinical manifestations of cocaine intoxication such as hypertension, diaphoresis, tachycardia, tremor, anxiety, and hyperthermia are also found as signs and symptoms of hyperthyroidism. Such symptoms may occur, in part, due to thyroid axis stimulation by cocaine. Conversely, the cocaine "crash" shares many signs and symptoms with the hypothyroid state, such as anergia, depression, bradykinesis, weight gain, and hypersomnia.

Addicts withdrawing from cocaine have normal T_3, T_4, and TSH levels, but significantly blunted TSH responses to thyrotropin-releasing hormone (TRH) infusions (Dackis *et al.*, 1985d). This finding of TRH receptor down-regulation is consistent with compensated hyperthyroidism, probably in response to chronic activation of the thyroid axis by cocaine. Since TRH is stimulated by NE circuits (Reichlin, 1975), NE re-uptake blockade during cocaine intoxication might be the mechanism of thyroid activation. With the sudden withdrawal of cocaine, TRH stimulation would not only cease, but be diminished due to functional NE depletion. Reduced TRH stimulation would occur in the face of desensitized pituitary TRH receptors, further amplifying the hypothryroid state. Since even mild forms of hypothyroidism have been shown to be associated with anergia and depressive symptoms, thyroid axis inhibition may contribute significantly to the cocaine "crash".

THE COCAINE "CRASH"

Although cocaine craving and euphoria are the primary reinforcers that drive cocaine addiction, secondary reinforcers also contribute to the cycle of addiction. Probably the most important of these is cocaine withdrawal syndrome, or the cocaine "crash". Cocaine withdrawal was largely overlooked or minimized for years, reinforcing the false notion that only "psychological" dependence occurs with this agent. Central stimulant withdrawal was first described with amphetamines (Kilbey and Ellinwood, 1977) as persisting 1 to 3 d and including symptoms opposite to those found during intoxication. Symptoms of cocaine withdrawal typically include anergia, depression, amotivation, hypersomnia, decreased libido, irritability, poor concentration, and hyperphagia. Other somatic symptoms such as headache and nausea are sometimes present. Interestingly, the cocaine "crash" does not usually include craving for cocaine, which tends to precede and follow the withdrawal period (Gawin and Kleber, 1986).

Cocaine withdrawal is not as dramatic as abstinence associated with opiates or sedatives. Hypertension, tachycardia, diaphoresis, panic, and seizures are not seen in cocaine withdrawal because its neurochemical basis involves catecholamine depletion rather than rebound hyperactivity (Gold *et al.*, 1980). Stimulant withdrawal also varies with the route, frequency, and amount of cocaine use, much in the way opiate withdrawal varies with severity of opiate addiction.

Physical withdrawal symptoms contribute to the perpetuation of drug addiction, but the extent of their contribution is probably overstated. For instance, it is commonly recognized that alcoholics and heroin addicts often relapse after completing detoxification unless rehabilitation is also pursued. Similarly, with cocaine addicts, craving is a more powerful reinforcer than the wish to avoid the "crash". Furthermore, since cocaine withdrawal is relatively mild, resolves quickly, and is not life threatening, detoxification is not critical from a medical perspective. Pharmacological treatments that address craving and euphoria will therefore have a more important role than those that reverse the brief and often transient cocaine "crash". To the extent that craving results from DA depletion, it may in fact be viewed as a form of protracted withdrawal. Defining addiction in terms of "physical" withdrawal syndromes is really a misconception resulting from the notion that craving and euphoria are merely "psychological" phenomena of secondary importance. Actually, the neurochemical bases of euphoria and craving are "physical" events in the brain, and comprise the most essential dynamics of addiction.

PHARMACOLOGICAL APPROACHES TO COCAINE ADDICTION

Successful treatment of severe cocaine addiction often requires a supervised period of abstinence and detoxification. This allows the intoxication, "crashing", and protracted withdrawal effects of cocaine to wear off while the patient's denial is addressed through confrontation and peer-based treatment. Protracted withdrawal appears to include craving and depressive symptoms that might result from neurotransmitter and neuroendocrine alterations already reviewed in this chapter. "Crashing" symptoms comprising cocaine withdrawal serve to perpetuate the addiction through negative reinforcement and might be amenable to detoxification. In addition, since cocaine addicts are frequently dependent on other agents, detoxification from alcohol, opiates, benzodiazepines, and other sedative/hypnotic drugs is often necessary. Pharmacological approaches to cocaine withdrawal and craving are only in the initial stages of investigation, but appear to be useful. It should be emphasized, however, that pharmacological treatments are adjunctive and do not represent treatments of addiction in and by themselves. Rehabilitation and behavioral changes remain the most crucial ingredients of successful recovery.

Understanding cocaine-induced neurochemical disruptions can translate into the formulation of effective pharmacological interventions. Clonidine serves as an excellent example of this process. Once opiate withdrawal was understood to involve noradrenergic rebound hyperactivity, an effective treatment for this condition was identified by researching nonopiate agents like clonidine that suppress NE activity (Gold *et al.*, 1980). In the case of cocaine addiction, we have reviewed the evidence that DA mediates cocaine reward, and the findings of DA depletion with chronic cocaine exposure. These findings suggest that DA agents are most likely to be effective in the pharmacotherapy of cocaine addiction (Dackis *et al.*, 1985).

Preliminary research indicates rapid efficacy of DA agonists for cocaine craving and withdrawal. Bromocriptine, a specific DA receptor agonist, was first administered to cocaine addicts based on the DA depletion model of cocaine craving. We reported rapid anti-craving efficacy in open trials (Dackis and Gold, 1985; Dackis and Gold, 1985b) and a double-blind, placebo-controlled study using a single dose of bromocriptine (Dackis *et al.*, 1987). This finding has been replicated in open (Extein *et al.*, 1986; Kosten *et al.*, 1988) and double-blind studies (Tennant and Sagherian, 1987; Giannini *et al.*, 1987; Extein *et al.*, 1988). To date, no negative studies using bromocriptine for cocaine craving or withdrawal have been reported. Tennant and Sagherian (1987) found bromocriptine to be effective against craving and cocaine withdrawal, but encountered side effects using total oral doses of 15 mg daily. Side effects included nausea, headaches, and orthostatic hypotension. We (Dackis and Gold, 1985b) and Extein *et al.* (1986) found total daily doses of

Table 4
Bromocriptine Protocol

Day	Bromocriptine dose, mg	Frequency
1–2	1.25	b.i.d.
3–4	1.25	t.i.d.
5–6	2.5	b.i.d.
7–11	2.5	t.i.d.
12–14	5	b.i.d.
15–20	Decrease by approximately 50% every 2 d	

Note: Titration of the dose is necessary with anti-craving and anti-withdrawal effects weighed against side effects. Lower doses than listed above are often sufficient. Maintenance bromocriptine treatment may be necessary if symptoms recur when discontinuation is attempted.

2–5 mg to be effective without encountering significant side effects. DA receptor supersensitivity (Taylor *et al.*, 1979) could account for the need to use extremely low doses of bromocriptine in cocaine addicts with this DA agonist. It should be emphasized that bromocriptine reverses the major DA alterations caused by cocaine, namely DA receptor supersensitivity, DA depletion, and hyperprolactinemia. Its rapid and specific symptom reversal may result from its ability to normalize these neurochemical imbalances.

Tennant and Sagherian (1987) also reported efficacy against cocaine craving and withdrawal with amantadine, which releases DA into the synapse. This finding is consistent with the DA depletion hypothesis. Theoretically, amantadine should correct DA depletion acutely, but might worsen the imbalance chronically because its indirect agonist activity depends on the release of depleted DA stores. Further research should clarify the efficacy of bromocriptine and other DA agonist treatment.

The duration requirements of bromocriptine treatment in cocaine addicts has not been fully researched to date. Maintenance bromocriptine treatment may be necessary if craving or depressive symptoms recur after the medication is discontinued, and the time course of these clinical phenomena remain poorly understood. Bromocriptine may also block cocaine euphoria (Tennant and Sagherian, 1987). If this observation proves to be true, bromocriptine treatment would address both of the primary reinforcers of cocaine addiction, with craving alleviated and euphoria blocked. In this case, a long-acting, injectable form of the drug would be optimal to ensure continued compliance, particularly for the blocking of euphoria.

Tricyclic antidepressants have been reported by one research group (Gawin *et al.*, 1988) to be effective in the treatment of cocaine withdrawal. Beneficial effects occur after several weeks, making this treatment more appropriate for patients with persistent withdrawal or depression. In fact, given the high prevalence of depression in these patients, desipramine efficacy may be largely due to its antidepressant effect. Desipramine efficacy against craving was not confirmed by another group (Tenant and Rawson, 1983), although it was reported to have antidepressant action in cocaine addicts (Giannini *et al.*, 1986). It remains unclear whether the reported efficacy of desipramine represents improvement in co-existing affective illness or actual antiwithdrawal action. Interestingly, tricyclic antidepressants produce some of the same neurochemical disruptions as cocaine, such as blockade of DA and NE re-uptake and increased postsynaptic DA sensitivity (Spyraki and Fibiger, 1981). Since these cocaine-induced DA and NE neurochemical imbalances would be aggravated by desipramine, its clinical efficacy should be carefully scrutinized. In fact, a recent study has found increased rates of recidivism in patients treated with desipramine (Weiss *et al.*, 1988). Confirmation of desipramine efficacy in cocaine addicts awaits further research.

NEW AVENUES FOR PHARMACOTHERAPY

Blockade of Cocaine Euphoria

DA antagonists (neuroleptics) are often indicated during cocaine intoxication for the treatment of cocaine-induced psychotic symptoms such as paranoia and hallucinations. Since DA mediates cocaine reward, the blockade of DA receptors with neuroleptics may prove to have a role in the management of certain cocaine addicts. This strategy has been utilized with the opiate receptor antagonist, naltrexone (Gold *et al.*, 1982), which can serve as a deterrent for certain motivated opiate addicts during early recovery. Two studies (Gunne *et al.* 1972; Angrist *et al.*, 1974) have found that neuroleptics block amphetamine euphoria in human subjects. Although cocaine has not been studied in this regard, it closely shares neurochemical reward characteristics with amphetamine (Wise and Bozarth, 1987).

Even if neuroleptics are proven effective as euphoria-blocking agents, their use in cocaine abusers must be weighed against a number of complications. We have reported that neuroleptics increase craving when administered to cocaine addicts during their first several weeks of abstinence (Dackis *et al.*, 1985e). This effect is consistent with the DA depletion hypothesis of craving. Recently abstinent cocaine addicts also appear to be very sensitive to neuroleptic side effects (Dackis *et al.*, 1985e), perhaps due to a compromised DA system. The risk of tardive dyskinesia is another complication mitigating against neuroleptic treatment. Since these patients may experience significant side effects and craving when treated with DA antagonists, the efficacy and safety of neuroleptic treatment for the blockade of cocaine euphoria must be carefully researched. One study has reported euphoria blockade with bromocriptine (Tennant and Sagherian, 1987). Although this observation is only anecdotal, it is an exciting possibility since both craving and euphoria could be addressed and treated by one agent. Further research should examine whether maintenance treatment with DA agonists like bromocriptine may reduce cocaine's ability to produce euphoria.

Reversal of Sensitization

Most addictive agents produce tolerance, defined as a physiologically based requirement for increased dosage to maintain the drug's same intensity and duration of effect. With regard to certain effects, such as epileptogenesis, cocaine actually induces reverse tolerance, or sensitization. For instance, daily administration of a fixed dose of cocaine will begin to produce seizures after several days. This phenomenon, known as "kindling", occurs because the brain becomes sensitized to cocaine (Post *et al.*, 1976). Sensitization may also occur with psychotic symptoms and its mechanism has been proposed to involve the supersensitivity of adrenergic receptors in response to presynaptic depletion (Dackis and Gold, 1985). To the extent that supersensitivity involves DA receptor supersensitivity, bromocriptine might reverse the process since it down-regulates DA receptors (Roehrich *et al.*, 1987). Similarly, direct NE agonists may have a similar role with respect to the NE neuronal system.

Acute Cocaine Overdose

Although deaths by cocaine overdose continue on the uptrend, there is no current antidote for this life-threatening condition (Cregler and Mark, 1986). The lack of a "naloxone equivalent" frustrates clinicians attempting to resuscitate or maintain patients in cardiorespiratory arrest or status epilepticus from cocaine overdose. Recent work in animals has suggested that calcium channel blockers might be effective in managing certain cardiac effects of cocaine (Nahas *et al.*, 1985). Animal research also points toward hyperthermia as a major cause of death in cocaine overdose (Gold *et al.*, 1988). This finding suggests that ice baths or pharmacological treatments that reverse cocaine hyperthermia could prevent mortality in overdosed patients. As the mechanisms of cocaine overdose become better understood, more refined treatments for this condition should emerge. The treatment of acute overdose remains the most critical target, in terms of medical severity, and probably the least researched. Further work in this area is imperative.

CONCLUSIONS

This paper has reviewed neurochemical and neuroendocrine disruptions caused by the acute and chronic administration of cocaine, with emphasis on clinical relevance and symptomatology. It is clear that cocaine produces significant alterations of the pleasure centers, and these abnormalities may lead to withdrawal and craving states. Disruptions of endogenous reward centers are consistent with this drug being "physically" addictive. Furthermore, neurochemical effects of cocaine point toward pharmacotherapies that might have important roles in the clinical approach to this serious and expanding disorder.

Pharmacological treatments for cocaine addiction have been proposed to treat symptoms associated with protracted cocaine withdrawal. Two preliminary treatments, bromocriptine and desipramine, have emerged as possible medications for these conditions. It is important to conceptually separate withdrawal, craving, and depression because these may be distinct syndromes that respond to different treatments. Based on current research, bromocriptine appears to be most effective in treating acute withdrawal and craving, and is the only agent with rapid efficacy. Desipramine may be more effective in treating persistent depression, and its clinical effect occurrs after several days or weeks. Other pharmacological applications might be developed to treat cocaine euphoria, overdose, and sensitization. Agents that restore neurochemical and neuroendocrine balance should be further researched for use in the management of cocaine addiction.

REFERENCES

B. Angrist, H.K. Lee, and S. Gershon. The antagonism of amphetamine-induced symptomatology by a neuroleptic. *Am. J. Psychiatry* **131**: (1974).

S.P. Banerjee, V.K. Sharma, L.S. Kung-Cheung, et. al. Cocaine and d-amphetamine induce changes in central beta-adrenoceptor sensitivity: effects of acute and chronic drug treatment. *Brain Res.* **175**: 119 (1979).

A. Carlsson, N.A. Hillarp, and B. Waldeck. Analysis of the Mg^{++}-ATP dependent storage mechanism in the amine granules of the adrenal medulla. *Acta Physiol. Scand. Suppl.* **215**: 1 (1963).

J.A. Cocores, C.A. Dackis, and M.S. Gold. Sexual dysfunction secondary to cocaine abuse in two patients. *J. Clin. Psychiatry* **47**: 384 (1978).

D. Corbett and R.A. Wise. Intracranial self-stimulation in relation to the ascending dopaminergic systems of the midbrain: a movable electrode mapping study. *Brain Res.* **185**: 1 (1980).

L.L. Cregler and H. Mark. Medical complications of cocaine abuse. *N. Engl. J. Med.* **315**: 1495 (1986).

C.A. Dackis, M.S. Gold, T.W. Estroff, and D.R. Sweeney. Hyperprolactinemia in cocaine abuse. *Soc. Neurosci. Abst.* **10**: 1099 (1984).

C.A. Dackis and M.S. Gold. New concepts in cocaine addiction: the dopamine depletion hypothesis. *Neurosci. Biobev. Rev.* **9**: 469 (1985).

C.A. Dackis and M.S. Gold. Bromocriptine as a treatment of cocaine abuse. *Lancet* **1**: 1151 (1985a).

C.A. Dackis and M.S. Gold. Pharmacological approaches to cocaine addiction. *J. Substance Abuse Treatment* **2**: 139 (1985b).

C.A. Dackis, M.A. Pasternak-Dackis, D. Martin, A.L.C. Pottash, and M.S. Gold. Platelet serotonin transporter in cocaine patients. *National Institute of Drug Abuse Monograph Series* **55**: 164 (1985c).

C.A. Dackis, T.W. Estroff, D.R. Sweeney, A.L.C. Pottash, and M.S. Gold. Specificity of the TRH test for major depression in patients with serious cocaine abuse. *Am. J. Psychiatry* **142**: 1097 (1985d).

C.A. Dackis, M.S. Gold, R.K. Davies, and D.R. Sweeney. Bromocriptine treatment for cocaine abuse: the dopamine depletion hypothesis. *Int. J. Psychiatry Med.* **15**: 125 (1985e).

C.A. Dackis, M.S. Gold, D.R. Sweeney, J.P. Byron, and R. Climko. Single dose bromocriptine reverses cocaine craving. *Psychiatry Res.* **20**: 261 (1987).

G.A. Deneau, T. Yanagita, and M.H. Seevers. Self-administration of psychoactive substances by the monkey. *Psychopharmacologia* **16**: 30 (1969).

H. DeWit and R.A. Wise. A blockade of cocaine reinforcement in rats with the dopamine receptor blocker pimozide but not with the noradrenergic blockers phentolamine or phenoxybenzamine. *Can. J. Psychol.* **31**: 195 (1977).

A.M. DiGiulio, A. Groppetti, F. Cattabeni, *et al.* Significance of dopamine metabolites in the evaluation of drugs acting on dopamine neurons. *Eur. J. Pharmacol* **16**: 171 (1971).

G.G. Dimijian. Contemporary drug abuse. In: *Medical Pharmacology: Principles and Concepts.* 7th ed., A. Groth, ed., C.V. Mosby, St. Louis, p. 313 (1974).

I.L. Extein, D.A. Gross and M.S. Gold. Cocaine detoxification using bromocriptine. *Am. Psych. Assoc. Abstr.* **NR 70**: (1986).

I. Extein, W.Z. Potter, M.S. Gold, P. Andre, W.A. Rafuls, and D.A. Gross. Persistent neurochemical deficit in cocaine abuse. *Am. Psych. Assoc. Abstr.* **NR 61**: (1987).

I. Extein and M.S. Gold. The treatment of cocaine addicts: bromocriptine or desipramine. *Psych. Ann.* **18**: 535 (1988).

G. Fouriezos and R.A. Wise. Pimozide-induced extinction of intracranial self-stimulation: response patterns rule out motor or performance deficits. *Brain Res.* **103**: 37 (1976).

E. Friedman, S. Gershon, and J. Rotrosen. Effects of acute cocaine treatment on the turnover of 5-hydroxytryptamine in the rat brain. *Br. J. Pharmacol.* **54**: 61 (1975).

F.H. Gawin and H.D. Kleber. Abstinence symptomatology and psychiatric diagnosis in cocaine abusers. *Arch. Gen. Psychiatry* **43**: 107 (1986).

F.H. Gawin. Chronic neuropharmacology of cocaine: progress in pharmacotherapy. *J. Clin. Psychiatry* **49**: 11 (1988).

A.J. Giannini, P. Baumgartel, and L.R. DiMarzio. Bromocriptine therapy in cocaine withdrawal. *J. Clin Pharmacol.* **27**: 267 (1987).

A.J. Giannini, D.A. Malone, M.C. Giannini, W.A. Price, and R.H. Loiselle. Treatment of depression in chronic cocaine and phencyclidine abusers with desipramine. *J. Clin. Pharmacol.* **26**: 211 (1986).

M.S. Gold, A.L.C. Pottash, I. Extein, *et al.* Clonidine in acute opiate withdrawal. *N. Engl. J. Med.* **302**: 1421 (1980).

M.S. Gold, C.A. Dackis, A.L.C. Pottash, *et al.* Naltrexone, opiate addiction, and endorphins. *Med. Res. Rev.* **2(3)**: 211 (1982).

M.S. Gold, J.R.L. Ehrenkranz, J.M. Jonas, and C.A. Dackis. Treatment of cocaine toxicity. *Soc. Neurosci. Abst.* **14**: (1988).

N.E. Goeders and J.E. Smith. Cortical dopaminergic involvement in cocaine reinforcement. *Science* **221**: 773 (1983).

N.E. Goeders and J. Kuhar. *Alcohol Drug Res.* **7**: 207 (1987).

L.M. Gunne, E. Anggard, and L.E. Jonsson. Clinical trials with amphetamine blocking drugs. *Psychiatr. Neurol. Neurochir.* **75**: 225 (1972).

L. Gyermek and E. Bindler. Blockade of the ganglionic stimulant action of 5-hydroxytryptamine. *J. Pharmacol. Exp. Ther.* **135**: 344 (1962).

B.T. Ho, D.L. Taylor, V.S. Estevez, *et al.* Behavioral effect of cocaine — metabolic and neurochemical approach, in: *Cocaine and Other Stimulants.* E.H. Ellinwood and M.M. Kilbey, eds., Plenum Press, New York, p. 229 (1977).

T. Hokfelt, K. Fuxe, M. Goldstein, *et al.* Immunohistochemical evidence for the existence of adrenaline neurons in the rat brain. *Brain Res.* **66**: 235 (1973).

M.M. Kilbey and E.H. Ellinwood, eds. Chronic administration of stimulant drugs: response modification, in: *Cocaine and Other Stimulants,* Plenum Press, New York p. 410 (1977).

S. Knapp and A.J. Mandell. Narcotic drugs: effects on serotonin biosynthetic systems of the brain. *Science* **177**: 1209 (1972).

T.R. Kosten, B. Schumann and D. Wright. Bromocriptine treatment of cocaine abuse in patients maintained on methadone (letter). *Am. J. Psychiatry* **145**: 381 (1988).

W.H. Lyness, N.M. Friedle, and K.E. Moore. Destruction of dopaminergic nerve terminals in nucleus accumbens: effect of D-amphetamine self-administration. *Pharmacol. Biochem. Behav.* **11**: 553 (1979).

R.M. MacLeod. Regulation of prolactin secretion, in: *Frontiers in Neuroendocrinology.* L. Martini and W.F. Ganong, eds., Raven Press, New York, p. 169 (1976).

J.H. Mendelson, S.K. Tesh, U. Lange, N.K. Mello, R. Weis, and A.S.T. Skupny. Hyperprolactinemia during cocaine withdrawal, in: *Proceedings, 49th Annual Scientific Meeting. Committee on Problems of Drug Dependence,* (1987).

J.E. Morley, R.B.Shafer, M.K. Elson, M.F. Slag, M.J. Raleigh, G.L. Brammer, A. Yuwiler, and J.M. Hershman. Amphetamine-induced hyperthyroxinemia. *Ann. Intern. Med.* **93**: 707 (1980).

G. Nahas, R. Trouve, J.F. Demus, and M. Von Sitbon. A calcium-channel blocker as antidote to the cardiac effects of cocaine intoxication. *N.Engl. J. Med.* **313**: 519 (1985).

J. Olds. Pleasure centers in the brain. *Sci. Am.* **195**: 1505 (1956).

R.L. Patrick and J.D. Barchas. Potentiation by cocaine of the stimulus-induced increase in dopamine synthesis in rat brain striatal synaptosomes. *Neuropharmacology* **16**: 327 (1977).

R. Pickens and W.C. Harris. Self-administration of d-amphetamine by rats. *Psychopharmacologia* **12**: 158 (1968).

D.K. Pitts and J. Marwah. Effects of cocaine on the electrical activity of single noradrenergic neurons from the locus coeruleus. *Life. Sci.* **38**: 1229 (1986).
R.M. Post. Cocaine psychosis: a continuum model. *Am. J. Psychiatry* **47**: 384 (1975).
R.M. Post, R.T. Kopanda, and K.E. Black. Progressive effects of cocaine on behavior and central amine metabolism in the rhesus monkey: relationship to kindling and psychosis. *Biol. Psychiatry* **11**: 403 (1976).
S.N. Pradhan, A.K. Battachargya, and S. Pradhan. Serotonergic manipulation of the behavioral effects of cocaine in rats. *Commun. Psychopharmacol.* **2**: 481 (1978a).
S. Pradhan, S.N. Roy, and S.N. Pradhan. Correlation of behavioral and neurochemical effects of acute administration of cocaine in rats. *Life Sci.* **22**: 1737 (1978b).
D.E. Redmond, Y.H. Huang, D.R. Snyder, *et al.* Behavioral effects of stimulation of the locus coeruleus in the stumptail monkey *(Marcaca Arctoides). Brain Res.* **116**: 502 (1976).
S. Reichlin. Regulation of the hypophysiotropic secretions of the brain. *Arch. Intern. Med.* **135**: 1350 (1975).
D.C.S. Roberts and G.F. Koob. Disruption of cocaine self-administration following 6-hydroxydopamine lesions of the ventral tegmental area in rats. *Pharmacol. Biochem. Behav.* **6**: 615 (1977).
H. Roehrich, C.A. Dackis, and M.S. Gold. Bromocriptine. *Med. Res. Rev.* **7**: 243 (1987).
S.B. Ross and A.L. Renyi. Uptake of some tritiated sympathomimetic amines by mouse brain cortex *in vitro. Acta Pharmacol. Toxicol.* **24**: 297 (1966).
R.K. Seigel. Cocaine and sexual dysfunction: the curse of mama coca. *J. Psychoactive Drugs* **14**: 71 (1982).
R.K. Seigel. New patterns of cocaine use: changing doses and routes. *National Institute on Drug Abuse Monograph Series* **61**: 204 (1985).
C. Spyraki and H.C. Fibiger. Behavioral evidence for supersensitivity of postsynaptic dopamine receptors in the mesolimbic system after chronic administration of desipramine. *Eur. J. Pharmacol.* **74**: 195 (1981).
D. Taylor and B.T. Ho. Neurochemical effects of cocaine following acute and repeated injection. *J. Neurosci. Res.* **3**: 95 (1977).
D. Taylor, B.T. Ho, and J.D. Fagen. Increased dopamine receptor binding in rat brain by repeated cocaine injections. *Commun. Psychopharmacol.* **3**: 137 (1979).
F.S. Tennant. Effect of cocaine dependence on plasma phenylalanine and tyrosine levels and on urinary MHPG excretion. *Am. J. Psychiatry* **142**: 1200 (1985).
F.S. Tennant and R.A. Rawson. Cocaine and amphetamine dependence treated with desipramine. *National Institute on Drug Abuse Monograph Series* **61**: 351 (1983).
F.S. Tennant and A.A. Sagherian. Double-blind comparison of amantadine hydrochloride and bromocriptine mesylate for ambulatory withdrawal from cocaine dependence. *Arch. Intern. Med.* **147**: 109 (1987).
R.D. Weiss, S.M. Mirin and J.L. Michael. Relapse to cocaine abuse after initiating desipramine treatment. *JAMA* **260**: 2545 (1988).
R.A. Wise. Catecholamine theories of reward: a critical review. *Brain Res.* **152**: 215 (1978).
R.A. Wise. Neural mechanisms of the reinforcing action of cocaine. *National Institute on Drug Abuse Monograph Series* **50**: 15 (1985).
R.A. Wise and M.A. Bozarth. A psychomotor theory of addiction. *Psychol. Rev.* **94**: 469 (1987).
R.J. Wyatt, F. Karoum, R. Suddath, and R. Fawcett. Persistently decreased brain dopamine levels and cocaine. *JAMA,* **259**: 2996 (1988).

INDEX

C

D

E

F

G

H

I

K

L

M

T

U

V

W

Y

Z